Solid-State Chemistry of Inorganic Materials III

MATERIALS RESEARCH SOCIETY
SYMPOSIUM PROCEEDINGS VOLUME 658

Solid-State Chemistry of Inorganic Materials III

Symposium held November 27–30, 2000, Boston, Massachusetts, U.S.A.

EDITORS:

Margret J. Geselbracht
Reed College
Portland, Oregon, U.S.A.

John E. Greedan
BIMR, McMaster University
Hamilton, Ontario, Canada

David C. Johnson
University of Oregon
Eugene, Oregon, U.S.A.

M.A. Subramanian
Dupont Central Research & Development
Experimental Station
Wilmington, Delaware, U.S.A.

Materials Research Society
Warrendale, Pennsylvania

This material is based upon work supported by the National Science Foundation under Grant No. NSF: DMR-0086290. Any opinions, findings, and conclusions or recommendations expressed in this material are those of the author(s) and do not necessarily reflect the views of the National Science Foundation.

Single article reprints from this publication are available through
University Microfilms Inc., 300 North Zeeb Road, Ann Arbor, Michigan 48106

CODEN: MRSPDH

This book has been registered with Copyright Clearance Center, Inc. For further information, please contact the Copyright Clearance Center, Salem, Massachusetts.

Published by:

Materials Research Society
506 Keystone Drive
Warrendale, PA 15086
Telephone (724) 779-3003
Fax (724) 779-8313
Web site: http://www.mrs.org/

Library of Congress Cataloging-in-Publication Data

Solid-state chemistry of inorganic materials III : symposium held November 27–30, 2000, Boston, Massachusetts, U.S.A. / editors, Margret J. Geselbracht, John E. Greedan, David C. Johnson, M.A. Subramanian
        p.cm.—(Materials Research Society symposium proceedings ; v. 658)
        Includes bibliographical references and indexes.
        ISBN 1-55899-568-4
        I. Geselbracht, Margret J. II. Greedan, John E. III. Johnson, David C. IV. Subramanian, M.A. V. Materials Research Society symposium proceedings ; v. 658
2001

Manufactured in the United States of America

CONTENTS

Preface xv

Materials Research Society Symposium Proceedings xvi

CRYSTAL CHEMISTRY OF
COMPLEX SYSTEMS

Structure Determination of $Ba_8CoRh_6O_{21}$, a New Member of the
2H-Perovskite Related Oxides GG1.4
    H.-C. zur Loye, M.D. Smith, K.E. Stitzer, A. El Abed,
    and J. Darriet

The Formation of Striped Phases by Charge Localization in
Vanadium Phosphates GG1.7
    Ranko P. Bontchev, Junghwan Do, and Allan J. Jacobson

Synthesis and Crystal Chemistry of New Transition Metal
Tellurium Oxides in Compounds Containing Lead and Barium GG1.10
    Boris Wedel, Katsumasa Sugiyama, Kimio Itagaki,
    and Hanskarl Müller-Buschbaum

SESSION IN HONOR OF
J.M. HONIG

Itinerant Vibrons and High-Temperature Superconductivity GG2.2
    John B. Goodenough

(Rare Earth/Li) Titanates and Niobates as Ionic Conductors GG2.3
    A. Morata-Orrantia, S. García-Martín, E. Morán, U. Amador,
    and M.A. Alario-Franco

Determination of T_N for $Ba_6Mn_4MO_{15}$ (M = Cu, Zn) by Neutron
Diffraction GG2.4
    P.D. Battle and E.J. Cussen

Powder Diffraction Refinements of the Structure of Magnetite
(Fe_3O_4) Below the Verwey Transition GG2.6
    Jon P. Wright, J. Paul Attfield, and Paolo G. Radaelli

*Invited Paper

DIELECTRICS, CRYSTAL CHEMISTRY, GLASSES, ELECTRICAL TRANSPORT

Ferroelectric Tunability Studies in Doped $BaTiO_3$ and $Ba_{1-x}Sr_xTiO_3$ GG3.4
Dong Li and M.A. Subramanian

Modification of Ferroelectric Properties of TGS Crystals Grown
Under a dc Electric Field GG3.5
G. Arunmozhi, E. Nogueira, E. de Matos Gomes,
S. Lanceros-Mendez, M. Margarida R. Costa, A. Criado,
and J.F. Mano

Mechanism of Quality Factor Compensation by Nb_2O_5 Addition
for Dielectric Properties of Low-T Sintered ZST Microwave
Ceramics GG3.14
Yong H. Park, Moo Y. Shin, Hyung H. Kim, and Kyung H. Ko

Grain Orientation of Aluminum Titanate Ceramics During
Formation Reaction GG3.17
Yutaka Ohya, Zenbe-e Nakagawa, Kenya Hamano,
Hiroshi Kawamoto, and Satoshi Kitaoka

Solid-State NMR Investigation of Mixed-Alkali Distribution in
Phosphate Glasses GG3.20
S. Prabakar and K.T. Mueller

[207]Pb NMR of $PbZrO_3$ and $PbZr_{1-x}Ti_xO_3$ Solid Solutions GG3.21
Paolo Rossi, Matthew R. Dvorak, and Gerard S. Harbison

Investigation of Lead Borosilicate Glass Structure With [207]Pb and
^{11}B Solid-State NMR GG3.22
James M. Gibson, Frederick G. Vogt, Amy S. Barnes,
and Karl T. Mueller

Interference Effect Between Electron and Ion Flows in
Semiconducting $Fe_{3-\delta}O_4$ GG3.25
Jeong-Oh Hong and Han-Ill Yoo

Stability and Structural Characterization of Epitaxial $NdNiO_3$
Films Grown by Pulsed Laser Deposition GG3.27
Trong-Duc Doan, Cobey Abramowski, and Paul A. Salvador

Isotopes in Neutron Diffraction—Detailed Structural Analysis at
the Metal-Insulator Transition in $SmNiO_3$ GG3.31
Mark T. Weller, Paul F. Henry, and C.C. Wilson

Crystal Chemistry of Colloids Obtained by Hydrolysis of Fe(III)
in the Presence of SiO_4 Ligands GG3.36
    Emmanuel Doelsch, Armand Masion, Jérôme Rose,
    William E.E. Stone, Jean Yves Bottero, and Paul M. Bertsch

TRANSPORT PROPERTIES/ METAL-INSULATOR SYSTEMS

Electronic and Magnetic Properties of d^1 Pnictide-Oxides:
$Na_2Ti_2Pn_2O$ (*Pn* = As, Sb) GG4.1
    Tadashi C. Ozawa, Mario Bieringer, John E. Greedan,
    and Susan M. Kauzlarich

Fine-Tuning the Physical Properties of Perovskite Related
La-Ti-Oxides by Merely Altering the Oxygen Stoichiometry GG4.2
    Olav Becker, Stefan Ebbinghaus, Bernd Renner,
    and Armin Reller

Quantum Critical Phenomena in the $NiS_{2-x}Se_x$ System GG4.3
    J.M. Honig

Slater Transition in the Pyrochlore $Cd_2Os_2O_7$ GG4.4
    D. Mandrus, J.R. Thompson, and L.M. Woods

Synthesis, Structure and Magnetic Properties of Monoclinic
$Nb_{12}O_{29}$ GG4.5
    J.E.L. Waldron and M.A. Green

Transport Properties and Crystal Chemistry of Ba-Sr-Bi Oxides GG4.7
    Oya A. Gökçen, James K. Meen, Allan J. Jacobson,
    and Don Elthon

Bonding of Guest Molecules in the Tubes of Nanoporous Cetineite
Crystals GG4.9
    E.E. Krasovskii, O. Tiedje, S. Brodersen, W. Schattke,
    F. Starrost, J. Jockel, and U. Simon

MAGNETISM AND MANGANATES

Metal Cyanide Networks Formed at an Air-Water Interface:
Stucture and Magnetic Properties GG5.2
    Jeffrey T. Culp, A. Nicole Morgan, Mark W. Meisel,
    and Daniel R. Talham

*Invited Paper

* Theoretical Analyses of Spin Exchange Interactions in Extended
Magnetic Solids Containing Several Unpaired Spins per Spin Site............................GG5.3
  D. Dai, H.-J. Koo, and M.-H. Whangbo

Micromagnetic and Magnetoresistance Studies of Ferromagnetic
$La_{0.83}Sr_{0.13}MnO_{2.98}$ Crystals GG5.6
  Guerman Popov, Sergei V. Kalinin, Rodolfo A. Alvarez,
  Martha Greenblatt, and Dawn A. Bonnell

Resistive Anomaly Relevant to Nd Moments in the
Antiferromagnetic Phase of the Bandwidth-Controlled
Manganites GG5.9
  H. Kuwahara, K. Noda, and R. Kawasaki

Temperature Dependence of Magnetic Compton Profile in $DyCo_5$ GG5.10
  Hayato Miyagawa, Yasuhiro Watanabe, Akihisa Koizumi,
  Nobuhiko Sakai, Masaichiro Mizumaki, Yoshiharu Sakurai,
  Tetsuya Nakamura, and Susumu Nanao

NEW MATERIALS,
MESO/NANOPOROUS MATERIALS

Effect of Polymerization of Precursor Solutions on Crystallization
and Morphology of $Ce_{0.9}Gd_{0.1}O_{1.95}$ Powders GG6.1
  Shuqiang Wang, Masanobu Awano, and Kunihiro Maeda

Decoration of Silicon Carbide Nanotubes by $CoFe_2O_4$ Spinel
Nanoparticles GG6.4
  Claude Estournès, Cuong Pham-Huu, Nicolas Keller,
  and Marc J. Ledoux

A New Three-Dimensional Lanthanide Framework Constructed
by Oxalate and 3,5-Pyridinedicarboxylate GG6.12
  Michael A. Lawandy, Long Pan, Xiaoying Huang, Jing Li,
  Tan Yuen, and C.L. Lin

Synthesis of Mesophase Cerium(IV) Oxides via Surfactant
Templating GG6.14
  Masahiko Inagaki, Atsushi Hozumi, Yoshiyuki Yokogawa,
  and Tetsuya Kameyama

Electronic and Optical Studies of Coexisting 5- and 6-Atom Rings
in Tetrahedral a-C GG6.15
  R.M. Valladares, A.G. Calles, M.A. Mc Nelis,
  and Ariel A. Valladares

*Invited Paper

Modified Sol-Gel Synthesis of Vanadium Oxide Nanocomposites
Containing Surfactant Ions and Their Partial Removal GG6.16
Arthur Dobley, Peter Y. Zavalij, Jürgen Schulte,
and M. Stanley Whittingham

Two Pillared Three-Dimensional Inorganic/Organic Hybrid
Structures Composed of Bimetallic Layers Bridged by Exo-
Bidentate Ligands GG6.17
Long Pan, Xiaoying Huang, and Jing Li

A Structural Analysis Method for Graphite Intercalation
Compounds GG6.18
Tatsuo Nakazawa, Kyoichi Oshida, Takashi Miyazaki,
Morinobu Endo, and Mildred S. Dresselhaus

A Novel Route for the Synthesis of $LiAl_xCo_{1-x}O_2$ Battery Materials
and Their Structural Properties GG6.19
M.S. Tomar, A. Hidalgo, P.S. Dobal, R.S. Katiyar, A. Dixit,
R.E. Melgarejo, and K.A. Kuenhold

Reducibility Study of the $LaM_xFe_{1-x}O_3$ (M = Ni, Co) Perovskites GG6.22
C. Estournès, H. Provendier, L. Bedel, C. Petit, A.C. Roger,
and A. Kiennemann

Conversion of an Aurivillius Phase $Bi_2SrNaNb_3O_{12}$ Into Its
Protonated Form via Treatment With Various Mineral Acids GG6.24
Masashi Shirata, Yu Tsunoda, Wataru Sugimoto,
and Yoshiyuki Sugahara

ESCA and Vibrational Spectroscopy of Alkali Cation Exchanged
Manganic Acids With the 2x2 Type Tunnel Structure Synthesized
via Different Chemical Routes GG6.25
Masamichi Tsuji, Hirofumi Kanoh, and Kenta Ooi

A New Microporous Silicate With 12-Ring Channels GG6.28
Jacques Plévert, Yoshihiro Kubota, Takahisa Honda,
Tatsuya Okubo, and Yoshihiro Sugi

Deposition and Characterization of $Y_3Al_5O_{12}$ (YAG) Films and
Powders by Plasma Spray Synthesis GG6.29
Sujatha D. Parukuttyamma, Joshua Margolis, Haiming Liu,
John B. Parise, Clare P. Grey, Sanjay Sampath, Perina Gouma,
and Herbert Herman

The Synthesis and Characterization of a Novel Gallium
Diphosphonate GG6.31
Martin P. Attfield and Howard G. Harvey

Synthesis of CoFe$_2$O$_4$ Nanoparticles via the Ferrihydrite Route GG6.32
A. Manivannan, A.M. Constantinescu, and M.S. Seehra

Study of Calcium/Lead Apatite Structure Type for Stabilizing
Heavy Metals..GG6.33
Z.L. Dong, B. Wei, and T.J. White

MICRO/MESO/NANOPOROUS
MATERIALS: INORGANIC-ORGANIC HYBRIDS

Transition Metal Based Zeotypes: Inorganic Materials at the
Complex Oxide - Zeolite Border...GG7.2
Paul F. Henry, Robert W. Hughes, and Mark T. Weller

A Computational Study of the Translational Motion of Protons in
Zeolite H-ZSM-5 ....GG7.4
M.E. Franke, M. Sierka, J. Sauer, and U. Simon

Disordered Mesoporous Silicates Formed by Templation of a
Liquid Crystal (L$_3$) GG7.5
Abds-Sami Malik, Daniel M. Dabbs, Ilhan A. Aksay,
and Howard E. Katz

Frameworks of Transition Metals and Linkers With Two or More
Functional Groups...GG7.8
Slavi C. Sevov

SYNTHESIS, NEW METHODS,
NEW MATERIALS

Synthesis of Large ZSM-5 Crystals Under High Pressure GG8.1
Xiqu Wang and Allan J. Jacobson

Synthesis of Tetramethylammonium Polyoxovanadates GG8.3
Nathalie Steunou, Laure Bouhedja, Jocelyne Maquet,
and Jacques Livage

Assembly of Metal-Anion Arrays Within Dion-Jacobson-Type
Perovskite Hosts...GG8.5
Thomas A. Kodenkandath, Marilena L. Viciu, Xiao Zhang,
Jessica A. Sims, Erin W. Gilbert, François-Xavier Augrain,
Jean-Noël Chotard, Gabriel A. Caruntu, Leonard Spinu,
Weilie L. Zhou, and John B. Wiley

Preparation of New Bismuth Oxides by Hydrothermal Reaction GG8.7
N. Kumada, T. Takei, N. Kinomura, and A.W. Sleight

Chemical Synthesis and Properties of Layered $Co_{1-y}Ni_yO_{2-\delta}$ Oxides
(0 ≤ y ≤ 1) GG8.12
  A. Manthiram, R.V. Chebiam, and F. Prado

SOLID-STATE IONICS, BATTERY MATERIALS,
THERMO POWER, OPTICAL MATERIALS

Preparation, Electrical Properties, and Water Adsorption
Behavior of $(1-x)Sb_2O_5 \cdot xM_2O_3 \cdot nH_2O$ (M = Al, Bi, and Y; 0 ≤ x ≤ 1) GG9.1
  Kiyoshi Ozawa, Yoshio Sakka, and Muneyuki Amano

High Oxidation State Alkali-Metal Late-Transition-Metal Oxides GG9.5
  David B. Currie, Andrew L. Hector, Emmanuelle A. Raekelboom,
  John R. Owen, and Mark T. Weller

In Situ XAFS Study on Cathode Materials for Lithium-Ion
Batteries GG9.6
  Takamasa Nonaka, Chikaaki Okuda, Yoshio Ukyo,
  and Tokuhiko Okamoto

Chemical Delithiation, Thermal Transformations and
Electrochemical Behavior of Iron-Substituted Lithium Nickelate GG9.7
  Pedro Lavela, Carlos Pérez-Vicente, and José L. Tirado

Lithium Deintercalation in $LiNi_{0.30}Co_{0.70}O_2$: Redox Processes,
Electronic and Ionic Mobility as Characterized by ^7Li MAS NMR
and Electrical Properties GG9.9
  D. Carlier, M. Ménétrier, and C. Delmas

^7Li MAS NMR Studies of Lithiated Manganese Dioxide Tunnel
Structures: Pyrolusite and Ramsdellite GG9.11
  Younkee Paik, Young J. Lee, Francis Wang, William Bowden,
  and Clare P. Grey

A Composite Ionic-Electronic Conductor in the (Ca,Sr,Ba)-Bi
Oxide System GG9.14
  James K. Meen, Oya A. Gökçen, I.-C. Lin, Karoline Müller,
  and Binh Nguyen

Manganese Vanadium Oxide Compounds as Cathodes for
Lithium Batteries GG9.16
  J. Katana Ngala, Peter Y. Zavalij, and M. Stanley Whittingham

Microporous Ruthenium Oxide Films for Energy Storage
Applications GG9.17
  J.P. Zheng and Q.L. Fang

Two Different Interactions Between Oxygen Vacancies and
Dopant Cations for Ionic Conductivity in CaO-Doped CeO$_2$
Electrolyte Materials G9.19
    Yuanzhong Zhou and Xi Chen

Nickel Determination by Complexation Utilizing a Functionalized
Optical Waveguide Sensor GG9.29
    Erin S. Carter and Klaus-H. Dahmen

Competitive Adsorption of O$_2$ and H$_2$O at the Neutral and
Defective SnO$_2$ (110) Surface GG9.33
    Ben Slater, C. Richard A. Catlow, David E. Williams,
    and A. Marshall Stoneham

NO Decomposition Property of Lanthanum Manganite Porous
Electrodes GG9.36
    Kazuyuki Matsuda, Takao Kanai, Masanobu Awano,
    and Kunihiro Maeda

SOLID-STATE IONICS, BATTERY MATERIALS, ENERGY STORAGE

Bismuth Contribution to the Improvement of the Positive
Electrode Performances in Ni/Cd and Ni/MH Batteries GG10.1
    V. Pralong, A. Delahaye-Vidal, B. Beaudoin, and J.-M. Tarascon

Li-Insertion Behavior in Nanocrystalline TiO$_2$-(MoO$_3$)$_z$
Core-Shell Materials.. GG10.4
    Gregory J. Moore, Annie Le Gal La Salle, Dominique Guyomard,
    and Scott H. Elder

Vanadium Oxide Frameworks Modified With Transition Metals GG10.7
    Peter Y. Zavalij and M. Stanley Whittingham

Study of Fluoride Ion Motions in PbSnF$_4$ and BaSnF$_4$ Compounds
With Molecular Dynamics Simulation and Solid State NMR
Techniques.. GG10.9
    Santanu Chaudhuri, Michael Castiglione, Francis Wang,
    Mark Wilson, Paul A. Madden, and Clare P. Grey

*Invited Paper

THERMO POWER, THERMAL EXPANSION, OPTICAL MATERIALS

* "Open Structure" Semiconductors: Clathrate and Channel
Compounds for Low Thermal Conductivity Thermoelectric
Materials..GG11.1
George S. Nolas

Spectral Properties of Various Cerium Doped Garnet Phosphors
for Application in White GaN-Based LEDs GG11.8
Jennifer L. Wu, Steven P. Denbaars, Vojislav Srdanov,
and Henry Weinberg

Origin of Ferroelectricity in Aurivillius Compounds GG11.9
Donají Y. Suárez, Ian M. Reaney, and William E. Lee

ADDENDUM

Formation and Erosion of WC Under W^+ Irradiation of Graphite...........O5.31
J. Roth, U. v. Toussaint, K. Schmid, J. Luthin, W. Eckstein,
R.A. Zuhr, and D.K. Hensley

Author Index

Subject Index

*Invited Paper

PREFACE

Solid-state chemistry is an interdisciplinary field, attracting investigators from a diverse set of backgrounds including materials science and engineering, ceramics, chemistry, chemical engineering, mineralogy/geology, and condensed-matter physics. Researchers share the common challenge of understanding, controlling, and predicting the structures and properties of solids at the atomic level. Symposium GG, "Solid-State Chemistry of Inorganic Materials," held November 27–30 at the 2000 MRS Fall Meeting in Boston, Massachusetts, was the third in a biennial series, all held at the MRS Fall Meetings. This symposium continues to provide an international, interdisciplinary forum for the presentation and discussion of recent fundamental advances in the solid-state chemistry of inorganic materials and the impact of these advances on the development of practical applications.

Symposium GG was the third largest symposium at the 2000 MRS Fall Meeting with a total of 175 contributed papers, including 73 oral presentations and 102 poster presentations. The symposium attracted a large gathering of scientists, including a significant number of international speakers, working in this core discipline of inorganic materials research. The program lasted four days and included the presentations of 14 invited speakers and three evening poster sessions. The topics covered included novel synthetic methods leading to new materials, solving complex crystal structures using four dimensional crystallography, magnetic interactions in superconductors and colossal magnetoresistive solids, solid-state ionics, microwave dielectrics, microporous and hybrid inorganic/organic materials, negative thermal expansion and thermoelectric materials. A highlight of the symposium was an afternoon session dedicated to Professor J.M. Honig in recognition of his many contributions to the discipline of solid-state chemistry and his stewardship of the Journal of Solid State Chemistry.

The organizers would like to thank all of the people who attended and participated in this symposium and those of you who chose to contribute to these proceedings. This first experience with exclusively electronic handling of manuscripts presented a number of challenges to both authors and reviewers, and we appreciate your patience and cooperation as we navigated these challenges. Your enthusiastic participation in this symposium made it a resounding success and underscored the pivotal role that this forum plays in the solid-state chemistry community. The success of this symposium would not have been possible without the financial support received from Academic Press, DuPont Central Research & Development, Elsevier Science, and the National Science Foundation. We would like to express our sincere gratitude to David Nelson, program director of Solid-State Chemistry in NSF's Division of Materials Research, for his enthusiastic and generous financial support of this symposium.

Margret J. Geselbracht
John E. Greedan
David C. Johnson
Mas A. Subramanian

September 2001

MATERIALS RESEARCH SOCIETY SYMPOSIUM PROCEEDINGS

Volume 609— Amorphous and Heterogeneous Silicon Thin Films—2000, R.W. Collins, H.M. Branz, M. Stutzmann, S. Guha, H. Okamoto, 2001, ISBN: 1-55899-517-X

Volume 610— Si Front-End Processing—Physics and Technology of Dopant-Defect Interactions II, A. Agarwal, L. Pelaz, H-H. Vuong, P. Packan, M. Kase, 2001, ISBN: 1-55899-518-8

Volume 611— Gate Stack and Silicide Issues in Silicon Processing, L.A. Clevenger, S.A. Campbell, P.R. Besser, S.B. Herner, J. Kittl, 2001, ISBN: 1-55899-519-6

Volume 612— Materials, Technology and Reliability for Advanced Interconnects and Low-k Dielectrics, G.S. Oehrlein, K. Maex, Y-C. Joo, S. Ogawa, J.T. Wetzel, 2001, ISBN: 1-55899-520-X

Volume 613— Chemical-Mechanical Polishing 2000—Fundamentals and Materials Issues, R.K. Singh, R. Bajaj, M. Moinpour, M. Meuris, 2001, ISBN: 1-55899-521-8

Volume 614— Magnetic Materials, Structures and Processing for Information Storage, B.J. Daniels, T.P. Nolan, M.A. Seigler, S.X. Wang, C.B. Murray, 2001, ISBN: 1-55899-522-6

Volume 615— Polycrystalline Metal and Magnetic Thin Films—2001, B.M. Clemens, L. Gignac, J.M. MacLaren, O. Thomas, 2001, ISBN: 1-55899-523-4

Volume 616— New Methods, Mechanisms and Models of Vapor Deposition, H.N.G. Wadley, G.H. Gilmer, W.G. Barker, 2000, ISBN: 1-55899-524-2

Volume 617— Laser-Solid Interactions for Materials Processing, D. Kumar, D.P. Norton, C.B. Lee, K. Ebihara, X.X. Xi, 2001, ISBN: 1-55899-525-0

Volume 618— Morphological and Compositional Evolution of Heteroepitaxial Semiconductor Thin Films, J.M. Millunchick, A-L. Barabasi, N.A. Modine, E.D. Jones, 2000, ISBN: 1-55899-526-9

Volume 619— Recent Developments in Oxide and Metal Epitaxy—Theory and Experiment, M. Yeadon, S. Chiang, R.F.C. Farrow, J.W. Evans, O. Auciello, 2000, ISBN: 1-55899-527-7

Volume 620— Morphology and Dynamics of Crystal Surfaces in Complex Molecular Systems, J. DeYoreo, W. Casey, A. Malkin, E. Vlieg, M. Ward, 2001, ISBN: 1-55899-528-5

Volume 621— Electron-Emissive Materials, Vacuum Microelectronics and Flat-Panel Displays, K.L. Jensen, R.J. Nemanich, P. Holloway, T. Trottier, W. Mackie, D. Temple, J. Itoh, 2001, ISBN: 1-55899-529-3

Volume 622— Wide-Bandgap Electronic Devices, R.J. Shul, F. Ren, W. Pletschen, M. Murakami, 2001, ISBN: 1-55899-530-7

Volume 623— Materials Science of Novel Oxide-Based Electronics, D.S. Ginley, J.D. Perkins, H. Kawazoe, D.M. Newns, A.B. Kozyrev, 2000, ISBN: 1-55899-531-5

Volume 624— Materials Development for Direct Write Technologies, D.B. Chrisey, D.R. Gamota, H. Helvajian, D.P. Taylor, 2001, ISBN: 1-55899-532-3

Volume 625— Solid Freeform and Additive Fabrication—2000, S.C. Danforth, D. Dimos, F.B. Prinz, 2000, ISBN: 1-55899-533-1

Volume 626— Thermoelectric Materials 2000—The Next Generation Materials for Small-Scale Refrigeration and Power Generation Applications, T.M. Tritt, G.S. Nolas, G.D. Mahan, D. Mandrus, M.G. Kanatzidis, 2001, ISBN: 1-55899-534-X

Volume 627— The Granular State, S. Sen, M.L. Hunt, 2001, ISBN: 1-55899-535-8

Volume 628— Organic/Inorganic Hybrid Materials—2000, R. Laine, C. Sanchez, C.J. Brinker, E. Giannelis, 2001, ISBN: 1-55899-536-6

Volume 629— Interfaces, Adhesion and Processing in Polymer Systems, S.H. Anastasiadis, A. Karim, G.S. Ferguson, 2001, ISBN: 1-55899-537-4

Volume 633— Nanotubes and Related Materials, A.M. Rao, 2001, ISBN: 1-55899-543-9

Volume 634— Structure and Mechanical Properties of Nanophase Materials—Theory and Computer Simulations vs. Experiment, D. Farkas, H. Kung, M. Mayo, H. Van Swygenhoven, J. Weertman, 2001, ISBN: 1-55899-544-7

Volume 635— Anisotropic Nanoparticles—Synthesis, Characterization and Applications, S.J. Stranick, P. Searson, L.A. Lyon, C.D. Keating, 2001, ISBN: 1-55899-545-5

Volume 636— Nonlithographic and Lithographic Methods of Nanofabrication—From Ultralarge-Scale Integration to Photonics to Molecular Electronics, L. Merhari, J.A. Rogers, A. Karim, D.J. Norris, Y. Xia, 2001, ISBN: 1-55899-546-3

MATERIALS RESEARCH SOCIETY SYMPOSIUM PROCEEDINGS

Volume 637— Microphotonics—Materials, Physics and Applications, K. Wada, P. Wiltzius, T.F. Krauss, K. Asakawa, E.L. Thomas, 2001, ISBN: 1-55899-547-1

Volume 638— Microcrystalline and Nanocrystalline Semiconductors—2000, P.M. Fauchet, J.M. Buriak, L.T. Canham, N. Koshida, B.E. White, Jr., 2001, ISBN: 1-55899-548-X

Volume 639— GaN and Related Alloys—2000, U. Mishra, M.S. Shur, C.M. Wetzel, B. Gil, K. Kishino, 2001, ISBN: 1-55899-549-8

Volume 640— Silicon Carbide—Materials, Processing and Devices, A.K. Agarwal, J.A. Cooper, Jr., E. Janzen, M. Skowronski, 2001, ISBN: 1-55899-550-1

Volume 642— Semiconductor Quantum Dots II, R. Leon, S. Fafard, D. Huffaker, R. Nötzel, 2001, ISBN: 1-55899-552-8

Volume 643— Quasicyrstals—Preparation, Properties and Applications, E. Belin-Ferré, P.A. Thiel, A-P. Tsai, K. Urban, 2001, ISBN: 1-55899-553-6

Volume 644— Supercooled Liquid, Bulk Glassy and Nanocrystalline States of Alloys, A. Inoue, A.R. Yavari, W.L. Johnson, R.H. Dauskardt, 2001, ISBN: 1-55899-554-4

Volume 646— High-Temperature Ordered Intermetallic Alloys IX, J.H. Schneibel, S. Hanada, K.J. Hemker, R.D. Noebe, G. Sauthoff, 2001, ISBN: 1-55899-556-0

Volume 647— Ion Beam Synthesis and Processing of Advanced Materials, D.B. Poker, S.C. Moss, K-H. Heinig, 2001, ISBN: 1-55899-557-9

Volume 648— Growth, Evolution and Properties of Surfaces, Thin Films and Self-Organized Structures, S.C. Moss, 2001, ISBN: 1-55899-558-7

Volume 649— Fundamentals of Nanoindentation and Nanotribology II, S.P. Baker, R.F. Cook, S.G. Corcoran, N.R. Moody, 2001, ISBN: 1-55899-559-5

Volume 650— Microstructural Processes in Irradiated Materials—2000, G.E. Lucas, L. Snead, M.A. Kirk, Jr., R.G. Elliman, 2001, ISBN: 1-55899-560-9

Volume 651— Dynamics in Small Confining Systems V, J.M. Drake, J. Klafter, P. Levitz, R.M. Overney, M. Urbakh, 2001, ISBN: 1-55899-561-7

Volume 652— Influences of Interface and Dislocation Behavior on Microstructure Evolution, M. Aindow, M. Asta, M.V. Glazov, D.L. Medlin, A.D. Rollet, M. Zaiser, 2001, ISBN: 1-55899-562-5

Volume 653— Multiscale Modeling of Materials—2000, L.P. Kubin, J.L. Bassani, K. Cho, H. Gao, R.L.B. Selinger, 2001, ISBN: 1-55899-563-3

Volume 654— Structure-Property Relationships of Oxide Surfaces and Interfaces, C.B. Carter, X. Pan, K. Sickafus, H.L. Tuller, T. Wood, 2001, ISBN: 1-55899-564-1

Volume 655— Ferroelectric Thin Films IX, P.C. McIntyre, S.R. Gilbert, M. Miyasaka, R.W. Schwartz, D. Wouters, 2001, ISBN: 1-55899-565-X

Volume 657— Materials Science of Microelectromechanical Systems (MEMS) Devices III, M. deBoer, M. Judy, H. Kahn, S.M. Spearing, 2001, ISBN: 1-55899-567-6

Volume 658— Solid-State Chemistry of Inorganic Materials III, M.J. Geselbracht, J.E. Greedan, D.C. Johnson, M.A. Subramanian, 2001, ISBN: 1-55899-568-4

Volume 659— High-Temperature Superconductors—Crystal Chemistry, Processing and Properties, U. Balachandran, H.C. Freyhardt, T. Izumi, D.C. Larbalestier, 2001, ISBN: 1-55899-569-2

Volume 660— Organic Electronic and Photonic Materials and Devices, S.C. Moss, 2001, ISBN: 1-55899-570-6

Volume 661— Filled and Nanocomposite Polymer Materials, A.I. Nakatani, R.P. Hjelm, M. Gerspacher, R. Krishnamoorti, 2001, ISBN: 1-55899-571-4

Volume 662— Biomaterials for Drug Delivery and Tissue Engineering, S. Mallapragada, R. Korsmeyer, E. Mathiowitz, B. Narasimhan, M. Tracy, 2001, ISBN: 1-55899-572-2

Prior Materials Research Society Symposium Proceedings available by contacting Materials Research Society

Crystal Chemistry of
Complex Systems

Mat. Res. Soc. Symp. Proc. Vol. 658 © 2001 Materials Research Society

Structure Determination of $Ba_8CoRh_6O_{21}$, a New Member of the 2H-Perovskite Related Oxides

H.-C. zur Loye*, M. D. Smith, K. E. Stitzer
Department of Chemistry and Biochemistry, University of South Carolina, Columbia, SC, 29208, USA, email: zurloye@sc.edu

A. El Abed[†] and J. Darriet
Institut de Chimie de la Matière Condensée de Bordeaux (ICMCB-CNRS), Avenue du Dr. Schweitzer, 33608 Pessac Cedex, France
† Permanent Address: Mohamed I Univ., Facultédes Sciences, Oujda, Morocco.

ABSTRACT

Single crystals of $Ba_8CoRh_6O_{21}$ were grown out of a potassium carbonate flux. The structure was solved by a general method using the superspace group approach. The superspace group employed was $R\bar{3}m(00\gamma)0s$ with $a = 10.0431(1)$ Å, $c_1 = 2.5946(1)$ Å and $c_2 = 4.5405(1)$ Å, $V = 226.60(1)$ Å^3. $Ba_8CoRh_6O_{21}$ represents the first example of an $m = 5$, $n = 3$ member of the $A_{3n+3m}A'_nB_{3m+n}O_{9m+6n}$ family of 2H hexagonal perovskite related oxides and contains chains consisting of six consecutive RhO_6 octahedra followed by one distorted CoO_6 trigonal prism. These chains in turn are separated from each other by $[Ba]_\infty$ chains.

INTRODUCTION

Low-dimensional magnetic systems have attracted much interest historically due to the presence of magnetic behavior unique to structurally highly anisotropic systems. [1-3] Insights into such behavior can be gained from structural families where it is possible to systematically vary either the structure or the composition independently. For this reason, perovskite and perovskite-related oxides in particular have long provided excellent candidates for structural and physical property studies, due to the compositional and structural flexibility of this huge extended oxide family. Recently, much interest has been focussed on a large and varied group of oxides closely akin to the pseudo-one-dimensional 2H hexagonal perovskites, with the general formula $A_{3n+3m}A'_nB_{3m+n}O_{9m+6n}$ (n, m = integers, A = alkaline earth; A', B = large assortment of metals including alkali, alkaline earth, transition, main group and rare earth metals). [4-25] An early general structural classification of these materials based on the filling of interstitial sites generated by the stacking of $[A_3O_9]$ and $[A_3A'O_6]$ layers was developed by Darriet and Subramanian. [26-27] This approach easily describes the structural composition of all the commensurate members of this family of structures and can be extended to encompass members that form incommensurately modulated (aperiodic) structures.

An alternate, complementary description that highlights the low-dimensional nature of these compounds is the composite structure approach. In this structural description, these oxides consist of two crystallographically independent sub-structures, $[A]_\infty$ chains and $[(A',B)O_3]_\infty$ columns made up of distinct ratios of face-sharing octahedra and trigonal prisms. In many cases, the ratio of the repeat distances of the two chains is not a rational number and, consequently, the structure is incommensurately modulated along the chain direction. As shown previously, a better structural formulation of such composites is $A_{1+x}(A'_xB_{1-x})O_3$, where $x = n/(3m+2n)$ and ranges continuously between 0 and 1/2, corresponding to chains containing all face-shared octahedra and alternating face-sharing octahedra and trigonal prisms, respectively. [27-29] For simple fractional values of x such as 1/5, 2/7, or 1/3, the structure is commensurate and the end-members, the 2H perovskite (BaNiO$_3$, $x = 0$) and the K$_4$CdCl$_6$ ($x = 1/2$) structure type, are well known.

Most structural and physical property measurements of these compounds (and oxides in general) have been carried out on polycrystalline samples, as oxide single-crystal growth is often unsuccessful. Recently, however, the application of single-crystal flux-growth techniques has enabled the growth of large, high-quality single crystals of this family of oxides, [30-32] which has made possible precise structural determination of both commensurate and incommensurate compounds, and promises to offer a deeper insight into these materials. We have discussed a 4 dimensional superspace group approach in recent papers, and further detailed explanations of the composite structure approach and its description using the superspace formalism can be found in several papers and references therein. [33-37]

EXPERIMENTAL

Synthesis and Crystal Growth. Black hexagonal rod-like single crystals of Ba$_8$CoRh$_6$O$_{21}$ ranging in size from sub-millimeter to ~ 7 mm in length were grown from molten potassium carbonate flux. Rh powder (0.20 g, 1.94 mmol; Engelhard, 99.5%), Co$_3$O (0.24 g, 1.00 mmol; Johnson-Matthey, 99.99%), BaCO$_3$ (1.16 g, 5.88 mmol; Alfa, 99.99%) and K$_2$CO$_3$ (15.8 g, 114 mmol; Fisher, reagent grade) were mixed thoroughly and placed in an alumina crucible. The filled crucible was covered and heated in air from room temperature to the reaction temperature of 1050°C at 600° h^{-1}, held at 1050°C for 12 h and cooled at 12°C h^{-1} to 875°C, at which point the furnace was turned off and the system allowed to cool to room temperature. The flux was removed with water and the crystals isolated manually. Figure 1 is an image of several crystals obtained from the flux, showing the variability in size and aspect. For all crystals, the presence of Ba, Co, and Rh was verified by SEM.

Data Collection. For the structure determination, a small crystal was carefully chosen. Data collection was performed on an Enraf-Nonius CAD4 diffractometer (Mo-Kα) in the supercell approach. The unit cell parameters of both subsystems were determined precisely and refinements led to the following values : $a = 10.0431(1)$ Å, $c_1 = 2.5946(1)$ Å for the [(Co, Rh)O$_3$] subsystem and $a = 10.0431(1)$ Å, $c_2 = 4.5405(1)$ Å for [Ba] subsystem. With the [(Co, Rh)O$_3$] subsystem as the reference system, the modulation wave vector is defined by $\mathbf{q} = \gamma c_1{}^*$ where γ $c_2{}^*/c_1{}^* = c_1/c_2 = 0.57143(1)$. The γ value of 0.57143(1) is a rational fraction (4/7) and corresponds to a commensurate chain sequence of 1 trigonal prism and 6 octahedra.

The X-ray intensity data were collected in the supercell approach. Data reduction, absorption corrections (Psi - scan) and transformation of the indices were carried out using the JANA 2000 program package. [38] The atomic coordinates for the two subsystems are given in Table I. The first and second subsystems are related to the (3+1)D superspace group by the application of the W$_1$ and W$_2$ transformation matrices respectively:

$$W_1 = \begin{vmatrix} 1 & 0 & 0 & 0 \\ 0 & 1 & 0 & 0 \\ 0 & 0 & 1 & 0 \\ 0 & 0 & 0 & 1 \end{vmatrix} \qquad W_2 = \begin{vmatrix} 1 & 0 & 0 & 0 \\ 0 & 1 & 0 & 0 \\ 0 & 0 & 0 & 1 \\ 0 & 0 & 1 & 0 \end{vmatrix}$$

The two possible superspace groups that are compatible with the observed extinction conditions of $-h + k + l \neq 3n$ for (h, k, l, m) and $m \neq 2n$ for $(h, 0, l, m)$ are R3m(00γ)0s and the corresponding centrosymmetric superspace group $R\overline{3}m(00γ)0s$. $R\overline{3}m(00γ)0s$ was chosen initially and confirmed by the successful solution of the structure. Final R values: R = 0.0537 (Rw = 0.0598) for all reflections; R = 0.0417 (Rw = 0.0463) for 501 main reflections; R = 0.0735 (Rw = 0.0718) for 934 satellites of order 1; R = 0.0652 (Rw = 0.0894) for 174 satellites of order 2.

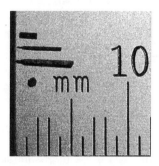

Figure 1: Flux-grown single crystals of $Ba_8CoRh_6O_{21}$. The specimens indicate the range in aspect ratio that is achievable, where the small crystal shown is of a typical size used for the structure determination.

RESULTS AND DISCUSSION

For $Ba_8CoRh_6O_{21}$, the subsystem $[(Co, Rh)O_3]$ was chosen as the reference system with the superspace group $R\overline{3}m(00γ)0s$. The fractional atomic average coordinates and thermal parameters are given in Table I. The symmetry of the superspace group generates six equivalent oxygen atoms surrounding the axis of the $[(Co, Rh)O_3]$ chains. Since all six positions cannot be simultaneously occupied, two types of oxygen positions, O_a and O_b, are generated that are each half occupied. Each set of three oxygens forms an equilateral triangle corresponding to the shared face of the polyhedra in the transition metal chain. A Crenel function, which runs along the x_4 axis and is characterized by the Crenel width, Δ, and the Crenel midpoint, $\hat{x}$, is used to model this occupational modulation. Thus two adjacent O_a or O_b triangles create a trigonal prism, O_a-O_a or O_b-O_b, while a sequence of O_a-O_b or O_b-O_a creates an octahedra, thus determining the polyhedral chain sequence for the transition metal chain. [22,27]

From the value of γ = 0.57143(1) (or 4/7), the value of x can be calculated according to the following relationship γ = (1+x)/2, thereby x = 1/7. For this compound, x is a rational fraction indicating a commensurate structure. From the value of x and γ, it is now possible to predict the polyhedra sequence for the $[(Co, Rh)O_3]$ column. [27] Figure 2 illustrates the ability to predict the sequence of polyhedra by carrying out 7 translations of 4/7 along x_4 that produces a

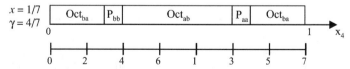

Figure 2: Schematic of the sequence of octahedra and prisms for x = 1/7 visualized along the internal coordinate x_4 for a column of $[(A', B)O_3]$.

Table I: Fractional atomic average coordinates and equivalent isotropic displacement factors, and atomic positional and DWF modulation coefficients.

Subsystem [(Co, Rh)O$_3$]: $R\bar{3}m(00\gamma)0s$

atom	x_0	y_0	z_0	$U_{eq}(\text{Å}^2)$
Co	0	0	0	0.0312(7)
Rh	0	0	0	0.0078(1)
O	0.148(1)	0.148(1)	1/2	0.049(2)

Subsystem [Ba]: $P\bar{3}c1(001/\gamma)$

atom	x_0	y_0	z_0	$U_{eq}(\text{Å}^2)$
Ba1	0.32593(5)	0	1/4	0.0118(2)
Ba2	0.34939(9)	0	1/4	0.0069(3)

Co: Amplitude = -0.0135 $\hat{x} = 1/4$ $\Delta = 0.0714$

Rh: $U_{z,1}^{Rh} = -0.0058(2)$ $U_{z,3}^{Rh} = -0.0050(2)$

Amplitude = -0.0805 $\hat{x} = 0$ $\Delta = 0.4286$

$U_{U11,2}^{Rh} = U_{U22,2}^{Rh} = 2U_{U12,2}^{Rh} = -0.0032(1)$ $U_{U33,2}^{Rh} = -0.0114(2)$

O: $U_{x,1}^{O} = U_{y,1}^{O} = 0.0079(6)$ $U_{x,2}^{O} = -U_{y,2}^{O} = 0.0003(2)$

$U_{z,2}^{O} = 0.0101(6)$

$U_{x,3}^{O} = -U_{y,3}^{O} = 0.0009(1)$ $U_{z,3}^{O} = 0.0030(5)$

$U_{x,1}^{O} = U_{y,1}^{O} = -0.00001(7)$

Amplitude = -0.0939 $\hat{x} = 1/4$ $\Delta = 1/2$

$U_{U11,1}^{O} = U_{U22,1}^{O} = -0.032(2)$ $U_{U33,1}^{O} = -0.024(2)$

$U_{U12,1}^{O} = -0.007(2)$ $U_{U13,1}^{O} = -U_{U23,1}^{O} = -0.006(1)$

$U_{U11,2}^{O} = -U_{U22,2}^{O} = -0.0011(3)$ $U_{U13,2}^{O} = U_{U23,2}^{O} = 0.0025(3)$

Ba1: $2U_{x,1}^{Ba1} = U_{y,1}^{Ba1} = -0.0063(2)$ $U_{z,1}^{Ba1} = -0.0021(1)$

$U_{x,2}^{Ba1} = -0.0011(2)$

Amplitude = -0.0029 $\hat{x} = 1/2$ $\Delta = 0.2083$

$U_{U11,1}^{Ba1} = U_{U12,1}^{Ba1} = 0.0013(4)$ $U_{U13,1}^{Ba1} = -0.0034(3)$

$U_{U11,2}^{Ba1} = -0.0018(4)$ $U_{U22,2}^{Ba1} = 2U_{U12,2}^{Ba1} = 0.0087(7)$

$U_{U33,2}^{Ba1} = 0.00002(24)$ $2U_{U13,2}^{Ba1} = U_{U23,2}^{Ba1} = 0.0057(4)$

Ba2: $2U_{x,1}^{Ba2} = U_{y,1}^{Ba2} = -0.0019(2)$ $U_{z,1}^{Ba2} = 0.0021(2)$

$U_{x,2}^{Ba2} = -0.0029(1)$

Amplitude = -0.0032 $\hat{x} = 0$ $\Delta = 0.1250$

$U_{U11,1}^{Ba2} = U_{U12,1}^{Ba2} = -0.0004(5)$ $U_{U13,1}^{Ba2} = -0.0095(4)$

$U_{U11,2}^{Ba2} = -0.0092(5)$ $U_{U22,2}^{Ba2} = 2U_{U12,2}^{Ba2} = 0.0033(8)$

$U_{U33,2}^{Ba2} = 0.0082(4)$ $2U_{U13,2}^{Ba2} = U_{U23,2}^{Ba2} = 0.0026(4)$

sequence of Oct_{ab}-Oct_{ba}-Oct_{ab}-P_{bb}-Oct_{ba}-Oct_{ab}-Oct_{ba} or six octahedra followed by one trigonal prism.

To distinguish the trigonal prismatic sites from the octahedral sites, a saw-tooth function was employed to account for the displacive modulation along x_4 of the atoms. [22] A graphical representation of the functions utilized is shown in Figure 3. Note how well the saw-tooth function effectively models the modulation of cobalt. The modulation of oxygen, rhodium, and barium deviate slightly from the ideal saw-tooth function, and so, consequently, an additional modulation was added to the saw-tooth function; a second order modulation for barium and a third order modulation for both rhodium and oxygen.

(a)   (b)

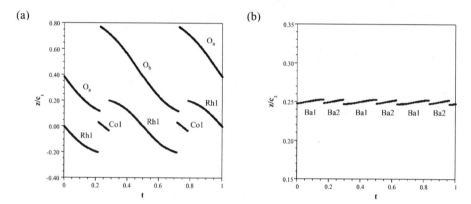

Figure 3: *Graphical representation of the internal z coordinate of (a) O_a, O_b, Rh1, and Co1 and (b) Ba1 and Ba2 versus the internal coordinate t ($t = x_4 - q \cdot r$).*

An approximate [110] view of the composite structure of $Ba_8CoRh_6O_{21}$ is shown in Figure 4. The repeat sequence in the face-shared polyhedral $[(A',B)O_3]_\infty$ subsystem consists of six consecutive RhO_6 octahedra followed by one distorted CoO_6 trigonal prism. The metal-

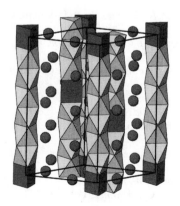

Figure 4: *Approximate [110] view of the structure of $Ba_8CoRh_6O_{21}$. Light gray: RhO_6 octahedra; Dark gray: CoO_6 trigonal prisms; Gray spheres: Ba atoms.*

oxygen bond distances (Co-O = 2.161(1) Å - 2.241(1) Å; Rh-O = 1.896(1) Å - 2.092(1) Å) are typical for oxides of this type. Intrachain Co-Rh (2.780(1) Å) and Rh-Rh (2.496(1) Å - 2.543(1 Å) distances are essentially non-bonding.

CONCLUSION

Successful structure solution of $Ba_8CoRh_6O_{21}$ using the superspace formalism indicate the efficiency of the method for commensurate as well as incommensurate structures. Thi compound represents the first example of an $m = 5$, $n = 3$ member of the $A_{3n+3m}A'_nB_{3m+n}O_{9m+6}$ family and one of only a few compositions other than $n = 1$ and $m = 0$ that crystallize in commensurate form. In addition, this compound displays some very interesting and highl anisotropic magnetic behavior that will be reported elsewhere. [39]

ACKNOWLEDGEMENT

Financial support from the National Science Foundation through Grant DMR:9873570 i gratefully acknowledged.

REFERENCES

1.   Schlenker, C.; Dumas, J. *Crystal Chemistry and Properties of Materials with Quasi-On Dimensional Structures, A Chemical and Physical Approach*; Rouxel, J. (ed), D. Reid Publishing Co.: Boston 1986, p. 135.
2.   de Jongh, L. J.; Miedema, A. R. *Adv. Phys.* **1974**, *23*, 1.
3.   Day, P. *Solid State Chemistry Compounds*, Cheetham, A.K.; Day P. (eds), Clarendc Press: Oxford 1992, Chp. 2.
4.   Nguyen, T. N.; Giaquinta, D. M.; zur Loye, H.-C. *Chem. Mater.* **1994**, *6*, 1642.
5.   Nguyen, T. N.; Lee, P. A.; zur Loye, H.-C. *Science* **1996**, *271*, 489.
6.   Fjellvåg, H.; Gulbrandsen, E.; Aasland, S.; Olsen, A.; Hauback, B.C. *J. Solid State Chen* **1996**, *124*, 190.
7.   Kagayama, H.; Yoshima, K.; Kosuge, K.; Mitamura, H.; Goto, T. *J. Phys. Soc. Jp* **1997**, *66*, 1607.
8.   Campá, J.A.; Gutiérrez-Puebla, E.; Monge, M.A.; Rasines, I.; Ruíz-Valero, C. *J. Sol State Chem.* **1994**, *108*, 203.
9.   Harrison, W.T.A.; Hegwood, S.L.; Jacobson, A.J. *J. Chem. Soc., Chem. Commun.* **1995** 1953.
10.   Battle, P.D.; Blake, G.R.; Darriet, J.; Gore, J.G.; Weill, F.G. *J. Mater. Chem.* **1997**, *7*, 1559.
11.   Strunk, M.; Müller-Buschbaum, Hk. *J. Alloys.Comp.* **1994**, *209*, 189.
12.   Dussarrat, C.; Fompeyrine, J.; Darriet, J. *Eu. J. Solid State Inorg. Chem.* **1995**, *32*, 3.
13.   Campá, J.; Gutierrez-Puebla, E.; Monge, A.; Rasines, I.; Ruíz-Valero, C. *J. Solid State Chem.* **1996**, *126*, 27.
14.   Layland, R. C.; Claridge, J. B.; Adams, R. D.; zur Loye, H.-C. *Z. Anorg. Allg. Chem.* **1997**, *623*, 1131.
15.   Reisner, B. A.; Stacy, A. M. *J. Am. Chem. Soc.* **1998**, *120*, 9682.
16.   Blake, G.R.; Sloan, J.; Vente, J.F.; Battle, P. D. *Chem. Mater.* **1998**, *10*, 3536.

17. Battle, P.D.; Blake, G.R.; Darriet, J.; Gore, J.G.; and Weill, F.G. *J. Mater. Chem.* **1997**, *7*, 1559.

18. Huvé, M.; Renard, C.; Abraham, F.; Van Tendeloo, G.; Amelinckx, S. *J. Solid State Chem.* **1998**, *135*, 1.

19. Boulahya, K.; Parras, M.; González-Calbet, J. M. *J Solid State Chem.* **1999**, *142*, 419.

20. Smith, M. D.; zur Loye, H.-C. *Chem. Mater.* **2000**, *12*, 2404.

21. Beauchamp, K. M.; Irons, S. H.; Sangrey, T. D.; Smith, M. D.; zur Loye, H.-C. *Phys. Rev. B.* **2000**, *61*, 11594.

22. Zakhour-Nakhl, M.; Claridge, J. B.; Darriet, J.; Weill, F.; zur Loye, H.-C.; Perez-Mato, J.-M. *J. Am. Chem. Soc.* **2000**, *122*, 1618.

23. Layland, R. C.; zur Loye, H.-C. *J. Alloys Comp.* **2000**, *299*, 118.

24. Layland, R. C.; Kirkland, S. L.; zur Loye, H.-C. *J. Solid State Chem.* **1998**, *139*, 79.

25. Smith, M. D.; zur Loye, H.-C. *Chem. Mater.* **1999**, *11*, 2984.

26. Darriet, J.; Subramanian, M. A. *J. Mater. Chem.* **1995**, *5*, 54.

27. Perez-Mato, J. M.; Zakhour-Nakhl, M.; Weill, F.; Darriet, J. *J. Mater Chem.* **1999**, *9*, 2795.

28. Evain, M.; Boucher, F.; Gourdon, O.; Petricek, V.; Dusek, M.; Bezdicka, P. *Chem. Mater.* **1998**, *10*, 3068.

29. Gourdon, O.; Petricek, V.; Dusek, M.; Bezdicka, P.; Durovic, S.; Gyepesova, D.; Evain, M. *Acta Cryst.* **1999**, *B55*, 841

30. Claridge, J. B.; Layland, R. C.; Henley, W. H.; zur Loye, H.-C. *Chem. Mater.* **1999**, *11*, 1376.

31. Henley, W. H.; Claridge, J. B.; zur Loye, H.-C. *J. Cryst. Growth* **1999**, *204*, 122.

32. zur Loye, H.-C.; Layland, R. C.; Smith, M. D.; Claridge, J. B. *J. Cryst. Growth* **2000**, *211*, 452.

33. Van Smaalen, S. *Phys. Rev. B* **1991**, *43*, 11330.

34. Janner, A.; Janssen T. *Acta Crystallogr. A* **1980**, *36*, 399.

35. Janner A.; Janssen, T. *Acta Crystallogr. A* **1980**, *36*, 408.

36. Perez-Mato, J. M.; Madariaga, G.; Zuniga, F. J.; Garcia Arribas, A. *Acta Crystallogr. A* **1987**, *43*, 216.

37. Van Smaalen, S. *Crystallogr. Rev.* **1995**, *4*, 79.

38. Dusek, M.; Petricek, V.; Wunschel, M.; Dinnebier, R.E.; van Smaaleen, S. *J. Appl. Cryst.*, Submitted.

39. zur Loye, H.-C.; Stitzer, K.E.; Smith, M.D.; El Abed, A.; Darriet, J. *Chem Mater.*, Submitted.

Mat. Res. Soc. Symp. Proc. Vol. 658 © 2001 Materials Research Society

The Formation of Striped Phases by Charge Localization in Vanadium Phosphates

Ranko P. Bontchev, Junghwan Do and Allan J. Jacobson
Department of Chemistry, University of Houston,
Houston, TX 77204-5641, USA.

ABSTRACT

The compound $(H_3O)_{1.5}\{[VO(H_2O)]PO_4\}_3 \cdot 2.5H_2O$ is a new member of a series of vanadium phosphates containing vanadium in both plus four and five oxidation states. Charge ordering occurs in these mixed vanadium phosphate layered compounds forming striped phases with different periodicities. The $VOPO_4$ layers undulate as a consequence of differences in the O-O atom distances in the ordered $V(V)O_6$ and $V(IV)O_6$ octahedra.

INTRODUCTION

Many aspects of the chemistry of layered vanadium phosphates have been studied since Ladwig's early work on the synthesis and properties of $VOPO_4 \cdot 2H_2O$ [1]. The structure of $VOPO_4 \cdot 2H_2O$ was determined by Tietze using single crystal X-ray diffraction [2]. Layers are formed by connecting distorted $VO_5(H_2O)$ octahedra with PO_4 tetrahedra. Adjacent $VO_5(H_2O)$ octahedra are related by an inversion center within the layer so that the orientation of the vanadyl V=O groups alternates above and below the layer (see Figure 1). The interlayer space is occupied by an additional water molecule that is hydrogen bonded to the coordinated water molecule. Intercalation reactions of both neutral molecules and ions have been reported [3-7]. Mixed valence compounds are formed by insertion of cations into the interlayer space and reduction of V(V) to V(IV) [3,5]. A large number of compounds based on $VOPO_4$ layers were subsequently synthesized in single crystal form by using mild hydrothermal conditions [8-12]. The crystal chemistry, including the interlayer cation and solvent molecule locations, are now well understood from the detailed structural studies. Here we note only that phases are known with a water molecule coordinated to the vanadium atom (distorted octahedral) [8,12] and without (square pyramidal) [8-10] and that only a small number of mixed valence systems have been characterized by single crystal X-ray diffraction [13-17]. The known mixed valence systems are summarized in Table I. A small number of mixed valence compounds in which the layers are cross linked by an additional phosphate group are also included [18].

Here we discuss the evidence for charge ordering in the mixed valence compounds and describe a new example with the composition $(H_3O)_{1.5}\{[VO(H_2O)]PO_4\}_3 \cdot 2.5H_2O$.

EXPERIMENTAL

Synthesis: a) The compound $Na[VO(H_2O)PO_4]_2 \cdot 2 H_2O$ (1) was prepared by hydrothermal synthesis in a Teflon lined steel autoclave at 130°C for 3 days: $NaVO_3$ (0.122 g, 1 mmol), BPO_4 (0.317 g, 3 mmol) and 3 mL H_2O. The reaction products were prismatic dark green crystals of (1), average size 0.1-0.3 mm, yield approximately 80% based on vanadium, together with some unidentified brownish powder. b) $(H_3O)_{1.5}\{[VO(H_2O)]PO_4\}_3 \cdot 2.5H_2O$ (2): V_2O_5 (1.819 g, 10 mmol), $H_2C_2O_4$ (0.480 g, 5 mmol), 1.20 mL H_3PO_4 (85 wt% solution in H_2O, 20 mmol) and 30 mL H_2O were mixed in a 100 mL flask and refluxed for 10 h at 70°C in air. After cooling the pH was adjusted to 1.33 by adding dropwise HNO_3 and the resulting blue solution was left at room temperature in air. Dark green plate-like crystals began forming after several days and the crystallization process was completed within a week, yield approximately 80% based on vanadium.

Table I. *Mixed valence layered vanadium phosphates.*

Compound	BVS [20]	Fig. No.	o × d*	Ref.
$Na_{0.5}VOPO_4 \cdot 2H_2O$	4.55, 4.56	Fig. 3c	0 × 1	[13]
	4.59, 4.69			[14]
1	4.57			this work
$K_{0.5}VOPO_4 \cdot 1.5H_2O$	4.57	Fig. 3c	0 × 1	[15, 13]
$Rb_{0.5}VOPO_4 \cdot 1.5H_2O$	4.51	Fig. 3c	0 × 1	[15]
$N_2C_2H_{10}(VO)_2(PO_4)_2 \cdot H_2PO_4$	4.97, 5.15	Fig. 3d	1 × 0	[19]
	4.04, 4.08			
$(C_4H_{12}N_2)(H_3O)[(VOPO_4)_4(H_2O)H_2PO_4] \cdot 3H_2O$	4.07	Fig. 3d	1 × 0	[18]
	5.09			
$(NH_4)_{0.5}VOPO_4 \cdot 1.5H_2O$	3.94	Fig. 3b	1 × 1	[17, 16]
	5.27			
	4.55			
$(H_3O)_{1.5}\{[VO(H_2O)]PO_4\}_3 \cdot 2.5H_2O$ **2**	4.07	Fig. 3a	2 × 1	this work
	5.04			
	4.46			

* o × d refer to the numbers of ordered (o) and disordered (d) 'stripes', BVS ≡ Bond Valence Sum.

Structure Determination: Single crystal X-ray diffraction data were obtained using a Bruker SMART 1K CCD diffractometer with Mo-K_α radiation and a graphite monochromator.

a) $Na[VO(H_2O)PO_4]_2 \cdot 2.5\ H_2O$ (**1**) at T = 293K: triclinic, space group P-1 (#2), a = 6.2883(10), = 6.2901(10), c = 6.8465(11) Å, α = 107.199(2), β = 92.453(3), γ = 90.102(3)°, V = 258.43(7) Å^3, Z = 1, ρ_{calcd} = 2.685 gcm^{-3}; μ = 2.069 mm^{-1}, θ_{max} = 39.96°, 831 measured reflections, 472 independent reflections; corrections for Lorentz factor, polarization, air absorption and absorption due to variations in the path length through the detector faceplate, Psi-scan absorpt correction. The structure was solved by direct methods and refined by full matrix least squares on F^2 (SHELXTL)[21], all non-H atoms anisotropic; GOF = 1.278, R values (I > 2σ(I)): R1 = 0.0218, wR2 = 0.0635; max./min. residual electron density 0.211 and -0.202 e Å^{-3}.

b) $(H_3O)_{1.5}\{[VO(H_2O)]PO_4\}_3 \cdot 2.5H_2O$ (**2**) at T = 293K: orthorhombic, space group Pbca (#61) a = 8.8453(6), b = 14.1484(9), c = 26.401(2) Å, V = 3304.0(4) Å^3, Z = 8, ρ_{calcd} = 2.403 gcm^{-3} = 2.069 mm^{-1}; θ_{max} = 35.98°, 7169 measured reflections, 1138 independent reflections; corrections as above. The structure was solved by direct methods and refined by full matrix least squares on F^2 (SHELXTL)[21], all non-H atoms anisotropic; GOF = 0.961, R values (I > 2σ(I)) R1 = 0.0709, wR2 = 0.1315; max./min. residual electron density 0.497 and -0.487 e Å^{-3}.

RESULTS

The mixed valent vanadium phosphates form an interesting series of systems with respect to the distribution of V(V) and V(IV) ions within the layers. All of the compounds in Table 1 have close to equal amounts of V(V) and V(IV) but have very different structures. In the alkali metal series, the structures of $K_{0.5}VOPO_4 \cdot 1.5H_2O$ and $Rb_{0.5}VOPO_4 \cdot 1.5H_2O$ are well determined and contain a single inequivalent vanadium atom. Some disagreement, however, exists in the literature concerning the structure of $Na_{0.5}VOPO_4$ $2H_2O$. Wang *et al.* solved the structure in space group P-1 with Z=4 [13], whereas Yamase and Makino [14] solved the structure in P1 with Z=1. We have synthesized and redetermined the structure of this phase (**1**) and obtained an excellent refinement in P-1 with Z=1 with only one inequivalent vanadium atom. All of the reported structure solutions for $Na_{0.5}VOPO_4 \cdot 2H_2O$ are similar and it is possible that small composition variations are responsible for the differences. The bond valence sum values have been calculated for each structure and are given in Table I. The values indicate that the V(IV) and V(V) sites are disordered in **1**. The structure of (**1**) is shown in Figure 1.

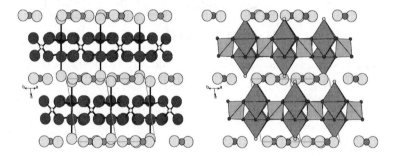

Figure 1. *The structure of Na[VO(H₂O)PO₄] 2H₂O* (**1**) *Selected bond lengths (Å): V(1)-O(1) = 1.578(3), V(1)-O(6) = 1.957(3), V(1)-O(5) = 1.958(3), V(1)-O(4) = 1.961(3), V(1)-O(3) = 1.962(3), V(1)-O(2) = 2.357(4). Bond valence sum (BVS): V(1) = 4.45.*

Single crystal X-ray diffraction studies revealed that the structure of (**2**) is also based on $VOPO_4 \cdot H_2O$ layers and contains vanadium in both V(V) and V(IV) oxidation states. The unit cell contents of the structure are shown in Figure 2. The bond distances and angles are typical for hydrated oxovanadium phosphates. In contrast to the parent $VOPO_4 \cdot 2H_2O$ and the mixed valence alkali metal phases, the structure contains three crystallographically distinct vanadium atoms, which bond valence sums indicate have different oxidation states. The atoms V1 and V2 are V(IV) and V(V) respectively and V3 is a disordered mixture of V(IV) and V(V). As a consequence of the vanadium charge ordering the layers are no longer flat but undulate periodically along the *c* axis. Similar behavior was observed previously in $(NH_4)_{0.5}VOPO_4 \cdot 1.5H_2O$ [16,17] and is discussed below. Hydronium cations and water molecules occupy interlayer sites in a complex and disordered arrangement.

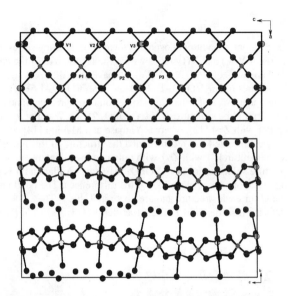

Figure 2. *Two projections of the unit of cell contents of* $(H_3O)_{1.5}\{[VO(H_2O)]PO_4\}_3·2.5H_2O$. *Selected bond lengths* (Å): V(1)-O(1) = 1.602(10), V(1)-O(4) =1.946(10), V(1)-O(6) =1.957(10), V(1)-O(11) =1.984(11), V(1)-O(12) =1.999(13), V(1)-Ow1 =2.273(12), V(2)-O(2) =1.546(11),V(2)-O(5) =1.873(10), V(2)-O(7) =1.884(10), V(2)-O(=1.930(11), V(2)-O(8) =1.935(11), V(2)-Ow2 =2.305(11),V(3)-O(3) =1.598(12),V(3)-O(10) =1.899(13), V(3)-O(14) =1.912(11), V(3)-O(13) =1.926(12), V(3)-O(15) =1.923(12), V(3)-Ow3 =2.315(13). Bond valence sums (BVS): V(1) = 4.07, V(2) = 5.04, V(3) = 4.46.

DISCUSSION

The structures of the compounds in Table 1 other than those of the alkali metal compou contain ordered arrangements of all or some of the V(IV) and V(V) atoms. The ordering apparent on inspection of the bond lengths or by calculation of bond valence sums [20]. ordering is shown schematically in Figure 3. In the horizontal direction in the figure, not only the oxidation states alternate but the V(IV) and V(V) octahedra alternate up and down so that the V(V) octahedra are on one side of the layer and the V(IV) octahedra are on the other. consequence of the charge ordering is that the $VOPO_4·H_2O$ layers must bend to accommodate larger and smaller V(IV) and V(V) octahedra on opposite sides of the layer. The relative num of the ordered and disordered rows ('stripes') is also indicated in Table I.

The origin of the undulation can be seen in more detail by inspection of the edge-on view the $VOPO_4$ layer in $(NH_4)_{0.5}VOPO_4·H_2O$ (Figure 4). The V1 and V2 octahedra are V(V) and V(IV) respectively, and the V3 octahedra are disordered. In the $V1O_6$ octahedron the O1-O1 ar O2-O2 distances are 2.851 and 2.858 Å and in the $V2O_6$ octahedron the O5-O5 and O3-O3 separations are 2.653 and 2.614 Å. Because V1, V2, and V3 share oxygen atoms from a comm tetrahedron, the consequence of the difference in oxygen-oxygen distances is to force the V3O octahedra to tilt. In the $V3O_6$ octahedra the O-O distances are similar (O7-O9 is 2.694 Å and (O8 is 2.739 Å). The orientation of the next pair of V1 and V2 octahedra in the horizontal direction in Figure 4 is inverted and consequently along this direction the layers undulate in a regular way. Similar behavior is observed for the other ordered structures.

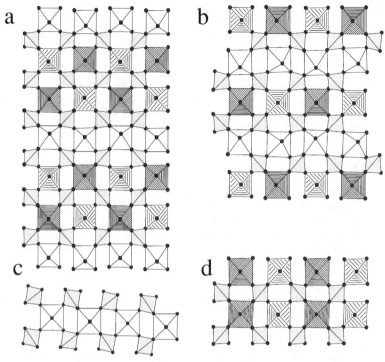

a b c d

Figure 3. *Ordered arrangements of V(IV) and V(V) in vanadium phosphate layered compounds. Phosphate tetrahedra are grey, V(V), V(IV) and disordered V(IV,V) octahedra are hatched, lightly hatched, and open, respectively. The arrangements correspond to different numbers of ordered(o) and disordered rows (d) and are represented by the symbol (o ×d). Figures 3a, b, c, d, correspond to the arrangements 2 ×1, 1 ×1, 0 ×1, 1 ×0.*

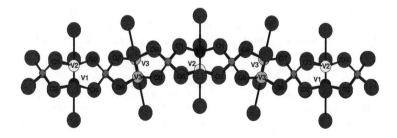

Figure 4. *Edge on view of the layer in $(NH_4)_{0.5}VOPO_4 \cdot 1.5H_2O$.*

The origin of the charge localization is more difficult to describe. Most likely it arises from the distribution of the interlayer ammonium and/or hydronium cations in the interlayer space an the hydrogen bonds that they form with the oxygen atoms in the layers. A detailed discussion of this is beyond the scope of this paper. We note here only that very strong hydrogen bonds (O-O distances 2.7Å) are formed with the water molecules coordinated to the V(V) atoms and that analysis of the hydrogen bonding is difficult because of the disorder of the interlayer species.

ACKNOWLEDGEMENTS

This work was supported in part by the National Science Foundation under Grant DMR–9805881 and by the Robert A. Welch Foundation. The work made use of MRSEC/TCSUH Shared Experimental Facilities supported by the National Science Foundation under Award Number DMR-9632667 and the Texas Center for Superconductivity at the University of Houston.

REFERENCES

1. G. Ladwig, *Z. Anorg. Allg. Chemie*, **338**, 266 (1965).
2. H. Tietze, *Aust. J. Chem.*, **34**, 2035 (1981).
3. J. W. Johnson, A. J. Jacobson, J. F. Brody, and S. M. Rich, *Inorg. Chem.*, **21**, 3820 (1982).
4. J. W. Johnson, and A. J. Jacobson, *Angew. Chem. Int. Ed. Engl.*, **22**, 412 (1983).
5. A. J. Jacobson, J. W. Johnson, J. F. Brody, J. C. Scanlon, and J. W. Lewandowski, *Inorg. Chem.*, **24**, 1782 (1985).
6. P. Amoros, M. Dolores Marcos, A. Beltran-Porter, and D. Beltran-Porter, *Current Opinions in Solid State and Materials Science*, **4**, 123 (1999).
7. J. Kalousova, J. Votinsky, L. Benes, K. Melanova, V. Zima, *Collect. Czech. Chem. Commun.*, **63**, 1 (1998).
8. H. Y. Kang, W. C. Lee, S. L. Wang, and K.-H. Lii, *Inorg. Chem.*, **31**, 4743 (1992).
9. K.-H. Lii, L.-S. Wu, and H.-M. Gau, *Inorg.Chem.*, **32**, 4153 (1993).
10. R. C. Haushalter, V. Soghomonian, and Q. Chen, *J. Solid State Chem.*, **105**, 512 (1993).
11. Y. Zhang, A. Clearfield, and R. C. Haushalter, *J. Solid State Chem.*, **117**, 157 (1995).
12. M. Roca, M. Dolores Marcos, P. Amoros, J. Alamo, A. Beltran-Porter, and D. Beltran-Porte *Inorg. Chem.*, **36**, 3414 (1997).
13. S. L. Wang, H. Y. Kang, C. Y. Cheng, and K.-H. Lii, *Inorg. Chem.*, **30**, 3496 (1991).
14. T. Yamase, and H. Makino, *J.Chem. Soc. Dalton Trans.*, 1143 (2000).
15. D. Papoutsakis, J. E. Jackson, and D. G. Nocera, *Inorg. Chem.*, **35**, 800 (1996).
16. L. Liu, X. Wang, R. Bontchev, K. Ross, and A. J. Jacobson, *J. Mat. Chem.*, **9**, 1585 (1999).
17. J. Do, R. P. Bontchev, and A. J. Jacobson, *Inorg. Chem.*, **39**, 3230 (2000).
18. J. Do, R. P. Bontchev, A. J. Jacobson, *J. Solid State Chem.*, in press (2000).
19. W. T. A. Harrison, K. Hsu, A. J. Jacobson, *Chem. Mater.*, **7**, 2004 (1995).
20. N. E. Brese, M. O'Keeffe, *Acta Crystallogr.*, **B47**, 192 (1991).
21. G. M. Sheldrick, Program SADABS, University of Gottingen, **1995**; b) G. M. Sheldrick, SHELXTL, Version 5.03, Siemens Analytical X-ray Instruments, Madison, WI, **1995**; c) G. M. Sheldrick, SHELX-96 Program for Crystal Structure Determination, **1996**.

Mat. Res. Soc. Symp. Proc. Vol. 658 © 2001 Materials Research Society

Synthesis and Crystal Chemistry of New Transition Metal Tellurium Oxides in Compounds Containing Lead and Barium

Boris Wedel, DOWA Mining Company Ltd., Central Research Laboratory, Hachioji, Japan,
Katsumasa Sugiyama, Kimio Itagaki, Tohoku University, Institute for Advanced Materials
Processing, Sendai Japan.
Hanskarl Müller-Buschbaum, Christian Albrechts University, Institute for Inorganic
Chemistry, Kiel, Germany.

ABSTRACT

During the past decades the solid state chemistry of tellurium oxides has been enriched
by a series of quaternary metallates. Interest attaches not only to the chemical and physical
properties of these compounds, but also to their structure, which have been studied by modern
methods. The partial similarity of earth alkaline metals and lead in solid state chemistry and their
relationships in oxides opens a wide field of investigations. Eight new compounds in the systems
Ba-M-Te-O (M= Nb, Ta) and Pb-M-Te-O (M = Mn, Ni, Cu, Zn) were prepared and structurally
characterized: $Ba_2Nb_2TeO_{10}$, $Ba_2M_6Te_2O_{21}$ (M = Nb, Ta) and the lead compounds $PbMnTeO_3$,
$Pb_3Ni_{4.5}Te_{2.5}O_{15}$, $PbCu_3TeO_7$, $PbZn_4SiTeO_{10}$ and the mixed compound $PbMn_2Ni_6Te_3O_{18}$. The
structures of all compounds are based on frameworks of edge and corner sharing oxygen
octahedra of the transition metal and the tellurium. Various different channel structures were
observed and distinguished. The compounds were prepared by heating from mixtures of the
oxides, and the single crystals were grown by flux method or solid state reactions on air. The
synthesis conditions were modified to obtained microcrystalline material for purification and
structural characterizations, which were carried out using a variety of tools including powder
diffraction data and refinements of X-ray data. Relationships between lead transition metal
tellurium oxides and the earth alkaline transition metals tellurium oxides are compared.

INTRODUCTION

So-called 'lone pair' cations such as Pb^{2+} or Te^{4+}, contain two valence electrons in their
outer occupied s^2-orbital. This s^2-orbital can be stereoactiv. The influence of the stereoactiv s^2-
orbital becomes obvious by a distortion the local coordination environment. In crystal chemistry,
the lone pair can be treated as a ligand of the cation. The shift of the lone pair cation out of the
center of its coordination polyhedron results in an asymmetric coordination environment, which
can lead to a variety of interesting physical properties.

The tellurium oxide compounds are more and more attractive to the researcher, because
of the interesting structural chemistry of Te(IV) and as alternative material for tungsten or 4th an
5th group transition metals as solid state device materials.

Quaternary earth alkaline tellurium oxides with 4th period transition elements crystallize
in numerous variants of the perovskite type. The earth alkaline metal can be substituted by lead
in these systems and the perovskite structure is retained. Pure lead quaternary 4th period
transition elements tellurium oxides were hitherto unknown.

Our study includes high temperature preparations of quaternary and quinternary tellurium
oxide compounds with special regard to lead (with 4th period transition elements) and barium
(with 5th group transition metals). Stability studies of the perovskite structures in the system of
quaternary lead tellurium oxides with 4th period transition metals and the influence of the stereo

active s^2-orbital (lone pair) have been conducted. Finally, the stability of the Te(IV)/Te(VI) in the temperature range from 873 to 1273 K has been studied.

The hitherto unknown compounds [1-8] were synthesized as a polycrystalline powder and as single crystals in a solid state reaction on air. The reaction temperatures in this study were between 873 and 1273 K.

After cooling, the single crystals were checked by energy dispersive X-ray spectroscopy (Hitachi X650S, EDX-system KEVEX 7000Q). The crystals were stable on air and they were suitable for single crystal X-ray diffraction analysis. The crystals for this study were selected under a polarized light microscope.

The symmetry and the lattice parameters were determined with oscillation, Weissenberg photographs and imaging plate measurements (Rigaku Image plate CS, modified for the Institute of Material research (IMR)). The crystals used for the final intensity measurement were mounted on the top of glass fibers. Diffraction measurements at 293 K were made on a Rigaku AFC6S four-circle diffractometer using MoKα radiation. The cell parameters were refined by the least-square procedure.

The structure was solved by a combination of direct methods and difference Fourier syntheses (SHELX-Package [9, 10]). All positions were refined with anisotropic thermal parameters. The space group was detected by using the ABSEN program [11]. All structure plot appearing in this paper were made with the programs ORTEP [12] and STRUPOL [13].

THE Pb-M-Te-O SYSTEM (M= Mn, Fe, Co, Ni, Cu, Zn)

The results of our investigations and an overview over the relevant quaternary earth alkaline tellurium oxides are shown in Table I. It has to be pointed out that in the earth alkaline series only perovskite structures are observed. The lead compounds with a stereo active s^2-orbital compounds crystallize in a different crystal structure.

Table I. Overview of 4th period transition metals tellurium oxide combined with earth alkaline metals or lead.

	Mn	Fe	Co	Ni	Cu	Zn
EA						
Sr	Sr_2MnTeO_6 c.P.[1.)]	$Sr_3Fe_2TeO_9$ c.P.	$Sr_3Co_2TeO_9$ c.P.	Sr_2NiTeO_6 monoclinic P.	Sr_2CuTeO_6 Elpasolith	———
Ba	Ba_2MnTeO_6 hex. P.	$Ba_3Fe_2TeO_9$ c.P.	$Ba_3Co_2TeO_9$ c.P.	Ba_2NiTeO_6 hex.P.	Ba_2CuTeO_6 triclinic P.	———
Pb	$Pb(Mn,Te)O_3$ c. P.	$Pb_3Fe_2Te_2O_{12}$ hex. Bronze type	Pb_2CoTeO_6 Elpasolith	$Pb_3Ni_{4.5}Te_{2.5}O_{15}$ $Pb_3Mn_7O_{15}$- type	$PbCu_3TeO_7$ New cryst. structure	$PnZn_4SiTe$ New crystal structure

$^{1.)}$ P = perovskite

Because of the limited length of this paper, only a short summary of the structural features is given and finally a mixed quinternary compound will be discussed. The compounds are discussed in the order of the atomic weight of the transition metal.

The quaternary manganese compound [5] crystallizes in a cubic disordered variation of the perovskite type Figure 1. The tellurium and manganese share the octahedral position in the anionic part.

Figure 1. Crystal structure of Pb(Mn,Te)O₃.

The quaternary iron compound [4] is characterized by a network of octahedra occupied by Fe^{3+} and Te^{6+} in a disordered manner. The crystal structure is a monoclinic variant of the hexagonal bronze type.

The cobalt compound [4] belongs to the elpasolithe type, which is a tetragonal variation of the perovskite.

The typical features of the crystal structure of the quaternary nickel compound [6] are face sharing $TeNiO_9$ octahedra doubles and planes of connected $(NiO_6)_6$ and $(TeO_6)_3(NiO_6)_3$ hexagons Figure 2. The one sided coordination of lead by oxygen is complemented by the lone pair effect. The compound is isotopic to the $Pb_3Mn_7O_{15}$ type.

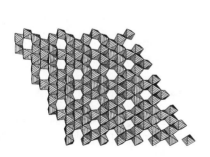 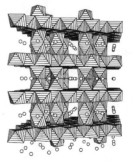

Figure 2. left: Plane of $Pb_3Ni_{4.5}Te_{2.5}O_{15}$ along [001].   right: Projection along [011].

The copper compound $PbCu_3TeO_7$ [3] crystallizes in a new crystal structure type, with interesting features. $Cu(1)O_5$ square pyramids and TeO_6 octahedra are members of the anionic part of the crystal structure and $Cu(2)$ atoms show tetrahedral coordination by oxygen. The $Cu(2)^{2+}$ and the Pb^{2+} cations are incorporated into a $[Cu(1)_2TeO_7]$ network.

In present stage, it was not possible to prepare a quaternary zinc compound in indicated temperature range. Only a quinternary compound has been synthesize so far. To knowledge, the $PbZn_4SiTe_{10}$ [7] compound represents the first quinternary silicon telluri oxide. The crystal structure is dominated by a $^3_\infty$ $[Zn_4O_{10}]$ framework with isolated TeO_6^{6+} SiO_4^{4+} polyhedra. The Pb^{2+} ions are incorporated in the network

The fact that the quaternary nickel compound crystallizes in the $Pb_3Mn_7O_{15}$ type led u the idea to check the stability of this compound with substitution of manganese. Surprisingly, substitution of nickel by manganese caused the $Pb_3Ni_{4.5}Te_{2.5}O_{15}$ crystal structure to change to new crystal structure of $PbMn_2Ni_6Te_3O_{18}$ [8]. Ni^{2+} and Te^{6+} show octahedral and Mn^{2+} shows trigonal prismatic coordination by O^{2-}. feature of this crystal structure is a $^3_\infty$ $[Ni_6Te_3O_{18}]^{6-}$ network together with two kinds of chann along [001] with trigonal and hexagonal shape. Isolated face shared $[Mn_2O_9]^{14-}$ trigonal prism polyhedra doubles are located in the trigonal channels while one-dimensional infinite chains $[PbO_{6+6}]^{22-}$ oxygen cube dodecahedra are incorporated in the hexagonal channels (see figure 3

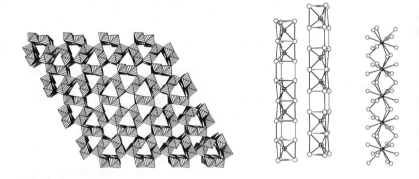

Figure 3. *right: Perspective view of the framework of $^3_\infty$ [Ni_6Te_3O_{18}]^{6-} along [001] left: $^1_\infty$ [Mn [empty]]^{8-} chain and $^1_\infty$ [PbO_9]^{16-} chain along [001] in PbMn_2Ni_6Te_3O_{18}.*

THE Ba-M-Te-O SYSTEM (M= Nb, Ta)

The barium compounds in these investigations are interesting as alternative materials molybdenum containing catalysts. We found two different stable compounds in the temperatu range between 873 k and 1273 K.

The Te(IV) compound $Ba_2M_6Te_2O_{21}$ (M = Nb, Ta) [1] crystallizes in a monoclinic symmetry. The channel crystal structure of is dominated by a $^3_\infty$ $[M_6O_{21}]^{12-}$ network. In this framework cross and diamond shaped channels exist along [010] (see figure 4). M^{5+} shows octahedral and Te^{4+} a one sided triangular coordination by O^{2-}. A triply capped BaO_8 cube ma describe the polyhedron around Ba^{2+}.

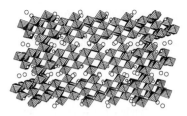

Figure 4. *Projection of the crystal structure along [010] in $Ba_2M_6Te_2O_{21}$ (M = Nb, Ta).*

Ba$_2$Nb$_2$TeO$_{10}$ [2] the Te(VI) compound crystallizes in a orthorhombic variant of the hexagonal bronze type. Planes of edge- and corner-sharing NbO$_6$- and TeO$_6$-octahedra characterize the crystal structure (see figure 5). The Barium environment can be described as an irregular polyhedron build from nine O^{2-}-ions.

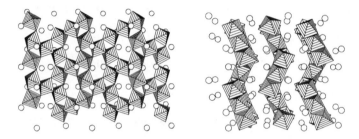

Figure 5. *Projection of the crystal structure along [010] (right) and [100] (left) in $Ba_2Nb_2TeO_{10}$.*

The tellurium rich Ba$_{1.5}$NbTe$_2$O$_9$ [14, 15] compound with nearly identical lattice constants and so far unknown crystal structure could not be identified in our experimental work. Under the conditions given in the publications [14, 15], we can only find the Ba$_2$Nb$_2$TeO$_{10}$ crystals described above.

CONCLUSIONS AND FUTURE WORK

We have systematically investigated the crystal structures of the Ba-M-Te-O system (M = Nb, Ta) and the Pb-M-Te-O system (M = Mn, Fe, Co, Ni, Cu, Zn) in the temperature range between 873 K and 1273 K using X-ray diffraction methods. Our investigations show that the tellurium is the minor part of the anionic lattice, whereas the 4th period or 5th group transition metals are major part. The structure of the tellurium oxide compound changes from the perovskite type to new structure types, if the s^2-orbital of the lead is stereo active.

The future work will include the combination of 4th period transition elements (e.g. Cu/Mn) with lead and tellurium oxides. The substitution of lead by rare earth metals or bismuth will be a promising opportunity to change the crystal structure again. Also the high temperature preparation of 6th transition group metals with earth alkaline metals, lead, bismuth and rare earth metals in combination with tellurium oxide seems to be an interesting research field with regard to materials for heterogeneous catalysis. Experimental work for the low temperature preparation of the presented systems and other quaternary tellurium oxides is in progress.

We will try to purify the observed phases for physical properties checks and tests for possible applications in the area of catalysis material or solid state devices.

ACKNOWLEDGEMENT

The generous support with a foundation given by the Japanese Ministry of Education and the German Research Society is gratefully acknowledged. A special thank you is given to Mr. Y. Sato for the EPMA analysis.

REFERENCE

1. B. Wedel, Hk. Mueller-Buschbaum, Z. *Naturforsch.* **51b**, 1407 (1996).

2. B. Wedel, Hk. Mueller-Buschbaum, Z. *Naturforsch.* **51b**, 1411 (1996).

3. B. Wedel, Hk. Mueller-Buschbaum, Z. *Naturforsch.* **51b**, 1587 (1996).

4. B. Wedel, Hk. Mueller-Buschbaum, Z. *Naturforsch.* **52b**, 35 (1997).

5. L. Wulff, B. Wedel, Hk. Mueller-Buschbaum, Z. *Naturforsch.* **53b**, 49 (1998).

6. B. Wedel, K. Sugiyama, Hk. Mueller-Buschbaum, Z. *Naturforsch.* **53b**, 527 (1998).

7. B. Wedel, K. Sugiyama, K. Hiraga, K. Itagaki, *Naturforsch.* **54b**, 469 (1999).

8. B. Wedel, K. Sugiyama, K. Hiraga, K. Itagaki, *Material Research Bulletin* **34**, 14/15, 21 (1999).

9. G. M. Sheldrick, SHELXS-86, Program for Crystal Structure Determination, Götting (1986).

10. G. M. Sheldrick, SHELXL-93, Program for Crystal Structure Refinement, Göttingen (1993

11. P. McArdle, *J. Appl. Cryst.* **29**, 306 (1996).

12. C. K. Johnson, ORTEP, Fortran Thermal Ellipsoid Plot Program for Crystal Structu Illustration, Report ORNL-3794, Oak Ridge National Laboratory, TN (1965).

13. R.X. Fischer, *J. Appl. Cryst.* **18**, 258 (1985).

14. S. Launay, P. Mahe, M. Quarton, *Powder Diffraction* **4**, 31 (1989).

15. S. Launay, P. Mahe, M. Quarton, *J. Less-Com. Met.* **147**, 161 (1989).

Session in Honor of
J.M. Honig

Mat. Res. Soc. Symp. Proc. Vol. 658 © 2001 Materials Research Society

ITINERANT VIBRONS AND HIGH-TEMPERATURE SUPERCONDUCTIVITY[*]

John B. Goodenough
Texas Materials Institute, ETC 9.102
University of Texas at Austin, Austin, TX 78712-1063

ABSTRACT

The $La_{2-x}Sr_xCuO_4$ phase diagram is interpreted within the framework of a transition from localized to itinerant electronic behavior. In the underdoped region $0 < x < 0.1$, holes in the $x^2 - y^2$ band are not small polarons; each occupies a mobile correlation bag of 5 to 6 copper centers at temperatures $T > T_F$, a spinodal phase segregation into the parent antiferromagnetic phase and a polaron liquid is accomplished below T_F by cooperative oxygen displacements. In the overdoped compositions $x > 0.25$, holes are excluded from strong-correlation fluctuations within a Fermi liquid. In the intermediate range $0.1 < x < 0.25$, the polaron liquid formed below room temperature changes character with increasing x and decreasing T. In the polaron liquid, mobile two-hole bags of four copper centers order with decreasing temperature into alternate Cu-O-Cu rows of a superconductive CuO_2 sheet at a critical composition $x_c \approx 1/6$. It is argued that hybridization of itinerant electrons with optical-mode phonons propagating along the Cu-O-Cu rows produces heavy electrons responsible for high-temperature superconductivity.

[*]Dedicated to J. M. Honig on his retirement in recognition of his long-time interest in the transition from localized to itinerant electronic behavior in solids.

INTRODUCTION

Of all the copper-oxide superconductors, the oxygen-stoichiometric system $La_{2-x}Sr_xCuO_4$, $0 \le x \le 0.3$, allows the most straightforward study of p-type high-temperature superconductivity in these materials. The number of holes in the CuO_2 sheets is given unequivocally by x; and in the solid-solution range $0 \le x \le 0.3$, the system changes with increasing x from an antiferromagnetic insulator to a superconductor and, finally, to a non-superconductive metal. Superconductivity appears below a critical temperature T_c in a phase having a narrow solid-solution range of intermediate composition $0.15 \lesssim x \lesssim 0.18$. The magnitude of T_c appears to vary with the volume of material that becomes superconductive.

Fig. 1 illustrates the structure of the parent compound La_2CuO_4; it consists of an intergrowth of LaO rock-salt layers and CuO_2 sheets. More complex p-type copper-oxide superconductors replace the LaO rock-salt layers with thicker $AO-\Phi-AO$ layers bounded by AO rock-salt sheets; moreover, the CuO_2 sheets may become layers of two or more CuO_2 sheets bound to one another by a plane of cations of intermediate size (*e.g.* Ca^{2+}, Y^{3+}, or a smaller rare-

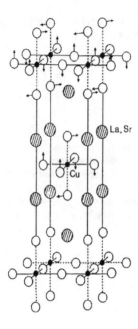

Figure 1. *The La₂CuO₄ structure. Arrows indicate direction of cooperative CuO₆/₂ rotations.*

earth) in eightfold oxygen coordination. The AO-Φ-AO layers may act as charge reservoirs, which makes determination of the number of holes in the CuO_2 sheets less straightforward; and with three or more CuO_2 sheets in a superconductive layer, the middle sheets have copper in fourfold coordination and the outer in fivefold coordination, which makes uncertain the distribution of holes between the sheets of a layer. In addition, the full range of compositions from antiferromagnetic insulator to the overdoped metallic phase is not realized in these comple structures. However, in all cases the superconductive components are CuO_2 sheets that contain every copper in a similar oxygen coordination, whether sixfold, fivefold, or fourfold; and the superconductive phase appears in all at a similar hole concentration in a CuO_2 sheet. The featu responsible for superconductivity in every copper-oxide superconductor appears to be the same even though T_c may vary from one structure to another by over 100 K. Therefore, the system $La_{2-x}Sr_xCuO_4$ should provide the key to an understanding of the phenomenon.

BACKGROUND

From Fig. 1, it is apparent that the mean equilibrium La-O bond length and the equilibrium Cu-O bond length of a CuO_2 sheet are matched only where the geometrical toleran factor

$$t = < \text{A-O} > / \sqrt{2} \ (\text{Cu-O}) \tag{1}$$

is $t = 1.0$. The factor t is calculated from the sums of ionic radii that have been obtained from room-temperature lattice parameters. Since the La-O bond has a larger thermal expansion than the Cu-O bond, a $t < 1$ for room-temperature La_2CuO_4 increases with temperature; it also increases with substitution of a larger Sr^{2+} ion for La^{3+}. A $t < 1$ places the CuO_2 sheets under a compressive stress; the stress is relieved by a cooperative rotation of the $CuO_{6/2}$ octahedra about a tetragonal [110] axis as is indicated by the arrows in Fig. 1. This rotation lowers the crystal symmetry from tetragonal to orthorhombic. Moreover, the compressive stress allows the removal of antibonding electrons to dope the material p-type, especially if the doping is done by substitution of a larger Sr^{2+} ion for La^{3+} or insertion of an interstitial O^{2-} ion between the LaO rock-salt sheets of the LaO rock-salt layers.

In La_2CuO_4, strong electron-electron interactions keep the Cu(II)/Cu(I) redox couple 2.0 eV above the Cu(III)/Cu(II) couple. The tetragonal ($c/a > 1$) distortion of the $CuO_{6/2}$ octahedra leaves the $(3z^2-r^2)$ orbitals filled; the operative bands are the antibonding (x^2-y^2) bands. The lower (x^2-y^2) band, which corresponds to the Cu(III)/Cu(II) couple, is pinned to the top of the $O^{2-}:2p^6$ bands because the states are antibonding with respect to the Cu-O interactions; they retain the (x^2-y^2) character because the O-$2p_\sigma$ orbitals are strongly hybridized with the Cu-3d x^2-y^2 orbitals. The crystal-field x^2-y^2 orbitals may be written as

$$\psi_{x^2-y^2} = N_\sigma(f_{x^2-y^2} - \lambda_\sigma\phi_\sigma - \lambda_s\phi_s) \tag{2}$$

where $f_{x^2-y^2}$ is the atomic 3d orbital, ϕ_σ and ϕ_s are the appropriately symmetrized O-$2p_\sigma$ and 2s orbitals of the four oxygen near neighbors in the CuO_2 sheet, and the covalent-mixing parameters are $\lambda_\sigma \equiv b_{p\sigma}/\Delta E_p >> \lambda_s \equiv b_{s\sigma}/\Delta E_s$, where ΔE_p and ΔE_s are the energies required to transfer an electron from an O-$2p_\sigma$ or 2s orbital to the empty Cu(II)/Cu(I) couple; $b_{p\sigma}$ and $b_{s\sigma}$ are the electron-energy-transfer (resonance) integrals for these electron transfers. In La_2CuO_4, the antiferromagnetic spin-spin coupling is given by the superexchange energy

$$\Delta\varepsilon^S \sim b^2/U \sim \varepsilon_\sigma\lambda_\sigma^4/U \tag{3}$$

where U is the on-site coulomb energy required to add an electron to a Cu(II) ion to make it Cu(I).

The crystal-field theory breaks down for a hole in the lower x^2-y^2 band because ΔE_p for electron transfer to a Cu(III)/Cu(II) couple becomes small or negative, which makes invalid the perturbation expansion for covalent mixing. A hole in the lower x^2-y^2 band hops between copper atoms in a time $\tau_h \approx \hbar/W_\sigma$, where the bandwidth W_σ increases with x. As x increases, a $\tau_h \approx \omega_o^{-1}$ decreases to $\tau_h < \omega_o^{-1}$, where ω_o^{-1} is the period of an optical-mode vibration of the oxygen atoms that would trap the hole as a small polaron. It is significant that the superconductive phase is stabilized at the cross-over from localized to itinerant electronic behavior where $\tau_h \approx \omega_o^{-1}$.

For central-force fields, the virial theorem states that a system of particles (the x^2-y^2 electrons in $La_{2-x}Sr_xCuO_4$) of mean kinetic energy <T> and potential energy <V> are constrained to the relation

$$2<T> + <V> = 0 \qquad (4)$$

If the volume occupied by the electrons increases discontinuously on going from localized to itinerant behavior, then $<T>$ decreases discontinuously and therefore so does $|<V>|$. For antibonding electrons, a decrease in $|<V>|$ is accomplished by a decrease in the equilibrium (Cu-O) bond length, and a discontinuous change in the equilibrium Cu-O bond length gives a first-order transition. Therefore, a

$$(Cu\text{-}O)_{loc} > (Cu\text{-}O)_{itin} \qquad (5)$$

may coexist at cross-over. A phase segregation associated with the first-order phase transition has been observed to set in below room temperature in $La_2CuO_{4+\delta}$ where the interstitial oxygen remain mobile to below 240 K. In $La_{2-x}Sr_xCuO_4$, there are no interstitial oxygen, but phase segregation may occur in the CuO_2 planes by cooperative atomic displacements where there is no atomic diffusion. If these cooperative displacements are dynamic, they are not detectable by a conventional diffraction experiment. Their existence may be deduced indirectly from the pressure dependence of the tolerance factor t. Normally a dt/dP < 0 is observed in AMO_3 perovskites because of a greater compressibility of the A-O bond [1], but the mean $<M\text{-}O>$ bond length, in our case $<Cu\text{-}O>$, becomes more compressible where a $(Cu\text{-}O)_{loc}$ and a $(Cu\text{-}O)_{itin}$ coexist, and a dt/dP > 0 has been observed. The prediction [2] of a $(Cu\text{-}O)_{loc}$ and a $(Cu\text{-}O)_{itin}$ coexisting in the range $0 < x < 0.3$ of $La_{2-x}Sr_xCuO_4$ has also been confirmed by direct measurement [3]; pair-distribution-function (PDF) analysis of neutron diffraction data has revealed the coexistence of two Cu-O bond lengths, one corresponding to that of the parent compound La_2CuO_4 and the other to that of the metallic phase $La_{1.7}Sr_{.03}CuO_4$. The consequences of locally cooperative fluctuations between two different equilibrium Cu-O bond lengths offer a key for understanding the phenomenon of high-temperature superconductivity in the copper oxides.

PHASE DIAGRAM

Fig. 2 is a tentative phase diagram for the system $La_{2-x}Sr_xCuO_4$. The Néel temperature T_N of the parent phase La_2CuO_4 drops precipitously with increasing x; a low-temperature spin-glass behavior is found for $0.02 < x < 0.12$ that suggests the antiferromagnetic phase is broken up into non-percolating clusters. The orthorhombic-tetragonal transition temperature T_t decreases sharply with increasing x, vanishing above $x \approx 0.22$. A temperature-dependent paramagnetic susceptibility exhibits a maximum at a T_{max} that also decreases sharply with increasing x as the concentration of localized spins on the copper atoms decreases.

Correlation Bags. The temperatures T_F and T_ρ are defined by the thermoelectric-power data of Fig. 3 for $0 < x \leq 0.1$. Above T_F, the thermoelectric power $\alpha(T)$ is temperature-independent, which is typical of polaronic conduction. However, the polaronic statistical term

$$\alpha = -(k/e) \ln [\beta(1\text{-}c)/c] \qquad (6)$$

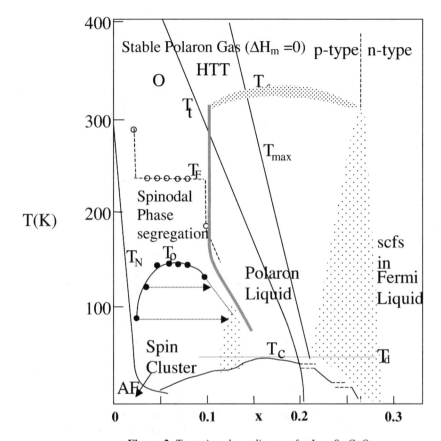

Figure 2. *Tentative phase diagram for $La_{2-x}Sr_xCuO_4$*

with spin-degeneracy factor $\beta=2$ requires a fractional occupancy $c = k_1 x$, where $k_1 = N/N^*$ is a measure of the average number of Cu centers in a polaron [4]. A $k_1 \approx 5.3$ gives a good fit for all compositions $0 < x \leq 0.1$, which means that the hole polarons do not collapse into Zhang-Rice [5] singlets, but occupy a molecular orbital (MO) of a cluster of five to six copper atoms [6]. Within a cluster, a $(Cu-O)_{itin}$ bond length delocalizes the x^2-y^2 electrons and captures a hole from the surrounding matrix having a $(Cu-O)_{loc}$. Therefore, these clusters represent *one-hole correlation bags*. Why the clusters do not collapse to a Zhang-Rice singlet is not yet established, but Bersuker [7] has calculated that a pseudo-Jahn-Teller deformation of the $CuO_{4/2}$ centers within a cluster can account for the size and mobility of these non-adiabatic, multicenter polarons.

One-hole bags of five to six copper centers are forced to make contact for $x > 0.1$. Where this happens, the locally cooperative oxygen displacements create a more complex interpenetration of fluctuating itinerant and localized domains. This "polaron-liquid" phase contains multihole bags. On cooling below T_F in the range $0 < x \leq 0.1$, the system appears to

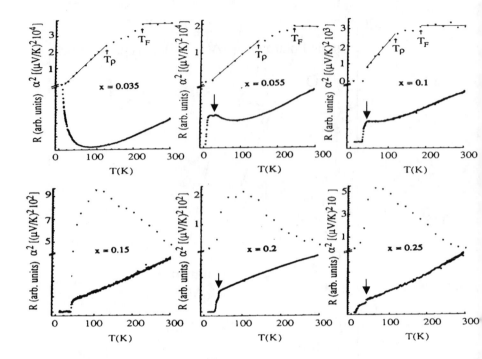

Figure 3. *Thermoelectric power α(T) and resistance R(T) for several values of x of La$_{2-x}$Sr$_x$CuO$_4$ after ref. [4]. The unmarked arrows on R(T) data indicate T$_d$.*

undergo a spinodal phase segregation into the parent antiferromagnetic and polaron-liquid phases, which results in a spin glass at low temperatures. I postulate that two principles guide the formation of the multihole bags: (1) they contain one hole for every two Cu atoms and (2) they form linear chains or chain segments along Cu-O-Cu rows as is illustrated in Fig. 4. The mobile two-hole bags containing four copper centers found near optimal doping by Egami *et al* [8] with PDF analysis of pulsed neutron data and the static stripes observed by Tranquada *et al* [9] in a non-superconductive $x = 1/8$ composition containing Nd in place of La both conform to these two principles. Moreover, the data of Egami *et al* suggest that optimal doping for high-T$_c$ superconductivity would produce a dynamic ordering of two-hole bags into alternate Cu-O-Cu rows where they are separated from one another by two Cu(II) atoms, see Fig. 4(d). Such a configuration would correspond to $x = 1/6$ with electron pairs within each two-hole bag. The change in α^2(T) at T$_\rho$ of Fig. 3 corresponds to the opening of a pseudogap (T$_\rho \approx$ T* of the pseudogap literature); stabilization of two-hole bags in the polaron liquid would open a gap between bonding and antibonding states within the bags. The R(T) data of Fig. 3 show a small

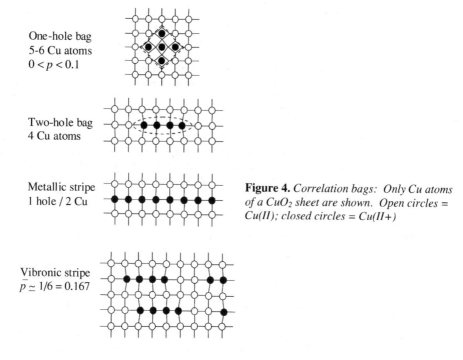

One-hole bag
5-6 Cu atoms
$0 < p < 0.1$

Two-hole bag
4 Cu atoms

Metallic stripe
1 hole / 2 Cu

Figure 4. *Correlation bags: Only Cu atoms of a CuO_2 sheet are shown. Open circles = Cu(II); closed circles = Cu(II+)*

Vibronic stripe
$\bar{p} \simeq 1/6 = 0.167$

anomaly at a $T_d = T_c(max)$, where $T_c(max)$ is the maximum value of the superconductive critical temperature T_c. This observation suggests that a change in the electron-lattice coupling from short-range to long-range order of the bond-length fluctuations occurs on cooling through T_d. I have suggested [10] that the long-range order corresponds to the formation of an itinerant vibronic state in which itinerant electrons hybridize with optical-mode phonons to produce heavy fermions. This hybridization would occur at a specific hole concentration x_c, and $T_d = T_c(max)$ at x_c would then correspond to a critical point.

Strong-correlation Fluctuations. To gain insight into the transition from a polaron liquid to a metal in the overdoped range $0.18 < x < 0.3$, the metallic system $Sr_{1-x}Ca_xVO_3$ proves instructive. As x increases, the π^* band of this system narrows until it approaches the Mott-Hubbard transition at $x = 1$. A seminal study with photoemission spectroscopy (PES) has revealed localized (strongly correlated) 3d electrons in a lower Hubbard band coexisting with itinerant 3d electrons [11]. Moreover, electrons were systematically shifted with increasing x from the itinerant-electron band to the lower Hubbard band. From this observation and the virial theorem, it would follow that a dynamic phase segregation is producing strong-correlation fluctuations in isolated clusters having $(V-O)_{loc}$ within a matrix having $(V-O)_{itin}$ and itinerant electrons. Therefore, we concluded that the conventional model of a homogenous Mott-Hubbard transition does not apply to these perovskites. We tested this conclusion by measuring $\alpha(T)$ for $CaVO_3$ under different hydrostatic pressures [12]. Strong-correlation fluctuations were found to suppress the low-temperature phonon-drag component of $\alpha(T)$; this component is partially restored as the π^* bands are broadened by the application of pressure. We have applied this signature to other single-valent perovskites, e.g. $La_{1-x}Nd_xCuO_3$ and $LnNiO_3$ (Ln = La, Pr, Nd, $Nd_{0.5}$, $Sm_{0.5}$), to show how strong-correlation fluctuations appearing on the approach to the Mott Hubbard transition can account for a smooth change from Pauli to Curie-Weiss paramagnetism [13, 14]. We have found that the phonon-drag component of $\alpha(T)$ is also suppressed in the overdoped compositions of the mixed-valent system $La_{2-x}Sr_xCuO_4$. The picture that emerges is that as x decreases from $x = 0.30$, strong-correlation fluctuations are first introduced as electron-rich clusters having $(Cu-O)_{loc}$ bond lengths; the clusters are isolated from one another in a matrix containing $(Cu-O)_{itin}$ bond lengths and itinerant electrons. Below a critical temperature, we should expect a spinodal phase segregation between the metallic phase and the polaron-liquid phase. Although this phase segregation appears to be more subtle than that on the underdoped side, evidence for two phases in the overdoped compositions is becoming available [6,15]. Moreover, recent experiments [16] with small concentrations of Zn substituted for Cu have indicated stabilization of a competitive non-superconductive phase at $x \approx 0.21$, which could correspond to $x = 5/24$ with a metallic stripe in every fourth Cu-O-Cu row (as in the non-superconductive $x = 1/8$ phase) and ordered two-hole bags in the middle row.

The Polaron Liquid. We turn finally to the polaron-liquid phase and its condensation at an $x_c \approx 1/6$ into the superconductive phase below a $T_d \geq T_c$. Measurement [6, 17] of $\alpha(T)$ in the range $0.1 < x < 0.25$ shows a weakly temperature-dependent value above $T_l \approx 300$ K and, as shown in Fig. 3, an enhancement below that reaches a maximum near 140 K. This enhancement is not due to phonon drag; it reflects the progressive transfer with decreasing temperature of spectral weight from the (π, π) to the $(\pi, 0)$ direction that has been observed directly by angle-resolved PES [18-20]. This remarkable observation has been found to occur in all the superconductive copper oxides, but not in the non-superconductive oxides. Given the coexistence of two equilibrium (Cu-O) bond lengths that fluctuate cooperatively to form hole-rich itinerant-electron domains with $(Cu-O)_{itin}$ interpenetrating hole-poor localized-electron domains with $(Cu-O)_{loc}$, it follows that the two domains would become more ordered spatially with decreasing temperature. Electrostatic, elastic, and magnetic-exchange energies drive any phase segregation and/or ordering of these domains. Strong magnetic-exchange interactions would drive a spinodal segregation of the parent antiferromagnetic phase from the polaron-liquid phase in the underdoped compositions, but the electrostatic and elastic energies would be critical for stabilizing long-range order within the polaron liquid. Moreover, strong coupling of the electrons to the bond-length fluctuations means that the ordering of the two Cu-O bond lengths must be reflected in the electronic states. Therefore, it is reasonable to assume that the transfer of spectral weight from the (π, π) to the $(\pi, 0)$ direction in reciprocal space, i.e. to heavy fermions having momenta oriented along the Cu-O-Cu bond axes, reflects a strong coupling of itinerant electrons to a progressive ordering of the Cu-O bond-length fluctuations from short-range cooperative to long-range cooperative. From the data of Egami et al [8], the low-temperature order at the optimal composition $x_c \approx 1/6$ appears to be that illustrated in Fig. 4(d) where evenly separated two-hole bags occupy alternate Cu-O-Cu rows. The formation of heavy fermions implies a hybridization of itinerant electrons with optical-mode phonons arising from long-range order of the bond-length fluctuations. Hybridization of electronic states with vibrational modes of the same symmetry gives a *vibronic state*; if the states are itinerant electrons and phonons, the *vibrons* are itinerant.

The Superconductive State. Five observations are of interest: (1) The phonon contribution to the thermal conductivity increases dramatically on cooling through a $T_c \approx 90$ K [21,22]. (The phonon contribution is too low at $T_c \approx 35$ K to give a dramatic increase in the $La_{2-x}Sr_xCuO_4$ system). (2) Electron-tunneling experiments [23] have revealed a superconductive-gap symmetry corresponding to (x^2-y^2) + ixy. (3) An Eliashberg analysis of the phonon spectrum for BCS-type superconductivity required addition of an electronic excitation [24]. (4) T_c increases with decreasing bending angle ϕ of the $(180^\circ - \phi)$ Cu-O-Cu bonds [25]. (5) Exchange of ^{18}O for ^{16}O has little effect on T_c at optimal doping, but a large effect near $x = 1/8$ and on the temperature $T^* \approx T_\rho$ where there is an opening of a pseudogap in the density of one-electron states [26, 27].

Zhou [28] has noted that short-range vibronic states in $La_{1-x}Sr_xMnO_3$ suppress the phonon contribution to the thermal conductivity. Similarly, short-range order of the Cu-O bond-length fluctuations suppresses the thermal conductivity in the normal state of the copper-oxide superconductors. Therefore, the observation of a dramatic increase in the thermal conductivity on cooling through T_c signals the introduction of phonons below T_c and therefore a long-range ordering of the (Cu-O) bond-length fluctuations. This observation is consistent with the formation of itinerant vibronic states below $T_d = T_c(max)$.

Hybridization of itinerant electrons with an optical-mode phonon arising from ordered fluctuations of Cu-O bond lengths would give a vibronic momentum vector $k' \sim \cos\theta + i\sin\theta$, where θ is the angle between the electronic vector **k** and a Cu-O-Cu bond axis. The vibronic energies would have the symmetry

$$E_{k'} \sim (\cos^2\theta - \sin^{-2}\theta) + i\sin2\theta \sim (x^2-y^2) + ixy \qquad (7)$$

and pairing of vibrons would give a superconductive energy gap of the same symmetry. Moreover, the electronic transition from molecular orbitals in a bag to localized orbitals at Cu(II) between bags would accompany the travelling charge-density wave represented by a vibron.

Stabilization of itinerant vibrons competes with retention of localized electrons in the electron-rich domains. Since itinerant fermions are stabilized by an increase in the bandwidth W_σ, $T_c(max)$ at x_c should increase with W_σ if $T_c(max) \approx T_d$ depends on the formation of itinerant vibronic states. Since W_σ increases with decreasing bending angle ϕ, an increase in T_c with decreasing ϕ is consistent with the formation of itinerant vibronic states as the origin of the heavy fermions responsible for high-T_c superconductivity in the copper-oxides.

The temperature below which the hole-rich domains would develop an ordered array of two-hole bags rather than metallic stripes depends on the structure, the mass of the oxygen atoms, and the concentration of holes in the CuO_2 sheets. A low-temperature-tetragonal (LTT) phase found at $x = 1/8$ in $La_{2-x}Ba_xCuO_4$ contains cooperative rotations of the $CuO_{6/2}$ octahedra around [100] and [010] axes in alternate CuO_2 sheets; the cooperative rotations are about a [110] axis in the low-temperature orthorhombic (LTO) phase. The rotations of the LTT phase tend to stabilize metallic stripes that are pinned, and stabilizing metallic stripes in a static charge-density/spin-density wave (CDW/SDW) suppresses superconductivity. In the $La_{2-x}Sr_xCuO_4$

system, a competition between the LTO and LTT phases persists near $x = 1/8$, which lowers T_c somewhat in the vicinity of $x = 1/8$. Exchange of ^{18}O for ^{16}O favors the LTT phase, which lowers T_c further. Therefore, T_c exhibits a big oxygen-isotope effect for composition near $x = 1/8$.

Two-hole bags containing two (x^2-y^2) electrons in MOs of a four-copper-center cluster would have an energy gap between the HOMO and LUMO of the cluster. Consequently, the formation of two-hole bags below a temperature T^* would introduce a pseudogap in the density of one-electron states. Since the formation of separated two-hole bags would become more difficult as x increases, T^* would decrease with increasing x and exhibit a large oxygen-isotope effect, as observed, if the opening of the pseudogap reflects formation of separated two-hole bag in stripes containing one hole for every three copper centers, Fig. 4(d).

Finally, the model calls for an order-disorder transition that need not be sensitive to the frequency of the oxide-ion vibrations along a Cu-O-Cu bond axis. Therefore, the formation of itinerant vibronic states at $T \geq T_c$ may depend more on the bandwidth than on the mass of the oxygen atoms. The model does not require that T_c exhibit a large oxygen-isotope effect.

SUPERCONDUCTIVE PAIRS

Given the small coherence length of the superconductive pairs, which is roughly the size of a two-hole bag, it remains an open question whether the pairs form a Bose-Einstein condensate or should be treated as Cooper pairs bound by a non-retarded potential rather than th retarded potential of conventional BCS theory. The formulation of itinerant vibrons would seem to favor the formation of Cooper pairs, but the semi-empirical reasoning presented here cannot resolve this question.

REFERENCES

1. J. B. Goodenough, J. A. Kafalas, and J. M. Longo, *Preparative Methods in Solid State Chemistry*, P. Hagenmuller, ed. (Academic Press, New York, N.Y. 1972), Chap. 1.
2. J. B. Goodenough, *Ferroelectrics* **130**, 77 (1992).
3. E. S. Bozin, G. H. Kwei, H. Tagaki, and S. Billinge, *Phys. Rev. Lett.* **84**, 5856 (2000).
4. J. B. Goodenough and J.-S. Zhou, *Phys. Rev. B* **42**, 4276 (1990).
5. F. C. Zhang and T. M. Rice, *Phys. Rev. B* **37**, 3759 (1988).
6. J. B. Goodenough, J.-S. Zhou, and J. Chan, *Phys. Rev. B* **47**, 5275 (1993).
7. G. I. Bersuker and J. B. Goodenough, *Physica C* **274**, 267 (1992).
8. T. Egami, Y. Petrov, and D. Louca, *J. Supercond.* (in press).
9. J. M. Tranquarda, N. Ichikawa, and N. Uchida, *Phys Rev. B* **59**, 14712 (1999).
10. J. B. Goodenough, *J. Supercond* (in press).
11. I. H. Inoue, I. Hase, Y. Aiura, A. Fujimori, Y. Haruyama, T. Maruyama, and Y. Nishihara, *Phys. Rev. Lett.* **74**, 2539 (1995).
12. J.-S. Zhou and J. B. Goodenough, *Phys. Rev. B* **54**, 13393 (1996).
13. J.-S. Zhou, W. Archibald, and J. B. Goodenough, *Phys. Rev. B* **57**, R2017 (1998).
14. J.-S. Zhou, J. B. Goodenough, B. Dabrowski, P. W. Klamut, and Z. Bukowski, *Phys. Re* *B* **61**, 4401 (2000).

15. H. H. Wen, X. Y. Chen, W. L. Yang, and Z. X. Zhao, *Phys. Rev. Lett.* **85**, 2805 (2000).
16. T. Kawamata, T. Adachi, T. Noji, and Y. Koike, *Phys. Rev. B* **62**, R11981 (2000); I. Watanabe, M. Aoyama, M. Akoshima, T. Kawamata, T. Adachi, Y. Koike, S. Ohira, W. Higemoto, and K. Nagamine, ibid, R11985.
17. J.-S. Zhou and J. B. Goodenough, *Phys. Rev. B* **51**, 3104 (1995).
18. Zhi-xun Shen, *Physica B* **197**, 632 (1994).
19. D. S. Dessau, Z.-X. Shen, D. M. King, D. S. Marshall, L. W. Lombardo, P. H. Dickinson, A. G. Loeser, J. DiCarlo, C.-H. Park, A. Kapitulmik, and W. E. Spicer, *Phys. Rev. Lett.* **71**, 2781 (1993).
20. M. R. Norman, H. Ding, M. Randeria, J. C. Campuzano, T. Yokoya, T. Takeuchi, T. Takahasi, T. Mochiku, K. Kadowski, P. Gutasarma, and D. G. Hinks, *Nature* **392**, 157 (1998); P. Coleman, ibid, 134.
21. J. L. Cohn, S. A. Wolf, and T. Vanderah, *Phys. Rev. B* **45**, 511, (1992).
22. K. Krishana, N. P. Ong, O. Li, G. D. Gu, and N. Koshizuka, *Science* **277**, 83 (1997).
23. G. Deutscher, Y. Dagan, and R. Kupke, *J. Supercond.* (in press).
24. W. Little, *J. Supercond.* (in press).
25. J.-S. Zhou, H. Chen, and J. B. Goodenough, *Phys. Rev B* **49**, 9084 (1994).
26. M. K. Crawford, W. E. Farneth, E. M. McCarron, III, R. L. Harlow, and A. H. Moudden, *Science* **250**, 1390 (1990).
27. D. Rubio Temprano, J. Mesot, S. Janssen, K. Conder, A. Furrer, H. Mutka, and K. A. Müller *Phys. Rev. Lett.* **84**, 1990 (2000).
28. J.-S. Zhou (unpublished).

Mat. Res. Soc. Symp. Proc. Vol. 658 © 2001 Materials Research Society

(Rare Earth/Li) Titanates and Niobates as Ionic Conductors

A. Morata-Orrantia[a], S. García-Martín[a], E. Morán[a], U. Amador[b] and M. A. Alario-Franco[a]

[a] Departamento de Química Inorgánica, Facultad de Ciencias Químicas, Universidad Complutense de Madrid, Madrid-28040, Spain.
[b] Departamento de Química Inorgánica y Materiales, Facultad de Ciencias Experimentales y Técnicas, Universidad San Pablo-CEU, Urb. Montepríncipe, 28668 Boadilla del Monte, Madrid, Spain.

ABSTRACT

The lithium ion conducting properties of materials of composition $La_{0.58}Li_{0.26}TiO_3$, $Nd_{0.58}Li_{0.26}TiO_3$, $La_{0.67}Li_{0.25}Ti_{0.75}Al_{0.25}O_3$ and $La_{0.29}Li_{0.12}NbO_3$ have been compared in relation with their microstructure. All the oxides have powder X-ray diffraction patterns characteristic of a perovskite-related structure with lattice parameters $a \sim \sqrt{2}a_p$, $b \sim \sqrt{2}a_p$, $c \sim 2a_p$ (p refers to cubic perovskite). However, some important differences are observed in their microstructure by SAED and HRTEM. Ordering between vacancies, Li^+ and La^{3+} or Nd^{3+} and twinning of the NbO_6 or TiO_6 octahedra tilting system are shown in $La_{0.29}Li_{0.12}NbO_3$ and $Nd_{0.58}Li_{0.26}TiO_3$, which are the materials having a lower ionic conductivity. The $La_{0.58}Li_{0.26}TiO_3$ and $La_{0.67}Li_{0.25}Ti_{0.75}Al_{0.25}O_3$ oxides do not show ordering between cations.

INTRODUCTION

Materials with perovskite-related structures are attractive candidates for a great variety of applications. The perovskite structure, ABO_3, tolerates total or partial substitution of the A and B cations by ions with different valence states, affecting both the structure and the physico-chemical properties of the corresponding oxides. Deviations from oxygen or/and cation stoichiometry are also allowed by chemical substitution, usually playing an important role in their electrical properties.

Research on lithium ion conducting materials is of much importance for the development of solid state lithium secondary batteries. Indeed, there has been a great interest in a family of Li^+ ion conductors of general formula $RE_{2/3-x}Li_{3x}TiO_3$ (RE=Rare earth elements) since Inaguma et al. reported a very high value of lithium conductivity in $La_{0.5}Li_{0.34}TiO_{2.94}$ ($\sigma= 1x10^{-3}$ Scm^{-1} at 25°C) [1]. These titanates have a perovskite-related structure with A-cation deficiency which seems to be favorable for the high lithium ion conductivity observed. The stoichiometry range of the solid solutions of these systems was soon determined [2-4] and the electrical properties of the materials well established [2-7]. However, controversy about their crystal structure still remains in the literature [1-13]. Besides, the large grain boundary resistances observed in the titanates and their electrochemical intercalation properties [14, 15], which limit their use as solid electrolytes, have lead to the search of new lithium ion conducting materials with this structure. In this sense, we have successfully carried out substitutions in the B cation sublattice, obtaining new oxides with perovskite-related structures which are good lithium-ion conductors.

We present, in this paper, some results covering the electrical properties of some niobates and titanates and related them to their microstructure.

EXPERIMENTAL

Different compositions of the $La_{2/3-x}Li_{3x}TiO_3$ (0.025<x<0.13), $Nd_{2/3-x}Li_{3x}TiO_3$ (0.016<x<0.12), $La_{1/3-x}Li_{3x}NbO_3$ ($0 \le x \le 0.06$) and $La_{2/3}Li_xTi_{1-x}Al_xO_3$ ($0.06 \le x \le 0.3$) systems were prepared from adequate stoichiometric amounts of Li_2CO_3, La_2O_3, Nd_2O_3, TiO_2, Nb_2O_5 and Al_2O_3. The mixtures were ground and heated in Pt boats at 700°C during 2 hours for decarbonation. Afterwards, the samples were ground, pelleted, covered with powder of the same composition to prevent lithia loss and fired at $1100^{\circ}C$ for one day followed by a further grinding, repelleting and refiring at $1250\text{-}1300^{\circ}C$ for another day to complete reaction. Samples were quenched from the synthesis temperature.

Crystalline phase identification was carried out by powder X-ray diffraction using a Philips X'PERT diffractometer with Cu $K_{\alpha1}$ radiation, a curved Cu monochromator and PEAPD (Philips) software.

Impedance measurements were performed in an impedance/gain-phase analyzer Solartron 1255A with a dielectric interface 1296 in a frequency range $1x10^{-3}\text{-}1x10^{6}$ Hz. Pellets of about 13mm diameter and 2mm thickness were prepared by pressing the powder samples and sintering at $1300^{\circ}C$. Electrodes were made by coating opposite pellet faces with platinum paste and heating at $850^{\circ}C$.

The microstructure of the samples was studied by selected area electron diffraction (SAED) and high resolution transmission electron microscopy (HRTEM). The samples were ground in n-butyl alcohol and ultrasonically dispersed. A few drops of the resulting suspension were deposited in a carbon-coated grid. SAED studies were performed in an electron microscope JEOL 2000FX (double tilt $\pm45^{\circ}$) working at 200 kV and HRTEM studies in an electron microscope JEOL 400 EX (double tilt $\pm25^{\circ}$) working at 400 kV.

RESULTS AND DISCUSSION

All the samples of the four different systems used for this study were single phase the materials whose powder X-ray diffraction patterns show a perovskite-related structure, the sometimes called diagonal cell, with unit cell parameters a~$\sqrt{2}a_p$, b~$\sqrt{2}a_p$, c~$2a_p$.

Table I shows the materials having the highest bulk ionic conductivities at 300K for each of the solid solutions and the corresponding activation energies.

Oxide	σ (Scm^{-1}) at 300K	E_A (eV)	Reference
$La_{0.58}Li_{0.26}TiO_3$	~$1x10^{-3}$	~0.3	5
$Nd_{0.58}Li_{0.26}TiO_3$	~$1x10^{-5}$	~0.3	3
$La_{0.67}Li_{0.25}Ti_{0.75}Al_{0.25}O_3$	~$9x10^{-5}$	~0.3	This work
$La_{0.29}Li_{0.12}NbO_3$	$4.2x10^{-5}$	~0.3	16

***Table I** Bulk ionic conductivities and activation energies corresponding to the four oxides*

The highest conductivity values are indeed observed for the La/Li/Ti system. It seems that the lower Li$^+$ ion conduction of the Nd/Li/Ti materials could be related to the smaller size of the Nd^{3+} ion that seems to compact somewhat the lattice making more difficult the Li$^+$ transport. Besides, the differences in the microstructure of both kinds of titanates (which we are going to comment in Figure 1) could also affect their properties. In the case of the La/Li/Ti/Al oxides, a lower concentration of vacancies in the A-sublattice per Lithium, in comparison with the La/Li/Ti system, could be responsible for the lower ionic conductivity of the former materials. The niobates, due to the higher oxidation state of niobium (Nb^{5+}) with respect to titanium (Ti^{4+}) in these perovskites, have more empty positions within the A sublattice, so that the lower amount of charge carriers coupled to a higher degree of microstructure ordering (Figure 1) might decrease the mobility of the Li ions.

Figure 1 shows the selected area electron diffraction patterns corresponding to the $[001]_p$ and $[\bar{1}10]_p$ zone axes of the four oxides.

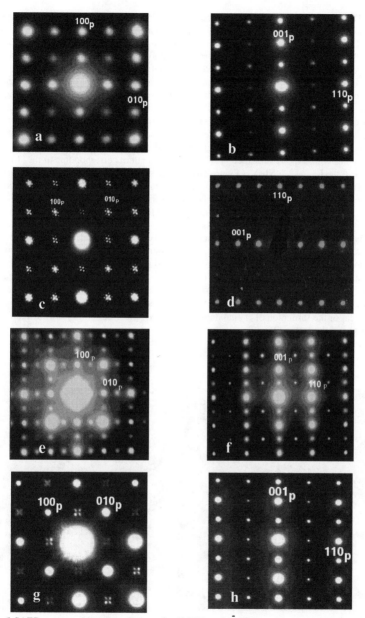

Figure 1 SAED patterns corresponding to the $[001]_p$ and $[\bar{1}10]_p$ zone axes of $La_{0.58}Li_{0.26}TiO_3$ (a and b), $Nd_{0.58}Li_{0.26}TiO_3$ (c and d), $La_{0.67}Li_{0.25}Ti_{0.75}Al_{0.25}O_3$ (e and f) and $La_{0.29}Li_{0.12}NbO_3$ (g and h).

The patterns show superlattice reflections which double the c-parameter of the basic cell as well as $(h/2\ k/2\ 1/2)_p$ reflections, giving place to a $\sim\sqrt{2}a_p$ x $\sim\sqrt{2}a_p$ x $\sim 2a_p$ unit cell, as observed by powder X-ray diffraction. This distortion of the perovskite structure is usually attributed to the tilting of the BO_6 octahedra network to accommodate the A (La^{3+} or Nd^{3+} and Li^+) cations. However, in the case of $Nd_{0.58}Li_{0.26}TiO_3$ and $La_{0.29}Li_{0.12}NbO_3$, the $(h/2\ k/2\ 1/2)_p$ reflections are split into four spots in the form of a cross. The splitting originates in the presence of a microdomain texture in which the system of the tilting of the octahedra is twinned across intersecting $(100)_p$ and $(010)_p$ domain boundaries, as it has been observed in other A cation deficient niobate perovskites [17- 19]. Apart from this, satellite reflections around the main $(h00)_p$ and $(0k0)_p$ Bragg reflections are observed in the patterns of the $Nd_{0.58}Li_{0.26}TiO_3$ oxide (Figure 1c), which can be attributed to an ordering between Li^+, Nd^{3+} and vacancies within the A sublattice [11]. Also, in the pattern corresponding to the $[\bar{1}10]_p$ zone axis of the $La_{0.29}Li_{0.12}NbO_3$, (Figure 1h), diffuse satellite reflections at odd multiples of $1/4\ d^*(001)_p$ and at about $1/7\ d^*(110)_p$ appear, suggesting a modulated structure along the $[001]_p$ and $[110]_p$ directions, that also may be due, by analogy to what happens in Thorium niobates and related materials to the ordering between Li^+, La^{3+} and A-vacancies within every another A-type plane [17-20]. Therefore, the $Nd_{0.58}Li_{0.26}TiO_3$ and $La_{0.29}Li_{0.12}NbO_3$ materials clearly have a more complicated microstructure than the other two oxides.

Another important feature observed in the patterns of the $[001]_p$ zone axis, common for the four materials, is the existence of $(h/2\ 0\ 0)_p$ and $(0\ k/2\ 0)_p$ reflections which double the a and b axes. The intensity of these reflections is quite strong in both $La_{0.58}Li_{0.26}TiO_3$ and $La_{0.67}Ti_{0.25}Ti_{0.75}Al_{0.25}O_3$ but very weak, almost invisible, in the other two materials. This apparent doubling takes us to consider two possibilities: a cell of lattice parameters $a\sim 2a_p$ $b\sim 2a_p$ $c\sim 2a_p$ or the formation of crystal domains of a $\sim\sqrt{2}a_p$ x $\sim\sqrt{2}a_p$ x $\sim 2a_p$ cell with different orientations of the c-axis.[21].

Figure 2 shows the HRTEM images corresponding to the $[001]_p$ zone axis of $La_{0.58}Li_{0.26}TiO_3$ and $La_{0.29}Li_{0.12}NbO_3$ oxides respectively.

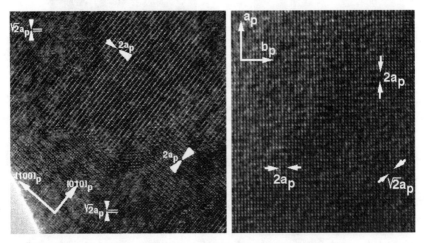

Figure 2 *HRTEM corresponding to the $[001]_p$ zone axis of $La_{0.58}Li_{0.26}TiO_3$ a) and $La_{0.29}Li_{0.12}NbO_3$ b).*

Both crystals present three types of domains: a set of domains with contrast differences showing $\sim 2a_p$ periodicity along the $[001]_p$ direction, another set having a $\sim 2a_p$ periodicity along the $[010]_p$ direction and a third one showing the $\sim\sqrt{2}a_p$ periodicity along $[110]_p$ direction. The domains are bigger in the case of the $La_{0.58}Li_{0.26}TiO_3$. Similar images were found for $La_{0.67}Li_{0.25}Ti_{0.75}Al_{0.25}O_3$ and $Nd_{0.58}Li_{0.26}TiO_3$. Therefore, the crystal structure of these materials is based on a perovskite-related cell with lattice parameters $a\sim\sqrt{2}a_p$ $b\sim\sqrt{2}a_p$ $c\sim 2a_p$, however, all the crystals are constituted by domains of this cell where in each domain the long c-axis is oriented along one of the three different crystallographic directions, x, y or z.

CONCLUSIONS

Besides the importance of the charge carriers and vacancy concentrations, ionic conductivity in materials with perovskite-related structure seems to be dependent on the degree of ordering of vacancies and cations within the A sublattice. Microstructure studies performed by TEM, show that the materials of the La/Li/Ti and La/Li/Ti/Al systems, which have the highest values of ionic conductivity, seem to be the most disordered. However, the oxides of the Nd/Li/Ti and La/Li/Nb systems have a more complex microstructure that is probably due to the ordering between cations and vacancies in the A positions and to the formation of domains of the octahedra tilting system. This ordering is likely to decrease the mobility of the Li ions. All the crystals of the four materials present a domain microstructure where in each domain the c-axis is oriented along one of the three crystallographic directions.

REFERENCES
1. Y. Inaguma, C. Liquan, M. Itoh, T. Nakamura, T. Uchida, H. Ikuto and M. Wakihara, *Solid State Commun., 86*, (1993), 689.
2. H. Kawai and J. Kuwano, *J. Electrochem. Soc., 141*, (1994), 278.
3. A. D. Robertson, S. García-Martín, A. Coats and A. R. West, *J. Mater. Chem., 5(9)*, (1995), 1405.
4. M. Morales, A.R. West, *Solid State Ionics 91* (1996) 33.
5. Y. Inaguma, L. Chen, M. Itoh and T. Nakamura, *Solid State Ionics, 70-71*, (1994), 196.
6. Y. Inaguma and M. Itoh, *Solid State Ionics, 86-88*, (1996), 257.
7. Y. Harada, T. Ishigaki, H. Kawai and J. Kuwano, *Solid State Ionics, 108*, (1998), 407.
8. A. Várez, F. García-Alvarado, E. Morán and M. A. Alario-Franco, *J. Solid State Chem., 118*, (1995), 78.
9. J. M. S. Skakle, G. C. Mather, M. Morales, R. I. Smith and A. R. West, *J. Mater. Chem. 5(9)* , (1995), 1807.
10. J. L. Fourquet, H. Duroy and M. P. Crosnier-López, *J. Solid State Chem., 127*, (1996), 283.
11. S. García-Martín, F. García-Alvarado, A. D. Robertson, A. R. West and M. Á. Alario-Franco, *J. Solid State Chem. 128*, (1997), 97.
12. H. T. Chung and D. S. Cheong, *Solid State Ionics, 120*, (1999), 197.
13. J. A. Alonso, J. Sanz, J. Santamaría, C. León, A. Várez and M. T. Fernández-Díaz, *Angew. Chem. Int. Ed. 39(3)*, (2000), 619.
14. Y. J. Shan, L. Chen, Y. Inaguma, M. Itoh and T. Nakamura, *J. Power Sources 54*, (1995), 397.
15. O. Bohnke, C. Bohnke and J. L. Fourquet, *Solid State Ionics 91*, (1996), 21.
16. S. García-Martín, J. M. Rojo, H. Tsukamoto, E. Morán and M. Á Alario-Franco, *Solid State Ionics 116*, (1999), 11.
17. M. A. Alario-Franco, I. E. Grey, J. C. Joubert, H. Vincent and M. Labeau, *Acta Cryst. A 38*, (1982), 177.

18. M. Labeau, I. E. Grey, J. C. Joubert, H. Vincent and M. A. Alario-Franco, *Acta Cryst. A* **38**, (1982), 753
19. M.Labeau *Thesis*, INPG, Grenoble, 1980.
20. S. García-Martín and M. A. Alario-Franco, *J. Solid State Chem.* **148**, (1999), 93.
21. M.A. Alario, M. Vallet, M. R. Henche, J. M. G. Calbet, J. C. Crenier, A. Wattiaux and P. Hagenmuller, *J. Solid State Chemistry*, **46**, 23, (1983).

Mat. Res. Soc. Symp. Proc. Vol. 658 © 2001 Materials Research Society

Determination of T_N for $Ba_6Mn_4MO_{15}$ (M = Cu, Zn) by Neutron Diffraction

P. D. Battle, E. J. Cussen

Inorganic Chemistry Laboratory, Oxford University, South Parks Road, Oxford, OX1 3QR, U.K.

ABSTRACT

The Néel temperatures of $Ba_6Mn_4MO_{15}$ (M = Cu, Zn) have been determined by neutron powder diffraction to be ~ 6 K. In both cases, short-range magnetic coupling within the individual chains of this pseudo-1D crystal structure produces a susceptibility maximum at a temperature somewhat higher than that at which the frustrated interchain interactions establish long-range magnetic order.

INTRODUCTION

There is currently a great deal of interest in the structural chemistry [1] and electronic properties [2, 3] of oxides having the general formula $A_{3n+3}A'_nB_{n+3}O_{6n+9}$. Their trigonal or rhombohedral crystal structures contain [001] chains consisting of a periodic sequence of $A'O_6$ trigonal prisms and BO_6 octahedra, the ratio of octahedra to prisms (o:p) being determined by the value of n; the A cations occupy sites in the interchain spaces. Much of the interest stems from the variation in the electronic properties that occurs as the value of n is changed, and recent work has focussed on the extent to which the electronic properties can be described as 1D [4], the degree of magnetic frustration present in the structure (which relates to the dimensionality of the structure) and the electronic consequences of the modulated structures that are observed in some compositions [5]. We have previously shown [6] that the n = 1, o:p = 4 structure (Figure 1) of $Ba_6Mn_4MO_{15}$ (M = Cu, Zn) is magnetically frustrated as a consequence of interchain interactions, although a long-range ordered, antiferromagnetic 120 ° spin structure was observed for both compositions in powder neutron diffraction experiments carried out at 1.7 K. The molar magnetic susceptibility of the Zn compound (Figure 2) was relatively low (~ 11 x 10 $^{-3}$ cm^3 mol^{-1}) throughout the temperature range 5 ≤ T/K ≤ 300, with a

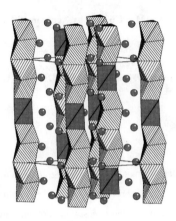

FIGURE 1 Crystal structure of n = 1 $A_{3n+3}A'_nB_{n+3}O_{6n+9}$. BO_6 octahedra are hatched, $A'O_6$ prisms are shaded, shaded circles represent A cations

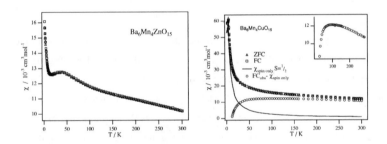

FIGURE 2 Temperature dependence of the molar magnetic susceptibility of $Ba_6Mn_4A'O_{15}$ (A' = Zn or Cu) [6]

broad maximum at 40 K; a small increase was observed below 16 K. Our original interpretation of these data was that the maximum at 40 K is a consequence of short-range spin coupling within the tetramers, and that the 3D Néel temperature lies at a lower temperature and is invisible in the susceptibility data. We argued that the change in the bulk magnetic moment of the sample on passing from a short-range ordered state to a long-range ordered state would be small and possibly undetectable in our experiment. The increase at low temperatures was assumed to arise as a result of Mn/Zn disorder over the two types of six-coordinate sites. The susceptibility of $Ba_6Mn_4CuO_{15}$ (Figure 2) increases slowly with decreasing temperature before showing a relatively strong paramagnetic tail in the range $8 \leq T/K \leq 30$. It reaches a sharp maximum at 5 K and hysteresis is apparent between field-cooled and zero-field-cooled data below this transition. We were able to show that the temperature dependent component of the susceptibility could largely be attributed to the presence of paramagnetic Cu^{2+} cations in the prismatic sites. The residual susceptibility after subtraction of this component bore a strong resemblance to that of the composition containing diamagnetic Zn^{2+}, albeit with the temperature of the broad maximum rising to 90 K. This feature was again ascribed to short-range spin ordering within the tetramers, and the 3D ordering temperature of the tetrameric spins was not identified directly. No long-range ordered moment was detected on the prismatic sites in the neutron diffraction experiment at 1.7 K and the 5 K transition was therefore taken to indicate the freezing (in spin-glass fashion) of the spins on these sites. In the case of both $Ba_6Mn_4CuO_{15}$ and $Ba_6Mn_4ZnO_{15}$ we have thus postulated that the maxima observed directly (Zn, 40 K) or indirectly (Cu, 90 K) in the susceptibility data do not correspond to 3D Néel points, and that a long-range ordered state is formed at lower, as yet unknown, temperatures. We describe below neutron diffraction experiments that were undertaken in order to test our hypothesis and to determine T_N.

EXPERIMENTAL

The preparation of the polycrystalline samples used in the experiments described below has been described previously [6]. Neutron diffraction data were collected from $Ba_6Mn_4CuO_{15}$ using the instrument D1b at the ILL, Grenoble. This relatively low-resolution diffractometer, has a stationary bank of 400 detectors covering the angular range $10 \leq 2\theta / ° \leq 90$, and operates with $\lambda = 2.522$ Å. The sample, contained in a vanadium can, was mounted in an ILL orange cryostat and data were recorded at fixed temperatures in the range $1.7 \leq T / K \leq 12.0$. During the course of each data collection the temperature varied by < 0.1 K. The data thus collected were subjected to a Rietveld analysis [7] using the GSAS suite of programs.[8] Neutron diffraction data were recorded on $Ba_6Mn_4ZnO_{15}$ using the backscattering detectors on the time–

of–flight (TOF) diffractometer OSIRIS at the ISIS facility, RAL. A pair of choppers was used to select a wavelength range of *ca.* 4 Å from the energetically broad neutron pulse exiting the moderator. Due to the limited d–spacing range sampled by the instrument the data were not used in a Rietveld refinement and the peak areas were instead determined by integration. The sample was again contained in a cylindrical vanadium can which was mounted in an ILL orange cryostat.

RESULTS

The data collected from $Ba_6Mn_4CuO_{15}$ contained insufficient Bragg peaks to allow a complete refinement of the crystal structure. The atomic parameters were therefore held at the values determined at 1.7 K using the high–resolution diffractometer D2b [6], and only a limited number of parameters were refined. These included a scale factor, 2 lattice parameters, 8 background parameters and a single parameter to describe the magnitude of the magnetic moment, the orientation of which was fixed in the 120 ° arrangement described previously. All of the data sets contained an instrumental feature in the angular range $71.8 \leq 2\theta / ° \leq 73.7$ that was excluded from the refinements. Each data set was analysed in this fashion and in every case a good fit was obtained. The typical quality of fit is illustrated by the data collected at 1.7 K shown in Figure 3 (R_{wp} = 3.80). The temperature dependence of the ordered magnetic moment is shown in Figure 4. The configuration of the OSIRIS TOF diffractometer meant that data could only be recorded on $Ba_6Mn_4ZnO_{15}$ over a limited d-spacing range, thus making a Rietveld refinement impossible. The integrated intensity of the strongest magnetic peak was therefore determined as a function of temperature. The effect of variation in the incident flux during the experiment was eliminated by comparing the intensity of the 103 magnetic reflection with that of the $20\bar{1}$ reflection, which occurs at a similar d-spacing to the magnetic reflection and has only a small magnetic contribution ($I_{mag}/I_{nuc} \approx 10^{-2}$). The intensity of the latter reflection is therefore temperature independent within the precision of the data and can be used to normalise the intensity of the magnetic scattering. We have previously determined (from a full Rietveld analysis of data collected using the diffractometer D2b) the value of the Mn magnetic moment at 1.7 K to be 0.6(1) μ_B[6], and this value was used to scale the OSIRIS data, leading to the temperature dependence of the ordered magnetic moment shown in Figure 5.

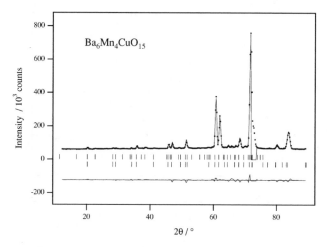

FIGURE 3 Observed (dots), calculated (line) and difference neutron powder diffraction patterns for $Ba_6Mn_4CuO_{15}$ at 1.7 K.

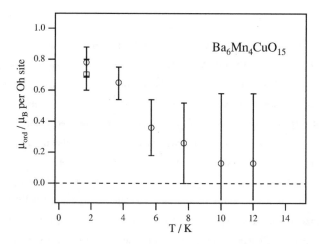

FIGURE 4 Ordered magnetic moment per octahedral site in $Ba_6Mn_4CuO_{15}$ as a function of temperature determined using data from D1b (circles) and D2b (square). The error bars represent 1 esd.

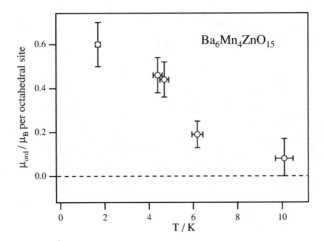

FIGURE 5 Ordered magnetic moment of $Ba_6Mn_4ZnO_{15}$ as a function of temperature. The error bars represent 1 esd.

DISCUSSION

The errors on the refined magnetic moments limit the precision with which the Néel temperatures of $Ba_6Mn_4CuO_{15}$ and $Ba_6Mn_4ZnO_{15}$ can be determined. Consideration of the error bars in Figures 4 and 5, together with a visual inspection of the diffraction patterns, suggests that both $Ba_6Mn_4CuO_{15}$ and $Ba_6Mn_4ZnO_{15}$ lose long-range magnetic order in the temperature range $6 \leq T_N/K < 10$, and that the non-zero values derived for the ordered moment of the former compound above this range result from attempts to fit noise in the neutron background. Whatever the precise value of T_N, it is clear that these results are consistent with the model which accounts for the susceptibility maxima observed at 40 and 90 K in terms of short-range ordering within the tetramers, rather than in terms of long-range order. Whilst the magnetic behaviour of these materials is clearly anisotropic, the occurrence of magnetic order below 6 K indicates that the interchain interactions are not negligible. We have recently shown [9] that the n = 0 composition 2H-BaMnO$_3$, in which the [001] chains are comprised only of MnO_6 octahedra, also shows a susceptibility maximum at a significantly higher temperature (150 K) than the Néel temperature (59 K). In all of these systems, magnetic frustration results in a value for the ordered moment that is much-reduced from the expected value (~ gS in the spin-only case). These studies demonstrate that the magnetic ordering temperature of the members of this structural family are not easily recognised in d.c. $\chi(T)$ data, and that neutron scattering has an important rôle to play in their study.

ACKNOWLEDGMENTS

We are grateful to EPSRC for financial support and to Thomas Hansen and Dennis Engberg for experimental assistance at ILL and ISIS respectively.

References

1. J. Darriet and M.A. Subramanian, *J. Mater. Chem.* **5**, 543 (1995).
2. C.A. Moore, E.J. Cussen, and P.D. Battle, *J. Solid State Chem.* **153**, 254 (2000).
3. S.H. Irons, T.D. Sangrey, K.M. Beauchamp, M.D. Smith, and H.C. zur-Loye, *Phys. Rev. B* **61**, 11594 (2000).
4. J.F. Vente, J.K. Lear, and P.D. Battle, *J. Mater. Chem.* **5**, 1785 (1995).
5. M. Zakhour-Nakhl, F. Weill, J. Darriet, and J.M. Perez-Mato, *Int. J. Inorg. Materials* **2**, 71 (2000).
6. E.J. Cussen, J.F. Vente, and P.D. Battle, *J. Amer. Chem. Soc.* **121**, 3958 (1999).
7. H.M. Rietveld, *J. Appl. Crystallogr.* **2**, 65 (1969).
8. A.C. Larson and R.B. von-Dreele, General Structure Analysis System (GSAS), *Los Alamos National Laboratories*, Report LAUR 86-748, 1990.
9. E.J. Cussen and P.D. Battle, *Chem. Mater.* **12**, 831 (2000).

Mat. Res. Soc. Symp. Proc. Vol. 658 © 2001 Materials Research Society

Powder Diffraction Refinements of the Structure of Magnetite (Fe_3O_4) Below the Verwey Transition.

Jon P. Wright,[1] J. Paul Attfield[1] and Paolo G. Radaelli[2]

[1]Department of Chemistry, University of Cambridge, Lensfield Road, Cambridge CB2 1EW, U.K.

[2]ISIS Facility, Rutherford Appleton Laboratory, Chilton, Didcot, OX11 0QX, U.K.

ABSTRACT

Magnetite is a classic example of a mixed-valent transition metal oxide, in which electronic conductivity and ferromagnetism result from electron hopping between octahedrally coordinated Fe^{2+} and Fe^{3+} states. Below the 122 K Verwey transition, the conductivity falls by a factor of ~100 and a complex monoclinic (or triclinic) superstructure of the high temperature cubic spinel arrangement is adopted. This is assumed to be the result of Fe^{2+}/Fe^{3+} charge ordering on the octahedral sites, but this has not been confirmed crystallographically, as single crystal refinements have been hampered by the extensive twinning that accompanies the Verwey transition. We have used very highly resolved powder diffraction data to attempt Rietveld refinements of the low temperature structure. The powder sample was prepared by grinding a single crystal of stoichiometric magnetite. Data were collected at 90 K on instruments HRPD at the ISIS neutron source, UK, and BM16 at the European Synchrotron Radiation Facility, France. The very high resolution of these data enables the monoclinic distortion to be observed, and the structure has been refined on the supercell proposed by Iizumi et al (Acta Cryst. B38, 2121 (1982)) with Pmca pseudosymmetry, giving parameters a = 5.94443(1), b = 5.92470(2), c = 16.77518(4) Å, β = 90.236(1)°. The mean octahedral site Fe-O distances differ from each other significantly, but the maximum difference between values is only 20% of that expected for ideal Fe^{2+}/Fe^{3+} charge ordering.

INTRODUCTION

Magnetite is the oldest known magnetic material, having been first described in ~800 B.C. At room temperature the iron spins order ferrimagetically, giving an overall bulk magnetisation. Formulated as $Fe^{3+}[Fe^{2+}Fe^{3+}]O_4$, magnetite has the inverse cubic spinel crystal structure, with Fe^{3+} at the tetrahedral A sites and Fe^{2+} and Fe^{3+} at the octahedral B sites. Rapid electron hopping occurs between the octahedral sites at ambient temperatures, but on cooling through the Verwey transition at T_v = 122 K, the resistivity rises sharply [1,2,3], which has been interpreted as a charge ordering of the Fe^{2+} and Fe^{3+} cations. Recent interest in magnetic conducting oxides for their magnetoresistive properties and associated phenomena such as charge ordering has given rise to a resurgence of interest in bulk and thin film samples of magnetite.

Despite much effort, no conclusive study of the crystal structure in the low temperature phase has emerged. A charge ordered structure with orthorhombic symmetry was originally proposed by Verwey et al [3] which was thought to have been confirmed by a single crystal neutron diffraction experiment [4]. However it was later shown that this experiment was flawed by multiple scattering effects [5]. A monoclinic $\sqrt{2}a \times \sqrt{2}a \times 2a$ supercell with Cc space group symmetry was proposed from experiments on a partially de-twinned crystal using neutron diffraction [6] and was supported by a single crystal x-ray study [7]. The structure below T_v w refined using neutron data from a mechanically de-twinned single crystal [8]. The authors use an $a/\sqrt{2} \times a\sqrt{2} \times 2a$ subcell of the unit cell identified above, and imposed orthorhombic symme constraints on the atomic positions, to obtain an approximation to the low temperature crystal structure. No charge ordered arrangement was identified in this refined structure.

The observation of a magnetoelectric effect [9] indicated that the symmetry in the low temperature phase is acentric and can be no higher than P1. Recent evidence for the triclinic distortion is provided by an x-ray topography study [10], which shows twins which were interpreted as being due to a triclinic distortion. However, a recent Raman and IR spectroscop investigation concluded that the structure is centric [11]. Additional evidence for a c-glide pla was found from the observation of dynamically forbidden electron diffraction lines [12], and a charge ordered arrangement has been proposed from a further electron microscopy study [13].

A recent NMR study [14] showed that the spectra are compatible with the Cc cell, which contains 16 different octahedral iron sites. However, these authors concluded that "below T_v th states of the iron sites on the B sublattice are so strongly mixed that the notion of 2+ and 3+ valency may lose its meaning". The most recent Mossbauer investigation [15] evidenced five iron sites below T_v; two octahedral Fe^{2+}, two octahedral Fe^{3+} and a single tetrahedral Fe^{3+}.

It is generally assumed that any charge ordered model must satisfy the Anderson criterion [16] which is that each tetrahedral cluster of four B sites must contain two Fe^{2+} and two Fe^{3+} ions. This leads to only ten possible fully charge ordered models in the above Cc symmetry ce However, the Anderson criterion can also be satisfied by short-range charge ordering leading t zero-point entropy [16].

The problems of multiple scattering, severe extinction and twinning effects, which beset single crystal studies on this material, have led us to study the structure of magnetite below T_v using powder diffraction. In a powder diffraction experiment only primary extinction has a significant influence on the observed intensities and twinning effects are dealt with by accounting for the overlap of the powder peaks. Also multiple scattering is much reduced in a powder sample, as the individual crystallites are much smaller. Results using constant wavelength neutron powder data have been published recently [17]. These showed that the powder pattern below the Verwey transition can be described to a fair approximation by a rhombohedral distortion of the cubic $Fd\bar{3}m$ spinel cell of magnetite. In this paper we summari results from a high resolution synchrotron X-ray and time-of-flight neutron powder study.

EXPERIMENTAL

A highly stoichiometric sample of $Fe_{3-3\delta}O_4$ with $\delta \approx 0.000$ was provided by Prof. J. Honig The powder was obtained by grinding single crystals grown by the skull melter technique. Hig resolution powder diffraction data were collected at 90 K and 130 K on the time-of-flight HRP instrument at the ISIS pulsed neutron source, UK. Powder synchrotron X-ray data were collec at 90 K and 130 K on the BM16 diffractometer at ESRF, Grenoble, France at a wavelength of 0.3251 Å.

RESULTS

The distortion below the Verwey transition is primarily rhombohedral, and elongates the cell along one of the [111] body diagonals, which decreases the angles between the cubic cell edges by ~0.2–0.3° and lowers the symmetry from $Fd\bar{3}m$ to $F\bar{3}m$ (a non-standard setting of $R\bar{3}m$). Doubling one of the cell vectors removes the threefold axis of the rhombohedral cell leading to monoclinic symmetry. Figure 1 indicates the relationship between the unit cell axes.

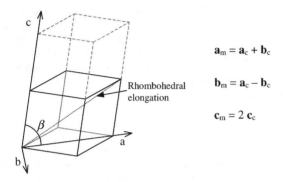

$$\mathbf{a}_m = \mathbf{a}_c + \mathbf{b}_c$$

$$\mathbf{b}_m = \mathbf{a}_c - \mathbf{b}_c$$

$$\mathbf{c}_m = 2\,\mathbf{c}_c$$

Figure 1. *The distortion of the cubic spinel unit cell below T_v in magnetite, Fe_3O_4. Face centering lattice points are not shown for clarity.*

The magnitude of the monoclinic distortion is smaller than that of the rhombohedral distortion, although this is frequently masked by the choice of axes used when reporting the monoclinic cell. The highest resolution data, from BM16 at ESRF in Figure 2, enable the two distortions to be compared. The rhombohedral distortion splits the (404) peak into two distinct components, while the monoclinic splitting broadens the left-hand component. The largest observed monoclinic splitting is that of the (800) peak.

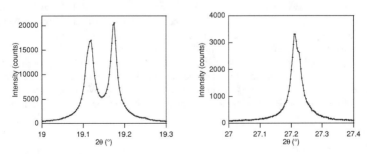

Figure 2. *Synchrotron powder X-ray diffraction peaks for Fe_3O_4 at 90 K. Left hand panel – (404) peak, right hand panel – (800) peak*

The principally rhombohedral nature of the distortion enables a reasonably good profile fit to be obtained with a simple $F\bar{3}m$ model as demonstrated in a recent paper [17]. This rhombohedral distortion splits the octahedral B positions into two sites in a 1:3 ratio. Refinements using this model lead to mean Fe-O distances for the two octahedral sites that diff slightly, e.g. 2.056(3) and 2.067(3) Å using the HRPD data.

Weak superstructure peaks are observed in both the HRPD and the BM16 profiles (Figure 3), and have been used for a combined refinement of the structure. Full details of the refinemen will be published elsewhere. The final model was refined in space group P2/c but uses orthorhombic Pmca symmetry constraints on the atomic positions, as were used previously in ref. 8. In this model, there are two crystallographically independent tetrahedral A sites and four octahedral B sites. The two tetrahedral sites are quite regular, with mean Fe-O bond distances A(1): 1.886(3) and A(2): 1.890(4) Å. The four octahedral sites are more distorted with some of the Fe atoms displaced off-center, and the mean Fe-O distances are B(1): 2.072(3), B(2): 2.043(4), B(3): 2.050(4), B(4): 2.068(4) Å.

CONCLUSIONS

These results show that high resolution powder X-ray and neutron diffraction experiments are able to resolve the monoclinic distortion of the magnetite structure below the Verwey transition, and observe the weak superstructure peaks. The quantitative Rietveld analysis of the data enables a refinement of the structure comparable to the previously reported single crystal neutron results to be obtained.

The refined model is very similar to that previously published [8], but the values of the Fe distances are slightly different, and this results in the average Fe-O bond distances falling into two sets, with two small and two large octahedral sites. However, as the ionic radii of Fe^{2+} and Fe^{3+} differ by 0.135 Å, the difference between the two types of B sites corresponds to only 20% of the difference between ideal charge ordered states (the same result is obtained from bond valence calculations). Nevertheless, these results provide evidence for at least partial charge ordering in magnetite below the Verwey transition.

The full Cc supercell of magnetite contains 16 independent octahedral sites so that each sit in our refinement averages over four distinct sites in the full structure. Hence, our small and larg sites could be viewed as being Fe^{3+} and Fe^{2+}, or ($0.75Fe^{3+} + 0.25Fe^{2+}$) and ($0.25Fe^{3+} + 0.75Fe^{2+}$). In the latter case, the difference between average distances would be half the above difference between Fe^{2+} and Fe^{3+} ionic radii, which is more in keeping with the observed distances. We note that even in well-defined, unfrustrated, charge ordered oxides, the difference in size are usually somewhat less than expected on the basis of idealised ionic radii for the two valence states. For example, in $Ca_2Fe^{3+}Fe^{5+}O_6$ the difference in Fe site charges calculated from the bond distances is only 55% of the ideal separation [18]. We are currently considering charge ordered models consistent with the refinement results and these will be published elsewhere.

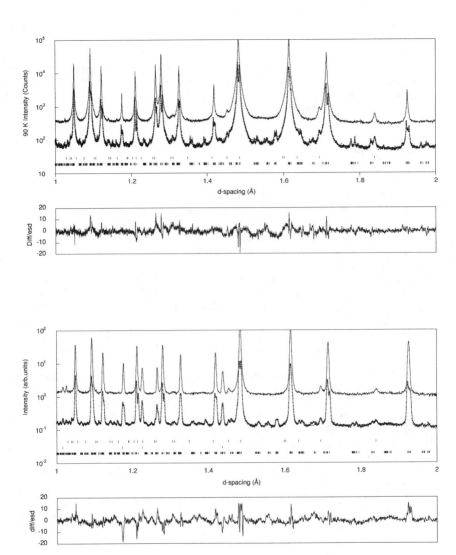

Figure 3. *Rietveld fits to part of the synchrotron (top) and time-of-flight neutron (bottom) powder diffraction data for magnetite at 90 K using logarithmic intensity scales. 130 K data are also plotted, one decade above the 90 K data, enabling the superstructure peaks that appear below the Verwey transition to be identified.*

ACKNOWLEDGMENTS

We thank Prof. J. Honig for kindly providing the sample of magnetite and for many useful discussions. We acknowledge EPSRC for providing beamtime and EPSRC and RAL for a studentship for J.P.W. We thank Drs. A. Fitch and E. Dooryhee (ESRF) and Dr. R. Ibberson (RAL) for assistance with data collection.

REFERENCES

1. E. J. W. Verwey, *Nature* (1939) **144**, 327.
2. E. J. W. Verwey and P. W. Haayman, *Physica (Utrecht)* (1941) **8**, 979.
3. E. J. W. Verwey, P. W. Haayman and F. C. Romeijan, *J. Chem. Phys* (1947) **15**, 181.
4. W. C. Hamilton, *Phys. Rev.* (1958) **110**, 1050.
5. G. Shirane, S. Chikazumi, J. Akimitsu, K. Chiba, M. Matsui and Y. Fuji, *J. Phys. Soc. Jpn.* (1975) **47**, 1779.
6. M. Iizumi and G. Shirane, *Solid State Comm.* (1975) **17**, 433.
7. J. Yoshido and S. Iida. *J. Phys. Soc. Jpn.* (1979) **47** 1627.
8. M. Iizumi, T. F. Koetzle, G. Shirane, S. Chikazumi, M. Matsui and S. Todo. *Acta. Cryst.* (1982) **B38**, 2121.
9. Y. Miyamoto and M. Shindo. *J. Phys. Soc. Jpn.* (1993) **62**, 1423
10. C. Medrano, M. Schlenker, J. Baruchel, J. Espeso and Y. Miyamoto. *Phys. Rev. B.* (1999) **59**, 1185.
11. L. V. Gasparov, D. B. Tanner, D. B. Romero, H. Berger, G. Margaritondo and L. Forro. *Phys.Rev.B.* (2000) **62,** 7939.
12. M. Tanaka, *Proceedings of the 45th EMSA,* edited by G. Bailey (San Francisco Press, San Francisco, 1987), p. 20.
13. J. M. Zuo, J. C. H. Spence and W. Petuskey. *Phys. Rev. B.* (1990) **42**, 8451.
14. P. Novak, H. Stepankova, J. Englich, J. Kohout and V. A. M. Brabers. *Pjys. Rev. B.* (2000) **61**, 1256.
15. F. J. Berry, S. Skinner and M. F. Thomas. *J. Phys: Condens Matter.* (1998) **10**, 215
16. P. W. Anderson. Phys. Rev. (1956) **102**, 1008.
17. J. P. Wright, A. M. T. Bell and J. P. Attfield. *Solid State Sci.*, in press.
18. P. M. Woodward, D. E. Cox, E. Moshopoulou, A. W. Sleight and S. Morimoto, Phys. Rev. B (2000) **62**, 844.

Dielectrics, Crystal Chemistry, Glasses, Electrical Transport

Mat. Res. Soc. Symp. Proc. Vol. 658 © 2001 Materials Research Society

Ferroelectric Tunability Studies in Doped BaTiO$_3$ and Ba$_{1-x}$Sr$_x$TiO$_3$

Dong Li and M. A. Subramanian
DuPont Central Research and Development, Experimental Station, Wilmington, DE 19880, USA

ABSTRACT

Acceptor and Donor codoped BaTiO$_3$ and Ba$_{1-x}$Sr$_x$TiO$_3$ are prepared. For Ba$_{1-x}$La$_x$Ti$_{1-x}$Fe$_x$O$_3$, BaTiO$_3$ remains as tetragonal phase up to about 5mol% LaFeO$_3$. For x $\geq$0.06, the structure changes to cubic at room temperature. The phase change shifts the Curie temperature to lower value and increases the tunability at room temperature. Doping of other acceptor (Al, Cr) and donor (Sm, Gd, Dy) ions has the same effect although with varying levels of tuning. BaTiO$_3$: 4%LaFeO$_3$ has the highest tunability among the studied systems, which is even higher than Ba$_{0.6}$Sr$_{0.4}$TiO$_3$. Co-doping of (La, Fe) and (La, Al) in Ba$_{1-x}$Sr$_x$TiO$_3$ also lowers the Curie temperature and increases the tunability of high Ba content samples at cryogenic temperature.

INTRODUCTION

The conventional tunable microwave devices for communication applications are made by mechanically tuned resonant structure, ferrite based waveguide, or semiconductor based voltage controlled electronics. Ferroelectric materials have received considerable attention recently for this application because of their advantages over the above mentioned materials. Ferroelectric devices can be tunned faster and at relatively lower voltage, consume lower power, and are smaller.

The two most important requirements of a ferroelectric material are low loss tangents (tanδ) and high dielectric tunability (Δε/ε). Device losses can be significantly reduced by using superconducting material instead of metal as electrodes. This idea is being actively pursued to produce novel voltage tunable microwave devices such as resonators, filters, and phase shifters [1-3]. Although tuning can be achieved in both the ferroelectric and paraelectric phases, the losses are usually higher below the Curie point (T$_C$) due to the additional hysteresis losses. To take fully advantage of the superconducting electrode, the T$_C$ of the ferroelectric materials should be lower than the operating temperature (usually 80K for high temperature superconductors). For room temperature application, the T$_C$ should be below 0°C.

The most often used ferroelectric material is Ba$_{1-x}$Sr$_x$TiO$_3$ (BSTO) [4]. Its Curie point can be tuned by varying the mole fraction x of Sr. The Curie temperatures are 40K and 393K for x=0.95 and pure BaTiO$_3$, respectively [5]. However the dielectric behaviors of high Sr content films depend on the substrate interaction.

It is widely known the dielectric properties of BaTiO$_3$ can be modified through partial substitution of isovalent or heterovalent cations at Ba (A site doping) or Ti sites (B-site doping). In general, there are two kinds of dopants: acceptor and donor [6,7]. Acceptor dopants are ions with a lower valence than the ions they replace (e.g. 1+ ions for Ba^{2+}, 3+ ions for Ti^{4+}). Donor dopants are ions with a higher valence than the ions they replace (e.g. 3+ ions for Ba^{2+}, 5+ ions for Ti^{4+}). Among the acceptor and donor doping, lanthanide ions [8] substituted into Ba sites (donor dopants) and transition metal ions [9, 10] substituted into Ti sites (mainly acceptor dopants) have been extensively studied. It has been found the doping can shift the Curie temperature. In most of these studies, the dopants are compensated by Ba vacancies, Ti vacancies, or O vacancies. Because the vacancy is one of the sources of dielectric loss, it is desirable to charge balance the dopants with other ions without generating vacancies. Few research groups studied the dielectric properties of acceptor-donor codoped $BaTiO_3$, especially tuning. Skapin and co-workers studied the phase change and Curie-temperature shift for the $BaTiO_3$-$LaAlO_3$ system [11]. Schwarzbach studied the semiconductive properties and Curie temperatures of $Ba_{1-n}La_{2n/3}TiO_3$-Fe_2O_3 system at very low content of La [12].

To search for $BaTiO_3$ and BSTO based ceramic material with high tuning and low loss properties, a wide range of acceptor-donor codoped $BaTiO_3$ were prepared in this lab and their dielectric properties were measured. High tunability is found in $BaTiO_3$-$LaFeO_3$ at room temperature. The study is extended to BSTO to search for high Ba content samples that can be used at cryogenic temperature. As an initial study, we only studied the samples doped with 4%La with 4%Fe, 4%La with 4%Al, and 6%La with 2%Al.

EXPERIMENTAL DETAILS

Solid solution oxide of the formula $Ba_{1-x}Ln_x^{3+}Ti_{1-x}M_x^{3+}O_3$, wherein Ln = La, Sm, Gd, Dy and M = Al, Fe, Cr, $(Ba_{1-y}Sr_y)_{0.96}La_{0.04}Ti_{0.96}Fe_{0.04}O_3$ [$(Ba_{1-y}Sr_y)TiO_3$: 4%La4%Fe], $(Ba_{1-y}Sr_y)_{0.96}La_{0.04}Ti_{0.96}Al_{0.04}O_3$ [$(Ba_{1-y}Sr_y)TiO_3$:4%La4%Al], and $(Ba_{1-y}Sr_y)_{0.94}La_{0.06}Ti_{0.97}Al_{0.02}\square_{0.01}O$ ($\square$: vacancy) [$(Ba_{1-y}Sr_y)TiO_3$:6%La2%Al], were prepared by the conventional powder sintering technique. The starting materials were Ln_2O_3 (Ln = lanthanide ion), M_2O_3 (M = Al, Fe, Cr), $SrCO_3$, $BaTiO_3$, and TiO_2 with purity of 99.9% or higher. Ln_2O_3 powders were heated at 1000°C before use. Appropriate amounts of starting oxides were weighed according to the stoichiometric ratios and mixed thoroughly in an agate mortar. The mixed powder was calcined at 1000°C for 8 hours. The calcined powder was reground and pressed to 12.7mm dia/1-2mm thick discs. The discs were sintered in air at 1325°C for 20 hours. The ramping rate was 200°C/hour and the cooling rate was 150°C/hour. X-ray powder diffraction patterns were recorded with Siemens D5000 diffractometer. The data showed all the samples were crystallized in a pseudo-cubic or cubic perovskite related structure. There are obvious $BaAl_2O_4$ impurities in 4%La4%Al doped samples. To minimize the impurities, 6%La2%Al doped BSTO samples were prepared.

The disc samples were polished to produce flat uniform surface and electroded with silver paint. The painted samples were dried at 70-100°C overnight. Capacitance and loss tangent measurements were taken on a HP-4275A LCR meter at room temperature and HP-4284A LCR meter below room temperature. Voltage up to 100V was applied to the sample by Keithley 228A voltage/current source. Cryogenic temperatures were achieved by using a closed cycle refrigerator with He. A heater was installed in the cold head to allow control of the sample temperature, which was measured using a Si diode thermometer.

RESULTS AND DISCUSSION

There is only limited solubility of La (up to 15%) at the Ba site [12] and Fe (up to 1.25%) at the Ti site [10] in $BaTiO_3$. However, our X-ray diffraction results indicate that the whole range solid solution can be formed between $BaTiO_3$ and $LaFeO_3$ under current preparative conditions. The data showed all the samples were crystallized in a pseudo-cubic or cubic perovskite related structure at room temperature. $LaFeO_3$ is well known to crystallizes in orthorhombic structure as one expect from tolerance factor (<1). The solid solution compositions remained as tetragonal up to 5% LaFeO$_3$ and for x values in the range 0.06-0.80 the compounds are cubic. For x ≥ 0.80, the compounds adopt orthorhombic LaFeO$_3$-type structure. The cell parameters of Ba$_{1-x}$La$_x$Ti$_{1-x}$Fe$_x$O$_3$ (x<0.08) is shown in Figure 1. At about 5% LaFeO$_3$, the phase changes from tetragonal to cubic. The structure study on doped BSTO shows they are all cubic at room temperature.

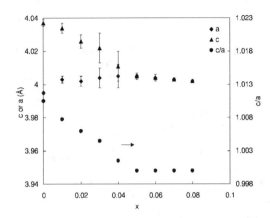

Figure 1. Unit cell parameters and c/a ratio vs. x for Ba$_{1-x}$La$_x$Ti$_{1-x}$Fe$_x$O$_3$ (x<0.08).

The temperature dependencies of relative dielectric constants and losses of BaTiO$_3$: x%LaFeO$_3$ are shown in Figure 2.

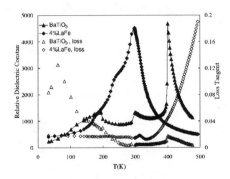

Figure 2. Dielectric properties of BaTiO3 and BaTiO3:4%LaFe.

BaTiO$_3$: 4%Ln,4%Fe (Ln = La, Sm, Gd, Dy) and BaTiO$_3$: 4%La,4%A (A=Al, Cr) are also prepared. T$_C$ decreases most when La is codoped with Fe compared to other lanthanide ions. When codoped with La, Fe, Cr, and Al all can decrease T$_C$ from about 120°C in pure BaTiO$_3$ to below room temperature at 4% doping level. The result of Al is in agreement with the literature

[11]. At higher concentration of LaFeO₃, the measured loss tangent is quite high. For doped BSTO samples, 6% La with 2% Al codoping shows the stronger effect to decrease the Curie temperature. La, Al codoped $Ba_{0.6}Sr_{0.4}TiO_3$ already shows the peak in figure of merit around 80K as shown in Figure 3. Figure of merit is used to show the combined effect of tuning and loss tangent. It is defined as:

$$[\varepsilon\ (v=0) - \varepsilon\ (v\neq0)] / [\varepsilon\ (v=0)] \times D\ (v=0)]\ (1)$$

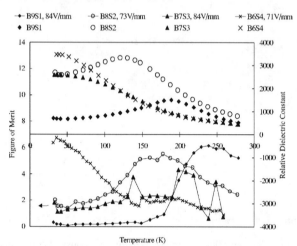

Figure 3. *Dielectric constant and figure of merit of La, Al doped BSTO (100k Hz).*

The tunability and loss of BaTiO₃: 4%LaFe are shown in Figure 4. Some selected tunabilit equations are shown in Table I.

Table I. Fitted tunability equation of selected materials.

Composition	Tunability Equation. (T: tunability in %, E: electric field in V/μm)
Room Temperature	
$Ba_{.96}La_{.04}Ti_{.96}Al_{.04}O_3$	T = 11.4 x (E)
$Ba_{.96}La_{.04}Ti_{.96}Cr_{.04}O_3$	T = 27.2 x (E)
$Ba_{.96}La_{.04}Ti_{.96}Fe_{.04}O_3$	T = 74.3 x (E)
$Ba_{.96}Sm_{.04}Ti_{.96}Fe_{.04}O_3$	T = 21.0 x (E)
$Ba_{.96}Dy_{.04}Ti_{.96}Fe_{.04}O_3$	No tuning observed
BaTiO₃	No tuning observed
$Ba_{0.6}Sr_{0.4}TiO_3$	T = 38.4 x (E)
80K	
$Ba_{0.2}Sr_{0.8}TiO_3$	T = 67.9 × (E)
$Ba_{0.4}Sr_{0.6}TiO_3$	T = 23.3 × (E)
$Ba_{0.6}Sr_{0.4}TiO_3$: 6%La, 2%Al	T = 71.4 × (E)
$Ba_{0.6}Sr_{0.4}TiO_3$: 4%La, 4%Al	T = 12.3 × (E)
$Ba_{0.6}Sr_{0.4}TiO_3$: 4%La, 4%Fe	T = 18.5 × (E)

Tunability, T = [K (v=0) – K (v≠0)] / K (v=0)]

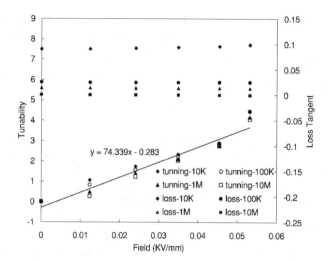

Figure 4. *Tunability and loss tangent versus applied electric field for BaTiO₃: 4%LaFeO₃ at room temperature.*

$BaTiO_3$: $4\%LaFeO_3$ has the highest tunability at room temperature among the materials we prepared. Its tunability is even higher than that of the $Ba_{0.6}Sr_{0.4}TiO_3$, a conventional material used for room temperature applications. Among the doped BSTO samples, 6%La with 2%Al codoped sample shows the best effect on increasing tunability of higher Ba content samples at 80K. Its tunability is comparable to that of the $Ba_{0.2}Sr_{0.8}TiO_3$. The effects of 4%La4%Al and 4%La4%Fe codoping are weaker. This might be caused by the impurity of $BaAl_2O_4$ and lower dielectric constant in LaFe codoped sample at low temperature. The loss tangents of the samples are not much influenced by the applied DC field. These doped BSTO samples provide much higher Ba content compared to the traditional sample composition used for cryogenic temperature region.

It is known that the T_C is higher when c/a ratio extends for perovskites with tetragonal symmetry [13]. So the reason of the decrease of T_C in $BaTiO_3$-$LaFeO_3$ system can be well explained by the results of cell parameter analysis. The decrease of T_C can also be understood from the microscopic point of view. The introduction of La^{3+} to replace Ba^{2+} will decrease the bond covalency, which decreases T_C. The same effect has been observed when F is substituted for O in $BaTiO_3$ [13]. When considering the Fe^{3+}-Ti^{4+} substitution, we reach the conclusion that T_C will decrease too. It is well known T_C (K) = 2.10^4 $(\Delta z)^2$ [13] where Δz denotes the movement along the polar axis. The effective ionic radius of Fe^{3+} (0.645Å) is a little bit bigger than Ti^{4+} (0.605Å) [14]. The increase in size will limit the displacement in the octahedron leading to a decrease in T_C.

CONCLUSIONS

The simultaneous doping of acceptor and donor ions in polycrystalline $BaTiO_3$ and $Ba_{1-x}Sr_xTiO_3$ can lowers the Curie point and increases the tunability at selected temperature. A whole range solid solution can be formed between $BaTiO_3$ and $LaFeO_3$. Up to 5mol% $LaFeO_3$, $BaTiO_3$ remains in tetragonal phase at room temperature. Above this level, $BaTiO_3$ changes to cubic. The phase and cell parameter changes of $BaTiO_3$ caused by La and Fe codoping is the reason of the shift of Curie temperature. The observed tunability for $BaTiO_3$: 4%LaFe is higher than that of $Ba_{0.6}Sr_{0.4}TiO_3$. The doping also increases the tunability of the high Ba content sample at the cryogenic temperature.

REFERENCES

1. S. Gevorgian, E. Carlsson, E. Wikborg, E. Kollberg, Integr. *Ferroelectr.*, **22** 245 (1998).
2. O. G. Vendik, L. T. Ter-Martirosyan, A. I. Dedyk, S. F. Karmanenko, R. A. Chakalo *Ferroelectrics*, **144** 33 (1993).
3. A. Hermann and V. Badri, *J. Supercond.*, **12** 139 (1999).
4. A. Outzourhit and J. U. Trefny, *J. Mater. Res.*, **10** 1411 (1995).
5. E. Hegenbarth, *Phys. Status Solidi*, **9** 191(1965).
6. S. B. Herner, F. A. Selmi, V. V. Varadan, V. K. Varadan, *Mater. Lett.*, **15** 317 (1993).
7. V. C. S. Prasad, L. G. Kumar, *Ferroelectrics*, **102** 141 (1990).
8. F. D. Morrison, D. C. Sinclair, J. M. S. Skakle, A. R. West, *J. Am. Ceram. Soc.*, **81** 19 (1998).
9. A. Inoue, M. Iha, I. Matsuda, H. Uwe, T. Sakudo, Jpn. *J. Appl. Phys.*, **30 [9B]** 2388 (1991).
10. H. Ihrig, *J. Phys. C: Solid State Phys.*, **11** 819 (1978).
11. S. Skapin, D. Kolar, D. Suvorov, *J. Solid State Chem.*, **129** 223 (1997).
12. J. Schwarzbach, Cezch. *J. Phys.*, **B18** 1322 (1968).
13. J. Ravez, M. Pouchard, P. Hagenmuller, *Ferroelectrics,* **197** 161 (1997).
14. R. D. Shannon, *Acta Crystallogr.*, **A32** 751 (1976).

Mat. Res. Soc. Symp. Proc. Vol. 658 © 2001 Materials Research Society

Modification of Ferroelectric Properties of TGS Crystals Grown under a dc Electric field

G. Arunmozhi[1], E. Nogueira[1], E.de Matos Gomes[1], S. Lanceros-Mendez[1], M. Margarida R. Costa[2], A. Criado[3] and J. F. Mano[4]
[1] Departamento de Física, Universidade do Minho, Campus de Gualtar 4710-057 Braga, Portugal.
[2] Departamento de Física, faculdade de Ciências e Tecnologia, Universidade de Coimbra, 3000 Coimbra, Portugal.
[3] Departamento de Física de la Materia Condensada, Universidad de Sevilla, Apartado 1065, 41080 Sevilla, Spain.
[4] Departamento de Engenharia de Polímeros, Universidade do Minho, Campus de Azurém 4800-058 Guimarães, Portugal.

ABSTRACT

Ferroelectric triglycine sulphate crystals have been grown under the influence of an intense electric field of $6x10^4$ V/m. Relative to crystals grown under ambient conditions (TGS) the crystals grown under the electric field (TGS-E) display a dielectric permittivity a factor of two lower. Significant differences are observed in the Curie-Weiss behavior of the ferroelectric phase, in the x-ray diffraction patterns and in the differential calorimetry measurements.

INTRODUCTION

Ferroelectric triglycine sulphate $(NH_2CH_2COOH)_3.H_2SO_4)$, (abbreviated TGS) was discovered by Mattias et al. [1] in 1956. It exhibits a second order phase transition with the Curie temperature at 49°C. Below and above the Curie point, the crystal has monoclinic symmetry. In the paraelectric phase (T > 49°C), it belongs to the centrosymmetrical crystal class 2/m and in the ferroelectric phase (T < 49°C) the mirror plane disappears and the crystals belongs to the polar point group 2 [2]. TGS can be grown in large sizes from aqueous solutions. The growth can be achieved by slow evaporation of the solvent at constant temperature or by slowly lowering the temperature at constant supersaturation.

Single crystals of triglycine sulphate (TGS) find wide application in pyroelectric detection. Depolarization of the glycine molecule with time in TGS seriously affects the device performance. Several dopants have been reported to stabilize the glycine molecule. Depolarization fields can be overcome if an internal bias field is created in the TGS lattice [3, 4]. In this paper we report preliminary results on the dielectric, structural ,SEM and thermal behaviour of TGS grown under a high dc electric field of $6.0x10^4$ V/m. To the best knowledge of the authors, no work has been reported in the literature on the effect of high dc fields applied during crystal growth.

EXPERIMENTAL DETAILS

Crystals of TGS were obtained by re-crystallizing a water solution under a dc electric field of roughly $6.0x10^4$ V/m. The electric field was generated by using a transformer to create a galvanic isolation as well as to provide an initial increase of the voltage. Subsequently a chain of voltage multipliers were used to both amplify and rectify the ac voltage source into a high dc voltage allowing us to eventually reach a differential of up to 40 kV. Rectangular glass cells containing the solution were then placed between condenser plates, the area of these condenser plates was much larger than the transverse cell dimensions while their spacing could be varied between 20 and 30 mm. With the highest voltage applied to the plates, an electric field of up to 1.5×10^6 V/m could be obtained in air between the condenser plates.

The results reported in this communication were obtained using a glass cell, with 2.5 mm walls and 10 mm width, of solution placed between condenser plates separated by 22 mm. A voltage difference of 35.8 kV was applied to the condenser plates giving rise to an electric field of approximately 6.0×10^4 V/m within the solution. (Taking the relative dielectric constant of glass to be 6.75 as is appropriate for Corning 0080 type [5] and assuming that the relative dielectric constant of the solution is essentially that of water, that is 71.99). Crystals were grown at room temperature using water as solvent, the crystal morphology being the same for crystals grown with and without the electric field (hereafter TGS-E and TGS respectively). There is no marked asymmetry in the growth rate along the b-axes and hence the macroscopic mirror plane normal to the b-axis is preserved. Well faceted good quality crystals of about $8 \times 8 \times 5$ mm^3 were grown in a week's time in the growth cells. The grown crystals were carefully removed from the rectangular growth cells and dried between filter papers to remove solvent on the surface.

Samples were cleaved from TGS and TGS-E grown crystals using a fine blade. The cleavage plane remained perpendicular to the polar b-axis. The cleaved plates were lapped and polished with fine alumina powder and diamond paste. Electrodes were made by painting electronic grade silver on the polished surfaces of the thin rectangular samples (thickness ~1.5 mm; area ~25 mm^2).

Dielectric measurements were carried out from room temperature up through the Curie point, at different frequencies with an applied electric field of 1V/cm. The sample thickness was always maintained at approximately 1.5mm. Using a Polymer Laboratories LCR thermal analyzer with a programmable temperature controller, the real and imaginary parts of the dielectric permittivity were measured from room temperature to 60°C at heating rate of 1°C/min for the following frequencies 0.3 kHz, 1 kHz, 10 kHz and 100 kHz.

Differential scanning calorimetry (DSC) studies were carried out in a Perkin-Elmer DSC7 apparatus, with a controlled cooling accessory. The usual calibration routines for temperature and heat flux were used. The heat capacity was calculated for the TGS and TGS-E samples using sapphire as reference. The scans were carried out between −20 to 85°C at 10°C min^{-1}. Only the first scan of each studied sample was considered.

X-ray powder diffraction was performed on both TGS and TGS-E samples using a PDS 12 powder diffractometer with the Debye-Scherrer geometry, Cu-K$_\alpha$ radiation (λ=1.5405 Å) and a linear position sensitive detector with an angular range of 120° (CPS 120).

Ferroelectric domains on freshly cleaved TGS and TGS-E samples were examined using a scanning electron microscope (Leica Cambridge) at low probe current (12pA) and accelerating voltage (2kV) to avoid charging of the surface.

RESULTS AND DISCUSSION

In TGS-E samples, as is shown in Figure 1, the real part of the dielectric permittivity value drops to more than half of the values of TGS grown under normal conditions, at the same frequencies of the applied field. The phase transition is less sharp than in TGS and is accompanied by a shift in the transition temperature from 49°C towards a higher value of 53°C

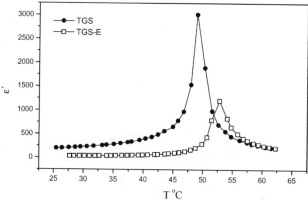

Figure 1. *Real part of the dielectric permittivity of TGS and TGS-E samples as a function of temperature for a frequency of 1kHz.*

The phase transition behavior is similar to the one observed in TGS when an external bias field is applied to a TGS crystal perpendicularly to the ferroelectric axis [6,7] and parallel [8] to the ferroelectric axis. As shown in Figure 2, the real part of the dielectric permittivity is quite constant for frequencies in the range of 300 Hz up to 100 kHz.

The Curie-Weiss temperature calculated for TGS and TGS-E, using the curves shown in Figure 3, was found to be 48.8°C and 51.3°C, respectively while the Curie Constant, C, was calculated to be 2720°C and 2270°C in the paraelectric phase. While for TGS the transition and Curie-Weiss temperatures are nearly the same, for TGS-E they differ by 2°C. In the ferroelectric phase the Curie constant of TGS and TGS-E have very different values, respectively 5263 °C and 788 °C. The ratios between the Curie constants in the paraelectric and ferroelectric phases are 0.5 for TGS, which agrees with literature [9] and 2.88 for TGS-E. These results might be an indication of domain clamping effects in the samples grown under the electric field.

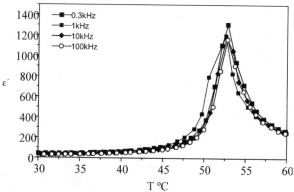

Figure 2. *Real part of the dielectric permittivity measured for TGS-E at different frequencies.*

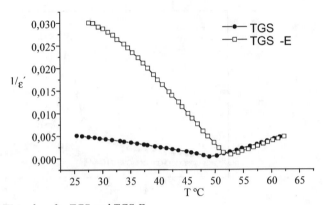

Figure 3. *Curie-Weiss law for TGS and TGS-E.*

The shift observed in the Curie-Weiss temperature is also confirmed by the DSC measurements performed as shown in Figure 4. The ferroelectric second order phase transition is well visible from the sudden drop of C_p at 49.2 °C for TGS and 50.1 °C for TGS-E. The higher transition temperature for TGS-E may be an indication of a thermodynamic stabilisation of the crystal when formed under the external electric field applied during recrystallization. During the transition the drop of C_p is similar: 0.2 J K^{-1} g^{-1}. However, the C_p of TGS-E is significantly lower than that of TGS. This difference is an indication of a higher number of degrees of freedom for molecular mobility for TGS compared to TGS-E. Some restriction in some normal modes could be caused by the crystal formation under the electric field. It is also interesting to call the attention for the λ shape of the TGS and TGS-E Cp curves; it has been reported that Cp for TGS does not have the characteristics of a lambda shape transition [4].

The X-ray diffraction patterns shown in Figure 5 indicate that there are in TGS-E, relative to TGS, shifts in 2-theta for some Bragg reflections which are more pronounced at higher angles and indicates that small modifications in the lattice parameters has occurred.

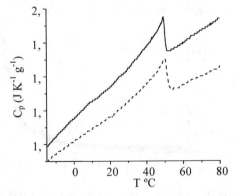

Figure 4. *Isobaric specific heat capacity of TGS and TGS-E within the studied temperature range.*

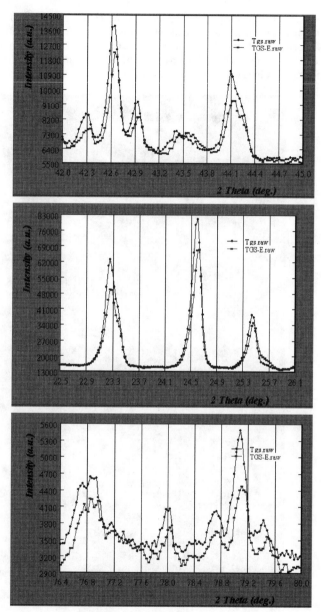

Figure 5. X-ray diffraction patterns of TGS and TGS-E with the TGS-E diffraction pattern being the lower of the two curves in each case.

Secondary electron image recorded on TGS and TGS-E is shown in Figure 6. Typical lens shaped (lenticular) domains with the major axes nearly perpendicular to the c-axis were observed on TGS crystals (a). On the electric field grown TGS crystals, the domain shape was found to be modified in the form of rods (b). Compared to the domains in TGS, in TGS-E, the major axis of the lenticular domain is found to be stretched.

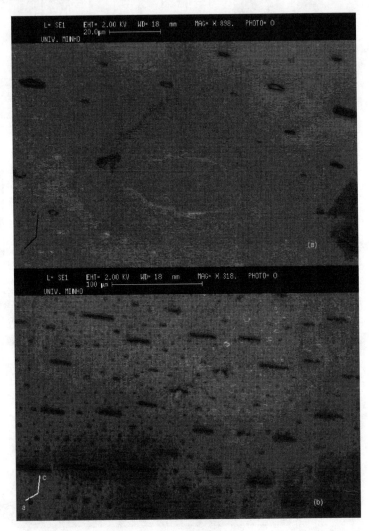

Figure 6. *Secondary electron image of (a) TGS (b) TGS-E*

CONCLUSIONS

In this paper we report for the first time the ferroelectric modifications of TGS crystals grown in a dc electric field. We conclude that growing TGS crystals under a high dc electric field induces dielectric and structural modifications that are responsible for a change in the physical properties.

Single crystal structure determination is underway for a better understanding of the structural subtleties induced by the electrical field. Hysteresis measurements will also be performed to study the ferroelectric behavior of the electric field grown crystals. A better understanding of the problems involved as well as detailed study on the influence of the electric field magnitude in the ferroelectric properties will be carried out in the near future.

ACKNOWLEDGEMENTS

The authors would like to thank Prof. M. Belsley for discussion of the manuscript. They also would like to thank IMAT- Institute of Materials and Accão Integrada Luso-Espanhola nº E-68/00 for supporting the research work.

REFERENCES

1. B. T. Mathias, C. E. Miller and J. P. Remeika, *Phys. Rev* **104** (1956) 849.
2. E. Wood and A. N. Holden, *Acta Cryst.* **10** (1957) 145.
3. C. S. Fang, M. Wang and H.S. Zhuo, *Ferroelectrics* **91**(1989) 349.
4. N. Nakatani, *Jp. J. Appl. Phys.* **29(10)** (1990) 2038.
5. *CRC Handbook of Chemistry and Physics*, **74** edition, (CRC Press, Boca Raton, 1993)
6. S.R. Stoyanov, M. P. Michailov, J. Stankowska, *Acta Phys. Polon.* **A65** (1984) 141.
7. K. Cwikiel, B. Fugiel, M. Mierzwa, *Physica B* **293(1)** (2000) 58.
8. Landoldt-Bornstein: *Ferro-und Antifferroelektrische sustanzen*, **Bd. 9**, (Springer Verlag, 1975)
9. F. Jona and G. Shirane, *Ferroelectric Crystals*, Dover Publications (1993).

Mat. Res. Soc. Symp. Proc. Vol. 658 © 2001 Materials Research Society

Mechanism of Quality Factor Compensation by Nb_2O_5 Addition for Dielectric Properties of Low-T Sintered ZST Microwave Ceramics

Yong H. Park *, Moo Y. Shin, Hyung H. Kim, Kyung H. Ko
Dept of MS&E, Ajou Univ., Suwon 442-749, Korea, *p021@chollian.net

ABSTRACT

It has been known that some additives such as Nb_2O_5 are efficient quality factor compensators to potential additives for low-T sintering of $(Zr_{0.8}Sn_{0.2})TiO_4$ such as ZnO. The compensation mechanism of Nb_2O_5 was analyzed in the light of its effects on the Zn incorporation in grains. ZST ceramics were prepared by conventional mixed oxide method and their microwave dielectric properties including quality factor, permittivity and temperature coefficient frequency (TCF) were measured at X-band. After sintering at 1350 and 1400°C, samples were post-annealed at 900~1100°C. It was found that as Nb_2O_5 was added, the quality factor of 6 mol % ZnO-ZST specimens increased from 24000 to 44000 without any sacrificing of other dielectric properties such as permittivity and TCF. According to TEM and XRD, Nb addition tends to enhance Zn diffusion toward grain boundaries while Nb moved in the opposite direction. Due to the fact that no second phases were formed, it was assumed that the redistribution of Zn and Nb could play a major role in the enhancement mechanism of quality factor. Post-annealing can also be a secondary booster for the quality factor of ZST. After annealing at 900~1100°C, the quality factor of a specimen with additives, increased again up to 48000 due to further decrease of oxygen vacancies resulting from further out-diffusion of Zn toward grain boundaries and incorporation of Nb into the grain.

INTRODUCTION

Zirconium tin titanate $((Zr_{0.8}Sn_{0.2})TiO_4$; ZST) ceramics are among the most popular commercial dielectric materials for microwave devices [1-6]. However, a major disadvantage of pure ZST ceramics is their poor sinterability at <1600°C. Therefore, extensive research [7-9] has focused on additives to the ZST system to lower the sintering temperature without significant loss of the required dielectric properties, such as the dielectric constant, the quality factor (Qf), and the temperature coefficient of the resonant frequency. Commercial ZST dielectric ceramics have been produced by additions of typical oxides, such as ZnO [7] or NiO [4], to enable sintering at <1400°C. However, the addition of sintering aids decreases the dielectric properties of ZST ceramics, especially the Qf value, so that other components must be incorporated to compensate the Qf.

Many reports [10,11] have focused on oxide compensators to improve the Qf in ZST ceramics sintered at low temperature (<1400°C). According to reports by Iddles et $al.$ [11], acceptor ions, such as La^{3+}, produced holes and increased dielectric loss, whereas donor ions, such as Nb^{5+}, reduced the number of oxygen vacancies and decreased dielectric loss. In addition, Yoon et $al.$ [12,13] noted that the unloaded Qf increased as the amount of additives (Nb_2O_5, Ta_2O_5, and Sb_2O_5) increased up to 1.0 mol%, because the oxygen vacancies in the ZST lattice decreased.

Conceivably, the Qf of dielectric ceramics such as ZST could be affected by synergistic factors among additives, especially when more than two components are incorporated into the main composition. In the present study, more detailed analysis has been applied to clarify this controversial point for ZST ceramics co-doped with ZnO and Nb_2O_5.

EXPERIMENT

Sample preparation

The $(Zr_{0.8}Sn_{0.2})TiO_4$ powders were prepared by conventional solid state synthesis from the simple oxides. The starting materials were reagent-grade ZrO_2, SnO_2, TiO_2, and ZnO. The mixtures were ball-milled for 24 hr in distilled water, filter pressed, dried and ground. After calcining at 1000°C for 2 hr, 6 mol % ZnO and 4 mol % Nb_2O_5 were added as additives. ZST powders prepared were pressed at 1000 kg/cm^2 pressure into disks (8 mm diameter and ~4 mm thickness). After sintering at 1350 and 1400 °C for 2 hr, samples were post-annealed at 900 ~ 1100°C for 5 hr in static air atmosphere.

Properties characterization

The sintered densities of the samples were measured by the Archimedes method. Dielectric properties in the microwave frequency range were measured, using the Hakki-colemann method. A network analyzer (HP 8720C, Hewlett-Packard Co.) was used for the microwave measurement system. X-ray powder diffraction (M18XHF-SRA, McScience, Japan) was used to determine the crystalline phases, using Cu Kα radiation. The ZST ceramics were processed by polishing, dimpling, and ion milling (Gattan, Model DuoMill 600) prior to transmission electron microscopy (CM 20, Philips Electronic Instruments, Inc., Amsterdam, Netherlands) and energy dispersive x-ray spectroscopy studies for microstructure and composition, respectively.

RESULTS

Figure 1 shows XRD patterns of ZST ceramics with 6 mol % ZnO and 6 mol % ZnO, 4 mol % Nb_2O_5 additives, sintered at 1350°C for 2h, respectively. In the sample (designated as ZST-S) with 6 mol % of ZnO, the second phases such as $ZnTiO_3$ and $Zn(Zn,Ti)O_4$ appeared. On the other hand, in the case of a sample (designated as ZST-C) co-doped with 6 mol % ZnO and 4 mol % Nb_2O_5, the second phase was not observed in the XRD measurement.

Dielectric properties of ZST-S and ZST-C are shown in Table I. Although Nb_2O_5 was added to ZST ceramics, relative density and dielectric constant changed only slightly. However, it was observed that the quality factor of ZST-C increased to 42000 from 25000 of ZST-S.

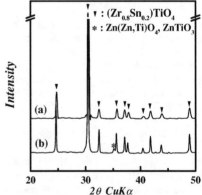

Figure 1. XRD patterns of ZST ceramics sintered at 1350°C for 2 hr; (a) 6 mol % ZnO and (b) 6 mol % ZnO, 4 mol % Nb_2O_5.

Figure 2 shows the TEM image near grain boundaries and the Zn concentration profiles obtained by X-ray microanalysis. Comparing data, the Zn content near the grain boundary remarkably decreased by Nb_2O_5 addition. Therefore, it is inferred that the addition of Nb_2O_5 can cure the side effects of Zn by out-diffusion as well as suppression of second phase formation.

Table I. *Dielectric properties of ZST ceramics with 6 mol % ZnO and 4 mol % Nb_2O_5 sintered at 1350^oC for 2 hr.*

	6 mol % ZnO	6 mol % ZnO + 4 mol % Nb_2O_5
Relative Density (%)	97.3	98.6
Dielectric Constant	36.6	37.4
Qf	25000	42000

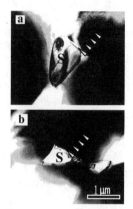
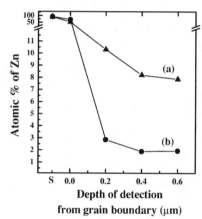

Figure 2. *TEM image and X-ray microanalysis of grain boundary area of ZST ceramics sintered at 1350^oC;(a) only 6 mol % ZnO and (b) 6 mol % ZnO and 4 mol % Nb_2O_5, ∇ detection point, S, second phase.*

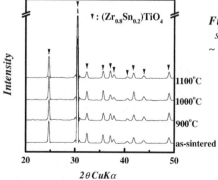

Figure 3. *XRD patterns of ZST-C ceramics sintered at 1300^oC and post-annealed at 900 ~ 1100^oC for 5 hr.*

Figure 3 shows the XRD patterns of ZST-C ceramics, which were sintered at 1350 and 1400°C for 2 hr and then post-annealed at 900 ~ 1100°C for 5 hr. Neither second phases nor changes in lattice parameter were observed at either sintering temperature, even after post-annealing. Also, as shown in Figure 4, it was found that while most dielectric properties were virtually unchanged by the post-annealing treatment, for the samples sintered at 1400°C, the quality factors increased up to 48000 after post-annealing.

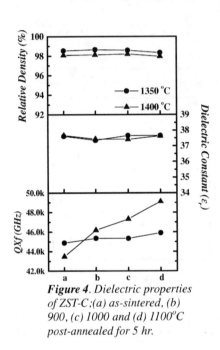

Figure 4. Dielectric properties of ZST-C;(a) as-sintered, (b) 900, (c) 1000 and (d) 1100°C post-annealed for 5 hr.

Figure 5 and Figure 6 show the TEM images and X-ray microanalyses of grains of the ZST-C ceramics sintered at 1350, 1400°C and post-annealed at 900 ~ 1100°C, respectively. In the TEM Figure 6, at the triple point of grain, Z rich phase were observed. It was inferred that these phases were responsible for the improvement of sinterability of ZST ceramics.

In the case of samples sintered at 1350°, the concentration profile of Zn and Nb inside t grain were not changed at all even after post-annealed at 900, 1100°C. On the other hand, in the samples sintered at 1400°C, the incorporati of Zn into the ZST grain decreased to 0.5 at % and that of Nb increased to 18 at % after post-annealing at 900 ~ 1100°C.

According to defect chemistry, the incorporation of Zn (valence of +2) into the ZS grain could generate the extrinsic oxygen vacancies in order to retain the charge neutrali By the same token, Nb ion (valence of +5) cou eliminate the oxygen vacancies in the ZST gra Because the post-annealing process encourage ionic diffusion, it could be therefore concluded that the synergistic effects of out-diffusion of Z and the incorporation of Nb resulted in further decrease of oxygen vacancies in the ZST grain and finally secondary enhancement of quality factors of ZST-C ceramics.

CONCLUSIONS

It was found that as Nb_2O_5 was added, the quality factor of 6 mol% ZnO-ZST specimens increased from 24000 to 44000 without any sacrifice of other dielectric properties such as permittivity and TCF. According to TEM and XRD, Nb addition tends to enhance Zn diffusion toward grain boundaries while Nb moved in opposite direction. Due to the fact that second phase formation was detected, it was assumed that the redistribution of Zn and Nb cou play a major role in the compensation mechanism of the quality factor. After annealing at 900~1100°C, the quality factor of ZST increased to 48000. In post-annealed samples, out-diffusion of Zn toward grain boundary and the incorporation of Nb into the ZST grain proceed thoroughly. Therefore, it is conceivable that co-addition of Nb with Zn could also enhance the post-annealing effects.

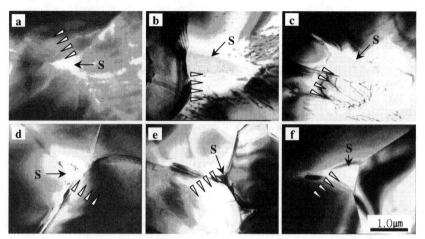

Figure 5. *TEM image of ZST-C ceramics;(a) as-sintered at 1350°C, post-annealed at (b) 900, (c) 1100°C, (d) as-sintered at 1400°C, post annealed at (e) 900, (f) 1100°C, ▽, detection point, S, second phase.*

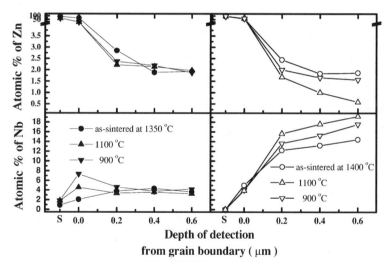

Figure 6. *X-ray microanalysis of ZST-C ceramics; Zn and Nb depth profile as-sintered at 1350 and 1400°C, post-annealed at 900 and 1100°C, S, second phase.*

REFERENCES

1. Yong H. Park, Moo Y. Shin, Ji M. Ryu, and Kyung H. Ko in *Chemical Processing of Dielectrics, Insulators and Electronic Ceramics*, ed. Anthony C. Jones, Janice Veteran, Dona Mullin, Reid Cooper and Sanjeev Kaushal, (Mater. Res. Soc. Pro. **606**, Pittsburgh, PA, 2000) pp. 293-298.
2. W. Wersing, in *Electronic Ceramics* ed. B. C. H. Steele (Elsevier A. A., London and New York, 1991), pp. 76-83.
3. K. Wakino, *Ferroelectrics*, **91**, 69-86 (1989).
4. K. Wakino, K. Minai, and H. Tamura, *J. Am. Ceram. Soc.*, **67[4]**, 278 (1984).
5. A. E. Mchale and R. S. Roth, *J. Am. Ceram. Soc.*, **66[2]**, C-18 (1983).
6. H. Tamura, *Am. Ceram. Soc. Bull.*, **73[10]**, 92 (1994).
7. K. R. Han, J. W. Jang, S. Y. Cho, D. Y. Jeong and K. S. Hong, *J. Am. Ceram. Soc.*, **81[5]**, 1209 (1998).
8. R. Kudensia, A. E. Mchale and R. L. Snyder, *J. Am. Ceram. Soc.*, **77[12]**, 3215 (1994).
9. T. Takada, S. F. Wang, S. Yoshikawa, S. J. Jang and R. E. Newnham, *J. Am. Ceram. Soc.*, **77[9]**, 2485 (1994).
10. N. Michiura, T. Tatekawa, Y. Higuchi and H. Tamura, *J. Am. Ceram. Soc.*, **78[3]**, 793 (199
11. D. N. Iddles, A. J. Bell and A. J. Moulson, *J. Mater. Sci.*, **27**, 6303 (1992).
12. K. H. Yoon, Y. S. Kim and E. S. Kim, *J. Mater. Res.*, **10[8]**, 2085 (1995).
13. K. H. Yoon and E. S. Kim, *Mater. Res. Bull.*, **30[7]**, 813 (1995).

Mat. Res. Soc. Symp. Proc. Vol. 658 © 2001 Materials Research Society

Grain Orientation of Aluminum Titanate Ceramics during Formation Reaction

Yutaka Ohya, Zenbe-e Nakagawa[1], Kenya Hamano[2], Hiroshi Kawamoto[3], and Satoshi Kitaoka[3]
Gifu Univ., Dept. of Chemistry, Gifu 501-1193, JAPAN
[1]Akita Univ, Akita 010-8502, JAPAN, [2]Professor Emeritus, Tokyo Inst. Tech., JAPAN, [3]JFCC, Nagoya 456-8587, JAPAN.

ABSTRACT

A microstructural change during the formation reaction of aluminum titanate from a mixture of rutile and corundum powders has been studied. The characterization was carried out using a polarization microscope, a scanning electron microscope and a micro-focus X-ray diffractometer. The formation of aluminum titanate was controlled by a nucleation step. The formation reaction proceeded to form spherically oriented regions of aluminum titanate grains among the matrix of rutile and corundum. At the end of the reaction, the specimen was entirely filled with the oriented region of consisting several hundred micrometers. The oriented region was composed of primary aluminum titanate grains of several micrometers and pores. Large cracks due to a thermal expansion anisotropy were formed at the boundaries of the orientated regions. The formation of the oriented region was caused by a small change in free energy, increasing elastic energy, and the endothermic nature of the reaction.

INTRODUCTION

The microstructure of aluminum titanate ceramics is very unique when they are fabricated from mixtures of alumina and titania, and the formation reaction occurs during the sintering [1, 2]. In these ceramics, small aluminum titanate grains oriented over a region of a few hundred micrometers. This structure is very interesting not only from the point of solid state chemistry, but also for practical applications of the ceramics. Aluminum titanate ceramics have been considered as refractory materials with a high thermal shock resistance due to their very small thermal expansion coefficient [3]. Since the microstructure of the ceramics affects the thermal expansion behavior and mechanical properties, studies on the microstructure and the grains' oriented structure are of importance. We have already reported the structure and discussed the reasons why the oriented structure is formed in part [2,4]. The keys of the formation of the oriented structure are a small change in thermodynamic free energy, increasing elastic energy by the formation reaction, and the endothermic nature of the reaction, which means that entropy controls the formation [5]. Among these, the small change in free energy and increasing elastic energy on the formation of aluminum titanate enhance epitaxial-like nucleation of aluminum titanate on the surface of the formed titanate. The endothermic reaction prevents a chain reaction of formation. These lead to the formation of the oriented region [4].

In the present work, we investigated the change in microstructure of the ceramics during the formation reaction of aluminum titanate and confirmed the argument above.

EXPERIMENTAL PROCEDURE

Starting materials were pure and uniform rutile (0.2 μm, TP-13, Fuji Titanium Co., Japan and corundum (0.5 μm, AKP-20, Sumitomo Chemical. Co., Japan). These powders were wet mixed in an alumina ball mill to get the powder mixture of an Al_2TiO_5 composition. In some case a small amount of pre-synthesized aluminum titanate powder (0.025 mass%, about 1 μm in size) was added into the mixture as seeds of the formation reaction. Rectangular specimens of 5x5x30 mm^3 were prepared by cold isostatic pressing at 98 MPa.

The specimens were fired at 1300 to 1500°C for various periods. For the observation using a polarization microscope, thin sections of 20 ~ 30 μm were prepared. An individual grain was observed on the polished and thermally etched surface using a scanning electron microscope (SEM). The change in crystal phases of the fired specimens was characterized using a micro-foc X-ray diffractometer. This XRD equipment can determine the crystalline phases in a small region of about 100 μm in diameter.

RESULTS AND DISCUSSION

Observation of Oriented Region

Figure 1 shows the polarization micrographs of a thin section of the specimen without seeds and fired at 1400°C for 4 hours. In this specimen, alumina and titania reacted completely to give the aluminum titanate phase. Figure 1 (a) was taken using single polarizer, *i.e.*, open nicol image and Figure 1 (b) was taken by placing the thin section between two polarizers whose directions of polarization were orthogonal with each other, *i.e.*, crossed nicols image. In Figure 1 (a), small grains of several micrometers and large cracks are observed. An enormous thermal expansion anisotropy causes these large cracks [3]. In Figure 1 (b), each grain in the area surrounded by the cracks showed the same contrast, or simultaneous extinction, i.e, each grain in this region has almost the same crystallographic orientation. To confirm the grain size, the polished and etched surface of the oriented region (specimen is fired at 1500°C for 4 hours) was observed using SEM and showed in Figure 2. Grain boundaries can be observed around each gra of 3 ~ 4 μm and coexisting with 1 ~ 2 μm pores. Therefore, the oriented region is not a single crystal but consists of primary oriented grains and pores.

200 μm

Figure 1. Polarization micrographs of sintered aluminum titanate ceramics,
(a):open nicol image and (b): crossed nicols images of the same area as (a).

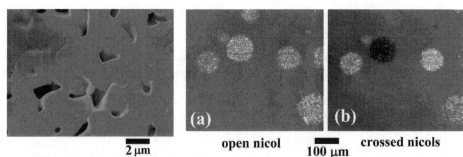

2 µm   **open nicol**   **crossed nicols**
100 µm

Figure 2. *SEM image of polished and etched surface of the oriented*

Figure 3. *Polarization micrographs of oriented aluminum titanate "island" in the matrix.*

Formation Process of the Oriented Region

In order to reveal a formation process of the oriented region, partially reacted specimens were fabricated and observed using the polarization microscope. The images of the specimen with seeds and heated at 1300°C for 18h are shown in Figure 3. In the sample, polycrystalline white circles, which means a transparent area, existed in a gray matrix as shown in an open nicol field image of Figure 3 (a). Each grain in the circle shows simultaneous extinction between crossed nicols image as shown in Figure 3 (b). Crystalline phases in the spherical region and the matrix were determined using the micro-focus X-ray diffractometer. Figure 4 shows the results. Position I, gray matrix, consists of rutile and corundum, and aluminum titanate existed only in position II, a spherical white area. Small amounts of corundum and rutile phase in the area II should be from outside the area of the aluminum titanate sphere, because the size of the measured area was almost the same as the diameter of the collimator of the diffractometer. Therefore the formation of aluminum titanate proceeds to form a sphere with the same crystallographic orientation as the grains inside the spherical region. The formation reaction occurred only on the surface of the aluminum titanate sphere. The difference in contrast between the white sphere and gray matrix should be caused by the difference in scattering of light. Since the matrix consisted of random oriented rutile and corundum grains with large differences in refractive indices, the boundaries of these grains scattered light and the gray area resulted. On the other hand, the oriented aluminum

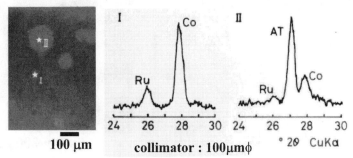

100 µm   **collimator : 100µmφ**   ° 2θ CuKα

Figure 4. *Micro-focus XRD patterns of area I, the matrix, and area II, the spherical "island", AT: Al$_2$TiO$_5$, Co: corundum, and Ru: rutile.*

titanate grains in the spherical area transmitted light to some extent and the white, transparent, region was formed.

The formation of spherical regions of aluminum titanate in the rutile and corundum matrix means that the reaction proceeds outward from a centered nucleus. We have already reported that the addition of 0.025 mass% aluminum titanate powder into a mixture of rutile and corundum enhanced the formation ratio to 67.5 % by firing at 1340°C for 10 minutes, while aluminum titanate was not found in blank specimens under the same firing condition [2]. These results indicate that the formation of aluminum titanate is controlled by a nucleation step and one oriented region corresponds to formation of one nucleus in the matrix.

Figure 5 shows the logarithmic plot of the oriented region size with time. Most of the data both with and without seeds fit on a straight line of slope 2 indicating that a growth mechanism of the region was neither a diffusion process nor a reaction process. Once the region becomes larger, the growth rate accelerated. This should be closely related to an increasing elastic energy to the reaction. The formation is accompanied with an increase of volume of 11%. This increment in the elastic energy retarded the nucleation and formation of aluminum titanate [4]. As the oriented region size increased, this volume increase during the formation reaction caused the occurrence of internal cracking as shown in a polarization micrograph in Figure 5. The size of the oriented region reached about 200 μm and cracking could be observed in the micrograph, while the sample in Figure 3 was about 100 μm with no cracks. Once the cracking occurred, the increase of elastic energy on the formation reaction diminished and the reaction rate was enhanced by great extent to give the slope of 2 in the logarithmic graph.

About the Formation of Oriented Region

We have already pointed out three factors for the formation of the oriented region, i.e., low free energy change, endothermic nature of the reaction, and increasing the elastic energy on formation reaction [4]. In this work we can see the spherical region of the initial stage of formation, which means an isotropic nature of the formation reaction. The large thermal expansion anisotropy and crystalline structure of aluminum titanate [6] suggested that many properties of the crystal strongly depend on the crystallographic direction. As an example, a needle like crystals grow by liquid phase sintering of aluminum titanate [7]. Consequently, the spherical growth is not

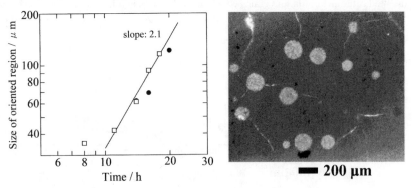

Figure 5. *Change in size of the oriented region (left), (open square: with seeds, closed circle: without seeds) and cracking of the matrix (right).*

a property of the crystal but a resultant morphology from the limited nucleation condition during formation. The growth rate in Figure 5 confirmed the importance of the elastic energy on nucleation of the aluminum titanate [4,8]. In solid state reactions, especially those reactions with large volume expansion, the elastic energy strongly influences the reaction rate to lower this.

The formation process of the oriented region is considered as follows: First an aluminum titanate nucleus forms and grows. The next aluminum titanate nuclei form on the surface of the first grain. This formation reaction should be epitaxial to minimize the interfacial energy, because the change in thermodynamic free energy is not so large. The reaction temperatures about 1300-1370°C are just a little higher than an equilibrium temperature of aluminum titanate, 1280°C [9]. Further the increase in the elastic energy also prefers the advantage of epitaxial nucleation. The formation of aluminum titanate is an endothermic reaction [5] and lowers its local temperature. This decrease in temperature leads to discrete growth of grains and temperature disturbance of epitaxy would also prevent the growth of one large single crystal. Finally the spheres impinge each other and fill the matrix as shown in Figure 1.

CONCLUSION

Structure and formation processes of orientated regions in aluminum titanate ceramics were studied. The oriented region consists of small grains of 3 ~ 4 μm size and pores of 1 ~ 2 μm. The formation rate of aluminum titanate is controlled by a nucleation step. An epitaxial reaction occurs on a surface of initial nucleus and forms a spherical oriented region in rutile-corundum matrix. As the reaction proceeds, these regions grow larger and impinge upon each other.

ACKNOLEDGEMENT

This study was supported in part by the New Energy and Industrial Technology Department Organization as a part of the Industrial Science and Technology Research and Development of Japan.

REFERENCES

1. K. Hamano, Z. Nakagawa, K. Sawano, and M. Hasegawa, *J. Chem. Soc. Jap.*, No.10, 1647-55 (1981).
2. K. Hamano, Y. Ohya, and Z. Nakagawa, *J. Ceram. Soc. Jpn.*, (*Yogyo-Kyokai-Shi*) **91**, 94-101 (1983).
3. W. R. Buessem, N. R. Thielke, and R. V. Sarakauskas, *Ceramic Age*, **60**, 38-40 (1952).
4. D-F. Qian, Y. Ohya, Z. Nakagawa, and K. Hamano, *J. Ceram. Soc. Jpn.*, **103**,1022-1026 (1995).
5. A. Navrotsky, *Am. Mineral.*, **60**, 249-56 (1975) .
6. B. Morrosin and R. W. Lynch, *Acta Cryst.*, **B28**, 1040-46 (1972).
7. Y. Ohya, M. Hasegawa, Z. Nakagawa, and K. Hamano, *Report of the Research Laboratory of Engineering Materials, Tokyo Institute of Technology*, No.12, 81-91 (1987).
8. B. Freudenberg and A. Mocellin, *J. Am. Ceram. Soc.*, **70**, 33-37 (1987).
9. E. Kato, K. Daimon, and J. Takahashi, *J. Am. Ceram. Soc.*, **63**, 355-56 (1980).

Mat. Res. Soc. Symp. Proc. Vol. 658 © 2001 Materials Research Society

Solid-State NMR Investigation Of Mixed-Alkali Distribution In Phosphate Glasses

S. Prabakar and K.T. Mueller
Department of Chemistry, The Pennsylvania State University
152 Davey Laboratory, University Park, PA 16802

ABSTRACT

The correlation between ^{31}P and ^{23}Na (or ^{133}Cs) nuclei in a mixed-alkali phosphate glass has been investigated by TRAnsfer of Populations by DOuble Resonance (TRAPDOR) NMR spectroscopy. The variation in spatial proximity between ^{31}P nuclei in phosphate tetrahedra and sodium (or cesium) modifier ions has been demonstrated in a phosphate glass with molar composition 25 Cs_2O: 31 Na_2O: 44P_2O_5. Interactions between ^{31}P and ^{23}Na (or ^{133}Cs) nuclei are shown to exist even at low dephasing times indicating the short-range nature of these couplings. The ratios of resonance intensities from Q^1 and Q^2 phosphate units show different trends depending on whether couplings to ^{23}Na or ^{133}Cs are investigated, demonstrating the differing structural roles of the modifier alkali cations within the glass.

INTRODUCTION

The "mixed-alkali effect" is a term coined to reflect the non-linear variation in glass properties (e.g. density and glass transition temperature) when one alkali in a multicomponent glass is gradually replaced by another alkali species [1]. Controversy remains over the atomic-level details (both structural and dynamical) that can provide such an effect. For example, the role of two dissimilar alkali ions in the same glass structure has been interpreted and modeled using both a random [2] or a clustered distribution [3] of the two alkali ions, depending on interpretations of the experiments used in different studies. Several experimental, theoretical, and computational approaches have been employed to understand the origins of the mixed-alkali effect, but the mixed-alkali distribution is not completely understood at this time [4-6]. We have proposed the use of multiple sets of dipolar "connectivity" experiments over a range of glass compositions in order to study the distributions of alkali atoms, and mixed-alkali phosphate glasses present a useful, and easily studied, model system.

Phosphate glasses are gaining increased attention [7] due to potential applications as glass-to-metal seals and laser glasses. Cesium-containing phosphate glasses may find applications in electron photomultiplier tubes and within containment systems for nuclear waste. Solid-state nuclear magnetic resonance (NMR) spectroscopy has been widely used as a tool to investigate complex microstructures in glasses [8], and ^{31}P magic-angle-spinning (MAS) NMR is known to be sensitive to the local environments of phosphorus atoms. For example, the number of bridging and non-bridging oxygens around phosphorus, the bond angles, and the degree of hybridization of phosphorus are all reflected in measurable NMR parameters [9]. Recently the dipolar interactions between ^{31}P and modifying cations have been exploited to probe structural correlations that exist between these nuclei [10]. In this article we investigate the alkali ion interaction with the phosphate network in a sodium-cesium phosphate glass system using the TRAPDOR NMR experiment.

EXPERIMENTAL

Solid-state NMR spectra were recorded using a home-built NMR spectrometer (9.4 T magnetic field) controlled by a Tecmag pulse programmer. MAS and TRAPDOR NMR spectra were acquired using a 5-mm triple-resonance MAS probe with rotor speeds of 9.0 kHz. The ^{23}Na, ^{133}Cs, and ^{31}P resonance frequencies were 105.805, 52.226 and 161.919 MHz, respectively at 9.4 T. Recycle delays for the ^{31}P acquisition experiments were 120 s. The chemical shift standard employed for ^{31}P was an 85% aqueous solution of H_3PO_4. The 90 degree pulse lengths were adjusted to 5 μs for ^{23}Na, ^{31}P, and ^{133}Cs, and therefore dephasing field strengths correspond to 50 kHz for both ^{23}Na and ^{133}Cs.

The pulse sequence used to acquire TRAPDOR spectra is shown in Figure 1. The TRAPDOR experiment actually consists of two separate experiments. In the first experiment a rotor synchronized ^{31}P spin-echo signal is collected. In the second, the same experiment is repeated with long dephasing pulses on the ^{23}Na or ^{133}Cs channels. The difference between the two spectra is the TRAPDOR spectrum. A saturation sequence on the ^{31}P channel before the TRAPDOR sequence ensures sampling of the same amount of magnetization during each scan. The typical number of scans per spectrum is 64. The intensity of a TRAPDOR difference spectrum is very sensitive to dipolar coupling, quadrupolar coupling constants (for ^{23}Na or ^{133}Cs), spinning speed, and the radio-frequency field strength used, as described in detail by Grey et al. [11,12].

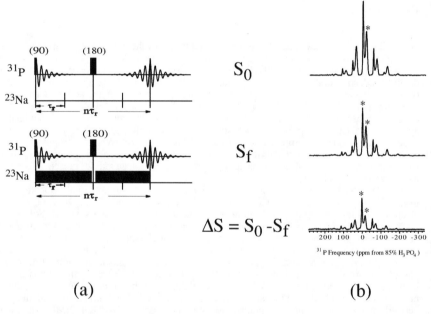

(a)   (b)

Figure 1. (a) Radio-frequency pulse sequence for the ^{31}P{^{23}Na}TRAPDOR experiment. The reference spectrum (S_0) is collected with the pulse sequence (top) with no dephasing pulses on the ^{23}Na channel; S_f is the spin-echo spectrum recorded with dephasing pulses on the ^{23}Na channel (bottom). τ_r is the rotor period. ΔS is the TRAPDOR difference spectrum. Asterisks (*) indicate the isotropic resonances, while all other peaks are spinning sidebands.

A phosphate glass with molar composition 25 Cs_2O: 31 Na_2O: $44P_2O_5$ was prepared starting from appropriate masses of Na_2CO_3, Cs_2CO_3, and $NH_4H_2PO_4$ (all obtained from Aldrich and used as received). The batched mixture was heated in a platinum crucible to approximately 400 °C for about 12 h to remove CO_2, H_2O, and NH_3. The resultant mixture was then melted at 650-700 °C. The sample was equilibrated for 30 min and quenched by pressing the melt between two steel plates. The glass was stored in a vacuum dessicator and subsequently powdered before the start of the experiment, for which a crushed sample was packed into zirconia rotors. A sodium phosphate glass of composition 56 Na_2O: $44P_2O_5$ was prepared in a similar manner.

RESULTS AND DISCUSSION

The ^{31}P spin-echo MAS spectrum (S_0) of the 56 Na_2O: $44P_2O_5$ glass sample is shown at the top of Figure 1(b). Two resonances are present, at 18 ppm and 1.8 ppm, corresponding to Q^2 and Q^1 phosphate units respectively. The phosphate units are represented in Q^n notation, where Q represents phosphate tetrahedra and n is the number of P-O-P (bridging) bonds to other phosphate species. The spectrum acquired with dephasing pulses on the ^{23}Na channel (S_f) is also shown in the middle of Figure 1(b), and the difference signal, ΔS, is due to phosphorous coupled by dipolar interactions with nearby sodium nuclei. The TRAPDOR spectra of the same glass as a function of

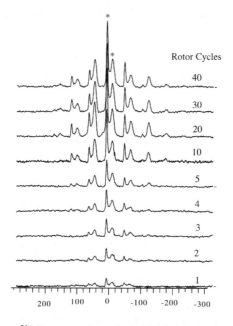

^{31}P Frequency (ppm from 85% H_3PO_4)

Figure 2. TRAPDOR difference spectra for the 56 Na_2O: $44P_2O_5$ sample are shown as a function of dephasing rotor periods (one rotor period corresponds to 111 μs of dephasing). Asterisks (*) indicates the isotropic resonances, while all other peaks are spinning sidebands.

dephasing times, given here in terms of number of rotor periods, are shown in Figure 2. The spinning speed used is 9.0 kHz, corresponding to 111 μs for one rotor period. Even at very short dephasing times, TRAPDOR signals are observed indicating a strong dipolar interaction between ^{23}Na and ^{31}P. As the ^{23}Na dephasing times are increased, longer-range couplings between ^{31}P and ^{23}Na nuclei will also contribute to the TRAPDOR difference signal.

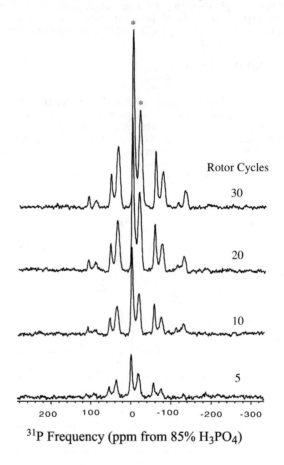

^{31}P Frequency (ppm from 85% H_3PO_4)

Figure 3. $^{31}P\{^{23}Na\}$ *TRAPDOR difference spectra of the 25 Cs_2O: 31 Na_2O: 44P_2O_5 glass are shown as a function of dephasing times. Asterisks (*) indicates the isotropic resonances, while other peaks are spinning sidebands.*

$^{31}P\{^{23}Na\}$ TRAPDOR spectra from the 25 Cs_2O: 31 Na_2O: 44P_2O_5 glass sample are shown in Figure 3, and as expected display two resonances corresponding to Q^2 and Q^1 phospha units. Again, the spectra show TRAPDOR signal even at low dephasing times indicating short-range correlation between ^{23}Na and ^{31}P atoms in this glass. There is no appreciable change in peak position with dephasing times. It has been shown [8] that dephasing times ranging from 0.5 to 3 ms gave satisfactory ^{31}P{^{23}Na} TRAPDOR signals for multicomponent glasses containing sodium and phosphorus.

Figure 4 shows the ^{31}P{^{133}Cs} TRAPDOR spectrum of the same 25 Cs$_2$O: 31 Na$_2$O: 44P$_2$O$_5$ glass sample, again displaying the expected two resonances corresponding to Q^2 and Q^1 units. It also shows considerable TRAPDOR signal at low dephasing times indicating short-range dipolar coupling between ^{133}Cs and ^{31}P.

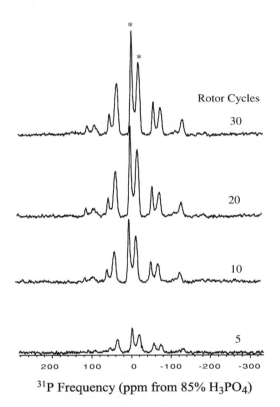

Figure 4. ^{31}P{^{133}Cs} TRAPDOR difference spectra of the 25 Cs$_2$O: 31 Na$_2$O: 44P$_2$O$_5$ glass are shown as a function of dephasing times. Asterisks (*) indicates the isotropic resonances, while all other peaks are spinning sidebands.

The presence of TRAPDOR signal for both $^{31}P\{^{23}Na\}$ and $^{31}P\{^{133}Cs\}$ spin pairs first indicates that ^{31}P is coupled to both ^{23}Na and ^{133}Cs via spin-spin dipolar interactions. The TRAPDOR signal at longer dephasing times may also reflect long-range couplings of ^{31}P with both ^{23}Na and ^{133}Cs. However, one important observation is that the Q^2/Q^1 peak ratios as a function of dephasing times for both $^{31}P\{^{23}Na\}$ and $^{31}P\{^{133}Cs\}$ TRAPDOR experiments are not similar. This fact indicates that the coupling of ^{23}Na with ^{31}P is substantially different from that of ^{133}Cs with ^{31}P. The Q^2/Q^1 ratio, calculated from the integrated intensities, is higher for the $^{31}P\{^{133}Cs\}$ spectra compared to $^{31}P\{^{23}Na\}$ spectra, indicating preferential distribution of cesium around Q^2 environments.

A study of the TRAPDOR ratios as a function of alkali oxide concentration will also reveal interesting insights and further information regarding the possible clustering of cations in these materials, and work in this direction is currently in progress. Other work in this laboratory has also shown that $^{23}Na\{^{133}Cs\}$ TRAPDOR NMR of this glass reveals dipolar correlation between these nuclei, another phenomenon that will be studied more fully on a range of glasses as an indication of cation interactions.

CONCLUSIONS

TRAPDOR NMR experiments involving ^{31}P and ^{23}Na (or ^{133}Cs) provide a measure of the complex role of cations in one mixed-alkali phosphate glass. These experiments probe the short- and long-range dipolar interactions between the alkali ions and the network forming nuclei, and indicate a preference for cesium cations to preferentially associate with Q^2 phosphate units in a glass with molar composition of 25 Cs_2O: 32 Na_2O: 44P_2O_5. Further investigation of the whole mixed-alkali series, x Cs_2O: (56-x) Na_2O: 44 P_2O_5 will provide more insight into the distribution alkali ions in these systems by tracking any changes in the roles of the glass modifiers as a function of concentration.

ACKNOWLEDGMENTS

KTM is an Alfred P. Sloan Research Fellow. This report is based upon work carried out with support from the National Science Foundation under Grant No. DMR-9458053 and with support from the Research Corporation (Cottrell Scholar Award to KTM).

REFERENCES

1. P. Maass, *J. Non-Cryst. Solids*, **255**, 35-46 (1999).
2. B. Gee and H. Eckert, *Ber. Bunsenges. Phys. Chem.*, **100**, 1610 (1996).
3. A.T- W. Yap, H. Forster, and S. R. Elliott, *Phys. Rev. Lett.*, **75**, 3946 (1995).
4. D.E. Day, *J. Non-Cryst. Solids*, **21**, 343 (1976).
5. M.D. Ingram, *Phys. Chem. Glasses*, **28**, 215(1987).
6. A. Hunt, *J. Non-Cryst. Solids*, **220**, 1 (1997).
7. R.K. Brow, *J. Non-Cryst. Solids*, **263&264**, 1-28 (2000).
8. H. Eckert, *Prog. NMR Spectroscopy*, **33**,131(1994).
9. R.J. Kirkpatrick and R.K. Brow, *Solid State NMR*, **5**, 9 (1995).
10. T. Schaller, C. Rong, M.J. Toplis, and H. Cho, *J. Non-Cryst. Solids*, **248**, 19 (1999).
11. C. P. Grey and A.J. Vega, *J. Am. Chem. Soc.*, **117**, 8232 (1995).
12. C.P. Grey, A.P.A.M. Eijkelenboom, W.S. Veeman, *Solid State NMR*, **4**, 113 (1995).

Mat. Res. Soc. Symp. Proc. Vol. 658 © 2001 Materials Research Society

^{207}Pb NMR of PbZrO$_3$ and PbZr$_{1-x}$Ti$_x$O$_3$ Solid Solutions

Paolo Rossi, Matthew R. Dvorak, and Gerard S. Harbison
Chemistry Dept., University of Nebraska, Lincoln, NE 68588-0304

ABSTRACT

PbZrO$_3$ undergoes a phase transition to a paraelectric phase at 230 °C. During this phase transition the unit cell changes from orthorhombic to cubic. Structural changes of PbZrO$_3$ have been monitored using solid state NMR by measuring the variation in the ^{207}Pb chemical shielding tensor as a function of temperature. The two distinct lead sites show rather different behavior as a function of temperature. The less shielded lead maintains an almost constant asymmetry parameter η = 0.2 from 0°C to 200 °C while the more shielded lead resonance becomes more axially symmetric as the temperature is raised going from η = 0.24 at 0 °C to η = 0.08 at 200°C. Powder pattern singularities become less distinct near the phase transitions, but the temperature dependence of the chemical shift tensor principal values remains continuous, and there is no evidence of an intervening higher-symmetry phase. At the phase transition, both resonances collapse into a single narrow line characteristic of a nucleus at a high-symmetry site. Results of a preliminary study of the PbZr$_{1-x}$Ti$_x$O$_3$ solid solutions by ^{207}Pb solid state NMR are also presented.

INTRODUCTION

Lead zirconium oxide PZ and mixed lead zirconium-titanium oxide PZT are perovskite-type materials of considerable technological importance [1]. The common methods of preparation of PZT based materials are ceramitization at high temperature and sol gel preparation techniques. In order to obtain materials that exhibit the desired properties, knowledge of condition-dependent structural changes is required. This is a particularly ambitious goal considering the fact that the even the structure determination of simple PZ is still being actively pursued and remains controversial. A large number of crystallographic studies of PZ have been reported in which the investigators arrived to different conclusion about the structure of PZ [2,3]. Lately the consensus appears to be that antiferroelectric PZ is orthorhombic with the *Pbam* space group [4,5]. The difficulties in determining the PZ originated largely because of structural disorder and the lack of a standard preparation method that gives uniform, defect-free PZ crystals with low Pb and O vacancies.

Given the sensitive nature of the ^{207}Pb nucleus to the surrounding environment, solid state NMR has often been used a tool to gain structural insights by correlating the values of ^{207}Pb chemical shift to the mean Pb-O distance in PZ and PZ-based compounds [6-9]. ^{207}Pb NMR has also been used for a qualitative examination of the changes in chemical shift as a function of preparation temperature [10]. However, because of the width of the spectrum, distorted and noisy spectra have been the rule rather than the exception, and these distortions have been in some cases sufficient to compromise the measurement of the principal values.. Moreover, the dependence of the chemical shift tensor principal values as a function of temperature has not been reported.

In this work ^{207}Pb NMR chemical shift tensor of PZ was measured as a function of temperature from 0 °C to 240 °C. A preliminary examination of a few PZT solid solutions at room temperature was also conducted.

EXPERIMENT

High purity $PbZrO_3$ (99.996%) and $PbTiO_3$ were purchased from Alfa Aesar and used without further purification. The $PbZr_{(1-x)}Ti_xO_3$ solid solutions with x = 0.2, 0.4, 0.6, 0.8 were prepared by mixing the appropriate amounts of $PbZrO_3$ and $PbTiO_3$, forming them into pellets, and sintering the pellets at 1100 °C for 24 hrs.

[207]Pb NMR spectra were recorded on a home-built spectrometer that uses a 7.1 T magnet. Static spectra were obtained with a home-built probe and magic angle spinning spectra (MAS) were recorded using a Doty DSI 782 supersonic probe equipped with a 4 mm silicon nitride rotor. The standard Hahn echo sequence $90° - \tau - 180° - \tau -$ was used for both static and MAS experiments; the 90° pulse lengths were 1.8 µs and 2.0 µs, respectively. The τ interval was chosen to be 60 µs in the case of static experiments, and was 1 rotor period long, typically around 80 - 120 µs, for the MAS experiments. Depending on temperature, a relaxation delay of between 10 s and 30 s was used. Static $PbZrO_3$ spectra were acquired at a temperature ranging between 0 and 240 °C; MAS spectra were acquired at room temperature at spinning speeds of 8.8 kHz and 12.2 kHz.

The [207]Pb NMR chemical shifts for the PZ and PZT solid solutions are reported in ppm from the 0.5 M aqueous $Pb(NO_3)_2$. The actual spectrometer frequency used during acquisition was 63.520 MHz for $Pb(NO_3)_2$ and 63.660 MHz for PZ and PZT solid solutions.

The principal components σ_{xx}, σ_{yy}, and σ_{zz} of the chemical shielding tensor σ are ordered using the convention [11]:

$$|\sigma_{zz}| > |\sigma_{xx}| > |\sigma_{yy}| \qquad (1)$$

The asymmetry parameter and the chemical shift anisotropy are defined as:

$$\eta = (\sigma_{yy} - \sigma_{xx})/ \sigma_{zz} \qquad (2)$$

and

$$\Delta\sigma = \sigma_{zz} - \sigma_{iso} \qquad (3)$$

Static lineshapes were fit by the method outlined by Mehring [12].

RESULTS

Variable-temperature [207]Pb NMR of lead zirconium oxide (PZ).

Static [207]Pb NMR spectra of PZ as a function of temperature are presented in Fig. 1. The experimental spectrum is composed of contributions from two inequivalent lead sites in the crystal; the intensity ratio of the two sites is expected from crystallography, and confirmed by simulation to be 1:1. The more anisotropic of the two powder patterns is tentatively assigned to crystallographic site 1 in the notation of Corker *at al.*[13], the more distorted of the two lead ions in the crystal lattice. The singularities in the powder patterns can be easily identified, except for the σ_{zz} values, which overlap at room temperature but become more distinct at higher temperature, and are difficult to assign by inspection.

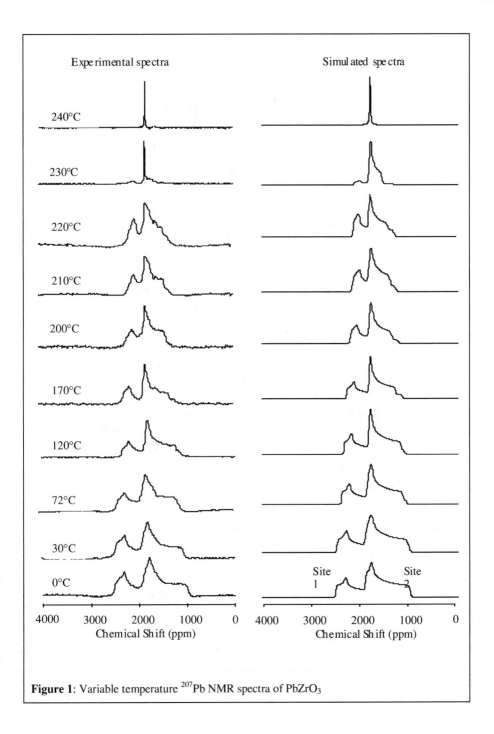

Figure 1: Variable temperature ^{207}Pb NMR spectra of PbZrO$_3$

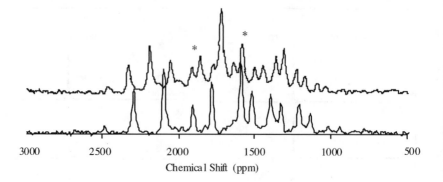

3000   2500   2000   1500   1000   500

Chemical Shift (ppm)

Figure 2: MAS NMR of PZ at 29°C. The isotropic lines for the two Pb sites are marked with an asterisk. Spinning speed for the top spectrum was 8.8 kHz; for the bottom, 12.2 kHz .

This ambiguity is resolved using MAS (figure 2) by extracting the value of σ_{iso} and using it to solve for the σ_{zz}| using σ_{yy} and σ_{yy} from the static spectrum. A second set of MAS experiments was obtained at 85°C to obtain a two-point fit of the σ_{iso} trends.

The three principal values of the chemical shift tensor are given for both lead sites in Table 1. It can be seen that there is a gradual reduction in the chemical shielding anisotropy of both sites with increasing temperature; this reduction presumably reflects reduced distortion of the coordination environment. However, the asymmetry parameters show rather different trends: the asymmetry parameter for site 2 decreases steadily with temperature, while site 1 remains highly axially symmetric.

The collapse of the spectrum into the single line of the cubic PZ phase occurs over a 4°C temperature range between 228° and 232° C; within this range the spectra appear to be a superposition of the lower temperature powder patterns and a single narrow line due to the cubic phase. This temperature dispersion may reflect temperature gradients across the sample. Above 232°C the spectrum exhibits a sharp single line at 1800 ppm.

Table 1

Isotropic chemical shift σ_{iso} , chemical shift anisotropy $\Delta\sigma$ and asymmetry parameter η as a function of temperature. All values are in ppm.

Temp(°C)	Lead site1			Lead site2		
	σ_{iso}	$\Delta\sigma$	η	σ_{iso}	$\Delta\sigma$	η
0	1915	-975	0.206	1562	-570	0.238
30	1930	-876	0.204	1598	-489	0.241
72	1898	-836	0.196	1595	-474	0.198
120	1875	-776	0.198	1605	-424	0.121
170	1844	-694	0.186	1625	-392	0.092
200	1836	-627	0.200	1655	-315	0.077
210	1812	-542	0.157	1660.	-280	0.142
220	1797	-532	0.122	1662	-252	0.020

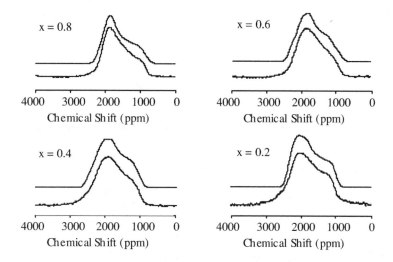

Figure 3: Room temperature ^{207}Pb NMR of PbZr$_{1-x}$Ti$_x$O$_3$ solid solutions.
Top spectrum: simulated. Bottom spectrum: experimental.

^{207}Pb NMR of PbZr$_{1-x}$Ti$_x$O$_3$ (PZT) solid solutions

Preliminary spectra of PZT solid solutions containing 20, 40, 60, and 80 mol percent zirconium are shown in Fig. 3. The spectra are strongly inhomogeneously broadened, and MAS NMR does not significantly reduce the broadening, suggesting that it has its origin primarily in a dispersion of isotropic chemical shifts. At high titanium content, the spectra appear to have a single lead site with very low asymmetry parameter, similar to PbTiO$_3$. However, 80 mole percent zirconium sample starts to show suggestions of a second zirconium site.

DISCUSSION

The results, while preliminary, have some significant consequences for the structure and phase transition behavior of PZ. The continuous variation in spectral parameters with temperature between 0° and 230Ç seems to rule out a major structural phase transition around 200°C, since the symmetries of the lead sites are not changed above this temperature. There is, however, evidence of more subtle change in the 200 – 230°C region. The lineshapes cannot be cleanly fit in this region, and the spectral singularities appear to be markedly broadened in this region. This broadening may indicate a marked increase in motion on a microsecond time scale, or I could also be explained by disorder in the lead sites. Measurement of T$_2$ relaxation rates in this region may resolve this question.

Below 200°C, the smooth decrease in chemical shift anisotropy with temperature resembles the trend we have previously observed in lead titanate [14], and we anticipate that an analysis of the changes in the geometry of the lead coordination environment with temperature at the two lead sites will generate a quantitative understanding of the temperature changes. Those studies are in progress.

ACKNOWLEDGEMENTS

This research was supported by NSF grant MCB-9604521.

REFERENCES

1. Y. Xu, Ferroelectric materials and their apllications, Elsevier, Amsterdam (1991).

2. H. D. Megaw, Proc. *Phys. Soc.*, **58**, 133 (1946).

3. F. Jona, G. Shirane, F. Mazzi, and R. Pepinsky, *Phys Rev.*, **105**, 899 (1957).

4. H. Fujishita, Y. Shiozaki, N. Achiwa, and E. Sawaguchi, *J. Phys. Soc. Jpn.*, **51**, 3583 (1982)

5. A. M. Glaser, K. Roleder, and J. Dec, *Acta Cryst.*, **B49**, 846 (1993).

6. A. D. Irwin, C. D. Chandler, R. Assnik, and M. J. Hampden-Smith, *Inorg. Chem.*, **33**, 1005 (1994).

7. G. Neue, C. Dybowski, M. L. Smith, M. A. Hepp, and D. L. Perry, *Solid State Nucl. Magn. Res.*, **6**, 241 (1996).

8. S. M. Korniyenko, I. P. Bykov, M. D. Glinchuk, V. V. Laguta, and L. Jastrabik, *Eur. Phys. AP*, **7**, 13 (1999).

9. P. Zhao, S. Prasad, J. Huang, J. J. Fitzgerald, and J. S. Shore, *J. Phys. Chem. B*, **103**, 106, 1 (1999).

10. J. Brieger, R. Merkle, H. Bertagnolli, and K. Muller, *Ber. Bunsenges. Phys. Chem.,* **102**, 1376 (1998)

11. Y.-S. Kye, S. Connolly, B. Herreros, and G. S. Harbison, *Main Group Metal Chem.,* **22**, 37 (1999).

12. M. Mehring, Principles of high resolution NMR in solids, Springer-Verlag, New York (1983).

13. D.L. Corker, A.M. Glazer, J.Dec, K. Roleder and R.W. Whatmore, *Acta Cryst.*, **B49**, 846 (1993).

14. D. A. Bussian, G. S. Harbison *Solid State Commun.*, **115**, 95 (2000).

Mat. Res. Soc. Symp. Proc. Vol. 658 © 2001 Materials Research Society

Investigation of Lead Borosilicate Glass Structure With ^{207}Pb and ^{11}B Solid-State NMR

James M. Gibson, Frederick G. Vogt, Amy S. Barnes and Karl T. Mueller
Department of Chemistry, The Pennsylvania State University
152 Davey Laboratory, University Park, PA 16802-6300

ABSTRACT

A series of three lead borosilicate glasses were synthesized and analyzed for structural information with both ^{11}B and ^{207}Pb solid-state nuclear magnetic resonance (NMR) spectroscopic methods. Results showed that increasing lead content caused lead to take a more active role in the network as a former and that the populations in these sites can be approximately quantified. ^{207}Pb phase-adjusted-spinning sidebands (PASS), ^{11}B magic-angle spinning (MAS), and ^{11}B multiple-quantum MAS (MQMAS) experiments were used to determine structural parameters for the two nuclei. The ^{207}Pb PASS experiment showed that at higher lead content, more covalent bonding was present. This principle was demonstrated in both an overall shift of the spectral resonances and a quantitative change in site ratios. The ^{11}B MAS experiment showed that the ratio of BO_3 to BO_4 units was dependent on the amount of lead and boron, consistent with previous studies. Preliminary ^{11}B MQMAS experiments failed to detect any BO_3^- units, previously hypothesized to exist in this system.

INTRODUCTION

Lead borosilicate glasses provide the basis for manufacturing of coatings, enamels, solder glasses, and glass-ceramic cements [1]. From prior analysis of nuclear magnetic resonance (NMR) data, x-ray photoelectron spectroscopy (XPS) results, and measurements of the thermal coefficient of linear expansion (TCLE) of lead silicates [2,3] and lead borosilicates [1,4,5], three major conclusions about lead borosilicate glasses can be drawn. Firstly, there are both BO_3 and BO_4 units in the glasses, and their ratio depends upon the molar composition of the glasses. Secondly, the lead has a dual structural role, in that it is both a network former and a network modifier within the glass. In small amounts the lead behaves as a modifier, while in greater amounts there is an increase in covalent bonds to lead, meaning that some of the lead is now better classified as a network former. Finally, it has been found that there is a broad immiscibility range where these systems can form two-phase glasses. The goals of this work are primarily to address the first two prior observations with newer solid-state NMR methods, as the resolution of both ^{207}Pb and ^{11}B NMR spectra have been significantly improved upon by techniques developed since the pioneering studies by Kim *et al.* [4]. One aim of this work is an attempt to quantify the amount of lead in the roles of network former or modifier using ^{207}Pb NMR. Another goal is to perform advanced ^{11}B NMR experiments in order to investigate the possibility of detecting BO_3^- units that were hypothesized to exist [4], as well as to quantify the ratio of BO_3 to BO_4 units in a range of glasses.

^{207}Pb NMR studies of a variety of lead compounds reveal spectral resonances with a wide chemical shift range. Due to its wide isotropic shift range and the often large anisotropy of the chemical shift interaction, the ^{207}Pb nucleus has previously been studied by two-dimensional PASS NMR [6,7]. PASS NMR is a quantitative technique developed originally for ^{13}C NMR studies, whereby isotropic resonances are separated from anisotropic sideband patterns encountered in a magic-angle-spinning NMR experiment [8]. The two-dimensional PASS experiment utilizes a 90° pulse followed by five variably spaced 180° pulses. When the 180° pulses are spaced evenly, the one-dimensional spectrum looks the same as a magic-angle-spinning spectrum, containing isotropic peaks and a number of corresponding sidebands at integer multiples of the spinning frequency. When the 180° pulses are spaced differently, based on a rigorous calculation [8], the sidebands change in phase strictly depending on the order of the sideband, while the isotropic peak retains the same constant phase. The pulse sequence and variations in pulse spacings for a PASS experiment are shown in Figure 1. The resulting series of spectra show modulation of sideband phases with frequencies depending on the sideband order. The spectra are then analyzed with a Fourier transform in this second dimension, leading to a two-dimensional spectrum containing isotropic peaks along one axis and the sidebands along the second axis. This spectrum is an anisotropic to isotropic correlation spectrum. In addition, isotropic peaks can then be isolated and summed for quantitative results in a one-dimensional isotropic projection spectrum. Quantitative PASS NMR results for ^{207}Pb were recently shown on a mixture of $PbSO_4$ and $Pb(NO_3)_2$, as well as a $PbZrO_3$ sample, all utilizing PASS NMR [6]. Glasses show much broader NMR resonances than crystalline compounds, making analysis much more difficult. However, a recent study demonstrated that a significant portion of the broadening in the ^{207}Pb MAS NMR spectrum from a lead phosphate glass was due to chemical shift anisotropy, and much of this broadening could be removed using ^{207}Pb PASS NMR [7].

One of the questions involving the role of lead as it moves from network modifier to network former, and the subsequent effect on glass structure, is whether or not there are appreciable numbers of BO_3^- units formed in the glass. Lower-resolution ^{11}B NMR techniques, such as those used by Kim [4], can not detect BO_3^- units even if they do exist in a system containing predominantly BO_3 and BO_4 units. ^{11}B is a quadrupolar nucleus (meaning that it has nuclear spin greater than 1/2, in this case 3/2). A technique, known as MQMAS NMR, was recently developed [9] for further narrowing and resolution of resonances from quadrupolar nuclei such as ^{11}B. It is useful for distinguishing resonances from multiple sites that are irresolvable with other one-dimensional techniques (including MAS). This technique introduces a new multiple-quantum dimension that separates sites that are not resolved in a conventional MAS experiment, and is therefore a good candidate for determining whether any BO_3^- sites exist in lead borosilicate systems.

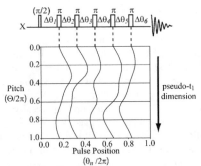

Figure 1: *The PASS pulse sequence used in the experiments described here. The positioning of the 180° pulses change as shown by the vertical curves.*

EXPERIMENTAL DETAILS

Lead borosilicate glasses were synthesized by melting a physical mixture of PbO, H_3BO_3, and SiO_2 at 1000°C for 30 minutes, according to the synthesis technique recommended by Johnson and Hummel [10]. The melts were then quenched between aluminum plates. Three different compositions were synthesized: sample A with a molar composition of 45% PbO, 32.5% B_2O_3, 22.5% SiO_2; sample B with 50% PbO, 30% B_2O_3, 20% SiO_2; and sample C with 55% PbO, 27.5% B_2O_3, 17.5% SiO_2. It is assumed that these three new samples had the same compositions as the batched compositions, consistent with previous elemental analyses of similar lead glasses synthesized by these methods in our laboratory. All of the glasses were examined with x-ray diffraction to ensure that samples were amorphous.

The NMR experiments were performed on an 11.7 T Chemagnetics/Varian Infinity NMR spectrometer system with Larmor frequencies at 160.297 MHz and 104.418 MHz for [11]B and [207]Pb respectively. The experiments were performed using a 2.5 mm double-resonance Chemagnetics magic-angle-spinning probe. The [11]B experiments used a spinning rate of $12,000 \pm 2$ Hz and the [207]Pb experiments were performed while spinning samples at $17,000 \pm 2$ Hz. Because the chemical shift of lead compounds is temperature dependent, the temperature was held constant at 25.0 ± 0.1°C with a Chemagnetics variable temperature accessory. Chemical shifts of [207]Pb compounds were referenced to tetramethyl lead at 0 ppm via a secondary reference, while chemical shifts of [11]B were referenced to boron trifluoride diethyl etherate at 0 ppm. For the [207]Pb experiments, radiofrequency pulse lengths were 1.85 μs for π/2 pulses and 3.7 μs for π pulses. For the [11]B experiments, the π pulses were 10 μs and pulse lengths of 2 μs were used in the one-dimensional MAS experiments to ensure quantitative results. For the MQMAS experiment [9], excitation pulses of 3.2 μs were used along with hard inversion pulses of 1.2 μs and selective z-filter pulses of 8 μs.

DISCUSSION

The [207]Pb PASS experiment was performed on three compositions of lead borosilicate glass. The isotropic projections from these data are shown in Figure 2. The centers of gravity (COGs) of the PASS isotropic projections are at −1000 ppm for glass A, −930 ppm for glass B and −645 ppm for glass C, demonstrating an overall more positive average shift as the amount of lead is increased in the system. Correlations between the lead-oxygen distances and the [207]Pb chemical shift have been suggested based on a prior study of several crystalline lead oxide compounds [11]. That study demonstrated that a shorter Pb-O distance leads to a higher [207]Pb chemical shift. Therefore, in this series of glasses the higher chemical shift would indicate that lead is taking more of a role as a network former on average, because these species will have shorter Pb-O internuclear distances than lead in the modifier role. From the isotropic PASS projections, it can be seen that on average the lead behaves most like a network modifier in glass A (the glass with the smallest amount of lead) and most like a network former in glass C (the glass with the greatest amount of lead). But without resolution of individual resonances in the [207]Pb spectra, this only reports on an average property of these systems.

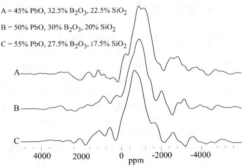

A = 45% PbO, 32.5% B$_2$O$_3$, 22.5% SiO$_2$

B = 50% PbO, 30% B$_2$O$_3$, 20% SiO$_2$

C = 55% PbO, 27.5% B$_2$O$_3$, 17.5% SiO$_2$

Figure 2: *PASS isotropic projections are shown here for three lead borosilicate glasses.*

For more information about the dual role of the lead in the individual glasses, it is necessary to examine the two-dimensional PASS spectra. An example from glass C is shown in Figure 3, and the isotropic projection shown in Figure 2 is obtained by a summation of anisotropic to isotropic correlation along the vertical axis. Although two sites are not distinct in the isotropic projection spectrum, when the two-dimensional PASS spectrum is examined a separation of the spectrum into resonances from at least two sites is observed. Similar spectra were obtained for glasses A and B, both also showing at least two isotropic sites. Using the two-dimensional spectra for identifying the average chemical shift values of the isotropic sites, the isotropic projection spectra can be deconvoluted into two peaks. The fits to the spectra can then be used to quantify the relative populations of lead in the sites. Each glass has a site that is more covalent and a site that is more ionic, likely indicating the presence of the lead as both network former and network modifier. Once again, the network former site would have shorter Pb-O bonds, meaning that it is more covalently bonded.

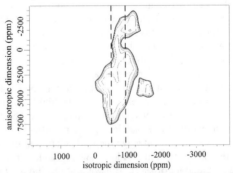

Figure 3: *The two-dimensional PASS spectrum of glass C shows a correlation of the anisotropic dimension to the isotropic dimension, and the increased spectral resolution suggests there are two isotropic sites. The COGs of the two peaks for glass C are at –900 ppm and –500 ppm (vertical lines).*

The resonances are referred to here as the "ionic" and "covalent" peaks, meaning the lower and higher frequency resonances. For glass A, the ionic peak is at –1200 ppm and the covalent peak is at –790 ppm. For glass B, the ionic peak shifts to –960 ppm and the covalent peak is at –640 ppm. Glass C has an ionic peak at –900 ppm and a covalent peak at –500 ppm. The ionic site in glass A is favored by about a 3:2 ratio, while in both glasses B and C the covalent site is favored by an approximately 2:1 ratio. In addition to having the overall COG move to a more positive chemical shift as the lead content increases, the individual covalent and ionic sites also have COGs that shift to higher frequency as lead content increases. These results suggest that both the former and modifier sites have more covalent bonding as the lead content in the sample increases. Kim [4] had also found that there were two sites and that the lead moved in its role from a network modifier to a network former as the lead content was increased, but previous studies were unable to provide direct spectral resolution that would result in a quantification for the two types of lead environments. In order to obtain a greater understanding of this system of lead borosilicate glasses, a larger variety of compositions will now be synthesized and studied by these methods.

^{11}B MAS NMR spectroscopy is used to measure the fraction of BO_4 units relative to total boron-containing units in a glass, as the BO_4 resonance is narrower and shifted (and hence resolved) when compared to other resonances in a ^{11}B MAS spectrum. The ^{11}B MAS spectra for all three glasses are shown in Figure 4. From these data, the percentage of BO_4 units is found to be 49% for glass A, 44% for glass B, and 41% for glass C, leading to similar conclusions as previous studies. Overall, the ratio of BO_4 units in a glass series decreases as the ratio of PbO to B_2O_3 is raised. Thus, it is clear that as a greater amount of PbO is introduced into the system, there are fewer BO_4 units. However, a relevant question arises regarding the potential for existence of BO_3^- units in these glasses.

Given favorable spectral parameters, MQMAS NMR is capable of separating peaks in a second (multiple-quantum) dimension, and therefore we have explored the possibility of using MQMAS to detect BO_3^- units in these systems. Although the MQMAS spectrum shows a very clear separation of the BO_3 and BO_4 units in these glasses, no new evidence for BO_3^- units was found using this technique. However, as of yet, this evidence is not conclusive in determining that such sites do not exist over the full compositional range, and these preliminary experiments will be continued in the future.

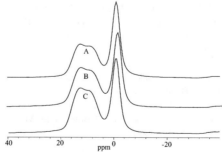

Figure 4: The ^{11}B MAS spectra of the three lead borosilicate glasses show peaks at approximately the same chemical shifts, but have different ratios of peak areas. Using a pulse width much less than that of a 90° pulse, quantitative data are obtained.

CONCLUSIONS

^{207}Pb PASS NMR was used to study the dual structural roles of lead in a short series of lead borosilicate glasses. In all of the glasses, both a "covalent" resonance and an "ionic" resonance are assigned. The covalent resonance can be assigned to lead acting primarily as a network former, while the ionic resonance can be assigned to lead in the role of a network modifier. Based on the overall center of gravity for the resonance lines, glasses with higher lead content correlated with shorter lead to oxygen bonds than those samples with lower lead content. Approximate quantitative ratios of lead atoms in the two environments were also obtained for all three glasses. The glass with the least amount of lead had the lowest ratio of former to modifier species. Finally, the individual covalent and ionic resonances showed an overall shift between the glasses, with both the former and modifier resonances moving to a higher frequency with each increase in lead content. Therefore, whether or not the lead is part of the network, the Pb-O bond distances are presumably shortened as the amount of lead in the system is increased.

^{11}B MAS and MQMAS NMR were also used to study the structural roles of boron in the same glasses. ^{11}B MAS was used to quantify the ratio of BO_4 units to all boron-containing units while MQMAS was used in an attempt to identify the presence of BO_3^- units in this system of glasses. The ^{11}B MAS spectra showed that the number of BO_4 units decreased as more lead was added to the system. Preliminary ^{11}B MQMAS experiment did not identify any BO_3^- units.

ACKNOWLEDGMENTS

KTM is an Alfred P. Sloan Research Fellow. The authors thank Prof. Domnique Massiot and Dr. Cindy Ridenour (Varian/Chemagnetics) for helpful discussions regarding the PASS experiment. Partial support for this research was provided by the National Science Foundation (Grant DMR-9458053) and the Research Corporation (Cottrell Scholar Award to KTM).

REFERENCES

1. T.S. Petrovskaya, *Glass and Ceramics*, **54**, 347-350 (1997).
2. J. Leventhal and P.J. Bray, *Phys. Chem. Glasses*, **6,** 113-116 (1965).
3. R. Dupree, N. Ford, and D. Holland, *Phys. Chem. Glasses*, **28**, 78-84 (1987).
4. K.S. Kim, P.J. Bray, and S. Merrin, *J. Chem. Phys.*, **64**, 4459-4465 (1976).
5. P.W. Wang and L. Zhang, *J. Non.-Cryst. Solids*, **194**, 129-134 (1996).
6. F.G. Vogt, J.M Gibson, D.J. Aurentz, K.T. Mueller, and A.J. Benesi, *J. Magn. Reson.*, **143**, 153-160 (2000).
7. F.Fayon, C. Bessada, A. Douy, and D. Massiot, *J. Magn. Reson.*, **137**, 116-121 (1999).
8. O.N. Antzutkin, S.C. Shekar, and M.H, Levitt, *J. Magn. Reson., Ser. A*, **115**, 7-19 (1995).
9. A. Madek, J.S. Harwood, and L. Frydman, *J. Am. Chem. Soc.*, **117**, 12779-12787 (1995).
10. D.W. Johnson and F.A. Hummel, *J. Am. Ceram. Soc.*, **51**, 196-201 (1968).
11. F. Fayon, I. Farnan, C. Bessada, J. Coutures, D. Massiot, and J.P. Coutures, *J. Am. Chem. Soc.*, **119**, 6837-6843 (1997).

Mat. Res. Soc. Symp. Proc. Vol. 658 © 2001 Materials Research Society

Interference Effect between Electron and Ion Flows in Semiconducting Fe$_{3-\delta}$O$_4$

Jeong-Oh Hong and Han-Ill Yoo
Solid State Ionics Research Lab., School of Materials Science and Engineering,
Seoul National University, Seoul 151-742, Korea

ABSTRACT

The effective valence, $z_{Fe}-\alpha_{Fe}^*$, of mobile Fe-ions (Fe^{2+}, Fe^{3+}) in semiconducting Fe$_3$O$_4$ was determined at elevated temperatures via the electrotransport experiment in association with the literature data on the cation diffusivity and total electrical conductivity. It has been found that the value for $z_{Fe}-\alpha_{Fe}^*$ varies systematically from below 2 up to 3 with oxygen partial pressure at a fixed temperature. The effective valence is determined not only by the mobility difference of Fe^{2+} and Fe^{3+} ions (z_{Fe}), but also by the cross effect between the cations and electrons upon their transfer (α_{Fe}^*). A value of $z_{Fe}-\alpha_{Fe}^*$ between 2 and 3 may be attributed to the mobility difference between Fe^{2+} and Fe^{3+} ions even in the absence of the cross effect, but the values of $z_{Fe}-\alpha_{Fe}^* < 2$ clearly indicate that the cross effect is in play in Fe$_3$O$_4$.

INTRODUCTION

In the treatment of charge/mass transport phenomena in a mixed ionic electronic conductor compound, independent migration of the charge carriers, electrons and mobile ions, has routinely been assumed. It has, however, been found by Yoo et al.[1-5] that the interference effect between ionic and electronic flows in a semiconducting oxide, Co$_{1-\delta}$O, is by no means negligible depending on its defect concentration. They employed, as a measure of this interference effect, the "charge of transport" of the mobile cations, α_{Co}^*, which phenomenologically corresponds to the number of electrons dragged by a cation upon its transfer [1]. In the system of CoO, α_{Co}^* takes a value in the range of 0 to 1.5 depending on the thermodynamic condition (T, Po$_2$) [4,5]. As a result of the interference effect, the effective valence of the mobile cations (Co^{2+}) is reduced from its formal valence $z_{Co}(=+2)$ to $z_{Co}-\alpha_{Co}^*$ [2]. It has since been concluded that the interference effect is not an exception but a rule for electronic conductor compounds. This work is aimed to see if the interference effect is in play in the system of magnetite, Fe$_{3-\delta}$O$_4$. To this end, we performed the Tubandt-type electrotransport experiment [6] on Fe$_{3-\delta}$O$_4$ at elevated temperature.

THEORETICAL BACKGROUND

When an electric field is applied across a mixed-conductor crystal, Fe$_{3-\delta}$O$_4$ with immobile oxide ions via a pair of identical reversible gas(O$_2$) electrodes, both electronic and cationic conduction takes place simultaneously through the oxide. The partial cationic conduction is often referred to as electrotransport [6]. During the electrotransport, the cations at the cathode react with surrounding oxygen gas to create the lattice molecules, whereas at the anode, the lattice molecules are destroyed, oxygen evolves into the gas phase, and cations move away. Thus, the electrotransport causes a shift of the crystal as a whole or a transfer of volume along the current direction. Since the electrotransported volume, ΔV is a measure of the flux of mobile cations, J_{Fe}, one can immediately write down the relation;

$$\frac{F(\Delta V / V_m)}{I \cdot \Delta t} = \frac{FJ_{Fe}}{i},$$
(1

where F is the Faraday constant, V_m the molar volume of $FeO_{4/3}(=Fe_3O_4/3)$, and Δt the time duration for which a constant current, I (current density, i) has been passed through the crystal.

The flux of Fe-ions under an electric field can be related to the transport properties of the Fe_3O_4 such as the self-diffusion coefficient of Fe ions, D_{Fe}, and the total electrical conductivity, σ. For Fe_3O_4, there may be three mobile charged components, two types of cations, Fe^{2+} and Fe^{3} and electrons, which are henceforth denoted by the subscript k=1, 2 and 3. According to the linear transport theory [1], the flux of each type of Fe ions can generally be expressed in terms of the chemical potential gradient of neutral component Fe, $\nabla\mu_{Fe}$ and the electrochemical potential gradient of electrons, $\nabla\eta_3$:

$$J_k = -(L_{k1} + L_{k2})\nabla\mu_{Fe} + (z_1 L_{k1} + z_2 L_{k2} - L_{k3})\nabla\eta_3, \qquad (k=1,2)$$
(2

where L_{kl}'s are the transport coefficients satisfying $L_{kl} = L_{lk}$ (Onsager reciprocity). Among L_{kl}'s, the cross coefficients, $L_{k3}(k=1,2)$ represent the interference between k-type ions and electrons. As a measure of this ion/electron interference effect, the charge of transport, α_k^* is defined as [1]

$$\alpha_k^* \equiv \frac{L_{k3}}{L_{kk}}. \qquad (k = 1, 2)$$
(3

In equation (2), $L_{k1}+L_{k2}$ may be identified in terms of self-diffusion coefficient of k-ions, D_k as $L_{k1}+L_{k2} = D_k c_k/RT$, c_k being the molar concentration of k-ions ($c_1=1/3V_m$, $c_2=2/3V_m$, neglecting the nonstoichiometry) [7,8]. It is noted that the diffusivity of each type of Fe-ions cannot be measured separately. Instead, they can be related to the self-diffusivity of Fe as a component, D_F through the jump balance

$$c_{Fe}D_{Fe} = c_1 D_1 + c_2 D_2 = RT[(L_{11} + L_{12}) + (L_{21} + L_{22})],$$
(4

where $c_{Fe}=c_1+c_2=1/V_m$. The electrotransport experiment is to be carried out under a uniform oxygen partial pressure ($\nabla\mu_O=0$). Then, $\nabla\mu_{Fe}=0$ due to Gibbs-Duhem equation for $FeO_{4/3}$, and $\nabla\eta_3=F \cdot i/\sigma$. By using equations (2)-(4), the overall flux of Fe-ions, $J_{Fe}(=J_1+J_2)$, which is observable in an electrotransport experiment, can be written as

$$J_{Fe} = \frac{D_{Fe}c_{Fe}}{RT}(z_{Fe} - \alpha_{Fe}^*)\frac{F \cdot i}{\sigma},$$
(5

where

$$z_{Fe} = \frac{z_1 c_1 D_1 + z_2 c_2 D_2}{c_1 D_1 + c_2 D_2}, \qquad \alpha_{Fe}^* = \frac{\alpha_1^* L_{11} + \alpha_2^* L_{22}}{L_{11} + L_{12} + L_{21} + L_{22}}.$$
(6

By combining equations (1) and (5), one can determine the effective valence of component

Fe, $z_{Fe}-\alpha_{Fe}^*$ from the result of the electrotransport experiment if the data on D_{Fe} and σ are available. For the system of Fe_3O_4, these data were well documented by Dieckmann and coworkers [9-12].

According to equation (6), the effective valence of Fe is determined not only by the mobility difference of Fe^{2+} and Fe^{3+} ions (z_{Fe}), but also by the cross effect between the cations and the electrons (α_{Fe}^*). Let us now examine the possible values for $z_{Fe}-\alpha_{Fe}^*$. By substituting $c_1=1/3V_m$, $c_2=2/3V_m$, $z_1=2$ and $z_2=3$ in equation (6), one has

$$z_{Fe} - \alpha_{Fe}^* = \left(\frac{2D_1 + 6D_2}{D_1 + 2D_2}\right) - \alpha_{Fe}^* . \tag{7}$$

As extreme cases, $z_{Fe}-\alpha_{Fe}^*=3-\alpha_{Fe}^*$, $(8/3)-\alpha_{Fe}^*$, $2-\alpha_{Fe}^*$ if $D_1<<D_2$, $D_1=D_2$, $D_1>>D_2$, respectively. Considering $\alpha_{Fe}^*\geq0$ by definition, therefore, if $z_{Fe}-\alpha_{Fe}^*=3$, then $D_2>>D_1$ and $\alpha_{Fe}^*=0$. If $2\leq z_{Fe}-\alpha_{Fe}^*$ <3, then one cannot unequivocally say whether $\alpha_{Fe}^*\neq0$. But, as long as $z_{Fe}-\alpha_{Fe}^*<2$, $\alpha_{Fe}^*>0$ for sure.

EXPERIMENTAL

The electrochemical cell as constructed for the electrotransport experiment is illustrated in Figure 1. A spherical Pt-bead with 0.45-0.60 mm diameter was employed as the anode to facilitate the determination of a volume transfer, ΔV after electrotransport. The cathode was a Pt-foil of 50 μm thickness. As the specimen, single crystal Fe_3O_4 was used, which is similar to that employed by Dieckmann et al. for their measurement of D_{Fe} and σ. The crystal was cut into square plates measuring $3\times3\times2$ mm^3. Both surfaces (3×3 mm^2) of a specimen, upon which the electrodes would be placed, were polished with diamond pastes of grit size down to 1μm. The assembly of electrodes and the specimen was mounted into a piston-and-cylinder type holder. It is noted that the electrical contact between the bead-anode and the specimen was kept by the weight ($\approx$1g) of a piston, 2-bore alumina tube, which was allowed to move freely through the cylinder. The details of the cell construction were described elsewhere [13].

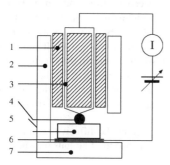

Figure 1. *Schematic of a piston-and-cylinder type sample holder [13] for the measurement of volume transfer induced by electrotransport: 1. two-bore alumina tubing(piston); 2. alumina tubing(cylinder); 3. Pt-lead wire; 4. Pt-bead anode; 5. Fe_3O_4 specimen; 6. Pt-foil cathode; 7. alumina stopper*

Electrotransport experiments were performed at 1200°C and 1300°C. The ambient oxygen partial pressures covered the interstitial dominating regime at 1200°C, and both the vacancy and interstitial dominating regimes at 1300°C. The specimen was first fully equilibrated at each of the prefixed thermodynamic conditions. Then, a constant current of no greater than 40 mA, which was found to be low enough not to cause any appreciable polarization effect [14], was passed for 8~72 hours depending on thermodynamic conditions. After switching the current off, the cell assembly was immediately pulled out of the furnace to prevent the anode bead from sticking to a specimen.

Figures 2(a) and (b) show the bead anode after its removal and the corresponding imprint on the oxide specimen, respectively. The anodic volume displacement was determined as the volume of the spherical indentation remaining into the oxide sample, ΔV, which can be evaluated from the radii of the bead, r_b and of the spherical indentation, r_s as [13]

$$\Delta V = \pi r_b^3 \left\{ (1-a) - \tfrac{1}{3}(1-a^3) \right\} \; ; \; a \equiv \sqrt{1-(r_s/r_b)^2} \; . \tag{8}$$

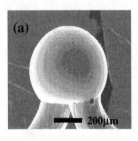 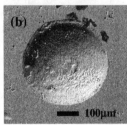

(a)   (b)

200μm   100μm

Figure 2. (a) An anode bead as separated from a Fe₃O₄ specimen and (b) the remaining anodic indentation after electrotransport run; I = 40 mA, Δt = 10.7 hours in Po₂ = 10⁻⁸·⁵ atm at T = 1200°C.

RESULTS AND DISCUSSION

The values for $z_{Fe}-\alpha_{Fe}^*$ are now evaluated via equation (1) and (5) from the measured value for ΔV and the reported values for σ and D_{Fe}. The results are plotted against oxygen partial pressure in Figure 3. As is seen, $z_{Fe}-\alpha_{Fe}^*$ decreases with decreasing oxygen partial pressures in each defect regime of vacancy and interstitial. One should note that $z_{Fe}-\alpha_{Fe}^*$ never goes higher than 3, but goes smaller than 2 depending on thermodynamic conditions. This experimental fact clearly indicates that there exists the cross effect between electrons and ions in Fe₃O₄.

In order to see the effect of the defect concentration on $z_{Fe}-\alpha_{Fe}^*$, we transform the oxygen partial pressure in Figure 3 into the concentration of defects per lattice molecule, $|\delta|$ by using nonstoichiometry data of Fe₃₋δO₄ [10]. The result is as shown in Figure 4. It appears that $z_{Fe}-\alpha_{Fe}^*$ decreases with increasing concentration in the interstitial regime. In the vacancy regime, on the other hand, $z_{Fe}-\alpha_{Fe}^*$ increases with increasing vacancy concentration. Furthermore, even though the data are available only at two different temperatures, it is indicated in the interstitial dominating region that $z_{Fe}-\alpha_{Fe}^*$ increases with increasing temperature at a fixed δ, in agreement with the reported on CoO [4,5].

According to equation (6) or (7), z_{Fe} is determined by the diffusivities of Fe²⁺ and Fe³⁺ ions. In the interstitial dominating region, the diffusivity of each kind of Fe-ions may be expressed as

$$x_1 D_1 = x_{Fe_i^{2+}} D_{Fe_i^{2+}} , \; x_2 D_2 = x_{Fe_i^{3+}} D_{Fe_i^{3+}} , \tag{9}$$

where Fe_i^{2+} and Fe_i^{3+} denote the interstitial Fe-ions, and x_k the mole fraction of species k. The defect concentrations in Fe₃₋δO₄ can be calculated defect-chemically:

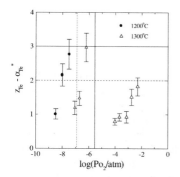

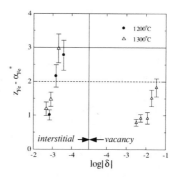

Figure 3. *The effective valence of Fe-ions, $z_{Fe}-\alpha_{Fe}{}^{*}$ in $Fe_{3-\delta}O_4$ against oxygen partial pressure, P_{O_2} at 1200°C and 1300°C. The vertical lines designate the P_{O_2}'s where $\delta=0$ at 1200°C(dotted),1300°C(solid).*

Figure 4. *The effective valence of Fe-ions, $z_{Fe}-\alpha_{Fe}{}^{*}$ against the nonstoichiometry, δ of $Fe_{3-\delta}O_4$ at different temperatures.*

$$3Fe_{Fe}^{2+}+\tfrac{2}{3}O_2(g)=2Fe_{Fe}^{3+}+V_{Fe}+\tfrac{1}{3}Fe_3O_4 \quad ; \quad K_1, \tag{10}$$

$$Fe_{Fe}^{2+}=V_{Fe}+Fe_I^{2+} \quad ; \quad K_2, \quad Fe_{Fe}^{3+}=V_{Fe}+Fe_I^{3+} \quad ; \quad K_3, \tag{11}$$

where $Fe_{Fe}{}^{n+}(n=2,3)$ and V_{Fe} denote regular Fe^{n+} ions and vacancy on Fe-sites, respectively, and K_i the relevant reaction equilibrium constant. Applying the mass action law, one obtains the number of each defect per lattice molecule as

$$[V_{Fe}]=\frac{[Fe_{Fe}^{2+}]^3}{[Fe_{Fe}^{3+}]^2}\cdot K_1 \cdot a_{Fe_3O_4}^{-1/3} \cdot P_{O_2}^{2/3}, \tag{12}$$

$$[Fe_I^{2+}]=\frac{[Fe_{Fe}^{3+}]^2}{[Fe_{Fe}^{2+}]^2}\cdot \frac{K_2}{K_1}\cdot a_{Fe_3O_4}^{1/3} \cdot P_{O_2}^{-2/3}, \quad [Fe_I^{3+}]=\frac{[Fe_{Fe}^{3+}]^3}{[Fe_{Fe}^{2+}]^3}\cdot \frac{K_3}{K_1}\cdot a_{Fe_3O_4}^{1/3} \cdot P_{O_2}^{-2/3}, \tag{13}$$

where $a_{Fe_3O_4}$ represents the activity of Fe_3O_4 and $[S]$ denotes the number of structure element S per lattice molecule. Substituting equation (13) into equation (9), one can obtain the expression for the diffusivity ratio R,

$$R=\frac{D_2}{D_1}=\frac{K_3}{2K_2}\cdot \frac{D_{Fe_I^{3+}}}{D_{Fe_I^{2+}}}. \tag{14}$$

Since all factors appearing on the right-hand side are determined only by temperature, the diffusivity ratio, R, and hence, z_{Fe} can be regarded as constant at a fixed temperature. In vacancy dominating regime, the diffusivity of each cation is proportional to the product of vacancy concentration and the jump frequency of each cation. Assuming the same jump distance, thus, R may be written in terms of the ratio of the jump frequencies of Fe^{2+} and Fe^{3+} ions, or

$$R = \frac{\Gamma_2}{\Gamma_1},$$

(1)

Because the jump frequencies, Γ_1, Γ_2 may be considered to be independent of oxygen partial pressure, R and hence z_{Fe} are regarded as constant again at a fixed temperature. It can therefore be said that the dependence of $z_{Fe}-\alpha_{Fe}^*$ on oxygen partial pressure or nonstoichiometry in Figure 3 and 4 stems from that of α_{Fe}^* on these variables. The α_{Fe}^* appears to increase with increasing interstitial concentration and with decreasing vacancy concentration.

The variation of $z_{Fe}-\alpha_{Fe}^*$ in Figure 3 suggests that it approaches 3 as the stoichiometric composition (or $\log Po_2$= -6.90, -5.55 at 1200°C, 1300°C, respectively) is approached in the interstitial dominating regime. Invoking that if $z_{Fe}=3$, then $\alpha_{Fe}^*=0$, this fact strongly suggests th $D_2>D_1$. Dieckmann [15] has also found that Fe^{3+} ions are more mobile than Fe^{2+} in Al_2O_3-dope MgO.

CONCLUSION

We now conclude that there clearly exists the cross effect between electrons and ions in Fe_3O_4. The effective valence of Fe, $z_{Fe}-\alpha_{Fe}^*$, increases with decreasing interstitial defect concentration, but with increasing vacancy defect concentration. From the fact that $z_{Fe}-\alpha_{Fe}^*$ approaches 3 as the stoichiometric composition is approached, Fe^{3+} seems more mobile than F in Fe_3O_4.

ACKNOWLEDGMENTS

The authors are grateful to Prof. K. D. Becker at Technical University of Braunschwei Germany, for providing the single crystal Fe_3O_4.

REFERENCES

1. H.-I. Yoo, H. Schmalzried, M. Martin and J. Janek, Z. Phys. Chem. N.F. **168**, 129 (1990).
2. H.-I. Yoo and M. Martin, Ceram. Trans. **24**, 103 (1991).
3. H.-I. Yoo, J.-H. Lee, M. Martin, J. Janek and H. Schmalzried, Solid State Ionics **67**, 317 (199
4. H.-I. Yoo and J.-H. Lee, J. Phys. Chem. Solids **57**, 65 (1996).
5. H.-I. Yoo and K.-C. Lee, Z. Phys. Chem. N.F. **207**, 127 (1998).
6. F. Morin, Solid State Ionics **12**, 407 (1984).
7. C. Wagner, Prog. Solid-State Chem. **10**, 3 (1975).
8. J.-H. Lee, M. Martin, and H.-I. Yoo, Electrochemistry **60**, 482 (2000).
9. R. Dieckmann and H. Schmalzried, Ber. Bunsenges. Phys. Chem. **81**, 344 (1977).
10. R. Dieckmann, Ber. Bunsenges. Phys. Chem. **86**, 112 (1982).
11. R. Dieckmann, C. A. Witt, and T. O. Mason, Ber. Bunsenges. Phys. Chem. **87**, 495 (1983)
12. R. Dieckmann and H. Schmalzried, Ber. Bunsenges. Phys. Chem. **90**, 564 (1986).
13. K.-C. Lee and H.-I. Yoo, Solid State Ionics **92**, 25 (1996).
14. J.-O. Hong and H.-I. Yoo, submitted to Solid State Ionics.
15 R. Dieckmann, J. Phys. Chem. Solids **59**, 507 (1998).

Mat. Res. Soc. Symp. Proc. Vol. 658 © 2001 Materials Research Society

Stability and Structural Characterization of Epitaxial NdNiO₃ Films Grown by Pulsed Laser Deposition

Trong-Duc Doan, Cobey Abramowski, and Paul A. Salvador
Carnegie Mellon University, Department of Materials Science and Engineering,
Pittsburgh, PA, 15213-3890

ABSTRACT

Thin films of $NdNiO_3$ were grown using pulsed laser deposition on single crystal substrates of [100]-oriented $LaAlO_3$ and $SrTiO_3$. X-ray diffraction and reflectivity, scanning electron microscopy, and atomic force microscopy were used to characterize the chemical, morphological and structural traits of the thin films. Single-phase epitaxial films are grown on $LaAlO_3$ and $SrTiO_3$ at 625°C in an oxygen pressure of 200 mTorr. At higher temperatures, the films partially decompose to Nd_2NiO_4 and NiO. The films are epitaxial with the (101) planes (orthorhombic *Pnma* notation) parallel to the substrate surface. Four in-plane orientational variants exist that correspond to the four 90° degenerate orientations of the film's [010] with respect to the in-plane substrate directions. Films are observed to be strained in accordance with the structural mismatch to the underlying substrate, and this leads, in the thinnest films on $LaAlO_3$, to an apparent monoclinic distortion to the unit cell.

INTRODUCTION

Thin film deposition of complex oxides having interesting properties, such as ferroelectricity, magnetoresistivity, superconductivity, optical activity, has attracted great attention for the development of advanced and novel devices (see for example [1]). Moreover, thin film deposition has also proven to be an important tool for synthesizing new and metastable phases and for allowing certain strain states to be accessed, which can lead to a variation in the film's structure and properties. Because advanced multilayer structures and complex device architectures require deposition of thin films with controlled structural and physical properties at very small length scales, it is imperative to understand the nature of materials as epitaxial thin films and their difference to their bulk counterparts if many of the potential applications are to be realized. In this communication we describe how the structure of the metastable perovskite oxide $NdNiO_3$ is affected by the growth conditions, underlying substrate, and overall film thickness.

The perovskite nickelates, of the stoichiometry $ANiO_3$ (A = Lanthanide, Y), are of interest because of their physical properties, which include a sharp, thermally driven metal-insulator transition. The temperature at which this transition occurs, T_{MI}, is known to be a strong function of the A cation and of external pressure[2-4]. Such electronic properties render these materials useful as metallic electrodes [5, 6] in ferroelectric devices, as optical switches, as bolometers, and as actuators. From a fundamental scientific perspective, the first order metal-insulator transition that occurs is of great interest because it is not a common feature of complex oxides and it is strongly coupled to the structural parameters (bond angles and distances) and weakly to the magnetic properties. For a review of these materials see Ref. [7]

One of the drawbacks to the applicability of the $RNiO_3$ compounds (not including $R = La$) is that the materials are unstable at elevated temperatures (> 800-900°C) at 1 atm of pressure. The most common approach to their synthesis involves the application of hydrostatic pressure at high temperatures to react the oxides in the presence of an oxidizing agent [2, 3, 7].

The entire series can be synthesized in this manner, although the pressure must be increased for the smaller lanthanides. Thin film deposition has been used as an alternate method of synthesis for R = La, Pr, Nd, and Sm. Extensive work has been carried out on the stable LaNiO$_3$ material as a metallic electrode. The high-pressure phases, R = Pr, Nd, Sm, were first synthesized by sputtering [8, 9], and they were more recently grown by pulsed laser deposition (PLD) [10, 11] and metal organic chemical-vapor deposition (MOCVD) [12].

Because the structural parameters play an important role in the determination of the physical properties, the T$_{MI}$ can be drastically varied by changing these parameters. For NdNiO$_3$ T$_{MI}$ has been decreased from 200 K to $\approx$100 K by application of hydrostatic pressure [13, 14]. In all cases where thin films have not been subjected to a post-deposition high-pressure anneal, the T$_{MI}$ has been suppressed from the bulk value of 200 K. It is clear that the biaxial stress that the films experience from the rigid substrate leads to a variation in the structure and properties of the films. This communication reports the evolution of the stability and structure of NdNiO$_3$ films [100]-oriented SrTiO$_3$ (STO) and LaAlO$_3$ (LAO) substrates. The physical properties of these films will be reported elsewhere.

EXPERIMENTAL

The NdNiO$_3$ PLD-target was synthesized using a standard powder synthesis route [15]. The target was black, dense, and consisted of a mixture of NiO and Nd$_2$NiO$_4$ according to the X-ray diffraction pattern. The overall stoichiometry was reconfirmed with energy dispersive spectroscopy to be 1:1 for Nd:Ni. The target was then mounted inside of the deposition chamber and used for film growth by pulsed laser ablation in a system described previously [15].

Polished, cleaned [15] single crystal substrates (Crystal GMBH, Berlin, Germany) of approximate dimensions $5 \times 5 \times 0.5$ mm were glued to the commercial heater support with silver paste. Samples were heated to the deposition temperature, which ranged from 575 to 800°C, in background pressure of $\leq 10^{-5}$ mTorr. A dynamic deposition atmosphere was established [15] with an oxygen pressure of 200 mTorr. The KrF laser (λ = 248 nm, pulse duration $\approx$ 20 ns) was operated at 3 Hz and was focused to an energy density of $\approx$ 2.25 J/cm^2 at the surface of the target. The target to substrate distance was $\approx$ 60 mm. After cleaning the target surface [15], films were deposited for an amount of time between 30 min and 2 hr. At the end of each deposition, the oxygen pressure was increased to 300 Torr and films were cooled at a rate of 20 °C/min in this static oxygen atmosphere.

X-ray diffraction (XRD) was carried out in Θ–2Θ, ω, and ϕ–scan modes using Rigaku diffractometers and a Philips MRD diffractometer, all equipped with a CuKα radiation. The substrate peaks were used as internal standards to correct for potential alignment inaccuracies and to determine instrumental resolution. The film thickness of thin samples was determined using the Philips MRD in reflectivity mode, and this was converted to a deposition rate. Scanning electron microscopy (SEM) was carried out using a Philips FEG-SEM equipped with an Oxford energy dispersive spectroscopy (EDS) analyzer. *Ex-situ* AFM measurements were carried out in air using an AutoProbe CP (Park Scientific Instruments) in contact mode with ultralever tips. To minimize surface contamination resulting from long exposure to air, AFM experiments were performed immediately after deposition.

RESULTS AND DISCUSSION

The Θ–2Θ XRD scans for $NdNiO_3$ films deposited on [100]-oriented STO as a function of deposition temperature are shown in Figure 1. These samples were deposited for 30 minutes and are $\approx$ 670 Å thick. The region between 2Θ = 27 and 45° has been magnified for each diffractogram (the magnification is listed above it) to highlight the presence or absence of impurity phases. All peaks observed in the diffractograms correspond to either the (00l) peaks from the substrate (S's), the ($h0h$) peaks from the perovskite $NdNiO_3$ (stars), NiO (triangles), and Nd_2NiO_4 (circles). At low temperatures, only the substrate and $NdNiO_3$ peaks are observed. At temperatures above 700°C, the films begin to develop second phases according to the decompositions reaction, 2 $NdNiO_3$ $\rightarrow$ NiO + Nd_2NiO_4 +1/2 O_2 (g). Further work on the $NdNiO_3$ films were, therefore, carried out on those samples grown at 625°C, where the relative intensities of the (101) and (202) are in best agreement with the expected bulk values.

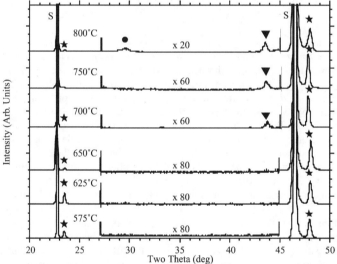

Figure 1. *X-ray diffractograms of NdNiO₃ films deposited upon [100]-SrTiO₃ at various temperatures. See text for explanation of symbols and of the central regions.*

Figure 2a-c shows the effect of the substrate on the film's structure for films deposited at 625°C for 30 minutes. Figure 2b shows the calculated pattern for a [101]-oriented crystal of $NdNiO_3$. For both substrates, the film's ($h0h$) peaks are shifted from the expected position. These shifts are in accordance with the type of strain imparted by the substrate, as depicted by the cartoons in the figure. The average mismatch [16], f_{avg}, for $NdNiO_3$ is 2.59 % (tensile) and –0.43 % (compressive) on STO and LAO, respectively. Hence, the biaxial substrate strain should lead to a decrease in the ($h0h$) spacing on STO and an increase on LAO, as observed. It appears that the peaks are shifted more on LAO than on STO, in spite of the larger mismatch on STO. This means that other relaxation mechanisms are important for films of this thickness on STO.

Film thickness was determined by x-ray reflectivity measurements in a manner similar to that reported elsewhere [17]. Films deposited for 30 minutes were $\approx$ 670 Å thick. This corresponds to a deposition rate of 22.3 Å/ min, or 0.12 Å / laser pulse. This is a relatively slow deposition rate, which allows for high-quality growth of these materials. SEM results indicated

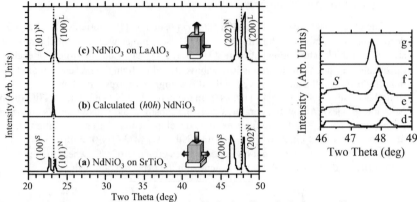

Figure 2. *a-f: XRD scans of NdNiO₃. Figs a-c highlight the effect of substrates on film structure. Cartoons indicate the strain state of films. Figs d-g highlight the effect of thickness for films grown on SrTiO₃. The films are 670 Å (d), 1270 Å (e), and 1970 Å (f). The pattern in g is a calculated pattern.*

that the films had very flat smooth surfaces with some standard defects associated with the PLD method [1]. A typical SEM is shown in Figure 3a for a 670 Å thick film on STO. Increased magnifications yielded no useful contrast as to the grain structure or surface morphology. Atomic Force Microscopy was therefore used to probe the surface morphology of these films. An image is shown in Figure 3b for a film of ≈ 670 Å thick on STO. The surface is characterized by a granular structure with lateral feature sizes on the order of a couple of hundred nanometers. Over the area of a square micron, however, the surface roughness is in the order of only a few nanometers. EDS indicated that the films had the appropriate nominal cationic stoichiometry, within the experimental error.

In order to ascertain the true epitaxial state of the films, we conducted rocking curves around the (202) peak and φ–scans for various peaks in the films. The rocking curves for the 670 Å thick films were 0.17° and 0.22° on LAO and STO, respectively. This implies good out-of plane alignment of the (101) planes for the NdNiO₃ films. The out-of-plane orientation was confirmed using φ–scans. Initially, both [101] and [010] variants were presumed to be present, but no intensity was found at any expected location corresponding to the [010] variants. On the

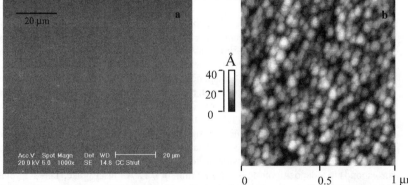

Figure 3. *Scanning electron micrograph (a) and atomic force micrograph (b) of NdNiO₃ on SrTiO₃.*

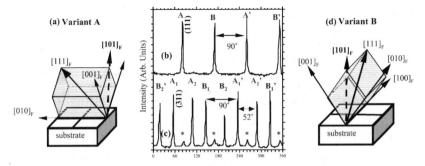

Figure 4. *Schematics of variant formation for a 90° rotation in plane, (a) and (d). The phi-scans demonstrate that four 90° variants exist for NdNiO₃ on SrTiO₃. The two variants A and B are observed in (b), a phi scan of the (111) for the film. The existence of all four potential variants, A and B and their 180° rotated versions, A' and B', are confirmed in (c), a phi-scan of the (311) for the film. Peaks marked by a star arise from residual substrate intensity.*

other hand, strong intensity was observed for peaks related to [101] variants. Phi-scans are given in Figure 4b and 4c for the {111} and {311} family of planes of the [101]-oriented films. Sharp, narrow diffraction peaks indicate that the films are highly epitaxial.

Only two peaks are expected in these phi scans from any single variant; the existence of more than two peaks implies that there are in-plane orientational variants in the film. These variants are related to each other by 90° rotations around the substrate normal. Two of the four possible variants are shown in Figure 4a and 4d. They are denoted as variant A and B, and peaks arising from these variants are marked accordingly in the (111) phi-scan. The (311) phi scan is more complex, because it discerns all four of the possible 90° variants, as denoted. The 52° separation between peaks of the A₁ and A₂ or B₁ and B₂ variants is expected in this experiment for the 180° rotations that relates them. Similar results were observed on the other films, including those deposited upon LaAlO₃.

Finally, using the observed two-theta and phi values of the distinct planes observed— (202), (111), (311), (210), (230), and (430)— it is possible to crudely refine the unit cell parameters of the thin films. A simple refinement program was written to determine the unit-cell parameters from the 11 observables recorded (there is no observable phi value for the (202) plane because it is perpendicular to the phi-rotation axis). It was observed that the thinnest films (670 Å) on LaAlO₃ were pseudomorphic in nature; this means that their in-plane periodicities matched that of the substrate. Because of this, it was difficult to match an orthorhombic unit cell to the observed diffraction values. On the other hand, a monoclinic cell fit the data quite well, with the following parameters: a = 5.392 Å, b = 7.590 Å, c = 5.408 Å, β = 88.82°. This large monoclinic distortion needs to be confirmed by electron diffraction, but it makes sense that the biaxial compressive stress leads to a closing of the angle between the [100] and [001] in the strained [101] oriented films (see Figures 4a and 4d). As the films grew thicker the lattice parameters relaxed towards the bulk structure, although the films were not completely relaxed even at 2000 Å (see Figures 2f and 2g). A monoclinic cell fit the data well for all films, with the monoclinic angle relaxing rather quickly to 90°. This monoclinic distortion could have important ramifications on the physical properties.

CONCLUSIONS

High-quality, epitaxial films of [101]-oriented $NdNiO_3$ were be grown under certain conditions using pulsed laser deposition on LAO and STO substrates. Although the films are epitiaxial, four degenerate variants exist corresponding to 90° rotations of the films crystalline axes. The built in strain decreases as the film thickness increases. The thinnest films on LAO experience a monoclinic distortion and are pseudomorphic with the substrate. Further work is being done to characterize better the unit cell parameters of these films, as well as their physical properties.

ACKNOWLEDGMENTS

This work was supported in part by the MRSEC program of the National Science Foundation under Award Number DMR-0079996. Partial funding for CA was provided by Carnegie Mellon's Undergraduate Research Initiative. TDD would like to thank the Lavoisier Program of the French Ministry of Foreign Affairs for Partial Support. The authors would like to thank J. Wolf and M. De Graef for their insightful comments and help with the diffraction analysis.

REFERENCES

1. D. B. Chrisey and G. K. Hubler. *Pulsed Laser Deposition of Thin Films* (Wiley, New York, 1994).
2. P. Lacorre, J. B. Torrance, J. Pannetier, A. I. Nazzal, P. W. Wang and T. C. Huang, *J. Solid State Chem.* **91**, p. 225 (1991).
3. J. A. Alonso, M. J. Martinez-Lope, M. T. Casais, M. A. G. Aranda and M. T. Fernández-Día *J. Am. Chem. Soc.* **121**, p. 4754 (1999).
4. J. B. Torrance, P. Lacorre, A. I. Nazzal, E. J. Ansaldo and C. Niedermayer, *Phys. Rev. B* **45**, p. 8209 (1992).
5. K. M. Satyalakshmi, R. M. Mallya, K. V. Ramanatham, X. D. Wu, B. Brainard, D. C. Gautier, N. Y. Vasanthacharya and M. S. Hegde, *Appl. Phys. Lett.* **62**, p. 1233 (1993).
6. K. M. Satyalakshmi and K. B. R. V. S. Hegde, *J. Appl. Phys.* **78**, p. 1160 (1995).
7. M. L. Medarde, *J. Phys. Condens. Matter* **9**, p. 1679 (1997).
8. J. F. DeNatale and P. H. Kobrin, , edited (Materials Research Society, Warrendale, 1997), p. 145.
9. P. Laffez, M. Zaghrioui, I. Monot, T. Brousse and P. Lacorre, *Thin Solid Films* **354**, p. 50 (1999).
10. G. Catalan, R. M. Bowman and J. M. Gregg, *J. Appl. Phys.* **87**, p. 606 (2000).
11. G. Catalan, R. M. Bowman and J. M. Gregg, *Phys. Rev. B* **62**, p. 7892 (2000).
12. M. A. Novojilov, O. Y. Gorbenko, I. E. Graboy, A. R. Kaul, H. W. Zandbergen, N. A. Babushkina and L. M. Belova, *Appl. Phys. Lett.* **76**, p. 2041 (2000).
13. P. C. Canfield, J. D. Thompson, S.-W. Cheong and L. W. Rupp, *Phys. Rev. B* **47**, p. 12357 (1993).
14. J. S. Zhou, J. B. goodenough, B. Dabrowski, P. W. Klamut and Z. Bukowski, *Phys. Rev. B* **61**, p. 4401 (2000).
15. A. J. Francis, A. Bagal and P. A. Salvador, in Innovative Processing and Synthesis of Ceramics, Glasses and Composites , edited by N. Bansal (The American Ceramic Society, Inc., Westerville, OH, 2000), p. 565.
16. M. Ohring, The Materials Science of Thin Films (Academic Press, San Diego, 1992).
17. P. A. Salvador, T.-D. Doan, B. Mercey and B. Raveau, *Chem. Mater.* **10**, p. 2592 (1998).

Mat. Res. Soc. Symp. Proc. Vol. 658 © 2001 Materials Research Society

Isotopes in Neutron Diffraction - Detailed Structural Analysis at the Metal-Insulator Transition in SmNiO$_3$

Mark T. Weller, Paul F. Henry and C.C. Wilson[1]
Department of Chemistry, University of Southampton, Highfield, Southampton, SO17 1BJ, UK.
[1]ISIS, Rutherford Appleton Laboratory, Chilton, Didcot, Oxon., OX11 0QX, UK.

ABSTRACT

The use of isotopically enriched materials on high intensity neutron powder diffractometers allows the derivation of much higher quality structural information than has hitherto been possible. The technique can be applied to materials such as ferroelectrics, superconductors and oxides exhibiting colossal magneto-resistance, where small structural changes associated with phase transitions need to be characterised. Three isotopically pure samples of samarium nickelate have been studied around the metal-insulator transition at 403 K. Simultaneous multi-histogram refinements permit extraction of very high quality structural information that shows smooth variations in bond lengths and angles with all nickel oxygen distances decreasing as the electrons localise.

INTRODUCTION

Neutron diffraction is a key technique in the characterisation of solids, often providing information unobtainable with any other method. Its importance has been amply demonstrated in recent years with investigations of high temperature superconductors [1], fullerenes and fullerides [2], compounds displaying colossal magneto-resistance [3], ferroelectrics [4], zero and negative coefficient of thermal expansion materials [5], zeolites [6] and molecular and organometallic solids [7]. Despite this widespread use, the method still has limitations deriving from the fundamental neutron scattering properties of many systems, coupled with (until now) relatively weak neutron count-rates, leading to the requirement for large sample sizes. With the advent of very high count-rate neutron diffraction instrumentation, such as D20 at the ILL, GEM at ISIS and the planned instruments at the SNS at Oakridge, incorporating large detector areas, sample sizes can be reduced dramatically whilst maintaining good data statistics. The work reported here is the first to address another aspect of powder neutron diffraction structural work that is possible with small sample sizes, namely the use of otherwise prohibitively expensive isotopes. By using isotopically pure materials with strongly contrasting scattering powers [8] for the elements present and multiple data set analysis techniques a new generation of powder neutron diffraction experiments, producing much higher quality structural information and structural data on systems, has become possible.

The potential of isotopes in neutron diffraction can be seen by consideration of their fundamental properties. For over half the elements in the periodic table there exists, at reasonable cost (<$20/mg commercially), two or more isotopes with strongly contrasting scattering lengths. In addition, for some elements there are suitable isotopes for overcoming absorption effects. With the need for only a few tens of milligrammes of sample for a neutron diffraction experiment on one of the new range of high count-rate instruments, the cost of producing an isotopically pure sample becomes reasonable – especially when compared with the costs of producing the neutrons themselves.

Specific areas where the use of isotopes may have an impact include the avoidance of absorption effects (through the use of non-absorbing isotopes for elements, such as Gd, Sm, E Cd and B, as has been undertaken previously including a study of ^{154}SmNiO$_3$ [9]), the determination of atomic distributions in materials containing multiple ions on various position (the use of several isotopes of species on the mixed ion sites allows extraction of the scatterin power from the site as a function of isotope quantity) and the observation of very low levels o dopant ions (through the selection of several strongly contrasting isotopes of the dopant ion). wide application will be the use of isotopes to allow the derivation of structural data with muc higher accuracy and precision through multi-histogram refinements, which minimise correlati between structural, instrumental, background and displacement parameters. In this article we demonstrate how isotopes can be used to markedly improve the determination of the structure SmNiO$_3$, as it undergoes a metal-insulator phase transition near 403 K, yielding positional an displacement parameters of an accuracy and precision hitherto unattainable, and discuss the w potential application of this technique in materials chemistry.

EXPERIMENTAL DETAILS

Polycrystalline samples of isotopic substituted SmNiO$_3$ were prepared as follows. Stoichiometric quantities of the elemental nickel isotope (Trace, 99.9% enriched) and ^{154}Sm$_2$((Trace, 99% enriched) were dissolved in the minimum amount of nitric acid and the solution diluted by addition of 50 ml water. An excess of citric acid was added along with 5 ml ethane to promote gel formation. The solution was heated, with constant stirring, until a thick gel wa formed. The gel was partially decomposed under air at 225 °C for 3 hours. The resulting mate was thoroughly ground and then heated in a furnace at 500 °C for 3 hours and then at 750 °C 12 hours to produce the precursor material for high-pressure annealing. The precursor was pressed into a 7 mm diameter pellet and annealed under 200 atm of flowing oxygen gas, using purpose built rig [10], at 1000 °C for 12 hours before cooling at 10 °C min^{-1} to room temperature. The high-pressure annealing procedure was repeated 3 times to ensure product homogeneity. Powder X-ray diffraction was used to monitor product formation, confirm phas purity and measure lattice parameters. Approximately 750 mg of ^{154}Sm^{58}NiO$_3$ and ^{154}Sm^{60}NiC and 350 mg of ^{154}Sm^{62}NiO$_3$ were produced.

Powder X-ray diffraction data were collected over the 2θ range 10 – 110 ° for 16 hour on each sample using a Siemens D5000 diffractometer operating in reflection mode (Cu$_{Kα1}$ λ 1.5406 Å). Rietveld refinement of the data was performed using GSAS [11]. Powder neutron diffraction data were collected on the POLARIS diffractometer at ISIS, Rutherford Appleton Laboratory at nine temperatures through the metal-insulator phase transition at approximately 403 K. Counting times were 45 minutes per temperature for ^{154}Sm^{58}NiO$_3$ and ^{154}Sm^{60}NiO$_3$ an 90 minutes per temperature for ^{154}Sm^{62}NiO$_3$. It is noteworthy that thermal equilibration of the sample is enhanced by the use of small quantities as compared with a typical neutron diffracti sample of several grams. A second series of experiments using the same samples, but over the temperature range 298-573 K at 50 K intervals, were carried out in order to confirm monotoni expansion of the structural and displacement parameters on either side of the transition temperature.

The data from the high-angle detector bank were then analysed by multi-profile, multi phase Rietveld refinement using GSAS. Before refinement, the isotopic data were corrected f absorption. The initial model for the Rietveld refinements was that of Rodriguez-Carvajal et a

[9]. In order to model the three different nickel isotopes, three identical phases were entered into the refinement program with the positional and displacement parameters constrained to be equivalent in each phase, but using the appropriate nickel neutron scattering length, calculated from the isotopic enrichment assay to be $b(^{58}Ni) = 1.438 \times 10^{-12}$ cm, $b(^{60}Ni) = 0.283 \times 10^{-12}$ cm and $b(^{62}Ni) = -0.847 \times 10^{-12}$ cm. The scattering length of ^{154}Sm was taken to be 0.93×10^{-12} cm. [8]. Final cycles of the refinement included all positional parameters as well as anisotropic temperature factors for all sites.

RESULTS AND DISCUSSION

Table I. *Refined atomic and displacement parameters for SmNiO₃ at selected temperatures between 388 K and 418 K. Space group Pbnm*

T / K	388	393	398	403	408	418
Sm (4c)						
$(xy^1/_4)$						
X	0.99027(31)	0.99074(32)	0.99099(33)	0.99027(32)	0.99064(32)	0.99122(33)
Y	0.05070(14)	0.05018(15)	0.05006(15)	0.04968(15)	0.04923(15)	0.04943(15)
$100 \times U_i/U_e$ (Å²)	0.83(4)	0.85(4)	0.89(4)	0.86(4)	0.90(5)	0.90(5)
Ni (4b)						
$(^1/_200)$						
$100 \times U_i/U_e$ (Å²)	0.65(4)	0.64(4)	0.63(4)	0.63(4)	0.63(4)	0.66(4)
O1 (4c)						
$(xy^1/_4)$						
X	0.07950(36)	0.07953(36)	0.07928(37)	0.07903(36)	0.07882(37)	0.07886(37)
Y	0.48472(35)	0.48487(35)	0.48475(36)	0.48459(36)	0.48510(36)	0.48563(36)
$100 \times U_i/U_e$ (Å²)	0.94(5)	0.95(5)	1.02(5)	1.00(6)	1.01(5)	1.02(6)
O2 (8d)						
(xyz)						
X	0.70802(26)	0.70808(27)	0.70812(27)	0.70821(27)	0.70840(27)	0.70829(27)
Y	0.29357(23)	0.29331(24)	0.29367(24)	0.29270(23)	0.29262(24)	0.29262(24)
Z	0.04122(17)	0.04130(17)	0.04100(17)	0.04090(17)	0.04077(17)	0.04064(18)
$100 \times U_i/U_e$ (Å²)	0.92(4)	0.93(5)	0.91(5)	0.93(5)	0.95(5)	0.97(5)
A (Å)	5.33590(9)	5.33596(9)	5.33613(9)	5.33625(9)	5.33658(9)	5.33740(9)
B (Å)	5.42922(9)	5.42852(10)	5.42753(10)	5.42586(9)	5.42484(9)	5.42470(9)
C (Å)	7.57462(12)	7.57414(12)	7.57381(12)	7.57327(12)	7.57301(12)	7.57402(12)
Volume (Å³)	219.435(6)	219.395(6)	219.353(6)	219.274(6)	219.239(6)	219.297(6)

Table I summarises the refined atomic positional and displacement factors at each of the temperatures investigated. Of direct importance in terms of the structure and physical properties of SmNiO₃ are changes in the derived bond-lengths and bond angles, particularly the superexchange Ni-O-Ni angles, as a function of temperature through the metal-insulator transition. This information is summarised in Figures 1-3 together with the previous data from Rodriguez Carvajal *et al.* [9].

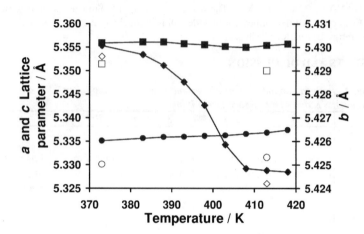

Figure 1. *Variation of the lattice parameters of SmNiO₃ between 373 and 418 K; a – •, b – ■ c - ♦; filled symbols this work – open symbols from reference [9], esds are within data points. Solid lines represent best fit trend lines to the data.*

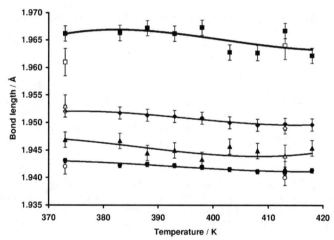

Figure 2. *Variation of the nickel – oxygen bond lengths between 373 and 418 K; Ni-O1 - •, Ni O2 ■ and ▲, average Ni-O ♦, esds are shown as error bars. Data from reference [9] are included for comparison as open symbols. Solid lines represent the best fit trend lines to the data.*

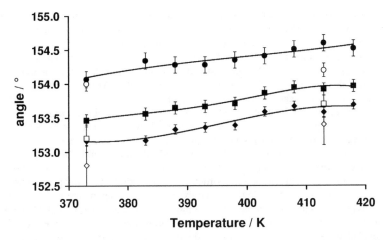

Figure 3. *Variation of O-Ni-O super-exchange angles between 373 and 418 K; Ni-O1-Ni - •, Ni-O2-Ni - ♦, average Ni-O-Ni - ■; esds are shown as error bars. Data from reference (9) are included for comparison as open symbols. The solid lines represent best fit trend lines to the data.*

It is notable that the correlation between various refined parameters is reduced markedly by the use of isotopes and multi- histogram refinement. In general, estimated standard deviations on positional and displacement factors are reduced by a factor of three. Extraction of anisotropic displacement parameters was also possible with the removal of the serious correlations between positional parameters and thermal displacements in the same coordinate direction that were apparent in single histogram refinements. In terms of derived bond lengths and angles this reduction in positional parameter uncertainty is reflected in very accurate values, with errors on bond lengths typically 0.0002 - 0.0004 Å. As the changes in such values around the transition are very small – of the order of 0.002 Å - it is obviously crucial that they are extracted with high precision in order to be able to monitor their evolution.

Lattice parameter changes in $SmNiO_3$ near the transition are in good agreement with those published previously – in particular the marked leveling and decrease in *b prior* to the transition to metallic behaviour at 403 K as observed in the transport properties. With the good resolution of POLARIS in comparison with the data obtained in the previous experiment [9], the correlations between the lattice parameters for the weakly distorted $SmNiO_3$ are also reduced and accuracy in these values is also increased.

Atomic positions, lattice parameters, bond lengths, angles and displacement parameters, extracted from the data collected over the wide temperature ranges before and after the M-I transition, studied in detail between 370 and 420 K, showed smooth increases in all values. For example the average nickel–oxygen distance increases by 1.1×10^{-5} Å/K, as expected through thermal expansion, up to around 30 K prior to the transition (403 K) and from 420 to 573K. From Figure 2 it can be seen that on approaching the transition temperature, the average nickel oxygen distance decreases at a rate of $\sim 7 \times 10^{-5}$ Å/K, before levelling off and then expanding again through the thermal effects. The contraction takes place roughly equally in *all* the nickel-oxygen distances, and the level of distortion of the NiO_6 octahedron, as measured as the standard deviation of the values from their average, remains constant. Over the same temperature range the average samarium-oxygen distances increase by 0.008 Å reflecting the expected continuous thermal expansion of this section of the lattice. The increase in bond lengths is mainly confined to two of the shorter Sm-O1 and Sm-O2 distances consistent with this ion moving to a more regular coordination geometry at the higher temperatures. As a result of the more detailed structural data that can be obtained with isotopes, plus the fact that several temperatures close both sides of the M-I transition could be studied on a reasonable experimental timescale, it is clear that the previous view of abrupt structural changes occurring in $SmNiO_3$, propelled by the delocalisation of the electrons and changes in the distortion parameter of the NiO_6 octahedron may be improved. The observed behaviour, with *smooth* variations in all the structural parameters over the temperature range from about 30 K below the M-I transition temperature, fits in with that expected for a first order transition with the high temperature metallic phase and the low temperature insulating phase co-existing over the temperature range 380-405 K. Such behaviour is clearly seen in the transport properties of the related $PrNiO_3$ [12]. Further discussion of the structural data presented in this article will be published elsewhere – where a more detailed analysis of parameters such as distortion of the NiO_6 octahedra, the level of deviation from the perfect perovskite structure and a full treatment of the displacement parameters will be discussed.

However it is clear from this work that full analysis of the diffraction data collected using several isotopic mixtures results in a marked improvement of the quality of the extracted structural parameters. For example, esds on positional parameters drop by a factor of about three while anisotropic temperature factors can also be extracted. For structural transitions, such as the first order M-I transition studied here, where the changes in bond lengths and angles are very small, and especially when derived parameters such as the level of distortion of an octahedral unit involves ratios of these values - compounding any errors - a much improved picture of the behaviour is obtained. For example, we have been able to show that the delocalisation results in decrease in the lengths of *all* the nickel-oxygen bonds leaving the distortion of the NiO_6 essentially unchanged.

The overall experiment time on a medium-high flux instrument such as POLARIS - 3 hours per temperature (combined time to collect all 3 data sets) - compares favourably with that required for a typical sample using [154]Sm alone - 36 hours on D1B (taken from Rodriguez-Carvajal et al. data [9]). Once very high count-rate instruments such as GEM [13] become generally available, experimental times will be reduced significantly - by a factor of 5 or more making the use of isotopes even more affordable and desirable. With sample sizes of a few milligrams, as has been investigated successfully on GEM, even extremely rare and expensive isotopes could be studied.

The improved structural information available using isotopes makes such methods ideal for studying subtle structural transformations and behaviour that are crucial to the understanding of many important effects in materials. Several further experiments have been completed or are planned using this technique. These will include a study of $ErBa_2Cu_3O_7$ and $ErBa_2Cu_4O_8$ (with Er, Ba and Cu isotopes) through their superconducting phase transitions in order to investigate whether predicted changes in anisotropic motion of the CuO_2 plane oxygens can be observed. Also, studies of negative and zero thermal expansion materials such as $M_2W_3O_{12}$ (where M = Sc or Yb using W and Yb isotopes) have allowed us to define accurately the thermal motion as a function of temperature [14]. We have also very recently undertaken a study of the ferroelectric

BaTiO$_3$ through the orthorhombic-tetragonal phase transition at 278 K using the new GEM diffractometer with its medium resolution 90° detector bank (Δd/d ~ 4 × 10^{-3}). The samples studied were isotopically pure ^{137}Ba^{46}TiO$_3$, ^{137}Ba^{48}TiO$_3$ ^{138}Ba^{46}TiO$_3$ and^{138}Ba^{48}TiO$_3$. Despite the only moderate resolution of this data bank, the extracted positional and displacement parameters showed significant improvement over the current best literature data [4]. Once high resolution data becomes available from the back scattering GEM detectors in 2000 [13] a full structural model for this key ferroelectric material at all its various phase transitions will be achievable.

Overall, it is clear that the use of isotopes offers the potential to markedly improve the structural data available from powders using neutron diffraction. This may be particularly so for instruments on time-of-flight sources where, as well as high precision positional parameters, excellent displacement parameters can now also be extracted, as the available data over a very large d-spacing range lead to a marked reduction in the reflection intensity to background correlation for the large numbers of grossly overlapping reflections.

ACKNOWLEDGMENTS

We thank the EPSRC for funding this work including access to neutron facilities at RAL under Grant GR/M21188. We also thank Ron Smith for assistance with the neutron experiments.

REFERENCES

1. W.I.F. David W.T.A. Harrison, J.M.F Gunn, O. Moze, A.K. Soper, P. Day, J.D. Jorgensen, D.G. Hinks, M.A. Beno, L. Soderholm, D.W. Capone, I.K. Schuller, C.U. Segre, K. Zhang, J.D. Grace, *Nature* **327**, 310 (1987).
2. K. Prassides, C. Christides, I.M. Thomas, J. Mizuki, K. Tanigaki, I. Hirasawa, T.W. Ebbesen, *Science* **263**, 950 (1994).
3. A. Moreo, S. Yunoki, E. Dagotto, *Science* **283**, 2034 (1999).
4. G.H. Kwei, A.C. Lawson, S.J.L. Billinge S.-W. Cheong, *J. Phys Chem.* **97**, 2368 (1993).
5. J.S.O. Evans, Z. Hu , J.D. Jorgensen, D.N. Argyriou, S. Short, A.W. Sleight, *Science* **275**, 61-65 (1997).
6. G. Vitale, C.F. Mellot, A.K. Cheetham, *J. Phys. Chem. B* **101**, 9886 (1997).
7. T. Vogt, A.N. Fitch, J.K. Cockcroft, *Science* **263**, 1265 (1994); R.Bau, M.H. Drabnis, L. Garlaschelli, W.T. Klooster, Z.W. Xie, T.F. Koetzle, S. Martinengo, *Science* **275**, 1099 (1997).
8. *Neutron News* **3(3)**, 29 (1992).
9. J. Rodriguez-Carvajal, S. Rosenkranz, M. Medarde, P. Lacorre, M.T. Fernandez-Diaz, F. Fauth, V. Trounov, *Phys. Rev. B* **57(1)**, 456 (1998).
10. B. Cleaver, D.B. Currie, *High Temp. High Press.* **22**, 623 (1990).
11. A.C. Larson, R.B. Von Dreele, *Generalised Structure Analysis System*, Los Alamos National Laboratory, (1994).
12. X. Granados, J. Fontcuberta, X. Obradors, J. B. Torrance, *Phys Rev B* **46**, 15683 (1992).
13. W.G. Williams, R.M. Ibberson, P. Day, J.E. Enderby, *Physica B* 241, 234, 1997.
14. M.T. Weller, P.F. Henry, C.C. Wilson, *J. Phys. Chem. B* (in press).

Mat. Res. Soc. Symp. Proc. Vol. 658 © 2001 Materials Research Society

Crystal Chemistry of Colloids Obtained by Hydrolysis of Fe(III) in the Presence of SiO_4 Ligands.

Emmanuel Doelsch[1], Armand Masion[1], Jérôme Rose[1], William E.E. Stone[1], Jean Yves Bottero[1] and Paul M. Bertsch[2]
[1]CEREGE, Europole Méditerranéen de l'Arbois, BP 80, 13545 Aix-en-Provence Cedex 04, France.
[2]AACES-SREL, University of Georgia, PO Drawer E, Aiken, SC 29802, USA.

ABSTRACT

The crystal chemistry of a number of Fe-Si systems (Si/Fe 0-4, pH 3-10) was investigated by combining local scale spectroscopic methods (EXAFS, FTIR and NMR) and at the semi local scale (SAXS). The Fe clusters within the precipitates have two growth regimes depending on the Si/Fe ratio: the growth is three and two dimensional for Si/Fe ≤ 1 and Si/Fe ≥ 1 respectively. The presence of Fe-O-Si bonds within the precipitated phases has been demonstrated. Their formation and relative proportion was found to be very dependent on the pH and Si concentration The size of silica domains within the precipitates was shown to increase with increasing Si/Fe and/or decreasing pH. The high fractal dimension (D_f) of the aggregates is attributed to the presence of the SiO_4 ligands, but the evolution of D_f linearly depends on the polymerization state of iron.

INTRODUCTION

Iron and silicon are two of the most abundant elements of the earth's crust. Fe-Si associations are of great relevance for materials science as well as environmental issues. Numerous studies concerning Fe-Si systems have focussed on the effects of Si on the crystallization of Fe phases [1-4]. It has been shown that, even at low Si levels, the crystallization of Fe was delayed or even inhibited. Most of the available literature on Fe-Si phases investigates systems that underwent some degree of "ripening" (i.e. heating and/or aging). Very few authors investigated the nucleation mechanisms of Fe-Si colloids on freshly prepared samples. Thus surprisingly little is known about the crystal chemistry of Fe-Si systems. The present study aims at investigating the local (i.e. up to 5 Å) and semi-local (up to 400 Å) structure of freshly prepared Fe-Si colloids or precipitates in order to elucidate the mechanisms that might influence the subsequent evolution of these systems upon heating and/or aging.

MATERIALS AND METHODS

Appropriate amounts of tetra ethylorthosilicate (TEOS) were added to a $FeCl_3$ stock solution so as to obtain a final Fe concentration of 0.2 M and Si/Fe molar ratios of 0, 0.25, 0.5, 1, 2 and 4. These systems were base hydrolyzed (NaOH) to reach pH values of 3, 5, 7 and 10. When required for further analysis, the suspensions were centrifuged at 40,000 rpm (2 hours) and freeze-dried.

X-ray diffraction (XRD) analyses were performed using a Philips PW 3710 diffractometer with Co Kα radiation at 40 kV and 40 mA. Fe K-edge extended X-ray absorption fine structure (EXAFS) measurements required the use of the synchrotron beamlines D42 (LURE, Orsay, France) and X23A2 (NSLS, Upton (NY), USA). Data reduction and modeling of the EXAFS data were accomplished as described previously [5, 6].

Solid state ^{29}Si nuclear magnetic resonance (NMR) spectra were recorded on a Bruker MSL 300 spectrometer at 59.63 MHz in the MAS mode (spinning rate 15 kHz). Spectra are referenced with respect to tetramethyl silane. Fourier transform infra-red (FTIR) spectra were obtained using a Nicolet 510 spectrometer with a spectral resolution of 4 cm^{-1}. Small angle X-ray scattering (SAXS) experiments were performed on beamline D24 of the LURE synchrotron. The wavelength was 1.89 Å and the covered Q range extends from 0.009 to 0.420 Å^{-1}. The detailed description of the data treatment can be found elsewhere [7, 8].

RESULTS AND DISCUSSION

The XRD diagrams reveal that the crystallinity of the formed species is sensitive to the Si concentration and the pH. The samples at pH = 3 display peaks originating from the presence of akaganeite (β-FeOOH). These peaks are broad, thus suggesting poor order within the precipitates. The amount of akaganeite steadily decreases with increasing Si/Fe ratio. At pH = 5, akaganeite is still detected by XRD but represents only a very minor phase as indicated by the very low intensity of the peaks. All the samples above this pH are amorphous to XRD.

A closer insight into the nature of the formed Fe species is given by EXAFS. For all the samples, only oxygen backscatterers are present in the first coordination shell around iron as opposed to less hydrolyzed FeCl$_3$ based systems (i.e. at lower pH) studied previously where up to 2 Cl atoms were detected [9]. Because of the large difference in atomic number between Fe and Si, it is very difficult to detect the presence of Si backscatterers in the vicinity of Fe atoms by Fe K-edge EXAFS [10, 6]. It has been demonstrated that the quality of the spectral fitting was not affected by whether Si shells were included in the calculation [10]. Consequently, only Fe-Fe contributions are considered here. The total number of Fe neighbors N_{tot} indicates the polymerization state of iron within the precipitated phases (Fig. 1). The highest number is observed for the samples prepared without Si. When Si is added, the values of N_{tot} are lower, thus demonstrating the inhibiting effect of the SiO$_4$ ligands on the polymerization of iron. Nevertheless, there is no linear trend between Fe polymerization and Si concentration: N_{tot} decreases for Si/Fe < 1 and re-increases for Si/Fe > 1 without, however, reaching the values determined for the samples without Si. There are major differences between the Si/Fe ratios concerning the number of neighbors by type of linkage (Fig. 2). The number of double corner linkages (Fe-Fe distance around 3.4 Å) decreases sharply from the lowest Si/Fe and remains nearly constant for Si/Fe ≥ 1. Single corner linkages (Fe-Fe around 3.8 Å) are detected only for Si/Fe ≤ 0.5. The evolution of edge linkages between Fe atoms is similar to N_{tot}. At low Si/Fe, the presence of edge and double corner linkages suggests a three dimensional growth of the Fe species. Indeed, three dimensional structures such as goethite(α-FeOOH) or akaganeite (β-FeOOH) include both

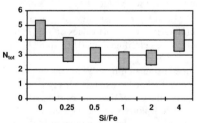

Figure 1: Total number of iron neighbors (Ntot) determined by EXAFS. Bars represent the lowest and highest value (all pH's).

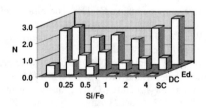

Figure 2: Number of iron neighbors (N) by linkage type. SC: single corner; DC double corner; Ed.: edge

types of linkages. For Si/Fe > 1, the presence of SiO_4 ligands inhibits the formation of corner linkages and the growth of Fe species occurs mainly by edge sharing. This corresponds to a two-dimensional growth. These trends are observed at all studied pH values.

FTIR and ^{29}Si NMR data were used to probe the status of Si in the formed precipitates. In the FTIR spectra, the major variations with pH and Si concentration were observed in the 800 to 1300 cm^{-1} region. At constant Si/Fe, this peak becomes narrower with increasing pH, and is broadened with increasing Si concentration at constant pH. This can be correlated to the IR spectra of silicate minerals. For Q_4 minerals (i.e. minerals in which each SiO_4 tetrahedron is linked to 4 others) such as quartz, the band is larger than for unpolymerized Q_0 minerals (800-1200 cm^{-1} and 820-1000 cm^{-1} respectively). Similarly, the Si polymerization in our Fe-Si samples can then be assumed to increase with increasing Si concentration and decreasing pH. The pH dependence of the Si polymerization could be expected from the higher solubility of silica for $pH \geq 7$ [11].

More detailed information is provided by the line fitting of the 800-1300 cm^{-1} region by gaussian peaks. The two main components considered here are the lines at 800 cm^{-1} (T-O-T symmetric stretching between SiO_4 tetrahedra), 1080 cm^{-1} (Si-O-Si asymmetric stretching) and 930 cm^{-1} which is associated with Si-O-Fe bonds. At low pH and up to Si/Fe = 1, the increase of the 1080 cm^{-1} contribution(Si-O-Si) is accompanied by an increase of the 930 cm^{-1} line (Si-O-Fe), thus showing that Si-O-Si and Si-O-Fe bonds are formed at the same time (Fig. 3). At higher Si/Fe, the 930 cm^{-1} line is not detected. It is possible that it is hidden by the larger amount of silica formed at these Si concentrations, as evidenced by the increase of the 800 cm^{-1} contribution. However, at higher pH (7 and 10), the 930 cm^{-1} peak is present even for the highest Si/Fe ratios. This is correlated with a lesser importance of the 800 cm^{-1} contribution at these pH values, which indicates less polymerized Si species.

The ^{29}Si NMR data give additional evidence for decreased Si polymerization with increasing pH. Since the proximity of the paramagnetic Fe atoms broadens the ^{29}Si NMR lines beyond detection, a signal is therefore indicative of a spatial separation between some Si and Fe atoms leading to silica domains within the solids. A signal was obtained only at low pH and high Si concentration (Si/Fe = 2, pH = 3; Si/Fe = 4, pH = 3 and 5). The intensity of the line decreased strongly between pH = 3 and pH = 5 and no signal was obtained at higher pH values. This is consistent with a reduced size of the silica domains as the pH increases and confirms the IR results.

At the semi-local scale, the shape of the SAXS curves obtained for Si/Fe 0.5, 1 and 2 at pH 3, 7 and 10, is in agreement with the predominantly amorphous nature of the

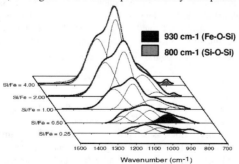

Figure 3:*Fit of the 800-1300 cm^1 region of the IR spectra (pH 3).*

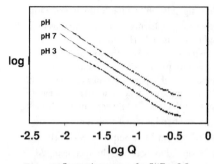

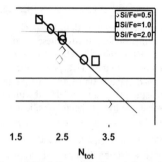

Figure 4 *Scattering curves for Si/Fe=0.5 systems. The curves are shifted in intensity for better clarity.*

Figure 5 *Evolution of the fractal dimension (Df) with Ntot, the total number of Fe neighbors.*

precipitates determined by XRD (Fig. 4). No correlation peak, which would indicate the presence of a privileged distance within the sample, was observed. The absence of an asymptotical behavior of the scattering curves at low Q shows that the size of the aggregates is higher than our detection limit, viz. 350 Å. The modeling of the outermost part of the curves (Q > 0.240 Å^{-1}, log Q > -0.62), performed as previously described [7], revealed that close to 100% of the iron is involved in species smaller than 7.5 Å without, however, providing precise information on the size, nature and relative proportion of the different subunits. Since the Fe-Si systems are amorphous, the description of their structure is done in terms of fractal dimension D_f. The D_f values range from 2 to 2.7. In the absence of complexing ligands, aggregates formed by hydrolysis of Fe(III) have a fractal dimension around 1.8 to 2 [12, 13]. The high values measured in the present study show that the presence of SiO_4 ligands results in more dense aggregation, possibly by bridging of the subunits. The evolution of D_f with pH and/or Si concentration cannot be correlated to the polymerization state of Si. However, it linearly depends on the polymerization state of iron (Fig. 5).

CONCLUSIONS

In Fe-Si systems, the growth regime of Fe species is not affected by the pH: it is three- or two-dimensional for low and high Si/Fe molar ratios respectively. The formation of Fe-O-Si bonds is very sensitive to both the pH and the Si concentration: it seems favored at low pH and Si/Fe as well as at high pH and Si/Fe. The size of silica domains within the precipitates was shown to increase with increasing Si/Fe and/or decreasing pH. The dense overall structure of the aggregates is due to the SiO_4 ligands, the variations of D_f with pH and Si concentration being however controlled by the polymerization state of Fe.

ACKNOWLEDGEMENTS

This work was partially supported by CNRS-NSF collaboration agreement number 7383. This research was carried out in part at the National Synchrotron Light Source, Brookhaven National Laboratory, which is supported by the U.S. Department of Energy, Division of Materials Sciences and Division of Chemical Sciences under contract number DE-AC02-98CH10886. P.M. Bertsch was supported by financial assistance award number DE-FC09-96SR18546 from the U.S. Department of Energy to the University of Georgia Research Foundation. The authors wish to thank the personnel at LURE and NSLS for their assistance during the experiments, and especially C. Bourgaux (D24, LURE), A. Traverse and F. Bouamrane (D42, LURE), and J. Woicik (X23A2, NSLS).

REFERENCES

1. R.K. Vempati and R.H. Loeppert, *Clays Clay Miner.*, **37**, 273-279 (1989).
2. K. Kandori, S. Uchida, S. Katoaka and T. Ishikawa, *J. Mater. Sci.*, **27**, 719-728. (1992).
3. T.D. Mayer and W.M. Jarrell, *Wat. Res.*, **30**, 1208-1214. (1996).
4. S. Glasauer, J. Friedl and U. Schwertmann, *J. Colloid Interface Sci.*, **216**, 106-115 (1999).
5. J. Rose, A. Manceau, J.Y. Bottero, A. Masion and F. Garcia, *Langmuir*, **12**, 6701-6707. (1996).
6. E. Doelsch, J. Rose, A. Masion, J.Y. Bottero, D. Nahon and P.M. Bertsch, *Langmuir*, **16**, 4726-4731. (2000).
7. A. Masion, D. Tchoubar, J.Y. Bottero, F. Thomas and F. Villiéras, *Langmuir*, **10**, 4344-4348. (1994).
8. A. Masion, J.Y. Bottero, F. Thomas and D. Tchoubar, *Langmuir*, **10**, 4349-4352. (1994).
9. J.Y. Bottero, A. Manceau, F. Villiéras and D. Tchoubar, *Langmuir*, **10**, 316-319. (1994).
10. A. Manceau, P. Ildefonse, J.L. Hazemann, A.M. Flank and D. Gallup, *Clays Clay Miner.*, **43**, 304-317. (1995).
11. R.K. Iler *The chemistry of silica. Solubility, polymerisation, colloid and surface properties and biochemistry.*; John Wiley & Sons: New-York (NY), (1979).
12. J.Y. Bottero, D. Tchoubar, M. Arnaud and P. Quienne, *Langmuir*, **7**, 1365-1369. (1991).
13. D. Tchoubar, J.Y. Bottero, P. Quienne and M. Arnaud, *Langmuir*, **7**, 398-402. (1991).

Transport Properties/
Metal-Insulator Systems

Mat. Res. Soc. Symp. Proc. Vol. 658 © 2001 Materials Research Society

Electronic and Magnetic Properties of d^1 Pnictide-Oxides: Na$_2$Ti$_2$Pn$_2$O (Pn = As, Sb)

Tadashi C. Ozawa[1], Mario Bieringer[2], John E. Greedan[2] and Susan. M. Kauzlarich[1]
[1]University of California
One Shields Avenue
Davis, CA 95616, U.S.A.
[2]Institute for Materials Research
McMaster University
Hamilton, Ontario
L8S 4M1, Canada

ABSTRACT

Bulk physical properties of layered pnictide-oxides, Na$_2$Ti$_2$Pn$_2$O (Pn = As, Sb) have been characterized. Both As and Sb compounds exhibit anomalous transitions in temperature dependent magnetic susceptibility and electrical resistivity. The powder neutron diffraction data show no superlattice reflections indicative of long-range order. The origin of the physical property anomalies and structure-property relationship between As and Sb compounds are described in terms of a charge-density-wave/spin-density-wave mechanism.

INTRODUCTION

Na$_2$Ti$_2$Pn$_2$O (Pn = As, Sb) are layered pnictide-oxide compounds that were first discovered by Adam et al. in the early 90's [1]. To our knowledge, this is the only pnictide-oxide containing Ti^{3+}, a d^1 transition metal. Geometrically, the structure of Na$_2$Ti$_2$Pn$_2$O (Figure 1) is the K$_2$NiF$_4$ type; however, positions of cations and anions are reversed from those in K$_2$NiF$_4$ type compounds. Thus, the chemical structure of this compound should be classified as the anti-K$_2$NiF$_4$ type. We have previously measured the temperature dependent magnetic susceptibility and electrical resistivity of the Sb compound, Na$_2$Ti$_2$Sb$_2$O, and have discovered that this compound exhibits a sharp drop in magnetic susceptibility and a metal-to-metal transition in the electrical resistivity at T$_c$ ~ 120 K reminiscent of CDW/SDW materials [2, 3]. In this paper, we report further investigations of the Na$_2$Ti$_2$Pn$_2$O system by synthesizing the bulk sample of the As compound, and comparing its physical properties to that of the Sb compound. For both Sb and As compounds, electrical resistivity and magnetic susceptibility were measured under various temperatures and magnetic fields. Also, temperature dependent powder neutron diffraction has been performed. The results from these experiments as well as the effect of Pn^{3-} on the physical properties of Na$_2$Ti$_2$Pn$_2$O system will be discussed.

EXPERIMENT

Bulk samples were prepared by solid state sintering of Na$_2$O, Ti and As or Sb [3]. The field and temperature dependent magnetization was measured on a commercial SQUID-magnetometer (MPMS XL, Quantum Design). The temperature dependence was measured at 10,000 Oe, and

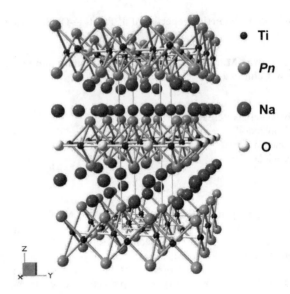

Figure 1. A perspective view of the structure of $Na_2Ti_2Pn_2O$ (Pn = As, Sb) showing alternation of $^2_\infty[Ti_{4/2}Pn_2O_{4/4}]^{2-}$ layers and double layers of Na^+.

the field dependent magnetization was measured above and below the T_c for both compounds. For all the magnetic property measurements, about 0.1 g of the sample was used, and it was sealed in a fused silica ampoule under a high vacuum in order to prevent its decomposition reaction with the air. The magnetic spin ordering and structural change of the $Na_2Ti_2Pn_2O$ system was investigated by powder neutron diffraction on the DUALSPEC diffractometer at the Chalk River Laboratories of AECL (Atomic Energy of Canada Limited). Diffraction data were acquired using $\lambda = 2.3692$ Å wavelength neutron beam. The electrical resistivity was measured by the four-probe method using the applied current of 0.10 mA.

RESULTS AND DISCUSSION

The powder X-ray diffraction peaks are indexed according to the calculated pattern based on the previously published $Na_2Ti_2Pn_2O$ structure [1]. For the As compound, all the peaks were indexed, except the two peaks at the two theta positions, 32° and 41°, arising from the presence of a small amount of an unidentified phase. Among these peaks, the peak at 32° is found to increase in intensity as the sample oxidizes. For the Sb compound, there was only one peak not indexed near 34°. When these samples are exposed to the air, they decompose to a darker gray phase and the volume of the sample increases by approximately a factor of 1.5.

Temperature dependent dc magnetism data show a drop in the susceptibility for both compounds; however, the Sb compound exhibits a sharper drop than the As compound. The As compound has the T_c^{onset} (onset transition temperature) = 330 K, and the susceptibility gradually decreases as temperature decreases down to T_c^{min} (minimum susceptibility temperature) = 195 K (Figure 2a). On the other hand, the Sb compound has the T_c^{onset} = 120 K, and the susceptibility decreases as temperature decreases down to T_c^{min} = 75 K (Figure 2b). Both compounds show no detectable difference between zero-field cooled and field cooled susceptibility, and the susceptibility decreases gradually above T_c^{onset} as temperature increases.

Field dependent magnetization at temperatures above and below the T_c of both As and Sb compounds is shown in Figure 3. For all the measurements, magnetization exhibits a linear dependence upon applied field indicating the presence of no ferromagnetic impurity in the sample. For both compounds, magnetization above T_c is about a factor of 3 larger than that below the T_c at all applied fields, and no hysteresis behavior was observed in a field loop measurement.

The powder neutron diffraction profiles are shown in Figure 4 for both As and Sb compounds. No superlattice reflections due to magnetic spin ordering or symmetry breaking were observed for either compound. Thus, these results indicate that the physical property anomaly is not likely arising from simple antiferromagnetic ordering or a spin-Peierls transition. Lattice parameters were refined by the least-squares method, and the results are summarized in the Table I. One thing noticeable from these data is that within their standard deviation, the lattice parameter a does not change above and below the T_c whereas the lattice parameter c decreases below T_c. This observation agrees with previously proposed $^2T_{2g}$ Jahn-Teller like distortion of Ti^{3+} coordination in the Sb compound [3] where the bond distance ratio of O-Ti-O/Sb-Ti-Sb increases below T_c. This distortion is commensurate, and the same symmetry is retained after the distortion.

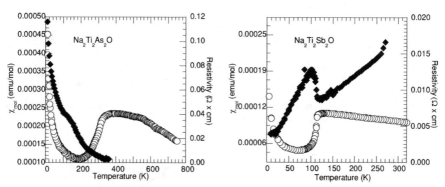

Figure 2. Temperature variation of dc magnetic susceptibility (marked as o) and electrical resistivity (marked as ♦) of (a) $Na_2Ti_2As_2O$ and (b) $Na_2Ti_2Sb_2O$.

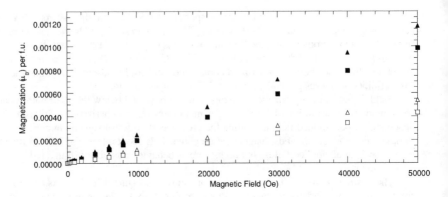

Figure 3. Magnetization versus applied field of $Na_2Ti_2As_2O$ and $Na_2Ti_2Sb_2O$ measured above and below their magnetic transition temperatures.

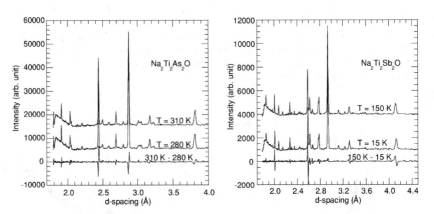

Figure 4. Powder neutron diffraction profiles above and below the T_c along with the difference of (a) $Na_2Ti_2As_2O$ and (b) $Na_2Ti_2Sb_2O$.

Table I. Lattice parameters of $Na_2Ti_2Pn_2O$ obtained from powder neutron diffraction

	a (Å)	c (Å)
$Na_2Ti_2As_2O$ at 280 K	4.0791(8)	15.316(2)
$Na_2Ti_2As_2O$ at 310 K	4.0810(9)	15.331(3)
$Na_2Ti_2Sb_2O$ at 15 K	4.161(2)	16.512(7)
$Na_2Ti_2Sb_2O$ at 150 K	4.160(2)	16.558(7)

The result of the temperature dependent electrical resistivity measurement exhibits the most pronounced difference between the As and Sb compounds. The As compound exhibits an insulator-to-insulator transition at T_c^{mid} = 135 K (inflection point)(Figure 2a). In contrast, the Sb compound exhibits a metal-to-metal transition at T_c^{mid} = 110 K with T_c^{onset} = 120K (Figure 2b), corresponding to the onset of the magnetic susceptibility anomaly. These behaviors are reminiscent of those in the quasi-two-dimensional CDW/SDW materials such as $TlMo_6O_{17}$ [4] and $P_4W_{14}O_{56}$ [5], respectively, for metal-to-metal and insulator-to-insulator transitions. Below T_c, the resistivity increases for both the As and Sb compounds.

The susceptibility decrease below T_c for both As and Sb compounds can qualitatively be explained in terms of gap opening at the Fermi surface caused by CDW/SDW instability [5, 6]. The higher susceptibility above T_c is due to Pauli susceptibility originated from conduction electrons. The gap opening at the Fermi surface due to CDW/SDW instability causes decrease in conduction electrons; thus, susceptibility decreases below T_c. Also, the increase in resistivity at the T_c for both As and Sb compounds can be explained in terms of the gap opening by CDW/SDW instability. The sample becomes insulating if the opening of the gap is complete. This is the case for the As compound. If the opening of the gap is incomplete, electrical transport retains metallic behavior with an increase in resistivity. This is the case for the Sb compound.

The property differences between the As and Sb compounds can be rationalized in terms of the dimensionality of the compounds. A CDW/SDW is known to be suppressed as the dimensionality of the system increases. In the case of the $Na_2Ti_2Pn_2O$ system, the As compound has a smaller inter-layer distance than the Sb compound. Thus, we can propose that the As compound has stronger inter-layer interaction and therefore more three-dimensional character than the Sb compound. The Sb compound has a more pronounced magnetic transition than the As compound due to its higher CDW/SDW amplitude arising from its lower dimensionality. Similarly, higher dimensionality quenches the electrical conductivity because electrical conductivity in these materials arises from the sliding motion of the CDW/SDW [7, 8]. Thus, the As compound, which has more three-dimensional character than the Sb compound, exhibits higher resistivity than the Sb compound. This CDW/SDW scenario is also supported by the previously published temperature dependent powder neutron diffraction results of the Sb compound [3]. The diffraction results show that there is a structural distortion at the same temperature as the T_c of resistivity and magnetic susceptibility providing the evidence for strong electron-phonon interaction. Furthermore, electronic structure calculation predicts that the Sb compound has a strongly "nested" box-like Fermi surface, which induces the CDW/SDW instability [9, 10]. Further characterization on single crystal sample should be performed in order to confirm the existence of the CDW/SDW.

ACKNOWLEGMENTS

We thank Warren E. Pickett for useful discussion, Robert N. Shelton for the use of the SQUID magnetometer and X-ray diffractometer, Peter Klavins for assistance with instrumentation. This research is supported by NSF DMR-9803074 and the Natural Sciences and Engineering Research Council of Canada. We also acknowledge the Neutron Program for Materials Research of the National Research Council of Canada, which operates the DUALSPEC instrument at the Chalk River Nuclear Laboratory.

REFERENCES

1. A. Adam and H.-U. Schuster, *Z. Anorg. Allg. Chem.*, **584**, 150 (1990).
2. E. A. Axtell, III, T. Ozawa, S. M. Kauzlarich and R. R. P. Singh, *J. Solid State Chem.*, **134**, 423 (1997).
3. T. C. Ozawa, R. Pantoja, E. A. Axtell III, S. M. Kauzlarich, J. E. Greedan, M. Bieringer and J. W. Richardson Jr., *J. Solid State Chem.*, **153**, 275 (2000).
4. K. V. Ramanujachary, B. T. Collins and M. Greenblatt, *Solid State Comm.*, **59**, 647 (1986).
5. M. Greenblatt, *Acc. Chem. Res.*, **29**, 219 (1996).
6. S. Kagoshima, H. Nagasawa and T. Sambongi, *One-Dimensional Conductors*, (Springer-Verlag, 1988) p.102.
7. J. Dumas, C. Schlenker, J. Markus and R. Buder, *Phys. Rev. Lett.*, **50**, 757 (1983).
8. S. Kagoshima, H. Nagasawa and T. Sambongi, *One-Dimensional Conductors*, (Springer-Verlag, 1988) p.205.
9. W. E. Pickett, *Phys. Rev. B.*, **58**, 4335 (1998).
10. F. Fabrizi de Biani, P. Alemany and E. Canadell, *Inorg. Chem.*, **37**, 5807 (1998).

Mat. Res. Soc. Symp. Proc. Vol. 658 © 2001 Materials Research Society

Fine-Tuning the Physical Properties of Perovskite Related La-Ti-Oxides by Merely Altering the Oxygen Stoichiometry

Olav Becker, Stefan Ebbinghaus, Bernd Renner and Armin Reller
Lehrstuhl für Festköperchemie, Universität Augsburg
Universitätsstrasse 1, 86135 Augsburg, Germany

ABSTRACT

The isostructural oxides $LaTiO_{3.5}$, $SrNbO_{3.5}$ and $CaNbO_{3.5}$ adopt layered structures made up of perovskite related slabs intermitted by layers of non-linking BO_6-octahedra (B=Ti, Nb). In these phases the transition metals Ti and Nb are in their highest oxidation states +IV and +V, respectively. They are insulators, in the case of the Ti phase a ferroelectric insulator with the highest known T_c (above 1700°C). By changing the oxygen stoichiometry, i.e. by controlled reduction or re-oxidation processes in the range of $2ABO_{3.5} \Leftrightarrow 2ABO_3 + ½ O_2$ (A=La or Sr, B=Ti or Nb) mixed valence phases are obtained. Accordingly, the physical properties of the different phases alter from insulating to semiconducting or to conducting. Detailed studies on the structural changes reveal, however, that the sublattice of the metal cations is basically conserved. In principle the reduction corresponds to a condensation of the perovskite layers leading to intermediate phases such as the semiconducting $LaTiO_{3.4}$, in which five TiO_6-octahedra thick perovskite slabs constitute the structural framework and 20% of the Ti^{+IV} cations are reduced to Ti^{+III}. Conductivity measurements using single crystals of the corresponding Ti and Nb phases reveal that these mixed valence oxides must be considered as one-dimensional conductors. The fully reduced phases $La^{+III}Ti^{+III}O_3$ and $Sr^{+II}Nb^{+IV}O_3$ adopt distorted perovskite structures. Extensive high resolution electron microscopic and light microscopic investigations have been carried out in order to characterize the structural mechanism of the reversible, highly topotactic reduction and re-oxidation processes. Thermogravimetric measurements in reducing and oxidizing atmospheres have been performed for the identification of the temperature ranges wherein the decisive mass changes take place. The results of the described experiments support that the properties of such metal oxides can be finely and reversibly tuned by merely changing the oxygen stoichiometries.

INTRODUCTION

Perovskite related structures are subject of interest long before High T_c superconductivity was discovered in cuprates. The possibilities of altering the corner sharing BO_6-framework and substituting the ions in the A and B position (as can be seen in Figure 1) allows one to synthesize materials with a huge variety of physical properties [1-3]. Our main subject is the alteration of physical properties by just changing the oxygen stoichiometry. Additionally the thermochemical behavior is studied by focusing on the redox mechanisms in the solid state.

It is known for a long time in the area of mineralogy that perovskite-related phases do exist over a wide range of oxygen stoichiometry (partly illustrated in Figure 2). The structure of oxygen deficient compounds is not exclusively built by BO_6-octahedra but consists of a sometimes highly ordered framework of corner sharing BO_6-octahedra, BO_5-pyramids and BO_4-tetrahedra etc. Already in the 1950s e. g. the brownmillerite structure, the mineral $Ca_2Fe^{+III}Al^{+III}O_5$, was examined by X-ray diffraction. On the other hand, in the range from ABO_3

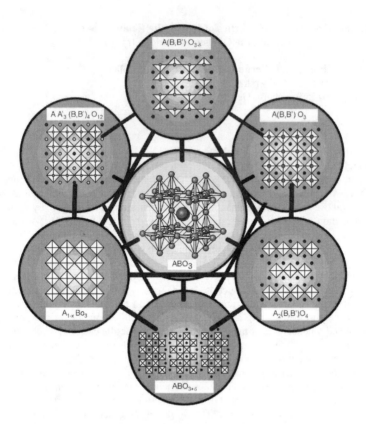

Figure 1: *Possible modifications of the ideal perovskite structure.*

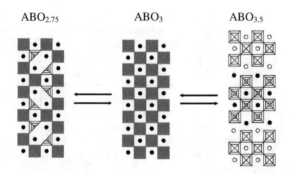

Figure 2: *Perovskite related structures as a function of oxygen content.*

to $ABO_{3.5}$ oxygen excess phases exist [4,5]. They are mainly characterized by additional oxygen layers separating perovskite slabs of different thicknesses and sequences. The basic perovskite structure is cut along the BO_6-octahedra edges (along [110]) by this extra oxygen. As shown for the system La/Ti/O macroscopic physical properties are a function of slab thickness and subsequently of the ratio of Ti^{+IV}/Ti^{+III} [6]. These properties are often anisotropic due to the crystal structure of the system and even one-dimensional systems can be stabilized.

EXPERIMENTAL

Single crystalline $LaTiO_{3+x}$-samples were prepared by zone melting crystal growth technique. Appropriate mixtures of La_2O_3, TiO_2 and TiO were ground, mixed with isopropanol to a stiff paste and formed to rods. These rods were sintered overnight at 1350°C in appropriate atmospheres, i.e. air for x=0.5 and $Ar/4\%H_2$ for x<0.5, respectively. The floating zone melting process was performed in the identical atmosphere. Details of this preparation procedure are published elsewhere [6,9]. Oxygen stoichiometries of the samples were checked by thermal analysis, namely by oxidizing the reduced phases to fully oxidized $LaTiO_{3.5}$.

RESULTS AND DISCUSSIONS

Reduction mechanisms

The system La/Ti/O has been studied extensively with respect to redox processes in the oxygen excess range between $ABO_{3.5}$ to ABO_3. It is well known that this system changes its crystal and electronic structure as a function of oxygen content. The question arose, at which oxygen contents stable phases exist during redox processes and whether one can observe the change of the crystal structures in-situ.

In the following, a structural reduction mechanism for $LaTiO_{3.5}$ is proposed. The parent material was exposed to a reducing atmosphere (4% H_2/Ar, 1400°C). In a first step an intermediate phase, i.e. $LaTiO_{3.44}$ is formed. The stabilization of this rather unusual phase can be described by the zipping mechanism depicted in Figure 3.

According to this mechanism, 2 slabs of corner sharing octahedra are zipped together by stepwise linking octahedra corners and by simultaneously removing a layer of oxygen. This type of mechanism was first observed by T. Williams et al. [7]. The emerging block of 8 corner sharing TiO_6-octahedra is subsequently divided into slabs with a thickness of 4 and 5 octahedra, which was observed in high resolution transmission electron microscope (HRTEM) as a sequence of 4-5-4-5- ... slabs of octahedra layers. Further reduction of this phase leads directly to the ideal perovskite structure $LaTiO_3$. This second step can also be described by a condensation / zipping mechanism between different slabs of TiO_6 octahedra as could be confirmed by HRTEM observations. Figure 3b shows clearly the effect of in-situ reduction of $LaTiO_{3.40}$ that was observed by removing the oxygen layer between two adjacent slabs which have been reduced before.

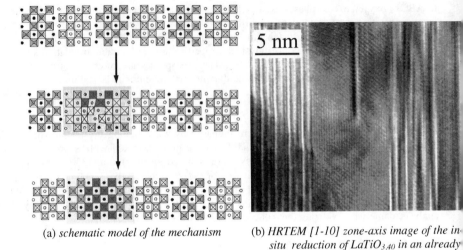

(a) *schematic model of the mechanism*

(b) *HRTEM [1-10] zone-axis image of the in situ reduction of LaTiO₃.₄₀ in an already reduced part of the sample*

Figure 3: *Zipping mechanism during the reduction of LaTiO₃₊ₓ.*

(Re-)Oxidation mechanism

In the reverse process, i.e. the oxidation of $LaTiO_3$ in air at 400°C no intermediate pha such as e.g. $LaTiO_{3.44}$ could be observed. X-ray diffraction intensities and TEM photogra gave evidence for an almost complete amorphisation of the sample. Only HRTEM stud revealed the presence of nanoscopic crystalline single phase product domains. This finding is to the non-topotactic structural transition from a three-dimensional perovskite structure to a t dimensional layered structure during oxidation. In order to obtain a detailed insight into structural mechanism of the oxidation in the range from 3.4 to 3.5 a sample of $LaTiO_{3.4}$ heated in a thermobalance and the mass change was determined as a function of temperature. can be seen in Figure 4 the weight gain is a very smooth process with no plateau denoting intermediate phase. The structure changes from the parent phase, made up of slabs contair five TiO_6-octahedra, to the product phase, made up of slabs containing four TiO_6-octahedra.

For this process we suggest a reverse mechanism as for the reduction of $LaTiO_{3.5}$. fact that no intermediate phase could be detected during the oxidation may be explained by low mobility of the cations at these remarkably low temperatures. The affinity of Ti^{3+} tow oxygen however, is the driving force for the oxidation. The details of the complete process be published elsewhere [8].

Weight gain [%]

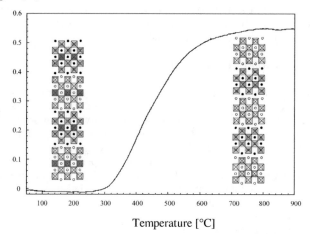

Temperature [°C]

Figure 4: Thermal analysis of the oxidation of LaTiO$_{3.4}$

FINE-TUNING OF PHYSICAL PROPERTIES

Based on the specified structural reduction and oxidation mechanisms the relation between actual structure and physical properties of these perovskite-related materials must be considered. In Figure 5 the structure of an intermediate, i.e. of the semi-conducting mixed-valence LaTiO$_{3.4}$ is depicted. It can be seen that along the c-axis the slabs made up of corner-sharing TiO$_6$-octahedra are separated. This intersection acts as insulating layer. Along the b-axis electrons have to move in a zigzag course through corner sharing oxygen octahedra. The conductivity along this direction is also very low. Only along the a-axis orbital overlap of infinite

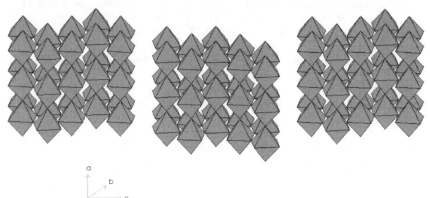

Figure 5: Octahedra structure of the pseudo 1D metal oxide conductor LaTiO$_{3.4}$.
(A-cations have been omitted for clearness)

titanium-oxygen-rows occurs, which is confirmed by a good electrical conductivity. T difference between the direction dependent conductivities is in the range of several orders magnitude (RT: $\rho_a \approx 10^{-3}\Omega$, $\rho_b \approx 10^{-1}\Omega$, $\rho_\perp \approx 10^2\Omega$ [9]). In summary the phase $LaTiO_{3.4}$ can described as a quasi one-dimensional conductor.

SUMMARY AND FUTURE WORK

It has been shown that crystal structures and physical properties of perovskite related I Ti-O- phases can be modified by solely changing the oxygen stoichiometry. The determination the mechanisms of the underlying reduction and/or oxidation processes are rather complicat However, by means of combined methods of investigation, detailed mechanistic insights may gained. In turn these insights allow the controlled generation of tailor made materials w specific properties. In these materials temperature dependent phenomena such as metal-insula and/or magnetic transitions will be studied in detail. Together with band structure calculations hope to achieve an improved concept for the description of structure-property-relations complex transition metal oxides.

ACKNOWLEDGEMENTS

The authors are indebted to the financial support of the ongoing studies by the Deuts Forschungsgemeinschaft DFG, Sonderforschungsbereich 484.

REFERENCES

1. F. S. Galasso, *Structure, Properties and Preparation of Perovskite-Type Compounds*, Oxfo Pergamon, 1986; *Perovskites and High-Tc Superconductors*, Gordon and Breach, 1990.
2. L. G. Tejuca and J. L. G. Fierro, *Properties and Applications of Perovskite-Type Oxides*, Marcel Denker, Inc., New York, 1993.
3. J.C. Grenier, M. Pouchard, and P. Hagenmuller, *Struct. Bonding*, **47**, 1 (1981).
4. M. Nanot F. Queyroux, J.-C. Gilles, A. Carpy and J. Galy, *Solid State Chem.* 11, 272 (1974
5. Gasperin, M., *Acta Cryst.*, **B31**, 2129 (1975)
6. F. Lichtenberg, D. Widmer, J.G. Bednorz, T. Williams and A. Reller, *Z. Phys. B* **82**, 211 (1991).
7. T. Williams, F. Lichtenberg, D. Widmer, J.G. Bednorz and A. Reller, *Solid State Chem.* **10.** 375 (1993).
8. O. Becker, *PhD Thesis*, University of Augsburg, Germany, 2000.
9. F. Lichtenberg, A. Herrnberger, K. Wiedemann, J. Mannhart, submitted to *Prog. Solid State Chem.*

Mat. Res. Soc. Symp. Proc. Vol. 658 © 2001 Materials Research Society

Quantum Critical Phenomena in the $NiS_{2-x}Se_x$ System

J.M. Honig
Department of Chemistry, Purdue University, West Lafayette, IN 47907-1393

ABSTRACT

Electrical conductivity experiments on $NiS_{2-x}Se_x$ (x = 0.45) subjected to moderate pressure in the 30 – 800 mK range permit the investigation of quantum critical phenomena in this system. Detailed data are presented in the context of standard theories of correlation effects in the vicinity of a critical point. The critical exponents pertaining to these effects have been evaluated; the significance of these findings require further study.

INTRODUCTION

There has been considerable interest in the electronic properties of the $NiS_{2-x}Se_x$ system because these characteristics can be changed drastically by adjustment of the S/Se ratio. Specifically, with increasing x one achieves a gradation of electrical properties leading from the insulating to a correlated metallic regime. An advantage of this system is that these changes can be brought about by isovalent substitutions in the anion sublattice, leaving the cation sublattice intact and the electron count fixed. The general properties of this system have been summarized in review articles [1,2] to which the reader is referred for details.

Here we emphasize the quantum critical fluctuations that the system displays under suitable conditions. This entails experimentation at cryogenic temperatures on a sample with x = 0.45, i.e., at a composition where the specimen is an insulator but very close to the phase boundary leading to the metallic regime. By application of moderate pressures the compound can be forced to cross the phase boundary; the degree of metallicity then encountered depends sensitively on the extent to which the applied pressure P exceeds the critical pressure range (P_c = 1.51 - 1.67 kbar) for these samples. Slightly above the critical pressure and at temperatures T < 1 K, one anticipates a display of quantum critical phenomena.

EXPERIMENTAL

The research was carried out on specimens grown at Purdue as single crystals in a Te flux, using procedures detailed in Ref. [3]. The advantage of this technique is that Te resists incorporation into the crystal lattice, in contrast to the conventional use of chemical vapor transport techniques, in which unintentional doping by the transporting agent cannot be avoided. The electrical measurements were performed by Rosenbaum and his collaborators at the University of Chicago [4,5] using a BeCu pressure cell mounted on a dilution refrigerator. Four-probe resistivity measurements were carried out with an ac bridge circuit. The relative measurements errors were below 0.01%, but the absolute values are uncertain within 25% because of the small size of the crystals (0.3 to 0.7 mm on a side) and the relatively large silver-paint contacts.

BACKGROUND INFORMATION

We first provide some important background information concerning quantum critic
fluctuations. The present discussion is necessarily very superficial; for a proper elementa
exposition the reader is referred to a very readable review [6].

Basically, a problem arises in the thermodynamic analysis of critical phenomena becau
the usual equations of state do not yield theoretical results in agreement with experiment. At t
outset one therefore resorts to a procedure that recognizes the important role of correlations ov
large distances, which become manifest near a critical point. While the methodology w
developed for thermodynamic variables, it is deemed applicable as well to transport parameters.

As implied above, the key properties of a system near its critical point involve correlatic
length scales ξ that become very large compared to atomic dimensions. Consider a variable K
interest; let its value at the critical point be K_c and define $\delta \equiv K - K_c$. The correlation length
assumed to diverge at the critical point as $\xi(K) \sim |\delta|^{-\nu}$, where $\nu > 0$ is the *correlation leng*
exponent whose value must be determined empirically. Associated with ξ there is as well
correlation time scale ξ_τ , such that $\xi_\tau \sim \xi^z$, where z is the so-called *dynamical scale exponen*
The introduction of ξ_τ reflects the fact that near the critical point long time periods are required
ensure the uniformity of properties of all parts of the system that are subject to the correlations.
z = 1, then the distance and time scales are on the same footing.

In considering ξ_τ we take note of an interesting analogy: the basic unit in statistic
thermodynamics is the partition function Z = tr exp (- βH), where $\beta \equiv k_B T$, k_B is Boltzmann
constant, H is the appropriate quantum mechanical operator that characterizes the properties
interest, and tr represents the trace operation. The quantity exp (-βH) closely resembles the tin
development of H in the Heisenberg operator formalism, exp (-iHτ/$\hbar$), where $\hbar \equiv$ h/2π, and h
Planck's constant. Such a juxtaposition suggests the establishment of the correspondence $\tau \leftrightarrow$
- i$\hbar\beta$. In other words, for our purposes time τ has as its analog an imaginary variable proportion
to 1/T. This seemingly formal identification has been properly substantiated in the literatur
Note that for T = 0, τ has no upper bound, whereas for T > 0 a finite upper bound on τ exists.

The analysis of quantum critical phenomena proceeds as follows: consider a quantu
mechanical operator $\hat{O}(k,\omega,K)$, representing an observable that depends on wave vector
frequency ω, and the experimental variable K. Then at T = 0 the following scaling law is assume
to hold:

$$\hat{O}(k,\omega,K) = \xi^{x_o} \hat{O}(k\xi,\omega\xi_\tau) \tag{1}$$

Here the variable K no longer appears explicitly on the right, but is subsumed through ξ; th
operator $\hat{O}$ on the right involves only dimensionless variables and is not of any immediate physic
interest. Eq. (1) clearly displays the important role played by correlation distances in criticality.

Where T > 0 one must be aware of the upper limit set on τ , namely the cutoff value L_τ
$\hbar\beta < \infty$. This fact reflects on ξ as well, since by the earlier argument one now has $L_\tau \sim \xi^z$, or
$\sim L_\tau^{1/z}$. We therefore replace Eq. (1) by

$$\hat{O}(k,\omega,K,T) = L_\tau^{x_o/z} \hat{O}(kL_\tau^{1/z},\omega L_\tau,L_\tau/\xi_\tau) \qquad (T > 0), \tag{2}$$

in which $L_\tau^{1/z}$ and L_τ are the fixed distance and time time scales to which k and ω are compare
and L_τ/ξ_τ measures the 'separation distance' from the case T = 0. Once again, the operator o

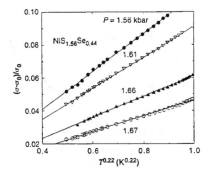

Figure 1. Electrical resistivity of Sample vs Temperature Slightly Above the Critical Pressure

the right involves dimensionless quantities and is of no direct interest.

This discussion provides the framework for the interpretation of the data.

EXPERIMENTAL MEASUREMENTS

Electrical resistivity (ρ) measurements were first carried out in the range $5 < T < 15$ K at pressures 3.5 and 5.0 kbar, well above P_c. Here ρ was found to be strictly proportional to T^2. This accords with the Baber-Pomeranchuk theory: electron correlation effects are predicted to give rise to this power law. The results are in keeping with the fact that at these pressures the $NiS_{2-x}Se_x$ system is barely metallic. As the temperature is lowered to the range $0.1 < T < 1$ K and pressures are reduced to the range 1.9 to 2.8 kbar, the conductivity ($\sigma \equiv 1/\rho$) obeys the power law $\sigma = T^{1/2}$. This is a regime where electron-electron interactions are dominated by disorder in the lattice. Fig. 1 shows what happens when the temperature is reduced to the range $35 < T < 800$ mK and the applied pressure approaches P_c. Here the influence of the critical point becomes apparent: now $\sigma \sim T^{0.22}$. In the present case Eq. (2) is applicable; thus, we set $x_\sigma/z = 0.22 \pm 0.02$.

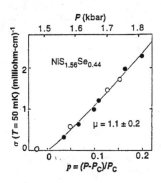

Figure 2. Conductivity of Sample at 50 mK as a Function of Applied Pressure p

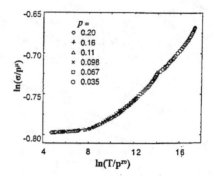

Figure 3. Collapse of Data onto a Single Curve; See Text for Details

Measurements in Fig. 2 show the variation of electrical conductivity with pressure at T = 50 mK for two samples of slightly different Se content. These data indicate that $\sigma \sim p^\mu$, where $p \equiv (P - P_c)/P_c$ and $\mu = 1.1 \pm 0.2$. T is deemed sufficiently low to render Eq. (1) applicable. We define $\mu \equiv x_\sigma \upsilon$; then, with $x_\sigma/z = \mu/z\upsilon$ together with the data from Fig. 1, one finds that $\upsilon z = 4.6$.

A fairly stringent test for T > 0 of the validity of the present approach involves taking the ratio

$$\sigma/p^\mu \sim T^{x_\sigma/z}/p^{x_\sigma \upsilon} \sim (T^{1/z}/p^\upsilon)^{x_\sigma}. \tag{3}$$

On taking logarithms one finds that a plot of $\ln (\sigma/p^\mu)$ vs. $\ln (T/p^{z\upsilon})$ should yield a universal curve. Indeed, as Fig. 3 shows, all of the data do collapse onto a single plot, thus validating the applicability of standard scaling theories.

It is possible to deconvolute z from υ by performing an additional experiment on the electrical conductivity in the nonohmic range. Consider carriers of charge e subjected to an external electric field E, whose response is limited to a mean free path l_ε. This quantity is regarded as the dominant length scale ξ of the experiment near the critical point; it has the associated time scale $l_\varepsilon^{\ z}$. The energy transferred to each carrier is $\Delta E = eEl_\varepsilon$. On account of the uncertainty principle it is reasonable to associate with this energy increment the quantity $\Delta E = \hbar/\xi_\tau \sim l_\varepsilon^{-z}$, so that $l_\varepsilon \sim E^{-1/(1+z)}$. We also define an effective temperature via $k_B T_{eff}$ which represents the electronic conductivity determined in the ohmic regime. Thus, near the quantum critical point,

$$T_{eff} \sim El_\varepsilon \sim E^{z/(1+z)}. \tag{4}$$

A plot of log T_{eff} vs log E should then lead to a single curve whose slope is given by $z/(z + 1)$. The experimental results, taken at pressure P = 1.56 kbar and at temperatures 350 to 800 mK are displayed in Fig. 4. Indeed, outside the ohmic range all the experimental points merge onto a single straight line of slope 0.73; very similar results were obtained at other pressures [4]. One thus finds that z = 2.7 ± 0.35, whence υ = 1.7 ± 0.25, and $x_\sigma = 0.64 \pm 0.13$.

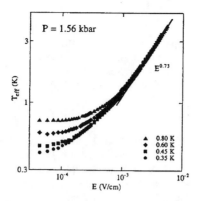

Figure 4. Resistivity Measurements in Nonohmic Region; Data Merge at High Electric Fields. See Text for Details.

The significance of these exponents has been the subject of some discussion [7-10]. On one hand the value of $\mu = 1$ is consonant with values encountered in four distantly related systems [7]. On the other hand, the product $\nu z = 4.6$ falls well outside the range $2 - 4$ in these systems, whose values also depend on the choice of tuning parameters for inducing a quantum phase transformation. Also, the hyperscaling relation $\mu = (d - 2)\nu$ is not properly obeyed in the present case. Schakel [10] cites values for z in the range 0.8 to 1 and ν values in the range 1.4 to 2.3 for materials undergoing transitions to the insulating state from superconducting, Hall liquid, or conducting phases. Clearly, the z value cited for $NiS_{2-x}Se_x$ falls well outside the normal range. Phillips [8] traces the deviations from expected values to an equipartitioning of the thermal activation energy between orbital and spin degrees of freedom.

CONCLUSIONS

It has been shown that quantum critical fluctuations can be induced in the $NiS_{2-x}Se_x$ system by investigating the alloy with x = 0.45 that is very close to the insulator-metal boundary. The passage from the insulating to the metallic regime at very low temperatures was effected at pressures barely exceeding the critical value required to reach the highly correlated metallic state. The three exponents characterizing the electronic transport properties in the critical regimes were determined. At present there is no clear consensus as to the proper interpretation of the experimental results. Further work on this system is necessary for an understanding of the physical phenomena in this regime.

ACKNOWLEDGMENTS

The author wishes to thank Professor Rosenbaum and his group for their expert accumulation of experimental data, and for many fruitful discussions of the results. This research was supported by NSF Grant 9612130 DMR

REFERENCES

1. J.M. Honig and J. Spałek,, Chem. Mater. **10**, 2910 (1998).
2. J.M. Honig, Acta Phys. Polon. B, **31**, 2857 (2000).
3. X. Yao and J.M. Honig, Mater. Res. Bull. **29**, 709 (1994).
4. A. Husmann, D.S. Jin, Y.V. Zastavker, T.F. Rosenbaum, X. Yao, J. M. Honig, Scie **274**, 1874 (1996).
5. A. Husmann, J. Brooke, T.F. Rosenbaum, X. Yao, J.M. Honig, Phys. Rev. Lett. **84**, 2 (2000).
6. S.L. Sondhi, S.M. Girvin, J.P. Carini, D. Shahar, Rev. Mod. Phys. **69**, 315 (1997).
7. E. Abrahams and G. Kotliar, Science **274**, 1853 (1996).
8. J.C. Phillips, Proc. Natl. Acad. Sci. USA **94**, 10532 (1997).
9. A.M.J. Schakel in *Correlations, Coherence, and Order*, edited by D.N. Shopova and ꓲ Uzunov (Plenum, New York, 1999), p. 295.
10. A.M.J. Schakel, J. Phys. Studies **3**, 337 (1999).

Slater Transition in the Pyrochlore $Cd_2Os_2O_7$

D. Mandrus, J. R. Thompson, and L. M. Woods
Solid State Division, Oak Ridge National Laboratory, Oak Ridge, TN 37831;
and Department of Physics, The University of Tennessee, Knoxville, TN 37996

ABSTRACT

$Cd_2Os_2O_7$ crystallizes in the pyrochlore structure and undergoes a metal-insulator transition (MIT) near 226 K. Here we present resistivity, heat capacity, and magnetization results on $Cd_2Os_2O_7$. Both single crystal and polycrystalline material were examined. We also present LAPW electronic structure calculations on $Cd_2Os_2O_7$. We interpret the results in terms of a Slater transition. In this scenario, the MIT is produced by a doubling of the unit cell due to the establishment of antiferromagnetic order. A Slater transition--unlike a Mott transition--is predicted to be continuous, with a semiconducting energy gap opening much like a BCS gap as the material is cooled below T_{MIT}.

INTRODUCTION

The synthesis and initial characterization of $Cd_2Os_2O_7$ were reported in 1974 by Sleight, *et al.* [1], but no subsequent publications have appeared on this compound. The physical properties reported in Ref. 1 are quite intriguing: $Cd_2Os_2O_7$ was found to crystallize in the pyrochlore structure and to undergo a continuous, purely electronic metal-insulator transition (MIT) near 225 K. It was also found that the MIT was coincident with a magnetic transition that the authors characterized as antiferromagnetic.

The idea that antiferromagnetic ordering can double the unit cell and for a half-filled band produce a metal-insulator transition goes back to Slater, who in 1951 proposed a split-band model of antiferromagnetism [2]. In this model, the exchange field favors up spins on one sublattice and down spins on another sublattice and for large U/W reduces to the atomic model of an antiferromagnetic insulator, with a local moment on each site. The thermodynamics of the metal-insulator transition were worked out in mean-field approximation by Matsubara and Yokota (1954) [3], and by Des Cloizeaux (1959) [4]. In both treatments a continuous metal-insulator transition was predicted, with a semiconducting gap opening much the way a BCS gap opens in a superconductor.

These results have, of course, been largely supplanted by modern SDW theory [5], but it must be remembered that SDW theory is unambiguously effective only in the weak coupling limit. For example, at half filling it should be possible to proceed smoothly from a weak coupling SDW state to a Mott insulating state as U/W is increased. Although strong coupling SDW theory gets some aspects of the Mott insulating state right, it is wrong in other respects and must be regarded as an incomplete description [5].

One of the major failings of strong coupling SDW theory is that the gap is predicted to disappear above the Neel temperature, whereas in true Mott insulators like CoO the gap persists despite the loss of long range magnetic order. In such materials a Mott-Hubbard description is clearly correct, but when the metal-insulator transition temperature and the Neel temperature coincide one should give careful thought as to whether a Mott-Hubbard or a Slater description is most appropriate. In such cases the thermodynamics of the MIT can provide valuable information about the underlying mechanism, because a Mott transition is expected to be discontinuous (first-order) whereas a Slater transition should be continuous [6].

EXPERIMENTAL DETAILS

Both single crystals and polycrystalline material were prepared. All preparations were performed under a fume hood because of the extreme toxicity of OsO_4. This oxide of osmium extremely volatile and melts at about 40°C. Exposure of the eyes to OsO_4 should be strictly avoided as permanent blindness can result. Even brief exposure to the vapor is dangerous. According to the MSDS: "If eyes are exposed to the vapor over a short period of time, night vision will be affected for about one evening. One will notice colored halos around lights."

Single crystals of $Cd_2Os_2O_7$ were grown following a method similar to Ref. 1 by sealing appropriate quantities of CdO, Os, and $KClO_3$ in a silica tube and heating at 800°C for 1 week. Shiny black octahedral crystals, up to 0.7 mm on an edge, grew on the walls of the tube. These crystals presumably grew from the vapor with OsO_4 as a transport agent. Crystals from several growths were used in the experiments reported here.

Polycrystalline material was synthesized from CdO and OsO_2 powders. These material were ground together thoroughly, pressed into pellets, and sealed in a silica tubes to which an appropriate amount of $KClO_3$ was added to provide the required oxygen. The tubes were then heated at 800 °C for several days. Dense polycrystalline material was produced using this method.

Resistivity measurements were performed using a standard linear 4-probe method. Silver epoxy and 25 μm Pt wires were used to make contacts to the samples. Heat capacity measurements were performed using a commercial heat-pulse calorimeter manufactured by Quantum Design. Magnetization measurements were performed in a SQUID magnetometer manufactured by Quantum Design.

The full-potential linearized augmented plane wave (LAPW) method in the WEIN97 code was used for the electronic structure calculations [7]. In this method the unit cell is partitioned into spheres (with a muffin-tin radius R_{mt}) centered on the atomic positions. The plane wave cutoff $R_{mt}K_{max} = 7.0$ and we used 56 k points in the irreducible wedge of the Brillouin zone.

ELECTRONIC STRUCTURE

Several pyrochlores with the formula $Cd_2M_2O_7$ are known where M^{5+} = Nb, Ta, Re, Ru, and C [1][8]. Insulating behavior is found for Nb^{5+} ($4d^0$) and Ta^{5+} ($5d^0$) as would be expected for a compound with no d electrons in its t_{2g} manifold. Metallic behavior is observed when M^{5+} is Re^{5+} ($5d^2$)[9] and Ru^{5+} ($4d^3$) [10], indicating that the t_{2g} manifold of d electrons is intimately involved in the conduction processes in $Cd_2M_2O_7$. Because Os^{5+} is in the $5d^3$ configuration and the t_{2g} manifold holds 6 electrons, it is likely that the conduction band in $Cd_2Os_2O_7$ is close to half filling. This condition, of course, is required for a Slater picture of the MIT to be valid.

Electronic structure calculations bear out this expectation. In Figure 1 we plot the total electronic density of states (DOS) of $Cd_2Os_2O_7$. These states have primarily Os t_{2g} character as expected, but there is considerable O $2p$ character as well, indicating the strong hybridization of the Os d electrons with their coordinating anions. The calculated DOS indicates that $Cd_2Os_2O_7$ is a metal. There is a sharp peak at the Fermi energy that reaches $N(E_F) = 25.4$ states/eV,

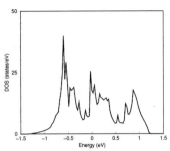

Figure 1: *Total DOS of $Cd_2Os_2O_7$ near E_F. The DOS is per unit cell, and the primitive cell used in the calculation contains 2 formula units.*

implying a Sommerfeld coefficient $\gamma = 29.9$ mJ/mol-K^2. This value is close to the Sommerfeld coefficient measured the related material $Cd_2Re_2O_7$ ($\gamma = 25$ mJ/mol-K^2) [8], and is also consistent with the value inferred from the analysis of the specific heat presented below.

RESISTIVITY

The resistivity of a single crystal of $Cd_2Os_2O_7$ appears in Figure 2. There was no indication of thermal hysteresis in the resistivity of either the single crystals or the polycrystalline material, consistent with expectations for a continuous phase transition. In a Slater picture of an MIT, we expect the resistivity to follow an expression of the form $\rho = \rho_0 exp(\Delta/T)$ as has been found for quasi-$1d$ SDW materials [11]. We also expect that the temperature dependence of a Slater insulating gap will behave in much the same way as a BCS gap [2,3]. In Figure 3 we plot the temperature dependence of the activation energy, Δ, calculated from the experimental data using $\Delta = T \, ln(\rho/\rho_0)$. Here ρ_0 is the resistivity just above the transition. Also in Figure 3 we plot a BCS gap function ($\Delta = 750$ K) for comparison. As is clear from the Figure, the resistivity of $Cd_2Os_2O_7$ is consistent with this picture at least for temperatures close to the transition. At lower temperatures the influence of extrinsic conduction mechanisms becomes increasingly important and masks the intrinsic behavior. This occurs at a higher temperature in the polycrystalline material as compared to the single crystal, consistent with the greater impurity concentration in the polycrystalline material. Assuming that $\Delta \approx 750$ K, we find that $2\Delta \approx 6.6 \, k_BT_C$. This is considerably higher than the $2\Delta \approx 3.5 \, k_BT_C$ predicted from weak coupling SDW theory, but, interestingly, quite similar to the $2\Delta/k_BT_C$ values found in cuprate superconductors.

MAGNETIZATION

In Figure 4 we plot M/H vs. T (ZFC, H = 20 kOe) for single crystal and polycrystalline samples and for comparison we plot the magnetic susceptibility data obtained on single crystals from Ref. 1. The single crystal results are in good quantitative agreement. The magnetic response of the polycrystalline material is somewhat weaker, but qualitatively the agreement is good. According

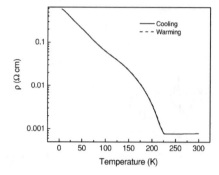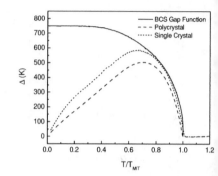

Figure 2 (left): *Resistivity of a single crystal of Cd₂Os₂O₇.*

Figure 3 (right): *Activation energy vs. temperature calculated for a single crystal and a polycrystalline sample Cd₂Os₂O₇. The activation energy was calculated by assuming Δ=exp(ρ/ρ₀) as described in the text. A BCS gap function, Δ(0) = 750 K, is also plotted for comparison.*

to Ref. 1, the data from 226-750 K do not obey a Curie-Weiss law. If, as we concluded from the specific heat analysis, the system above T_{MIT} is a moderately correlated metal, we do not expect the magnetic response to obey a Curie-Weiss law. What we expect is an exchange-enhanced Pauli paramagnetic response, and the data are fully consistent with that.

It is difficult to imagine that $Cd_2Os_2O_7$ is simply a local-moment antiferromagnet. In the first place, magnetically ordered Os compounds are extremely rare and $4d/5d$ materials in general tend to have low ordering temperatures. Secondly, the pyrochlore lattice is known to be geometrically frustrated, and for antiferromagnetic nearest-neighbor interactions no long range order is predicted in the absence of further-neighbor interactions [12]. In fact, antiferromagnetism is quite rare in pyrochlores [13]. Given all this, a Neel temperature of 226 is remarkably high and demands explanation.

SPECIFIC HEAT

Powder X-ray diffraction measurements in Ref. 1, and single-crystal refinements performed us above and below the MIT indicate no change in the symmetry or volume of the unit cell across the MIT. Therefore we associate the anomaly in the specific heat with the electronic, rather than the lattice, degrees of freedom. The estimated electronic contribution to the specific heat is plotted in Figure 5. This estimate was obtained by first *assuming* a Sommerfeld coefficient above T_{MIT} and subtracting off the assumed electronic contribution. Then a smooth polynomial was fitted to the data outside the region of the anomaly as an estimate of the lattice contribution. Then the lattice contribution was subtracted from the raw data. Sommerfeld coefficients between 0 and 30 mJ/mol-K² were explored, but the entropy analysis described below is relatively insensitive to the precise value of the assumed γ. A γ of 20 mJ/mol-K² was used to produce Figure 4. This value is in the range of the Sommerfeld coefficients reported o the metallic pyrochlores mentioned above and is meant to provide a reasonable estimate of the entropy of the itinerant electrons at temperatures above the MIT.

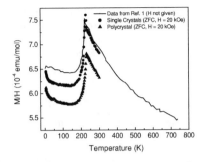 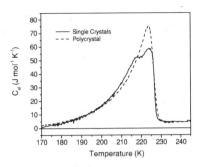

Figure 4 (left): *Effective susceptibility (M/H) obtained on single crystals and polycrystalline $Cd_2Os_2O_7$ under zero field cooled conditions with an applied field of 20 kOe. Data fom Ref. 1 obtained under unspecified conditions are also plotted.*

Figure 5 (right): *The electronic portion of the specific heat vs. temperature of $Cd_2Os_2O_7$ obtained as described in the text.*

Let us now consider how electronic entropy is eliminated as T → 0. A system of localized d electrons has a large electronic entropy of order $k_B ln(N)$ per magnetic ion, where N is the degeneracy of the ground state of the atomic d electrons subject to Hund's rule and crystal field interactions. In this case, entropy is typically eliminated as T→ 0 by a transition to a magnetically ordered state. In a system of itinerant electrons, on the other hand, the specific heat is given by C = γT, and the Pauli principle ensures that the entropy vanishes as T → 0.

In $Cd_2Os_2O_7$, we can estimate the entropy associated with the MIT by integrating C_{el}/T from 170 K to 230 K. Depending on the assumed value of γ above T_{MIT}, the answer ranges (in single crystals) from S_{MIT} = 5.4 J/mol-K (γ = 0 mJ/mol-K^2) to S_{MIT} = 6.6 J/mol-K (γ = 30 mJ/mol-K^2) and (in polycrystals) from S_{MIT} = 5.6 J/mol-K (γ = 0 mJ/mol-K^2) to S_{MIT} = 6.8 J/mol-K (γ = 0 mJ/mol-K^2). For an assumed γ of 20 mJ/mol-K^2 we have S_{MIT} = 6.2 J/mol-K (single crystal) and S_{MIT} = 6.5 J/mol-K (polycrystal). Localized Os^{5+} ($5d^3$) ions are expected to have spin-3/2 and to eliminate $2Rln(4)$ = 23.0 J/mol-K via a magnetic transition. This is clearly much higher than the observed value, and suggests that the transition does not involve ordering of localized $5d$ moments. Even if we assume that spin-orbit coupling breaks the degeneracy of the t_{2g} manifold making the ions effectively spin-1/2, we still expect an entropy of $2Rln(2)$ = 11.5 J/mol-K which is again significantly higher than the observed value. It seems more reasonable to identify S_{MIT} with the entropy of an itinerant electron system above the MIT. If we make this association, S(230 K) = 4.6 J/mol-K for γ = 20 mJ/mol-K^2 and S(230 K) = 6.9 J/mol-K for γ = 30 mJ/mol-K^2. These numbers are much closer to the observed values and support the notion that the entropy associated with the MIT in $Cd_2Os_2O_7$ is simply that of the itinerant electron system above T_{MIT}. This picture is consistent with what we expect from a Slater transition.

CONCLUSION

In this work we have characterized the metal-insulator transition in $Cd_2Os_2O_7$ using a variety of experimental techniques and have argued that a coherent picture of the transition emerges if the data are interpreted in terms of a Slater transition. Although the possibility of a

Slater transition has occasionally been mentioned in the study of metal-insulator transitions (e V_2O_3 under pressure [14-15]), $Cd_2Os_2O_7$ appears to be the first well-documented example of a pure Slater transition and therefore merits further study. Particularly needed are theoretical studies in the intermediate coupling regime ($U/W \approx 1$). This is the regime that characterizes m $4d/5d$ materials and promises to yield many intriguing discoveries in the years ahead.

REFERENCES

1. A. W. Sleight, J. L. Gillson, J. F. Weiher, and W. Bindloss, *Solid State Comm.* **14**, 357 (1974).
2. J. C. Slater, *Phys. Rev.* **82**, 538 (1951).
3. T. Matsubara and Y. Yokota, in *Proc. Int. Conf. Theor. Phys., Kyoto-Tokyo 1953* (Sci. Council Japan, Tokyo, 1954), p. 693.
4. J. Des Cloizeaux, *J. Phys. Radium*, Paris **20**, 606 (1959).
5. P. Fazekas, *Lecture Notes on Electron Correlation and Magnetism* (World Scientific, Singapore, 1999).
6. N. F. Mott, *Metal-Insulator Transitions* (Taylor and Francis, London, 1990).
7. P. Blaha, K. Schwarz, P. Sorantin, and S. B. Trickey, *Comput. Phys. Commun.* **49**, 399 (1990).
8. K. Blacklock and H. W. White, *J. Chem. Phys.* **71**, 5287 (1979).
9. P. C. Donohue, J. M. Longo, R. D. Rosenstein, and L. Katz, *Inorg. Chem.* **4**, 1152 (1965).
10. R. Wang and A. W. Sleight, *Mater. Res. Bull.* **33**, 1005 (1998).
11. G. Gruner, *Rev. Mod. Physics* **66**, 1 (1994).
12. J. N. Reimers, A. J. Berlinsky, and A.-C. Shi, *Phys. Rev. B* **43**, 865 (1991).
13. M. A. Subramanian, G. Aravamudan, and G. V. S. Rao, *Prog. Solid State Chem.* **15**, 55 (1983).
14. S. A. Carter, T. F. Rosenbaum, M. Lu, H. M. Jaeger, P. Metcalf, J. M. Honig, and J. Spale *Phys. Rev. B* **49**, 7898 (1994).
15. S. A. Carter, J. Yang, T. F. Rosenbaum, J. Spalek, and J. M. Honig, *Phys. Rev. B* **43**, 607 (1991).

Mat. Res. Soc. Symp. Proc. Vol. 658 © 2001 Materials Research Society

Synthesis, Structure and Magnetic Properties of Monoclinic $Nb_{12}O_{29}$

J. E. L. Waldron[1] and M. A. Green[1,2,*]

1. Royal Institution of Great Britain, 21, Albemarle Street, London. W1X 4BS. U. K.

2. University College London, 20 Gordon Street, London. WC1H 0AJ U. K.

ABSTRACT

The synthesis, structure and magnetic properties of monoclinic $Nb_{12}O_{29}$ are described. The synthesis of a pure bulk sample is difficult due to the large number of other similar phases. It is achieved by rapid reduction of $H-Nb_2O_5$ with Nb metal. The compound is shown to undergo a charge ordering transition at low temperature which provokes long range magnetic order in an intriguing one dimensional arrangement.

INTRODUCTION

Reduced d^0 oxides of early transition metal oxides often eliminate point defects by the formation of crystallographic shear planes, which creates a wide range of structures over a small range of oxidation states. $Nb_{12}O_{29}$ has a crystallographic shear structure composed of (3 x 4 x ∞) blocks of corner shared NbO_6 octahedra. These blocks can be connected to one another by edge sharing octahedra in two ways producing a monoclinic or an orthorhombic polymorph.[1] Extensive electron microscopy work has been done on $Nb_{12}O_{29}$, and other crystallographic shear structures, giving detailed information about their defects and stacking faults.[2,3] In contrast, very little is known about their electronic properties. This is primarily due to the difficulty in producing bulk samples of sufficient quality and quantity for measurement. Monoclinic $Nb_{12}O_{29}$ is particularly difficult to isolate as the orthorhombic polymorph seems to be energetically more favourable. Indeed, the only bulk property measurement reported on the monoclinic polymorph is a X-ray diffraction study performed on a twinned crystal.[1] Orthorhombic $Nb_{12}O_{29}$ has been reported to be metallic down to low temperature as well as containing localised electrons that show anti-ferromagnetic ordering at 12 K.[4] Here, we describe the synthesis in detail of a bulk sample (>5 g) of monoclinic $Nb_{12}O_{29}$ by rapid reduction of $H-Nb_2O_5$ using Nb metal and briefly describe its structure and electronic properties.

EXPERIMENTAL

H-Nb$_2$O$_5$ was formed by heating the commercially available Nb$_2$O$_5$ for 12 hours at 1100°C. Stoichiometric amounts of H-Nb$_2$O$_5$ and Nb metal were ground, pelletised and sealed in a quartz tube evacuated to 5 x 10^{-6} Torr. The reaction time and temperature required is dependent on the sample mass and is described below. Powder X-ray diffraction was carried out with a Siemens D500 diffractometer equipped with a primary monochromator giving Cu Kα1 radiation (λ=1.54056 Å). Magnetic susceptibility measurements were performed with a Quantum Design MPMS7 magnetometer. Conductivity measurements were performed with a Oxford Instruments Maglab 2000 system.

SYNTHESIS

Commercially available Nb$_2$O$_5$ contains a large number of polymorphs. H-Nb$_2$O$_5$ is the high temperature phase and it was obvious that formation of Nb$_{12}$O$_{29}$ was aided by its presence. This can be rationalised on inspection of the structure of H-Nb$_2$O$_5$ which is shown in Figure 1 (a). It is itself a crystallographic shear structure and can be derived from the ReO$_3$ structure through two nearly orthogonal crystallographic shears along the $(20\bar{9})$ and (601) axes. This results in a (3 x 4 x ∞) block arrangement connected via NbO$_4$ tetrahedra. A comparison of this structure with that of monoclinic Nb$_{12}$O$_{29}$ (shown in Figure 1 b) reveals a number of common features such as the presence of (3 x 4) blocks in a similar diagonal arrangement.

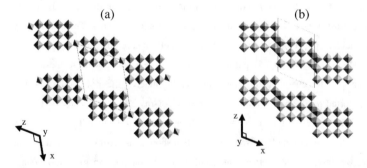

Figure 1 *(a) Structure of H-Nb$_2$O$_5$ and (b) structure of monoclinc Nb$_{12}$O$_{29}$.*

It was found that a direct reaction of H-Nb$_2$O$_5$ with Nb led to rapid formation of Nb$_{12}$O$_{29}$. However, the purity of the final product is highly dependent on the reaction time, temperature, grinding time of the starting materials and sample size. For lower temperatures (<1200 °C), the reaction does not go to completion even with long heating times and remnant amounts of Nb_2O_5 remained, shown in Figure 2a. For higher temperatures (>1300 °C), $Nb_{12}O_{29}$ is rapidly formed; any further heating causes a structural distortion which is manifested as a large increase in [h00] reflections, shown in Figure 2b, which is presumably caused by the creation of stacking faults.

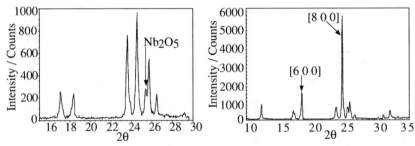

Figure 2 *Examples of the X-ray diffraction pattern of an under reacted (left) and an over reacted sample (right).*

A balance of reaction time and temperature between complete reaction of Nb_2O_5 and over reaction of the $Nb_{12}O_{29}$, was found by performing a series of reactions. One gram of starting material was mixed and ground for 10 minutes to give a homogeneous mixture and to standardise the reaction. This mixture was pelletised and heated at different combinations of temperature and reaction time. The purity of the sample was tested by Rietveld refinement of powder X-ray diffraction data. The atomic positions were constrained to the values reported by Norin[1] and only the lattice parameters, background function, peak shape parameters, zero point and scale factors were allowed to refine. The reaction was considered to have gone to completion when no $H-Nb_2O_5$ was detected in the X-ray diffraction pattern. The X-ray diffraction pattern of $H-Nb_2O_5$ is very similar to that of monoclinic $Nb_{12}O_{29}$, and therefore, impossible to detect without the use of Rietveld refinement. The most obvious indicator of remnant $H-Nb_2O_5$ is seen as a small peak at $2\theta \approx 25.5$ °. Visual inspection and the goodness of fit factors from the Rietveld refinement were used as indicators of the purity of the samples without remnant $H-Nb_2O_5$. In particular, the ratio of the intensities of the [300] and [400] reflections was found to be extremely sensitive to the reaction conditions.

Around 100 combinations of conditions were attempted and the optimum conditions for making a 1 g sample of monoclinic $Nb_{12}O_{29}$ were found to be 1200 °C for 30 minutes.

Larger samples synthesised under the same conditions were found to still contain H-Nb_2O_5 as an impurity. A series of reactions, as described for the 1 g samples, were performed for 5 g of starting material. It was found that the temperature and heating time must be increased to 1300 °C and 45 minutes respectively to obtain similar high purity samples.

STRUCTURE

The structure was solved by powder neutron diffraction, which will be described in detail elsewhere. It was found to be isostructural at high temperature with $Ti_2Nb_{10}O_{29}$, crystallising in the A2/m space group, but undergoes a charge ordering transition at low temperature, which lowers the symmetry to the space group Am. This occurs due to the very different geometry requirements of a d^0 and d^1 ion. Ions with the d^0 electronic configuration are susceptible to an out-of-centre distortion, which creates one long, one short and four roughly equal bonds within an octahedral unit, such as in Nb_2O_5, which displays bond lengths of 1.983Å, 1.949Å, 2.001Å, 1.983Å and one at 1.800Å and one at 2.270Å. In contrast, ions with a d^1 electronic configuration do not undergo the same distortion and all 6 bonds in an octahedra have very similar bond distances. At low temperature, a structure containing a mixture of d^0 and d^1 ions can undergo a charge ordering transition that locates the different ions onto crystallographically distinct sites. This can have important consequences on the electronic properties as one of these ions contains an unpaired spin, the other is diamagnetic, thus unusual magnetic or electrical properties can be formed. In the case of monoclinic $Nb_{12}O_{29}$, Rietveld refinement of the neutron data clearly indicates that the d^1 ions order on a central corner shared octahedra, as shown in Figure 3.

Figure 3 *Low temperature structure of $Nb_{12}O_{29}$ showing the Nb^{4+} (dark shaded) octahedra ordered on specific sites and surrounded by Nb^{5+} (lightly shaded) octahedra, forming one dimensional chains.*

ELECTRONIC PROPERTIES

The magnetic susceptibility and electrical conductivity measurements of monoclinic $Nb_{12}O_{29}$ (see Figure 4) show anti-ferromagnetic ordering at 12 K along with metallic behaviour over the entire temperature range. $Nb_{12}O_{29}$ can be written as $Nb^{4+}_2Nb^{5+}_{10}O_{29}$, therefore, the observation of long range magnetic order is a surprising result for such a magnetically dilute system.

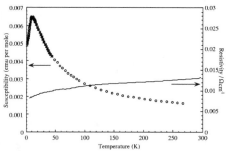

Figure 4 *Magnetic susceptibility and electrical conductivity of monoclinic $Nb_{12}O_{29}$.*

The above result can be rationalised by considering the charge ordering at low temperature, which was observed from the powder neutron diffraction data. The ordering of an unpaired electron onto a specific site forms chains along the b axis (see Figure 3). The magnetic order can then occur in an unusual one dimensional arrangement.[5]

This work was funded by the EPSRC. NIST and ILL are thanked for provision of powder neutron diffraction beam time.

REFERENCES

1. R. Norin Acta *Chem. Scand.* **20**, 871, (1966).

2. S. Iijima, S. Kimura and M. Goto *Acta Cryst.* **A29**, 632, (1973).

3. M. J. Sayagues and J. L. Hutchinson *J. Solid State Chem.* **124**, 116, (1996).

4. R. J. Cava, B. Batlogg, J. J. Krajewski, P. Gammel, H. F. Poulson, W. F. Peck and L. W. Rupp *Nature* **350**, 598, (1991).

5. J. E. L. Waldron, M. A. Green and D. A. Neumann, *J. Amer. Chem. Soc.* (in press, 2001)

Mat. Res. Soc. Symp. Proc. Vol. 658 © 2001 Materials Research Society

Transport Properties And Crystal Chemistry Of Ba-Sr-Bi Oxides

Oya A. Gökçen, James K. Meen, Allan J. Jacobson and Don Elthon

Department of Chemistry, Texas Center for Superconductivity, and Materials Research Science and Engineering Center, University of Houston, Houston, Texas 77204, U.S.A.

ABSTRACT

Rhombohedral (R) phases of the binary systems AE-Bi-O (AE = Ca, Sr, Ba) have low temperature (β_2) and high temperature (β_1) polymorphs, and the β_1 polymorph is a good oxygen ion conductor. In a previous study, we showed that in the ternary system Ba-Sr-Bi-O, conductivities of polycrystalline R samples are unaffected by Ba:Sr, whereas temperature ranges of the $\beta_2 \leftrightarrow \beta_1$ polymorphic transition are. Recently, we proved that the conductivity of both polycrystalline and single-crystal R samples is sensitive to spatial direction. For single crystals, the conductivity is highest perpendicular to the c-axis. Along the conduction plane, the conductivities are significantly higher than those of polycrystalline R samples; perpendicular to that plane they are significantly lower. The $BaBiO_3$ perovskite (P) phase, predominantly an electronic conductor, has conductivities of 10 to 10^2 S/cm at 350-800°C in air. Conductivity measurements of the R-P assemblages with different R:P showed that percolation threshold is around 35 volume % P.

INTRODUCTION

The β_1 high temperature polymorph of the R solid solution of the AE-Bi oxide binary systems (AE = Ca, Sr, Ba) and of the Ba-Sr-Bi oxide ternary system is a very good oxygen ion-conductor with conductivities much higher than those of the stabilized zirconias at around 700°C. The conductivities of the R phase of the AE-Bi oxide binary systems [1-3] and those of the R phase of the Ba-Sr-Bi oxide ternary system [4-6] have been investigated by different authors. A recent work [7] on phase equilibria of the R solid solution in the Ba-Sr-Ca-Bi oxide quaternary system reveals the polymorphic transition temperatures with respect to alkaline-earth content of this phase. In each case, the polymorphic transition from the β_2 polymorph to the β_1 polymorph was marked with sudden increase in conductivity. Mercurio *et al.* [8] studied the crystal structure of R specimens in AE-Bi oxide binary systems and suggested a plausible oxygen ion conductivity mechanism. The structure is characterized as a repetition of three fluorite-like sheets separated by an intersheet space free of anions for the β_2 polymorph with the oxygen ions located within the fluorite-like sheet side of the structure. The movement of the oxygen ions from the fluorite-like sheet to the intersheet space causes the $\beta_2 \leftrightarrow \beta_1$ transition. This transition is accompanied by an increase in the thickness of the intersheet space as well as an increase in the conductivity because the oxygen ions now occupy this corridor and can freely move along it. Thus, the main oxygen ion conductivity mechanism that is responsible for the high ionic conductivities of the β_1 phase is along the conduction planes perpendicular to the c-axis.

In the Ba-Sr-Bi oxide ternary, the R phase coexists with a $BaBiO_3$ P phase over a large range of temperatures and the P phase that can dissolve small amounts of Sr is an electronic conductor. Thus, any composition synthesized within the biphasic region of the phase diagram that contains both the P and the R phases will be a mixture of an oxygen-ion conductor and an electronic conductor. This allows us to fabricate a composite mixed ionic-electronic conductor that does not require application of an external potential to cause diffusion of oxygen ions.

In a previous publication [5], we discussed the results of our work on electrical conductivi ties of pure Ba-Sr-Bi oxide polycrystalline R samples, all with an $^{AE}/_{Bi}$ molar ratio of ¼. This $_{P}$ per, discusses results of a study on electrical conductivities of pure Ba-Sr-Bi oxide single-cryst R samples both along and perpendicular to the conduction planes. Electrical conductivities of a pure $BaBiO_3$ P polycrystalline sample as well as those of three R-P assemblages with different ratios of the two phases are also reported. Synthesis conditions of the samples and their effect $_{\bullet}$ the conductivities are explained.

EXPERIMENTAL

All polycrystalline samples were synthesized using solid-state reaction techniques. Pure samples were prepared by mixing pure, predried Bi_2O_3, BaO_2 and $SrCO_3$ in appropriate stoichiometric amounts. On the basis of microprobe analysis (below), starting materials are all believed to be > 99 % pure on a cation basis. The mixture was pressed into a pellet and sintere at 700°C for three consecutive days with grinding, mixing and repelletizing steps every 24 h. The final product was analyzed using Electron Microprobe Analysis (EMPA)(JEOL JXA-8600 Electron Microprobe), powder X-ray diffraction (Scintag XDS 2000 automated diffractometer, Cu K-α radiation λ = 1.54178 Å) and, in some cases, by Scanning Electron Microscopy (SEM) (JEOL JSM-6330F) for material characterization. After confirming the purity and the correct composition of the phases, final powder was pressed into a 13 mm pellet applying uniaxial pressure by a hydraulic pump. The pellet was sintered at 730°C for 10 h with heating and cooling rates of 2 to 3°C/min. A rectangular bar with an approximate cross section of 4 mm^2 a length of 12-13 mm was carefully cut from the sample pellet using a thin diamond saw and polished flat at two ends. Gold paste was applied to the ends of the pellet and dried at 600°C f 30 min. Electrical conductivity of the bar was measured by 2-probe AC impedance method in the 5 Hz-13 MHz frequency range using an HP 4192A Impedance Analyzer. The measuremen were made perpendicular to the uniaxial pressing axis of the bar and equilibration time at each experimental temperature was 1h. Impedance spectra were analyzed using the "Equivalent Circuit" program [9].

Electrical conductivities of a $BaBiO_3$ polycrystalline bar and those of R-P biphasic assemblages were measured in air as a function of temperature by the 4-probe DC resistivity technique using gold electrodes. A Keithley 220 programmable current source and a Keithley 196 digital multimeter were used for measurements. Two new steps were added to the synthes procedure. After mixing the starting oxides in appropriate proportions, the sample mixture was transferred to a gold crucible, partially melted to perovskite + liquid, kept at this temperature f 5 min and cooled down by air quenching. This step was introduced to the synthesis method to equilibrate Ba and Sr ions in the R phase. The crystalline chunk was then crushed to a coarse powder and ball-milled for 24 h. Sintering of sample pellets at 700°C was followed by materia characterization. The final sample pellet was uniaxially pressed as before and densified further by Cold Isostatic Pressing (CIP) under 40,000 psi pressure. After final sintering of the sample bar at 700°C for 10 h, the conductivities of the polycrystalline sample bars were measured perpendicular to the uniaxial pressing axis.

R single crystals were prepared by heating the sample in a Au crucible to 150°C above melting and by then cooling the melt at 0.15°C/min to room temperature. Following the material characterization steps, a crystal with a size near 5 mm x 3 mm x 0.1 mm was carefully picked from the crystal mass and studied by SEM to make sure it was free of any major defects. For conductivity experiments, Au electrodes were attached to the crystal using Au paste. Electrical conductivities were measured using the 2-probe AC impedance method under air or pure O_2 both along and perpendicular to the conduction planes of the crystals.

RESULTS

Our previous work on conductivities of pure Ba-Sr-Bi oxide R polycrystalline samples [5] shows that keeping the AE:Bi at 1:4 and interchanging the amounts of Ba and Sr present in the samples has little effect on conductivities [5]. However, the $\beta_2 \leftrightarrow \beta_1$ polymorphic transition temperature –that is marked by increase in the conductivity- changes with the Ba:Sr ratio. Transition temperature goes up almost in a linear way as the amount of Ba (the bigger cation) rises (Figure 1).

The planar mechanism of the oxygen ion conductivity suggests highest conductivities perpendicular to the c-axis of the structure. We, therefore, synthesized single-crystal samples with similar compositions to the pure R polycrystalline samples (AE:Bi near 1:4) and measured their conductivities parallel to the conduction plane (perpendicular to the c-axis) and perpendicular to the conduction plane (parallel to c). Four R crystals were grown from melts ($Ba_{1.64}Bi_{8.36}O_{14.18}$, $BaSrBi_8O_{14}$, $Sr_2Bi_8O_{14}$, $Sr_{2.1}Bi_{7.9}O_{13.95}$), their orientations determined by diffraction techniques, and the conductivities of one or more single crystals of each composition measured in both directions. Figure 2 summarizes these results. Electrical conductivities along the conduction plane are 4 to 5 orders of magnitude higher than those perpendicular to that plane. For most single-crystal samples, the $\beta_2 \leftrightarrow \beta_1$ polymorphic transition is poorly reflected in the conductivity profile and is not accompanied by a distinct increase in conductivity. This is probably due to the imperfection of these crystals as discussed below.

Melt grown single crystals with four different compositions were studied by SEM and GADDS X-ray diffraction methods. Back-scattered electron imaging shows uniform compositions of each crystal. However, SEM images of crystal faces perpendicular to the c-axis reveal presence of linear deformation features that might be kink bands. These are believed to be caused by thermal contraction during cooling that occurred after crystallization resulting in

Figure 1.
Electrical
conductivities
of Ba-Sr-Bi oxide
polycrystalline
rhombohedral (R)
samples vs. 1000/T.

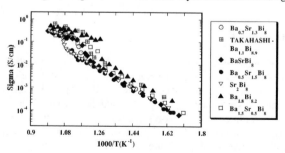

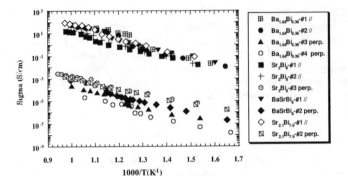

Figure 2. Electrical conductivities of Ba-Sr-Bi oxide rhombohedral (R) single crystals (measured parallel and perpendicular to the conduction plane) vs. 1000/T.

multiple diffraction maxima at any fixed 2θ, corresponding to individual lattice planes that are at high angles to the c-axis. The presence of several closely-spaced maxima rather than a single blurred maximum is consistent with formation of individual subgrains rather than a single distorted lattice.

Conductivities of a pure $BaBiO_3$ polycrystalline sample and those of three different R-P biphasic assemblages with R/P ≥ 1 have been measured. Results are shown in Figure 3. Electrical conductivities of the P increase from 6 to 40 S/cm on increasing the temperature from 300°C to 750°C. This increase is nearly linear compared to the pure R samples, the conductivities of which increase by almost 4 orders of magnitude between 350°C and 730°C. Hence, activation energies of the pure β_2 and pure β_1 phases are much higher than that of the pure $BaBiO_3$.

Among the R-P biphasic samples, the 20%P-80%R sample has similar conductivities to the pure R samples. The 35%P-65%R sample has higher conductivities than the pure R and 20%P-80%R samples in the β_2 region. In the β_1 region its conductivities are similar to those of the 20%P-80%R sample. The conductivity profile of that sample is almost parallel to that of pure P and the sharp increase in conductivity that marks the $\beta_2 \leftrightarrow \beta_1$ polymorphic transition is replaced by a small jump in conductivity. The 50%P-50%R sample has a conductivity profile subparallel to the pure $BaBiO_3$ with no sharp changes in conductivity near the $\beta_2 \leftrightarrow \beta_1$ polymorphic transition. Electrical conductivities of this composite are almost 2 orders of magnitude lower than the pure $BaBiO_3$ at each experimental temperature.

DISCUSSION

Comparing conductivities of pure R single crystals and those of the pure R polycrystalline samples, it was observed that conductivity curves of the polycrystalline specimens are in between the single-crystal conductivities along the conduction plane and single-crystal conductivities perpendicular to the conduction plane. This means that uniaxial pressing of the polycrystalline samples cause some R crystals to align with their c-axis parallel to the pressing axis. Supporting this conclusion, measured electrical conductivities of a polycrystalline sample parallel to the pressing axis are lower than in ones orthogonal to that.

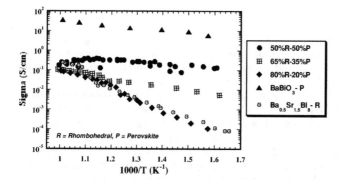

Figure 3. *Electrical conductivities of pure $BaBiO_3 (P)$, pure rhombohedral (R) and rhombohedral (R)-$BaBiO_3 (P)$ polycrystalline composites vs. 1000/T.*

Although different cooling rates and crucible sizes were used to grow crystals, the R single-crystals contained imperfections, presumably due to the stresses set up during cooling. The off-sets in conduction layers reduce the mobility of the oxygen ions through those layers to a degree depending on the offset in c. Therefore, conductivities of these crystals may differ from those of a perfect crystal. Methods to avoid such structural deformation are under investigation.

We can see from Figure 3 that the percolation threshold for conductivities of R-P biphasic assemblages is near 35%P (molar % are almost equal to volume %) as conductivities of the 35%P-65%R sample are distinctively different than the pure R samples. Percolation threshold is defined as the composition above which a single phase in a multi-component system starts behaving as a continuous phase. Thus, in our case, any biphasic assemblage that has more than 35 volume % P should behave as pure P phase. However, this assumption is valid for a dense sample. Our polycrystalline samples had densities around 85 % and it should be expected that pores which occupy 15 % of the sample volume will behave as a third phase and interrupt the continuity of the P phase within the composites. Although the P phase is isotropically distributed in composite samples, the connection between two adjacent P grains is sometimes a very narrow neck. This can also cause a barrier in obtaining highest conductivities from the composites and sintering methods to optimize the size of these necks are under investigation.

Wu and Liu [10] have modelled ambipolar transport properties of composite mixed ionic-electronic conductors for both dense and porous composites. Using their formulation for porous composites with similar densities, the theoretical conductivity profile of our composite materials almost overlaps that of experimental ones.

Two properties of these composites seem to have negative effects in manufacturing of samples to obtain highest conductivities. The first is the rather low melting temperatures of the composites that do not allow sintering sample pellets at higher temperatures to obtain higher densities. The second problem is structural. As explained before, the R phase has a planar structure and proper alignment of these planes is very important in optimizing the conductivities. The $BaBiO_3$ on the other hand has a more nearly isotropic structure. We believe that the presence of the P grains retards alignment of R grains with parallel conduction planes.

CONCLUSIONS

In the Ba-Sr-Bi oxide system, a rhombohedral phase (R) –an oxygen ion conductor- and $BaBiO_3$ (P) –an electronic conductor- coexist over a wide range of temperatures. That allows us to fabricate a composite mixed ionic-electronic conductor that does not require application of any external potential.

The conductivity of the pure R phase is very sensitive to spatial direction and is highest perpendicular to the c-axis. Keeping the Bi amount constant and interchanging the amount of th AE showed that increasing the amount of the bigger cation Ba lowers the temperature of the β_2 $\leftrightarrow \beta_1$ polymorphic transition so higher conductivities occur at lower temperatures.

The percolation threshold for the total conductivities of R-P biphasic assemblages is near 35% P. Methods to improve the structural properties of the composites to obtain higher conductivities above the percolation threshold are under investigation.

ACKNOWLEDGEMENTS

We would like to thank Sangtae Kim and Susan Wang for providing invaluable assistance in conductivity measurements and I-Ching Lin for valuable discussions. We also want to thank Kent Ross, Ranko Bontchev, and Robin Francis for their contribution in material characterizatio of the samples. Support from Robert Welch Foundation, Texas Center for Superconductivity, MRSEC Program of the National Science Foundation of the USA under Award Number DMR-9632667, and the Texas Advanced Research Program under Award Number 003652-886 are als acknowledged.

REFERENCES

1. T. Takahashi, H. Iwahara and Y. Nagai, *J. Appl. Electrochem.* **2**, 97-104 (1972).
2. T. Takahashi, H. Esaka and H. Iwahara, *J. Solid State Chem.* **16**, 317-323 (1976).
3. J.C. Boivin and D.J. Thomas, *Solid State Ionics* **5**, 523-526 (1981).
4. M. Drache, J.P. Wignacourt and P. Conflant, *Solid State Ionics* **86-88**, 289-295 (1996).
5. O.A. Gökçen, S. Kim, J.K. Meen, D Elthon and A.J. Jacobson, in: *Solid State Ionics V*, eds. G-A Nazri, C. Julien and A. Rougier, Materials Research Society Proceedings **548**, Boston MA, USA, 499-504 (1999).
6. O.A. Gökçen, J.K. Meen, and A.J. Jacobson, in *Mass and Charge Transport in Inorgani Materials: Fundamentals in Devices*, eds. P. Vincenzini and V. Buscaglia, Proceedings c the International Conference on "Mass and Charge Transport in Inorganic Materials; Fundamentals to Devices", Lido di Jesolo, Italy, 91-98 (2000).
7. J.K. Meen, O.A. Gökçen, I-C. Lin, K. Müller, and B. Nguyen, *[this volume]*.
8. D. Mercurio, J.C. Champarnaud-Mesjard, B. Frit, P. Conflant, J.C. Boivin and T. Vogt, ' *Solid State Chem.* **112**, 1-8 (1994).
9. B.A Boukamp, *Solid State Ionics* **18-19**, 136-140 (1986).
10. Z. Wu and M. Liu, *Solid State Ionics*, **93**, 65-84 (1997).

Mat. Res. Soc. Symp. Proc. Vol. 658 © 2001 Materials Research Society

Bonding of Guest Molecules in the Tubes of Nanoporous Cetineite Crystals

E.E. Krasovskii, O. Tiedje, S. Brodersen, W. Schattke, F. Starrost[1],
J. Jockel[2], U. Simon[2]
Institut für Theoretische Physik und Astrophysik, Christian-Albrechts-Universität Kiel,
Leibnizstraße 15, 24098 Kiel, Germany
[1] Department of Chemistry and Biochemistry, Box 951569, University of California, Los
Angeles, CA 90095-1569, USA
[2] Institut für Anorganische Chemie, RWTH Aachen, Professor-Pirlet-Str.1, 52074 Aachen,
Germany

ABSTRACT

A theoretical study of the optical excitation processes in cetineites is presented. This
new exciting class of crystals with tubular structures displays photoconductivity, which
strongly depends on the presence of guest molecules within the tubes. Based on
self-consistent electronic structure calculations we present calculated dielectric function and
photoelectron spectra of the (Na;Se) cetineite with the tube filled with different guest
molecules. Calculations are performed with the extended linear augmented plane wave **kp**
method. The configuration and the arrangement of the filling molecules is derived from
available crystallographic data and from heuristic arguments on chemical binding.

INTRODUCTION

The electronic structure of various pure cetineite compounds has been previously
studied extensively, see Ref. [1] and references therein. The general composition formula is
$A_6[Sb_{12}O_{18}][SbX_3]_2 \cdot (6 - mx)H_2O \cdot x[B^{m+}(OH^-)_m]$, where $A = Na^+$, K^+, Rb^+; $X = S^{2-}$,
Se^{2-}; and $B = Na^+$, Sb^{3+}. The crystal formulae are abbreviated by $(A;X)$. These
materials present an extremely attractive and promising subject owing to their single
crystal porous structures with tube diameter of about 0.7 nm (see Figure 1) and their
semiconducting properties. Potential applications are similar to those of the zeolites with
the additional feature of controling their function in an application by electric power. To
date, only in the case of $K_6[Sb_{12}O_{18}][SbSe_3]$ homogeneous single crystals can be grown
entirely free of guest molecules within the pores (in the form of needles of about 0.5 mm
diameter and 2 mm length along the axes of the pores).

Geometrical and electronic properties of the cetineites are discussed in Ref. [1]. In that
paper *ab initio* band structure calculations were presented for $A = Na$, K; $X = S$, Se. The
bonding mechanisms have been explained by the hexagonal aggregation of the tubes
consisting of SbO_3 tetrahedra – netted to form the tube walls. The tubes are stabilized by
the presence of singly ionized alkali atoms close to the walls inside the tubes and the
negatively charged SbX_3 pyramids between the tubes (see Figure 1). In Ref. [1] the
anisotropy of the photoconductivity has been explained in terms of the band structure of
the guest-free crystals. Two qualitatively different conduction processes can be deduced
from the bandstructure results, namely, hopping between the SbO_3 tetrahedra and

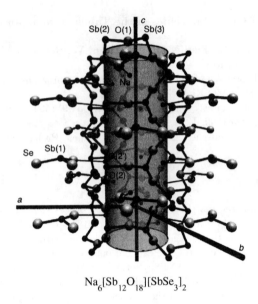

$$Na_6[Sb_{12}O_{18}][SbSe_3]_2$$

Figure 1: *Crystal structure of the cetineites. (Figure from Ref. [1].)*

traveling inside the tube. The latter mechanism is most interesting; by the condition of th orthogonality of the conduction band wave functions to the valence band ones, conductio band states can be formed that have negligible overlap with the orbitals of any of the atomic constituents. Owing to the nanoporosity the conditions for the formation of delocalized states of that nature exist in the cetineites, and we indeed observed an energetically well separated state confined to the tube in our calculation for (K;S). [1] (Similar nearly free electron like states have been observed in calculations on boron nitrid nanotubes [2].) It should be, however, noted that this feature of the electronic structure i very sensitive to the atomic composition (e.g., in other cetineites the state was more strongly hybridized with the wall orbitals) and can be easily destroyed by the guest molecules.

Because the guest free material possesses the most attractive features, it has to be understood why in the majority of cases of crystallization water and some other ions of th solution are relatively strongly bound to the interior of the tubes. A direct theoretical approach would require a geometry optimization procedure based on the *ab initio* total energy minimization within the Density Functional Theory (DFT). That would imply considering supercells of large volume and low symmetry, that is beyond our computation abilities.

In this paper we restrict ourselves to a few configurations with assumed geometry of water octahedra where adsorption data are available [3] and the configuration may be fixe by symmetry arguments. The actual *ab initio* calculation then covers the bandstructure determination, as well as optical and photoelectron spectra. It is hoped that by this procedure one obtains, first, a confirmation of the assumptions through a comparison with experiment, and second, an insight into the bonding properties of the guest molecules which may eventually lead to synthesizing guest free (dry) crystals.

COMPUTATIONAL METHOD

We have considered the following possibilities to fill the tubes with the water (or water-derived) molecules: 1) Inert water cluster $(H_2O)_6$ with symmetrically arranged oxygen atoms (their positions are deduced from x-ray diffraction measurements [4]), which are kept together by hydrogen bonds. 2) The oxygen atoms of the water molecules are connected to the oxygen atoms belonging to the wall via hydrogen bonds. This leads to weaker hydrogen bonds between the oxygen atoms of the water cluster. 3) The center of the cluster is occupied by an antimony atom and there are only hydrogen bonds to the wall, which leads to the $Sb_2(OH)_6$ configuration. The configurations keep the symmetry of the crystal the same as in the case of empty tubes, whereby the symmetry of the molecules might be higher than the actual one, and which has a technical advantage of strongly reducing the computer time. It should be also noted that in nature an incomplete filling of the tubes may occur.

The calculations were performed with the full potential extended linear augmented plane wave **kp** method [5]. The basis set of this direct variational method describes the wave functions in the interstitial by plane waves (the energy cutoff of 9 Ry was used in this work) and inside the muffin-tin spheres by linear combinations of orbitals with several radial functions per angular momentum channel. This leads to a Hamiltonian matrix of dimension 3615. The density-of-states (DOS) curves, dielectric function, and photoelectron spectra were obtained with the tetrahedron method by interpolating the electron eigenenergies, l-projected partial charges, and momentum matrix elements between 12 irreducible **k**-points, which corresponds to the division of the Brillouin zone into 324 tetrahedra, 42 of which are inequivalent.

The resulting self-consistent crystal potential and the band structure agree well with our previous calculations [1], however, our increased computer power and recent methodological progress have made it possible to achieve considerably better numerical accuracy at all stages of the calculation and to treat the more complicated unit cells. In particular, in evaluating the photoelectron spectra the explicit calculation of the optical interband transitions between ~ 150 occupied and ~ 300 unoccupied bands became feasible, owing to our using the plane wave (PW) expansion of the all-electron wave functions instead of the original augmented plane wave (APW) representation. The PW procedure (see Ref. [6]) scales as n^3 with the number of atoms in the unit cell in contrast to the straightforward calculation in terms of APWs, which scales as n^4.

RESULTS AND DISCUSSION

In order to visualize the implications of the guest molecules for the optical excitation processes, we present in Figure 2 the imaginary part of the dielectric function $\varepsilon(\omega)$ for the

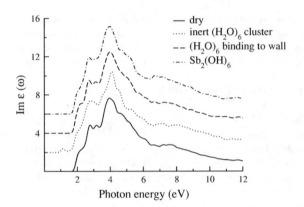

Figure *2: The imaginary part of the dielectric function of the (Na;Se) cetineite with different guest molecules.*

four crystals. It can be seen that the guest molecules have no dramatic effect on the optical absorption over the low photon energy region. In other words, the initial and the final states of important optical transitions have apparently the same nature irrespective of the filling of the tubes. However, that does not mean that there is no effect on the photoconductivity: the energy distribution of the mobility of the carriers changes more significantly with the filling of the tubes. (The mobility is proportional to the averaged group velocity of the electrons at a given energy and denoted by ω_p^2 in Ref. [1].) To conclude on the implications for the photoconductivity one needs a plausible approximation for the distribution function of the carriers – this work is in progress.

For the (Na;Se) crystal a photoemission experiment has been performed [1] for the photon energy 21.22 eV, but in Ref. [1] the measured spectrum has been compared only to the theoretical DOS function. In the present work we have calculated the angle-integrated photoelectron energy distribution curves (EDCs) in the framework of interband optical transitions between bulk states for a number of photon energies around 21.22 eV. The calculation is free from adjustable parameters and it takes into account the crystal momentum conservation and the dipole matrix elements. In the photon energy region considered the theoretical spectrum changed only slightly with the photon energy; thus, the uncertainties due to our using the DFT-derived one-particle solutions, which neglect self-energy effects, are apparently small. The EDCs for the pure (Na;Se) and for three different configurations of the guest molecules are shown in Figure 3.

The electron emission from the valence band is seen to be reproduced with excellent quality by the pure (Na;Se) calculation. However, the well-defined experimental maximum at −12 eV has no counterpart in the theoretical curve. The filling of the tubes has a strong effect on the EDC – our results suggest that none of the configurations we have tried is likely to occur in reality. On the other hand, it can be seen that the guest molecules may

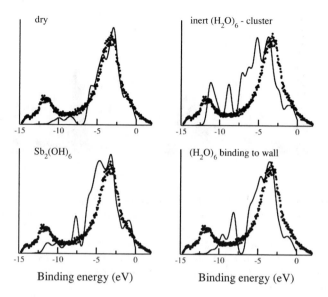

dry

inert $(H_2O)_6$ - cluster

$Sb_2(OH)_6$

$(H_2O)_6$ binding to wall

Binding energy (eV)   Binding energy (eV)

Figure *3: Comparison of the theoretical angle-integrated UPS spectra of the (Na;Se) cetineite with different guest molecules (solid lines) with the measurements of Ref. [1].*

give rise to relatively strong emission at binding energies deeper than -10 eV. The maximum at -12 eV may arise from the presence of alkali atoms in the tubes or may be the effect of contamination. However, in view of the surface sensitivity of the photoemission experiment, the possibility cannot be excluded that this structure comes from surface effects not described by the bulk band structure.

Another interesting effect is observed in the (K;S) compound in the presence of the water molecules in the inert water cluster $(H_2O)_6$ configuration: owing to their small binding energy the $3p$ states of potassium hybridize with the SbO_3 tetrahedra orbitals of the walls to form a satellite DOS maximum about 1 eV higher in energy (see the right panel of Figure 4).

CONCLUSIONS

We have performed *ab initio* calculations of the energy band structure, dielectric function, and photoelectron energy distribution curves for the (Na;Se) cetineite with various configurations of the guest molecules filling the nanotube. The guest molecules have been found to only slightly affect the optical absorption in the visible and UV photon energy range. At the same time, according to our calculations, their effect must be well seen in the photoemission experiments. The experimentally observed influence of guest molecules on the photoconductivity is apparently caused by changes in the mobility of the carriers rather than by changes in the optical excitation process.

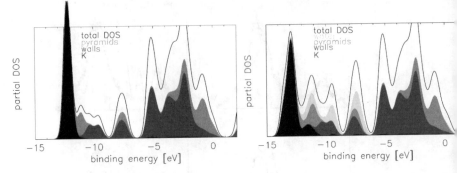

Figure *4: Atom-resolved partial DOS curves for the pure (K;S) compound (left panel) a with the tubes filled with inert water (H_2O)_6 clusters (right panel). The contributions fr (H_2O)_6-water (right panel only), SbS_3-pyramids, $Sb_{12}O_{18}$ tetrahedra, and potassium ato are shown by shaded areas, with the darkness increasing in that order. Thereby the wh region shows the contribution from the interstitial region and the solid line gives the t DOS. The DOS at the potassium atom is seen to have a satellite at about -12 eV in presence of water molecules.*

ACKNOWLEDGMENT

This work was supported by the Deutsche Forschungsgemeinschaft under contracts S 609/2-1, SCHA 360/14-2, and in part by Forschergruppe DE 412/21. One of us (FS) wo like to thank Prof. Emily A. Carter for her support.

REFERENCES

[1] F. Starrost, E. E. Krasovskii, W. Schattke, J. Jockel, U. Simon, R. Adelung, and L. Kipp, *Phys. Rev. B* **61**, 15697 (2000).

[2] X. Blase, A. Rubio, S. G. Louie, M. L. Cohen, *Europhys. Lett.* **28**, 335 (1994).

[3] U. Simon, F. Schüth, S. Schunk, F. Liebau, and X. Wang, *Angew. Chem. Intern. E Engl.* **36**, 1121 (1997).

[4] X. Wang, F. Liebau,*Eur. J. Solid State Inorg. Chem.* **35**, 27 (1998); X. Wang, F. Liebau, *Z. Kristallog.* **214**, 820 (1999).

[5] E. E. Krasovskii, F. Starrost, and W. Schattke,*Phys. Rev. B* **59**, 10504 (1999).

[6] E.E. Krasovskii and W. Schattke, *Phys. Rev. B* **60**, 16251 (1999).

Magnetism and Manganates

Mat. Res. Soc. Symp. Proc. Vol. 658 © 2001 Materials Research Society

Metal Cyanide Networks Formed at an Air-Water Interface: Structure and Magnetic Properties

Jeffrey T. Culp,[1] A. Nicole Morgan,[2] Mark W. Meisel,[2] and Daniel R. Talham[1]

[1] Department of Chemistry, University of Florida, Gainesville, FL
[2] Department of Physics, University of Florida, Gainesville, FL

ABSTRACT

Extended two-dimensional coordinate covalent networks have been formed at an air-water interface by condensing a Langmuir monolayer of an amphiphillic pentacyanoferrate(III) complex and aqueous nickel(II) ions. The cyanide-bridged arrays were transferred to solid supports by the Langmuir-Blodgett (LB) technique. The resulting thin films were structurally characterized by UV-Vis and FT-IR spectroscopies, X-ray diffraction, X-ray photoelectron spectroscopy (XPS), and SQUID magnetometry. Magnetic measurements on the thin films transferred to Mylar show evidence for ferromagnetic exchange interactions between the $S = 1/2$ Fe(III) and $S = 1$ Ni(II) centers.

INTRODUCTION

Supramolecular chemistry [1] is a synthetic method based on the spontaneous self-assembly of discreet molecular building blocks into supermolecules or supermolecular arrays. While the method has traditionally been applied to purely organic systems employing weak van der Waals interactions as structural directors, recent reports of metal-based clusters, [2] molecular squares [3], and networks [4] have shown the structural versatility achievable when coordination complexes are used as the building units. Supermolecular systems are currently being pursued for their potential host-guest, catalytic, optical, electronic, and magnetic properties. This synthetic strategy may eventually prove to be a new route to future nanoscale electronic and magnetic devices. In order to extend the potential applications of this class of materials, new synthetic methods for forming these structures at an interface need to be developed. We have synthesized a two-dimensional Fe-CN-Ni grid network at an air-water interface to demonstrate the potential for forming coordinate covalent networks at a surface. By employing paramagnetic transition metal ions in the construction of these networks, we were able to introduce ferromagnetic properties into the resulting inorganic arrays.

EXPERIMENTAL

The ligand 4-didodecylaminopyridine (DDAP) was prepared by reaction of 1-bromododecane with N-methyl-4-aminopyridinium iodide in acetonitrile. The ligand was deprotected in molten pyridinium chloride at 190° C. Formation of the amphiphillic complex pentacyano(4-didodecylaminopyridine)ferrate(III) (FeDDAP) was achieved by reaction of

DDAP with a methanolic suspension of $Na_3Fe(CN)_5NH_3$ (Aldrich) in air analogous to a previously reported procedure for the preparation of pentacyano(4-octadecylaminopyridine)-ferrate(III). [5] Langmuir monolayer studies and Langmuir-Blodgett film depositions were performed with a KSV Instruments KSV3000 system. Surface pressure of the reacting FeDDAP-Ni monolayers was measured using a Wilhelmy balance. All multilayer films were transferred over a 0.5 mM $Ni(NO_3)_2$ / 2 mM $Cs(NO_3)$ subphase at a surface pressure of 25 mN/m^2 by the Langmuir-Blodgett technique onto hydrophobic surfaces. Glass slides were made hydrophobic by treatment with octadecyltrichlorosilane.

RESULTS

Our strategy for forming two-dimensional Fe-CN-Ni networks is outlined in Scheme 1. The octahedral amphiphillic iron(III) complex (FeDDAP) provides the 90 degree bond angles needed for the assembly of a square grid network. The cyanide ligands provide the linear bridging units that can coordinate the Ni^{2+} ions in the subphase. By making the FeDDAP complex amphiphilic, the reaction can be confined to the water surface preventing growth in three dimensions. To form the network, a measured amount of the FeDDAP complex dissolved in chloroform was spread on the surface of a Ni^{2+} solution confined to a Teflon Langmuir trough. The mean molecular area of the monolayer film was monitored by the change in surface pressure during the compression of the films by a set of mobile Teflon barriers. When compared to the same monolayer spread over a Na^+ subphase, a shift to a smaller onset molecular area is seen in the isotherm over the Ni^{2+} subphase. This shift is an indication of a condensation reaction occurring between the FeDDAP complex and the aqueous Ni^{2+} ions. Extrapolation of the steepest part of the isotherm over Ni^{2+} yields a mean molecular area per FeDDAP of 52 $Å^2$.

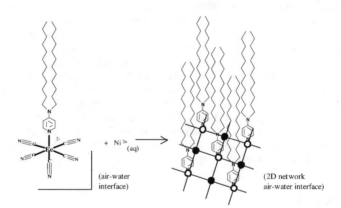

Scheme 1. *Method for assembling a two-dimensional Fe-CN-Ni grid network at the air-water interface.*

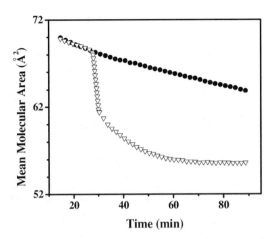

Figure 1. *Mean molecular area verses time for FeDDAP over a Na⁺ subphase (filled circles) and after injection of Ni²⁺ ions into the subphase (open triangles). Surface pressure was held constant at 10 mN/m².*

This is the expected value for the mean molecular area of FeDDAP when confined to a face centered square grid with a cell edged of 10.4 Å, a value typically observed in similar cyanide bridged systems. [6] The condensation reaction can be monitored in real time by measuring the change in the mean molecular area per FeDDAP complex at a fixed pressure of 10mN/m² over a Na⁺ subphase and comparing to the same experiment run on a Na⁺ subphase with injection of a Ni²⁺ solution under the monolayer. As shown in Figure 1, the mean molecular area slowly decreases over the sodium subphase due to the slight solubility of the FeDDAP complex. After injection of a Ni²⁺ solution, the mean molecular area rapidly collapses before stabilizing at an area of *ca.* 54 Å². These surface pressure measurements on the reacting FeDDAP complex indicate a condensation reaction is occurring at the air-water interface.

To further characterize the product of this reaction, the networks were transferred to solid supports by the Langmuir-Blodgett technique. The presence of Ni²⁺ in the FeDDAP film was verified by XPS. Integration of the nickel and iron peaks in the spectrum indicated an iron to nickel ratio of approximately 1:1 in the film. An iron to nickel ratio of 1:1 is expected in a face centered square grid network containing two Fe and two Ni per unit cell.

The lamellar nature of the multi-layer assemblies was confirmed by X-ray diffraction. The positions of the (001) and (002) Bragg reflections yielded an interlayer spacing of 35 angstroms. This value indicates the alkyl chains are tilted approximately 30 degrees to maximize their van der Waals contacts. This tilt is necessary to compensate for the size mismatch between

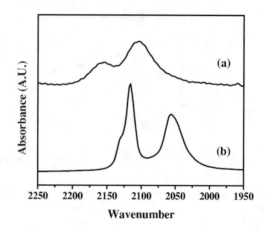

Figure 2. *FT-IR spectra showing the shift of the cyanide stretching energy in the (a) LB film containing the Fe-CN-Ni network compared to (b) pure FeDDAP (KBr pellet).*

the alkyl chains and the larger cyanide-bridged head group. Convincing evidence for the formation of a cyanide-bridged network is seen in the FT-IR spectrum of the FeDDAP-Ni film, shown in Figure 2. The stretching frequency of the C-N triple bond is known to be sensitive to the mode of binding of the CN⁻ ligand. The unreacted FeDDAP spectrum taken in a KBr pellet shows two dominant stretching bands centered at 2056 cm^{-1} and 2116 cm^{-1}. The same stretching bands in the FeDDAP-Ni film are shifted to 2105 cm^{-1} and 2156 cm^{-1}. This shift in CN stretching vibration is typically seen when cyanide is in a bridging mode and can been rationalized as both an electronic effect in which electron density is being removed from the C-N antibonding orbital and as a kinematics effect in which the structure has become more rigid. [7]

A possible side effect of performing reactions of transition metal complexes in aqueous media is hydrolysis of the metal complexes. The integrity of the FeDDAP complex can be verified by UV-Vis spectroscopy. Pentacyano(4-aminopyridine)ferrate(III) complexes possess a strong ligand to metal charge transfer band centered at 665 nm in aqueous solutions. The complexes are also known to be relatively inert to hydrolysis at room temperature. [8] This band is also seen in the reacted films of FeDDAP-Ni. The intensity of this band increases linearly with the number of layers transferred indicating the FeDDAP complex remains intact through the reaction and during the course of film transfer. The stability of the complex is also seen in the stability of the mean molecular area after reaction as seen in Figure 1.

The formation of an ordered array of paramagnetic ions opens the possibility of introducing magnetic behavior. Research on the three dimensional Prussian blue family of cubic metal cyanides has recently yielded molecular based magnets ordering above room temperature. [9] Magnetic exchange interactions can be predicted in these systems of octahedral metal ions

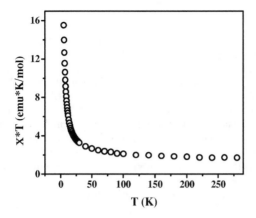

Figure 3. *Plot of the product of the molar susceptibility and temperature for the Fe-CN-Ni network formed at the air-water interface transferred to Mylar.*

by comparing the orbital symmetries of the occupied d orbitals. Unpaired electrons in d orbitals of the same symmetry undergo antiferromagnetic exchange via super exchange pathways whereas unpaired electrons in orthogonal d orbitals undergo ferromagnetic exchange. The networks prepared in the present study contain low spin $S = 1/2$ Fe(III) and $S = 1$ Ni(II) ions. The electron on the iron complex resides in a t_{2g} orbital which is orthogonal to the two electrons in the nickel e_g orbitals. These electrons have been shown to undergo a ferromagnetic exchange interaction in the three dimensional Fe(III)-CN-Ni(II) analog, ordering at a T_c of 23 K. [10] To find further evidence for the formation of an extended network in the FeDDAP-Ni system, the magnetic properties of a multi-layer sample transferred to Mylar was investigated. Evidence for a ferromagnetic exchange interaction is seen in the χT vs T plot for the iron-nickel two-dimensional network, Figure 3. The upturn in the χT product at low temperature is a sign of ferromagnetic exchange. The magnetization (M) verses field (B) data at 2 K showed a rapid rise in magnetization verses field and could not be fit to Brillion functions with various spin combinations. The M vs B and χT vs T data discriminate against a simple dimer or superpara-magnetic state, and suggest the network is coherent over several unit cells. However, no direct evidence for spontaneous magnetization was seen in the zero field cooled or field cooled magnetization data down to 5 K. The apparent lack of significant long-range order may be due to small domain sizes or to the low dimensionality of the material. Accurate determination of the Weiss constant by linear extrapolation of the high temperature data in the inverse susceptibility vs temperature plot was hampered by uncertainty in the background subtraction due to the Mylar substrate.

CONCLUSIONS

The reaction of a Langmuir monolayer of an amphiphillic pentacyanoiron(III) complex with aqueous Ni^{2+} ions yielded a two-dimensional square grid network at the air-water interface. The condensation reaction was observed on the water surface by monitoring the change in mean molecular area of the iron complex as Ni^{2+} ions were introduced into the subphase. The result networks can be transferred to solid supports by the Langmuir-Blodgett technique. Structural characterizations of the transferred films confirmed that the iron complex is resistant to hydrolysis, the cyanide ligands are in a bridging mode and the multilayer films are lamellar in nature. Further evidence for an extended inorganic network is seen in the χT vs T data. A rapid rise in χT at low temperatures indicates a ferromagnetic exchange interaction between the S = 1/2 Fe(III) and S = 1 Ni(II) ions. No evidence for spontaneous magnetization down to 5 K was observed in the films. The lack of long-range order in these networks may be due to a combination of small domain size and low dimensionality.

REFERENCES

1. *Comprehensive Supramolecular Chemistry*, edited by J. L. Atwood *et al*. (Pergamon, New York, 1996) vol. 1-11.
2. P. A. Berseth, J. J. Sokol, M. P. Shores, J. L. Heinrich, J. R. Long, *J. Am. Chem. Soc.* **12.** 9655 (2000).
3. M. Fujita, J. Yazaki, K. Ogura, *J. Am. Chem. Soc.* **112**, 5645 (1990).
4. B. F. Hoskins, R. Robson, D. A. Slizys, *J. Am. Chem. Soc.* **119**, 2952 (1997).
5. F. Armand, H. Sakuragi, K. Tokumaru, *New. J. Chem.* **17**, 351 (1993).
6. H. J. Buser; D. Schwarzenbach, W. Petter; A. Ludi, *Inorg.Chem.* **16**, 2704 (1977).
7. K. Nakamoto, *Infrared and Raman Spectra of Inorganic and Coordination Compounds*, ed. (Wiley, New York, 1978) p. 259.
8. N. V. Hrepic, J. M. Malin, *Inorg. Chem.* **18**, 409 (1979).
9. S. Ferlay, T. Mallah, R. Ouahes, P. Veillet, M. Verdaguer, *Nature* **378**, 701 (1995).
10. V. Gadet, M. Bujoli-Doeuff, L. Force, M. Verdaguer *et al*, in *Magnetic Molecular Materials*, edited by D. Gatteschi *et al* (NATO ASI Series E, 1991) vol. 198, pp. 281-295

Mat. Res. Soc. Symp. Proc. Vol. 658 © 2001 Materials Research Society

Theoretical Analyses of Spin Exchange Interactions in Extended Magnetic Solids Containing Several Unpaired Spins per Spin Site

D. Dai, H.-J. Koo and M.-H. Whangbo*

Department of Chemistry, North Carolina State University, Raleigh, North Carolina 27695-8204

ABSTRACT

For extended magnetic solids containing several unpaired spins per spin site, we reviewed briefly recent progress in quantitative and qualitative methods of describing their spin exchange interactions on the basis of electronic structure calculations for their spin dimers.

INTRODUCTION

In a magnetic field up to magnetic saturation, the magnetic energy states allowed for an extended magnetic solid may not have a gap (Fig. 1a) or may have a gap between the singlet ground state and the first excited state (i.e., a spin gap) (Fig. 1b) [1,2]. It is also possible that a magnetic solid has a spin gap as well as another gap in the middle of the excited states (Fig. 1c) (i.e., "mid-spin gap") [3]. A spin gap causes the magnetic susceptibility to vanish below a certain low temperature, while either a spin gap or a mid-spin gap causes a plateau in the magnetization $vs.$ magnetic field curve (i.e., magnetization plateau). For a Heisenberg chain of L spin sites per unit cell and spin S per spin site, a magnetization plateau is predicted to occur when the magnetization m per spin site satisfies the condition $L(S - m)$ = integer [4]. In recent years magnetic solids with spin gaps and magnetic plateaus have received much attention.

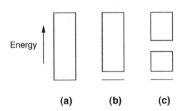

Figure 1. *Schematic diagrams showing the allowed magnetic energy states of an extended magnetic solid in a magnetic field up to magnetic saturation for the cases of (a) no forbidden energy gap, (b) a spin gap, and (c) a spin gap and a mid-spin gap.*

Allowed magnetic states of a magnetic solid are described in terms of a spin-Hamiltonian written as a sum of pair-wise spin exchange interactions $-J_{ij}\hat{S}_i \cdot \hat{S}_j$ between adjacent spin sites i and j. ($\hat{S}_i$ and $\hat{S}_j$ are the spin operators at the sites i and j, respectively, and J_{ij} is the spin exchange parameter associated with the spin exchange interaction between the two sites.) Thus a spin-Hamiltonian expresses low-energy excitation energies of a magnetic solid in terms of spin exchange parameters J_{ij}. Experimentally these excitations are probed typically by measuring magnetic susceptibility as a function of temperature or neutron inelastic scattering at a very low temperature (e.g., ~10 K). When the temperature-dependent magnetic susceptibility or angle-resolved inelastic neutron scattering data are analyzed in terms of a spin-Hamiltonian, the J_{ij} parameters act as numerical fitting parameters needed to reproduce the experimental data. What set of J_{ij} parameters to employ in this fitting process depends on our perception of what spin exchange pathways are important for the magnetic solid under investigation. In principle, therefore, experimental data can be fitted equally well with more than one set of J_{ij} parameters especially for magnetic solids of complex crystal structure. To relate such "experimental" J_{ij} parameters to the geometrical structure of a magnetic solid unambiguously, it is essential to perform appropriate electronic structure calculations for its spin dimers (i.e., the structural units containing two adjacent spin sites).

In the present work, we briefly review recent progress in how to calculate quantitatively the spin exchange parameters of an extended magnetic solid with several unpaired spins per spin site [5] and how to explain the trends in the spin exchange parameters of such magnetic solids on a qualitative basis [6].

QUANTITATIVE APPROACH

For a spin dimer consisting of a single unpaired spin at each spin site, its spin exchange parameter J is equal to the energy difference ΔE between the singlet and triplet states of the spin dimer, i.e., $J = \Delta E = {}^1E - {}^3E$. Quantitative evaluation of ΔE has been a challenging task mainly for two reasons. First, calculations of accurate ΔE values require state-of-the-art computational efforts on the basis of either configuration interaction (CI) wave functions or density functional theory (DFT) [7]. Second, a spin dimer of an extended solid is defined as a cluster obtained by breaking the bonds linking the spin dimer to the crystal lattice and then replacing each broken bond with an unshared electron pair belonging to the spin dimer, so that as an object for electronic structure calculations a spin dimer becomes a highly charged anion cluster.

For a general spin dimer that has M unpaired spin at one spin site and N spins at the other spin site, let us consider the energy difference between the highest-spin state Ψ_{hs} (Fig. 2a) and the low-spin state Ψ_{ls} (Fig. 2b) on the basis of the spin Hamiltonian $\hat{H} = -J\hat{S}_1 \cdot \hat{S}_2$. The spin exchange interaction between the sites 1 and 2 is ferromagnetic if the state Ψ_{hs} is more stable than the state Ψ_{ls}, and antiferromagnetic otherwise. If the states Ψ_{hs} and Ψ_{ls} are represented as pure spin states as depicted in Figs. 2a and 2b,

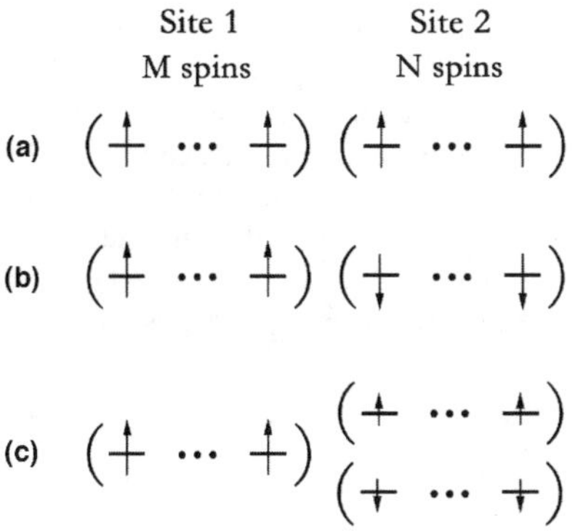

	Site 1	Site 2
	M spins	N spins

Figure 2. *Two spin states of a spin dimer that has M unpaired spins at the spin site 1 and N unpaired spins at the spin site 2: (a) the highest-spin state Ψ_{hs} and (b) the low-spin state Ψ_{ls}. Each spin site has the highest-spin arrangement, and the two spin sites interact ferromagnetically in Ψ_{hs}, and antiferromagnetically in Ψ_{ls}. In the transition state (c), each of the 2N magnetic orbitals (α- and β-spin) of the spin site 2 has half an electron while the M unpaired electrons of the spin site 1 are accommodated in the α-spin magnetic orbitals.*

respectively, then their energies (E_{hs} and E_{ls}, respectively) are given by the expectation values of the spin Hamiltonian $\hat{H} = -J\hat{S}_1 \cdot \hat{S}_2$. Thus we obtain [5]

$$E_{hs} = \langle \Psi_{hs} | \hat{H} | \Psi_{hs} \rangle = -\frac{MN}{4}J \qquad (1a)$$

$$E_{ls} = \langle \Psi_{ls} | \hat{H} | \Psi_{ls} \rangle = \frac{MN}{4}J \qquad (1b)$$

and the spin exchange parameter J is related to the energy difference between the spin states Ψ_{hs} and Ψ_{ls} as [5]

$$J = \frac{2(E_{ls} - E_{hs})}{MN} \qquad (2)$$

This expression provides a general relationship between the spin exchange parameter and the energy difference between the highest-spin and low-spin states of a spin dimer. Noodleman's relationship [8] between J and the energy difference between the highest-spin and the broken-symmetry states of a spin dimer is a special case of Eq. 2. The electronic states of a spin dimer corresponding to the pure spin states shown in Figs. 2a and 2b can be constructed in the CI wave function or DFT approach, so that the spin exchange parameter J is obtained by calculating the total energies associated with these electronic states (to be identified as E_{hs} and E_{ls}, respectively). Accordingly, the spin exchange parameter of a spin dimer can be determined by performing two separate total energy calculations. Thus this approach may be referred to as the total-energy-difference method.

Within DFT, the total energy of an electronic state can be written as a function of the electron occupation numbers and can be expanded in Taylor series with respect to the electron occupation numbers. Let us define the transition state as the state in which each of the $2N$ magnetic orbitals (α- and β-spin) of the spin site 2 has half an electron and the M unpaired electrons of the spin site I are accommodated in the α-spin magnetic orbitals (Fig. 2c). Then the total energy difference ($E_{hs} - E_{ls}$) is related to the magnetic orbital energies of the transition state as follows [5]

$$E_{ls} - E_{hs} = \sum_{j=1}^{N} (\varepsilon_j^\beta - \varepsilon_j^\alpha) \qquad (3)$$

where ε_j^γ ($\gamma = \alpha, \beta$) is the j-th magnetic orbital (spin γ) energy of the transition state. Thus Eqs. 2 and 3 lead to the following expression for J [5].

$$J = \frac{2}{MN} \sum_{j=1}^{N} (\varepsilon_j^\beta - \varepsilon_j^\alpha) \qquad (4)$$

The calculation of spin exchange parameters using Eq. 4 may be referred to as the transition-state method.

QUALITATIVE APPROACH

In general the spin exchange parameter J is expressed as $J = J_F + J_{AF}$, where the ferromagnetic term J_F favors the triplet state (i.e., $J_F > 0$), and the antiferromagnetic term J_{AF} favors the singlet state (i.e., $J_{AF} < 0$). Qualitative trends in the antiferromagnetic J parameters of magnetic solids containing one unpaired spin per spin site are well explained in terms of the one-electron spin orbital interaction energies Δe of spin dimers [9-15]. The spin orbital interaction energy of a spin dimer with one unpaired spin on each spin site refers to the energy difference Δe between the two singly occupied energy levels of the spin dimer when the two spin sites are equivalent (Fig. 3). For the interaction between two equivalent spins, the J_{AF} term is related to Δe by $J_{AF} \propto -(\Delta e)^2$ if the singly filled levels of a spin dimer are given as linear combinations of orthogonal spin orbitals at the two spin sites [16]. Alternatively, J_{AF} is related to Δe by $J_{AF} \propto -S\Delta e$, when the singly filled levels of a spin-dimer are given as linear combinations of non-orthogonal spin orbitals localized at the two spin sites [17]. Here S is the overlap integral between the two non-orthogonal spin orbitals. The two formulations are identical in nature because of the relationship $\Delta e \propto S$.

When the spin sites I and 2 of a spin dimer have M unpaired spins, the overall spin exchange parameter J of the spin dimer is described by Eq. 5 [18].

$$J = \sum_{\mu=1}^{M} \sum_{\nu=1}^{M} \frac{J_{\mu\nu}}{M^2} \qquad (5)$$

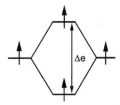

Figure 3. *Spin orbital interaction energy Δe associated with the interaction between two equivalent magnetic orbitals of a spin dimer.*

From the viewpoint of non-orthogonal spin orbitals localized at the spin sites, the antiferromagnetic contribution J_{AF} from each off-diagonal term $J_{\mu\nu}$ ($\mu \neq \nu$) is negligible because the overlap integral between two adjacent spin orbitals of different symmetry is either zero or negligible. Consequently for the discussion of antiferromagnetic spin exchange interactions, it is reasonable to assume that only the M diagonal $J_{\mu\mu}$ terms can contribute significantly to the antiferromagnetic term J_{AF} [6]. Consequently,

$$J \approx \sum_{\mu=1}^{M} \frac{J_{\mu\mu}}{M^2} \tag{6}$$

Therefore the trends in the antiferromagnetic spin exchange parameters J can be correlated with the average of the spin orbital interaction energies $< \Delta e >$ or that of the spin orbital interaction energy squares $<(\Delta e)^2>$ defined as follows [6],

$$< \Delta e > = \sum_{\mu=1}^{M} \frac{\Delta e_{\mu\mu}}{M^2} \tag{7a}$$

$$<(\Delta e)^2> = \sum_{\mu=1}^{M} \frac{(\Delta e_{\mu\mu})^2}{M^2} \tag{7b}$$

where $\Delta e_{\mu\mu}$ is the spin orbital interaction energy associated with the two singly filled molecular orbitals of a spin dimer that result from the spin orbitals μ from the two spin sites.

REPRESENTATIVE APPLICATIONS

Let us now examine the intrachain spin exchange interactions of MnF$_5$-chain containing systems on the basis of the quantitative and qualitative approaches described above. The MnF$_5$ chains are made up of *trans*-corner sharing MnF$_6$ octahedra (Fig. 4), and the Mn^{3+} cations each have four unpaired spins. The spin monomers and spin dimers of these compounds are (MnF$_6$)$^{3-}$ and (Mn$_2$F$_{11}$)$^{5-}$ clusters (Figs. 5a and 5b), respectively. Each (MnF$_6$)$^{3-}$ octahedron has a Jahn-Teller distortion such that the two axial Mn-F

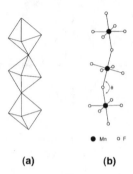

(a)   (b)

Figure 4. *Schematic diagrams of a MnF_5 chain present in $A_2MnF_5 \cdot H_2O$ (A = K, Rb, Cs, Tl), A'Mn $F_5 \cdot H_2O$ (A' = Sr, Ba) and A_2MnF_5 (A = Li, Na, NH_4, Rb, Cs): (a) Polyhedral representation and (b) ball-and-stick representation.*

bonds along the z-axis (Mn-F_{ax}) become longer than the four equatorial Mn-F bonds (Mn-F_{eq}) (Fig. 5a). Thus in the d-block levels of a distorted $(MnF_6)^{3-}$ octahedron, the x^2-y^2 level is empty while the remaining four d-levels are each singly filled. In all the MnF_5-chain containing compounds, the intrachain spin exchange interactions are antiferromagnetic. Table 1 lists the intrachain J values of A_2MnF_5 (A = Rb, Cs) and their ∠Mn-F_{ax}-Mn bridge angles.

To quantitatively determine the intrachain spin exchange parameter of A_2MnF_5 (A = Rb, Cs), DFT electronic structure calculations were performed for the spin dimer

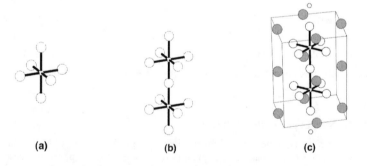

(a)   (b)   (c)

Figure 5. *Schematic representations of (a) the spin monomer $(MnF_6)^{3-}$, (b) the spin dimer $(Mn_2F_{11})^{5-}$, and (c) the spin dimer $(Mn_2F_{11})^{5-}$ plus its nearest-neighbor A^+ and Mn^{3+} cations of the MnF_5 chains of A_2MnF_5 (A = Rb, Cs). The small empty and the large empty circles represent the Mn and F atoms, respectively. The shaded circles represent the A^+ cations.*

$(Mn_2F_{11})^{5-}$ after surrounding it with appropriate point positive charges at the positions of the nearest-neighbor A^+ and Mn^{3+} cations (Fig. 5c) [5]. In first-principles methods (e.g., CI and DFT), the electronic structure of a system is strongly affected by the number of electrons it contains. The high negative charge on the spin dimer $(Mn_2F_{11})^{5-}$ is a consequence of defining a spin dimer as an object for electronic structure calculations.

In first-principles electronic structure calculations, this unphysical situation is corrected by surrounding an isolated spin dimer with a set of point positive charges (at the cation positions of the lattice around the spin dimer) such that the resulting attractive potential removes the unphysical effect of the high negative charge. For the electronic structure of the spin dimer $(Mn_2F_{11})^{5-}$ to be physically meaningful, the point charges on the surrounding cations should be chosen such that each F^- ion of a spin dimer has the net charge close to -1. Table 1 summarizes the intrachain spin exchange parameters J of A_2MnF_5 (A = Rb, Cs) calculated by the total-energy-difference and transition-state methods using the ADF program [19]. The J values calculated by the total-energy-difference method are about three times greater than the experimental values. The J values calculated by the transition-state method deviate from the experimental values by only about 20 % and hence are in near quantitative agreement with experiment [5].

Table 1. Mn-F_{ax}-Mn bridge angles and experimental/calculated intrachain spin exchange parameters J of A_2MnF_5 (A = Rb, Cs).

	Rb_2MnF_5	Cs_2MnF_5
$\angle$Mn-F_{ax}-Mn angle (°)	180.0	180.0
Experimental J/k_B (in K)	-22.6	-19.4
Calculated J/k_B (in K):		
Total-energy-difference method	-69	-61
Transition-state method	-27	-15

The trends in the intrachain J values [20] of $A_2MnF_5 \cdot H_2O$ (A = K, Rb, Cs, Tl), $A'MnF_5 \cdot H_2O$ (A' = Sr, Ba) and A_2MnF_5 (A = Li, Na, NH_4, Rb, Cs) were analyzed from the viewpoint of the qualitative approach [6] by calculating the $\Delta e_{\mu\mu}$ values of the $(Mn_2F_{11})^{5-}$ spin dimers using the extended Hückel method [21,22]. The electronic structure of a system calculated by this semi-empirical method does not depend on the number of electrons the system has [23], which makes it unnecessary to surround the spin dimers with point positive charges in calculating the $\Delta e_{\mu\mu}$ values. Nevertheless, for such calculations, it has been found to be crucial to employ double zeta Slater type orbitals for the transition-metal d- and the ligand s/p-orbitals (e.g., the d-orbitals of Mn and the 2s/2p orbitals of F) [9-15]. In a magnetic orbital of a magnetic solid that has unpaired spins on transition metal atoms, the major component is the transition-metal d-orbital, which combines out-of-phase with the ligand s/p-orbitals. Being the minor component, the ligand s/p-orbitals act as "tails" of a magnetic orbital. The overlap between two adjacent

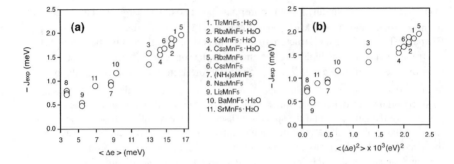

Figure 6. *Plots of (a) the intrachain J versus* $< \Delta e >$ *and (b) the intrachain J versus* $< (\Delta e)^2 >$ *for the MnF$_5$-chain compounds*

magnetic orbitals is primarily governed by how well their tails overlap. In order to adequately describe the tail parts of a magnetic orbital, both contracted and diffuse orbitals are needed for the ligand s/p-orbitals. This explains why the trends in the antiferromagnetic spin exchange parameters of various magnetic solids with one unpaired spin per spin site are explained well by the spin orbital interaction energies calculated with double zeta Slater type orbitals for the transition-metal d- and the ligand s/p-orbitals [9-15]. Fig. 6a shows the J vs. $< \Delta e >$ plot for the MnF$_5$-chain systems [6], and Fig. 6b the corresponding J vs. $< (\Delta e)^2 >$ plot. The magnitude of J increases with increasing $< \Delta e >$ or $< (\Delta e)^2 >$ [6], and both plots exhibit a reasonably good linear relationship. Thus, for the qualitative purpose of discussing the trends in the spin exchange parameters of magnetic solids containing several spins per spin site, it is useful to employ the $< \Delta e >$ or $< (\Delta e)^2 >$ values calculated for their spin dimers using the extended Hückel method.

DISCUSSION AND CONCLUDING REMARKS

For an extended magnetic solid containing several unpaired spins per spin site, the spin exchange parameter of its spin dimer is related to the energy difference between the two spin states Ψ_{hs} and Ψ_{ls} as given by Eq. 2. Thus the spin exchange parameter of a spin dimer can be determined by performing total energy calculations for the two states on the basis of CI wave functions or DFT. Within DFT the spin exchange parameter is related to the orbital energy differences of the transition-state as given by Eq. 4. For the magnetic solids A$_2$MnF$_5$ (A = Rb, Cs), self-consistent-field (SCF) convergence was achieved in the electronic structure calculations for both the total-energy-difference and the transition-state methods [5]. These calculations show that the intrachain spin exchange parameters obtained by the total-energy-difference method are greater than the experimental values by a factor of three approximately, while those obtained by the transition-state method are in near quantitative agreement with experiment. The derivation of Eq. 2 is based on the spin Hamiltonian and hence assumes that the magnetic orbitals representing the states Ψ_{hs} and Ψ_{ls} have identical spatial distributions. In the total-energy-difference method, this assumption cannot be rigorously satisfied because SCF calculations are carried out for two separate states. In the transition-state method the energy difference between the states Ψ_{hs} and Ψ_{ls} is obtained by SCF calculations for one state (i.e., the transition state) so that the assumption is satisfied rigorously. This might explain why the transition-state method leads to a better description than does the total-energy-difference method.

Quantitative calculations of spin exchange parameters using either the total-energy-difference or the transition-state method frequently lead to SCF convergence problems. This difficulty arises mainly from the fact that, as an object for electronic structure calculations, a spin dimer of an extended solid is a highly charged anion cluster. This artificial situation is in part corrected by surrounding a spin dimer with a set of point positive charges designed to provide the potential that the dimer feels in the crystal. Even with this correction, however, the lowest-energy magnetic state of an isolated spin dimer predicted by electronic structure calculations may not necessarily be the magnetic state appropriate for the spin dimer embedded in the crystal. Hence it is a difficult problem to force the SCF convergence to the physically meaningful electronic state of the spin dimer. To be able to carry out first-principles electronic structure calculations for various spin dimers, it is highly desirable to develop an efficient algorithm that ensures SCF convergence to their physically meaningful magnetic states.

It is possible to determine spin exchange parameters of an extended magnetic solid without using the spin dimer method. In this approach, electronic band structure calculations are carried out using DFT for various ordered spin arrangements of a magnetic solid, and then the electronic energy differences between these spin states are projected into the corresponding energy differences resulting from a model spin Hamiltonian that is expressed in terms of the spin exchange parameters to be determined [23]. Regardless of whether one employs this band structure approach or the spin dimer method, first-principles electronic structure calculations of spin exchange parameters become difficult to carry out for magnetic solids with large and complex unit cells. To analyze the spin exchange interactions of such solids, it is necessary to employ a qualitative approach.

The anisotropy in the spin exchange interactions of an extended magnetic solid is intimately related to its crystal structure, i.e., its atom arrangement. To understand this structure-property correlation for a magnetic solid of complex structure, it is necessary to go beyond simulating experimental data (e.g., temperature-dependent magnetic susceptibility or angle-resolved inelastic neutron scattering data) with a spin-Hamiltonian made up of several spin exchange parameters that one perceives as important invariably from the mere inspection of the crystal structure. It is essential to examine how the electronic structures of spin dimers depend on their atom arrangements by performing appropriate electronic structure calculations. The level of calculations one needs depends on the nature of the answer one hopes to find from the calculations. For the quantitative determination of spin exchange parameters, there is no substitute for state-of-the-art calculations on the basis of either CI wave functions or DFT [5,7]. However, in providing a conceptual framework of understanding, theoretical predictions need not be quantitative but provide a slight bias toward correct thinking, and the extended Hückel method has provided such a role for the past three decades and a half [24,25]. For magnetic solids containing one unpaired spin per site, the qualitative trends of their spin exchange interactions are well explained in terms of the spin orbital interaction energies calculated for their spin dimers using the extended Hückel method [9-15]. For magnetic solids containing several unpaired spins per spin site, the antiferromagnetic spin exchange interactions between spin sites result largely from the diagonal terms [6]. Therefore the average of the spin orbital interaction energies $< \Delta e >$ or that of the spin orbital interaction energy squares $< (\Delta e)^2 >$, defined in Eqs. 7a and 7b, respectively, provides a qualitative measure for the relative strength of antiferromagnetic spin exchange interactions of magnetic solids. In this qualitative approach, it is crucial to employ double zeta Slater type orbitals for the transition-metal d- and the ligand s/p-orbitals.

ACKNOWLEDGMENTS

This work was supported by the Office of Basic Energy Sciences, Division of Materials Sciences, U. S. Department of Energy, under Grant DE-FG05-86ER45259.

REFERENCES

1.   O. Kahn, *Molecular Magnetism* (VCH, Weinheim, 1993).

2.   F. D. M. Haldane, *Phys. Lett.* **93A**, 464 (1983); *Phys. Rev. Lett.* **50**, 1153 (1983).

3.   Y. Narumi, M. Hagiwara, R. Sato, K. Kindo, H. Nakano and M. Takahashi, *Physica B* **246-247**, 509 (1998).

4.   M. Oshikawa, M. Yamanaka I. Affleck, *Phys. Rev. Lett.* **78**, 1984 (1997).

5.   D. Dai and M.-H. Whangbo, *J. Chem. Phys.*, in press.

6.   H.-J. Koo, M.-H. Whangbo, S. Coste and S. Jobic, *J. Solid State Chem.*, in press.

7   F. Illas, I. de P.R Moreira, C. de Graaf, and V. Barone, *Theoret. Chem. Accts.* **104**, 265 (2000), and the references cited therein.

8.   L. Noodleman, *J. Chem. Phys.* **74**, 5737 (1981).

9.   K.-S. Lee, H.-J. Koo and M.-H. Whangbo, *Inorg. Chem.* **38**, 2199 (1999).

10   H.-J. Koo and M.-H. Whangbo, *Solid State Commun.* **111**, 353 (1999).

11.   M.-H. Whangbo, H.-J. Koo and K.-S. Lee, *Solid State Commun.* **114**, 27 (2000).

12.   H.-J. Koo and M.-H. Whangbo, *J. Solid State Chem.* **151**, 96 (2000).

13.   M.-H. Whangbo and H.-J. Koo, *Solid State Commun.* **115**, 115 (2000).

14.   H.-J. Koo and M.-H. Whangbo, *J. Solid State Chem.* **153**, 263 (2000).

15.   H.-J. Koo and M.-H. Whangbo, *Inorg. Chem.* **39**, 3599 (2000).

16.   P. J. Hay, J. C. Thibeault and R. Hoffmann, *J. Am. Chem. Soc.* **97**, 4884 (1975).

17. O. Kahn and B. Briat, *J. Chem. Soc. Faraday Trans. II* **72**, 268 (1976).

18. M. F. Charlot and O. Kahn, *Nouv. J. Chim.* **4**, 567 (1980).

19. Amsterdam Density Functional (ADF 2.01) program. (Theoretical Chemistry, Vrije Universiteit, Amsterdam, 1995).

20. P. Núñez and T. Roisnel, *J. Solid State Chem.* **124**, 338 (1996), and the references cited therein.

21. R. Hoffmann, *J. Chem. Phys.* **39**, 1397 (1963).

22. Our calculations were carried out by employing the *CAESAR* program package (J. Ren, W. Liang and M.-H. Whangbo, *Crystal and Electronic Structure Analysis Using CAESAR*, 1998, http://www.PrimeC.com/).

23. For example, see: A. Chartier, P. D'Arco, R. Dovesi and V. R. Saunders, *Phys. Rev. B*, **1999**, *20*, 14042.

24. M.-H. Whangbo, *Theoret. Chem. Accts.* **103**, 252 (2000).

25. R. Hoffmann, *Chem. Eng. News* **52**, 32 (1974) (No. 30, July 29).

Mat. Res. Soc. Symp. Proc. Vol. 658 © 2001 Materials Research Society

Micromagnetic and Magnetoresistance Studies of Ferromagnetic $La_{0.83}Sr_{0.13}MnO_{2.98}$ Crystals

Guerman Popov,[a] Sergei V. Kalinin,[b] Rodolfo A. Alvarez,[b] Martha Greenblatt,[a] and Dawn A. Bonnell [b]

[a] Department of Chemistry, Rutgers, the State University of New Jersey, Piscataway, New Jersey 08854, USA

[b] Department of Materials Science and Engineering, University of Pennsylvania, Philadelphia, Pennsylvania, 19104, USA

ABSTRACT

Magnetic force microscopy (MFM) and atomic force microscopy (AFM) were used to investigate the surface topography and micromagnetic structure of $La_{0.83}Sr_{0.13}MnO_{2.98}$ single crystals with colossal magnetoresistance (CMR). The crystals were grown by fused salt electrolysis and characterized by chemical analysis, X-ray diffraction, magnetic and transport measurements. The crystals are rhombohedral (R$\bar{3}$c). Magnetic and transport measurements indicate that the ferromagnetic ordering at 310 K is associated with an insulator-metal transition at the same temperature. A maximum negative magnetoresistance (-62 %) is observed at 290 K in an applied magnetic field of 5 T. The magnetoresistance increases in magnitude sharply (1.8 %), comparing to the rest of the change, with increasing magnetic field up to 20 G, and then it increases slowly with increasing field.

MFM and AFM were used to study the (110) surface as well as a number of unspecified surfaces. Surface topography of an as-grown crystal exhibits well-developed surface corrugations due to extensive twinning. The corrugation angle at twin boundaries can be related to the unit cell parameters, surface and twinning planes. Magnetic force microscopy images show that magnetic domain boundaries are pinned to the crystallographic twins; a small number of unpinned boundaries are observed. The statistical analysis of domain boundary angle distribution is consistent with cubic magnetocrystalline anisotropy with easy axis along [100] directions for this material. Unusual magnetization behavior in the vicinity of topological defects on the surface is also reported. MFM contrast was found to disappear above the ferromagnetic Curie temperature; after cooling a new magnetic structure comprised of Bloch walls of opposite chiralities developed.

INTRODUCTION

$LnMnO_3$-type rare earth manganates have received recent interest, because of the discovery of colossal magnetoresistance (CMR), the effect by which the electrical resistivity decreases by orders of magnitude upon the application of a magnetic field. CMR properties of polycrystalline materials and thin films have been studied extensively. It was shown that CMR behavior in many cases is associated with magnetic disorder at grain and domain boundaries [1]. Hence, the origins of CMR behavior of these materials can be best understood from the combination of spatially resolved local and single crystal studies.

$La_{1-x}Sr_xMnO_3$ is a typical perovskite CMR system [2-7]. Its structural symmetry changes from orthorhombic to rhombohedral with increasing x up to 0.8. Depending on x in

$La_{1-x}Sr_xMnO_3$ and oxygen stoichiometry the ferromagnetic transition may vary greatly [2, 7] Compositions with T_c above room temperature can be obtained.

In this work we undertake micromagnetic studies on $La_{0.83}Sr_{0.13}MnO_{2.98}$ single crystal While many single crystal growth techniques require special equipment and are in general difﬁ cult to utilize, recently the fused salt electrolysis technique was proven to be an easy method growing single crystals of doped rare earth manganates [2].

EXPERIMENTAL

The synthesis conditions were similar to those described in [2]. The crystals of $La_{0.83}Sr_{0.13}MnO_{2.98}$ were grown by electrolysis in a melt obtained from a mixture of Cs_2MoO_4 and MoO_3 to which $SrMoO_4$, $MnCO_3$ and La_2O_3 were added. The molar ratios were 2.00 : 1.00 0.390 : 0.740 : 0.280 respectively. Electrolysis was carried out in an yttria-stabilized zircon crucible using Pt electrodes, in air, at 975°C for 73 hours using an 18 mA current. Electrolys was terminated by lifting the electrodes from the melt. The use of a zirconia crucible and n alumina was necessary to prevent incorporation of aluminum that can be abstracted from the cru cible [8]. The anode product was washed in a warm solution of 5 % K_2CO_3 and 2.5 % disodiu ethylenediaminetetraacetate.

Chemical analyses for La, Sr and Mn were carried out with a Baird Atomic Model 20" inductively coupled plasma emission spectrometer (ICP). The sample was quantitatively di solved in HCl. The solution was diluted appropriately to bring the concentrations of the studie elements within the range of the available standards. Analyses are accurate to within 1-2 %. Th formal average oxidation state of manganese was established by iodometric titration employin an amperometric dead-stop end point method [9]. Lattice parameters were determined by Rie veld refinement employing the GSAS software package [10]. Data for Rietveld refinement w. collected on finely ground crystals using a Scintag PAD V diffractometer (CuK_α radiation) in th range $2\theta = 10° - 120°$.

The index of the face used in MFM study was determined with a CAD4 diffractomet using graphite monochromatized Mo K_α radiation ($\lambda = 0.71073$ Å). Electrical resistivity mea urements were made by a four-probe technique. Magnetic and magnetoresistance measuremen were performed on a piece containing intergrown crystals in a Quantum Design SQUID magn tometer (MPMS) between 4 K and 400 K.

The AFM and MFM measurements were performed on a commercial instrument (Digit Instruments Dimension 3000 NS-III). Unpoled CoCr coated Magnetic Force Etched Silic Probe (MESP) tips ($l \approx 225$ μm, resonant frequency ~60 kHz, $k \approx 1$-5 N/m) were used. The li height for the interleave scans in the MFM was usually 100 nm. The scan rate varied from 0.2 I for large scans (~60 μm) to 1 Hz for smaller scans (~10 μm). The low coercive field of the CM samples used precluded imaging at lower lift heights and application of strongly magnetized tip

RESULTS AND DISCUSSION

The product crystals grow directly on the anode in the form of batches of black inte grown highly reflecting crystals with cubic-like habits. The dimensions of the individual crysta lites were typically 0.5– 1.5 mm on an edge. The average manganese oxidation state was four to be +3.213. Based on the results of ICP and titration, the composition of the sample is $La_{0.83}Sr_{0.13}MnO_{2.98}$.

The powder X-ray diffraction pattern was indexed in rhombohedral symmetry. No evidence of non-uniform composition of the solid solution $La_{1-x}Sr_xMnO_{3-\delta}$, such as unusual broadening or doubling of peaks, was found. The lattice parameters of the studied crystals are $a = 5.4805(4)$ Å and $\alpha = 60.602(1)°$ in the rhombohedral setting. The lattice parameters in rhombohedral setting indicate that the structure is only slightly distorted from cubic symmetry ($\alpha \approx 60°$)Figure 1a shows the temperature dependence of magnetization at an applied field of 100 G and the temperature dependence of resistivity at zero field, as well as the magnetoresistance at 5 T. Magnetoresistance (MR%) is defined as $[(\rho(5T)- \rho(0))/\rho(0)]\times100$, where $\rho(0)$ and $\rho(5T)$ are the resistivities at an applied field of zero and 5 Tesla respectively. The sample shows typical behavior for the hole-doped manganese perovskites. The ferromagnetic ordering is accompanied by an insulator-metal transition at 310 K. $T_{c, onset}$ of the ordering is 334 K and it is complete at 280 K. Maximum magnetoresistance, –62 %, is observed at 290 K and 5 T (Figure 1a).

Figure 1b presents the hysteresis measurements taken at 293 K. The coercive field of this sample is 21 G, however, saturation is not fully achieved even at an applied field of 5 T. Figure 1c shows the dependence of magnetoresistance on the magnetic field at 300 K. The negative magnetoresistance increases in magnitude with increasing magnetic field, however, there is a

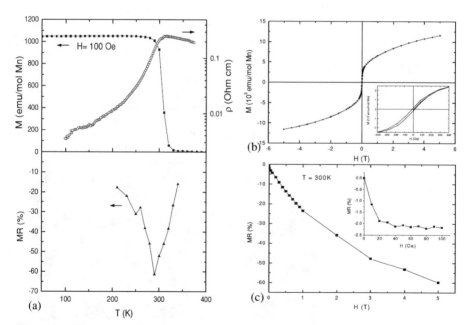

Figure 1. (a) Temperature dependent magnetization, resistivity, magnetoresistance, (b) hysteresis, and (c) field dependence of magnetoresistance of $La_{0.83}Sr_{0.13}MnO_{2.98}$.

sharp negative magnetoresistance change at low field of ~20 G (see Figure 1c inset). Taking into consideration the coercive field value of 21 G we conclude that this sharp change in MR is attributed to magnetic domain reorientation.

The magnetic parameters of $La_{0.83}Sr_{0.13}MnO_{2.98}$ were derived by fitting the magnetic susceptibility in the paramagnetic range to the Curie-Weiss law ($\chi = C/(T-\theta)$). The effective magnetic moment was found to be $\mu_{eff} = 5.61 \mu_B$ and $\theta = 308$ K. From the data of the average manganese oxidation state, the calculated spin-only magnetic moment is $\mu_{calc} = 4.70 \mu_B$. The experimental value of the magnetic moment is larger than expected (μ_{calc}), which can be attributed to the formation of superparamagnetic clusters [11].

The surface topography and the magnetic structure of the crystals were studied by various scanning probe microscopy (SPM) techniques [12]. The presence of crystallographic twin boundaries results in formation of surface corrugations on the topographical image (Figure 2). The images indicate a high degree of twinning with characteristic twin sizes of 5-20 μm, which is expected for this type of material [3]. The corrugation angles for the twins were $1.4° \pm 0.1°$ (Figure 2a) and $0.9° \pm 0.1°$ (Figure 2b). This corrugation angle is determined by bulk crystallographic structure, surface and twinning plane and it is unique for a given combination of the three.

Figure 3 shows the surface topography and the micromagnetic structure of the (110) surface plane. The magnetic domain boundaries are pinned to the twin boundaries, although there is a small number of unpinned boundaries. This behavior can also be observed in Fig. 4 along with various micromagnetic structures on different unspecified surface planes. Angles between domain walls on the (110) surface have well-defined bimodal distribution (Fig. 3c) consistent with cubic magnetocrystalline anisotropy. Simultaneously performed surface potential imaging (SSPM) indicates that the surface potential is uniform within the areas studied as expected on a conductive surface. This indicates the absence of a second phase on surfaces and evidences a purely magnetic origin of MFM images.

Topographic defects (unspecified surface) are associated with an unusual local distortion of micromagnetic structure (Fig. 5). Increased MFM contrast indicates larger out-of-plane magnetization. As confirmed by energy dispersive X-ray analysis, the defect is purely topographical and there is no associated impurity inclusion. Noteworthy is the characteristic labyrinthine magnetization pattern, which is typical for thin films and amorphous materials.

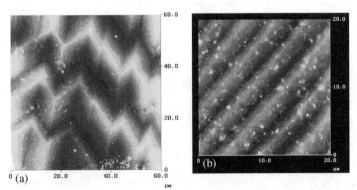

Figure 2. *Surface topography of two regions on as grown LSMO crystals surface.*

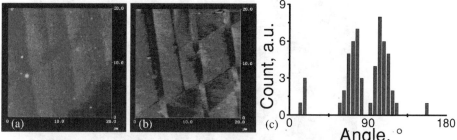

Figure 3. (a) Surface topography, (b) micromagnetic structure, and (c) domain angle distribution of the (110) surface plane (rhombohedral setting).

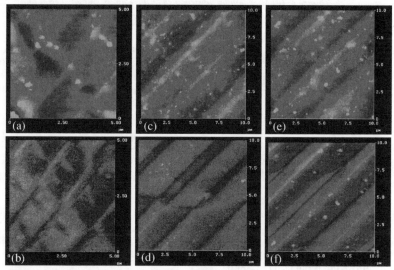

Figure 4. (a, c, e) Surface topographies and (b, d, f) corresponding micromagnetic structures of various unspecified surface planes.

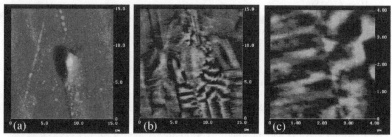

Figure 5. (a) Surface topography and (b, c) micromagnetic structure of the unspecified surface with a topographic defect on it.

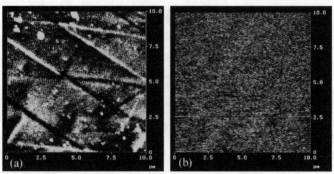

Figure 6. (a) Micromagnetic structure of the rapidly cooled crystal and
(b) micromagnetic structure observed above the Curie temperature.

Heating the crystal above the Curie temperature and quickly cooling it down to room
temperature resulted in a domain structure comprised of 180° Bloch walls with different chiral
ties (Figure 6a). Heating this crystal again above the Curie temperature resulted in the disappear
ance of domain structure (Figure 6b) in agreement with theoretical expectations.

ACKNOWLEDGEMENTS

The authors would like to thank Dr. W.H. McCarroll for discussions and help with IC
and Dr. T.J. Emge for single crystal X-ray diffraction measurements.

REFERENCES

1. J.Z. Sun, A. Gupta, *Annual Rev. Mater. Sci.* **28**, 29 (1998) and references therein.
2. W.H. McCarroll, K.V. Ramanujachary, I. Fawcett, and M. Greenblatt, *J. Solid State Chem.*
    145, 88 (1999).
3. M. Déchamps, A.M. de Leon Guevara, L. Pinsard and A. Revcolevschi, *Philosophical
    Magazine* A **80**(1), 119 (2000).
4. V.A. Cherepanov, L.Yu. Barkhatova, and V.I. Voronin, *J. Solid State Chem.* **134**, 38 (1997
5. A. Maignan, V. Caignaert, Ch. Simon, M. Hervieu and B. Raveau, *J. Mater. Chem.* **5**(7),
    1089 (1995).
6. M.S. Osofsky, B. Nadgorny, R.J. Soulen, Jr., P. Broussard, M. Rubinstein, J. Byers,
    G. Laprade, Y.M. Mukovskii, D. Shulyatev, and A. Arsenov, *J. Appl. Phys.* **85**, 5567 (1999
7. N. Abdelmoula, K. Guidara, A. Cheikh-Rouhou, and E. Dhari and J.C. Joubert, *J. Solid
    State Chem.* **151**, 139 (2000).
8. W.H. McCarroll, K.V. Ramanujachary, M. Greenblatt, and F. Cosandey, *J. Solid State
    Chem.* **136**, 322 (1998).
9. F. Licci, G. Turilli, P. Ferro, *J. Magn. Magn. Mater.* **164**, L268 (1996).
10. A.C. Larson and R.B. Von Dreele, General Structure Analysis System (GSAS), Los Alamo
    National Laboratories, Report LAUR 86-748 (1994).
11. J. Töpfer and J.B. Goodenough, *J. Solid State Chem.* **130**, 117 (1997).
12. Scanning Probe Microscopy: Theory, Techniques and Applications, Ed. D.A. Bonnell,
    Wiley VCH, 2000.

Mat. Res. Soc. Symp. Proc. Vol. 658 © 2001 Materials Research Society

Resistive Anomaly Relevant to Nd Moments in the Antiferromagnetic Phase of the Bandwidth-Controlled Manganites

H. Kuwahara[1,2*], K. Noda[1], and R. Kawasaki[1],
[1]Sophia University, 7-1 Kioi-cho, Chiyoda-ku, Tokyo 102-8554, Japan
[2]PRESTO, Japan Science and Technology Corporation (JST), Tokyo, Japan

ABSTRACT

We have investigated the electronic and magnetic properties of $(Nd_{1-y}Sm_y)_{0.45}Sr_{0.55}MnO_3$ ($0 \leq y \leq 1$) crystals, in which one-electron bandwidth (W) is systematically decreased from the parent compound ($y=0$) with increase of y. We have found remarkable magnetic phase transition concerning the change of rare-earth (RE) moments in low temperatures below 25 K. The subtle drop in resistivity superimposed upon the spin-valve like magnetoresistance (MR) was observed for the isothermal MR measurements, e.g. $\Delta\rho(H)/\rho(H') \approx 4.7\%$ at 3.5 T and 4 K. The phase transition fields corresponding to these concomitant magnetic and resistive changes monotonically decrease with temperature and disappear above ~ 25 K. It turned out that the resistive drop is due to the field-induced increase of magnetic moments of RE ions from magnetization measurements. The field-induced phase transition from the small moment state to the large one in RE ions can be explained in terms of energy level crossing between the crystal-field-split J multiplets by the Zeeman effect.

INTRODUCTION

The parent $Nd_{0.45}Sr_{0.55}MnO_3$ ($y=0$) compound, which shows the layered antiferromagnetic (AF) spin structure and the extremely large anisotropy in resistivity, has attracted much interest from the view point of application such as a spin-valve-type MR device and of fundamental science such as spin-charge coupled phenomena, orbital ordering, and so on [1, 2]. In this compound, the observed large anisotropy in spite of the nearly cubic crystal structure is thought to be due to confinement of the spin-polarized carriers within the ferromagnetic sheets induced by the magnetic-ordering as well as $3d_{x^2-y^2}$ type orbital-ordering [2]. Akimoto et al. have recently reported on AF metallic states in $(Nd_{1-y}La_y)_{0.46}Sr_{0.54}MnO_3$ by increase of W [3]. In the present work, we have investigated electronic and magnetic properties of $(Nd_{1-y}Sm_y)_{0.45}Sr_{0.55}MnO_3$ ($0 \leq y \leq 1$) crystals, in which W is systematically decreased with increase of Sm content y. We have found the metal-insulator phase boundary in the ground states of this system: The anisotropic metallic state with AF structure ($y=0$) has been transferred to the AF insulator ($0.3 \leq y$) in zero field (See the inset of Figure 1) [4]. This (Nd,Sm)-based compound gives a unique opportunity to investigate the influence of magnetic state of the RE ions upon the charge dynamics of conducting Mn $3d$ e_g electrons. Our motivation is to clarify the interaction between Mn $3d$ electrons and RE $4f$ spins. To the best of our knowledge, there is no report on the charge transport concerning the RE spins so far. Only specific heat and neutron scattering measurements have shown the magnetic states of RE ions, however resistivity measurements have not detected the resistive anomaly due to RE moments yet [5, 6]. In this study, we have observed a very small change in resistivity arises from the magnetic phase transition from the small-moment state to the large one in RE ions. We show evidence of the phase transition by systematic measurements of resistivity and magnetization as a function of temperature and applied magnetic fields.

*h-kuwaha@sophia.ac.jp

EXPERIMENT

Crystals of $(Nd_{1-y}Sm_y)_{0.45}Sr_{0.55}MnO_3$ ($0 \leq y \leq 1$) were grown by the floating zone metho with use of a lamp-image furnace; details are published in Ref. [7]. Powder X-ray diffrac tion (XRD) was used to check the crystal quality and to determine the lattice parameters Rietveld refinement of XRD data for the pulverized crystals indicated that all peaks ar indexed in the orthorhombic crystal symmetry ($Pbnm$) without impurity phases. There i no crystallographic phase boundary in this system over the whole Sm-substitution range Resistivity measurements were performed using the conventional four-probe method. Resis tivity in magnetic fields up to 8 T was measured by using a cryostat with a superconductin magnet equipped with liquid-He free Gifford-McMahon (GM) refrigerator. Magnetizatio was measured by using a SQUID magnetometer. Although anisotropy is important in th system, the resistivity and magnetization measurements were performed on randomly cu samples (the current direction is not specified), because the multi-domain structure cannc thoroughly be eliminated in the present nearly cubic crystal.

RESULTS AND DISCUSSION

First of all, we show in Figure 1 the $M-T$ curves for the whole $(Nd_{1-y}Sm_y)_{0.45}Sr_{0.55}MnC$ series of crystals. The Néel temperatures (T_N) and paramagnetic Curie temperatures (Θ_f for the crystals monotonically decrease with increase of Sm content y, as simply expecte from decrease of W with increase of y. As one can immediately notice from the figure, mag netization increases with lowering temperature below 20 K. This paramagnetic Curie-lik behavior is attributed to the RE moments, because the Mn $3d$ spins are well ordered antife romagnetically in this temperature region as evidenced by neutron scattering measuremen of the $y=0$ sample [1, 8]. We have fit the raw data to Curie-Weiss behavior and calculate the effective moments p_{eff} of RE ions by subtracting the Mn-spin contributions: the p_{eff} li early decreases from 2.84 μ_B/(Nd ion) ($y=0$) to 0.88 μ_B/(Sm ion) ($y=1$). These values ar qualitatively consistent with the estimated ones of the ground state of each free-ion: Nd^3 ($y=0$) is 3.62 μ_B (($4f$)^3, $^4I_{9/2}$) and Sm^{3+} ($y=1$) is 0.84 μ_B (($4f$)^5, $^6H_{5/2}$). The Weiss ten perature remains constant at around zero Kelvin, independent of y. We have paid attentic to this low temperature region, in which we have investigated the effect of $4f$ spins upon e_g conduction electrons as discussed later.

The inset in Figure 1 shows the conductivity, $\sigma_0(T=0)$, in zero field as a function of y. Th metal-insulator phase boundary exists in the vicinity of $y = 0.3$ when we define the insulate as σ is zero at a finite temperature. The anisotropic metallic state with AF structure ($y=$ has been transferred to the AF insulator ($0.3 \leq y$) in zero field. The critical compositio y, changes from 0.3 to 0.5 by application of a magnetic field of 12 T. This implies the the magnetic field induced the spin-canting and broadening-W and the resultant insulato to-metal transition [4]. Moreover, we have observed the metamagnetic transition from t layered AF state to the three dimensional ferromagnetic state in the magnetic field of 35 T l using a nondestructive long-pulse magnet [9]. In accord with the metamagnetic transition, large negative MR was also observed, e.g., $\rho(19T)/\rho(0)=0.13$ (10 K), which arises from t destruction of the $d_{x^2-y^2}$ type orbital-ordering and the complete mixture of the $d_{3z^2-r^2}$ ar $d_{x^2-y^2}$ orbitals induced by high magnetic fields. This indicates the dimensional crossover d to the field-induced canting of not only the spin but also the orbital.

We show in the left panels of Figure 2 the isothermal MR (upper) at fixed temper tures below 15 K. The observed large negative-MR background arises from the field-induce spin-canting, which gives rise to the revival of the electron hopping along the AF-coupli

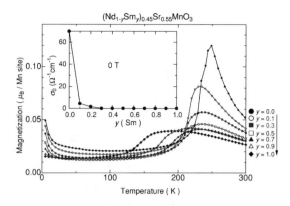

Figure 1: *The temperature dependence of magnetization for the $(Nd_{1-y}Sm_y)_{0.45}Sr_{0.55}MnO_3$ $(0 \leq y \leq 1)$ crystals. Magnetizations were measured on warming runs in the magnetic field of 0.5 T after the samples were cooled in zero magnetic field (ZFC). The $y=0$ sample shows the layered antiferromagnetic structure below T_N of 220 K [1, 8]. The inset shows the extrapolated conductivity, σ_0, in zero magnetic field and at zero Kelvin as a function of Sm content y. The values of σ_0 are determined by the linear extrapolation of $\sigma - T$ curves below 20 K.*

direction (c-axis) [2]. This is quite a different mechanism from the conventional CMR, i.e., the field-induced suppression of spin scattering in the paramagnetic phase near the ferromagnetic transition temperature (T_c). In the present case of the layered AF phase, a small ferromagnetic moment appears to induce such a large MR as observed. The normalized MR value $\rho(H)/\rho(0)$ measured in the wide temperature ranges below T_N is approximately scaled with an empirical function of the normalized magnetization M/M_s ($M_s \approx 3.45\mu_B$/Mn being the saturated magnetization): $\rho(H)/\rho(0) = 1 - C(M/M_s)^\alpha$, where the scaling parameter C was estimated to be ~ 7 and the critical exponent α was ~ 1.5.

As shown in the Figure 2(a), we have found another remarkable transition concerning the RE moments at low temperatures. The very subtle drop in resistivity superimposed upon the above mentioned spin-valve like MR was observed for the isothermal MR measurements, e.g., $\Delta\rho(H)/\rho(H') \approx 4.7$ % at 3.5 T and 4 K. The transition fields for these anomalies monotonically decrease with increase of temperature and disappear above ~ 25 K. (See downward open-triangles in lower panel of Figure 2(a)). To reveal the anomalous resistive drops, we have measured the $M - H$ curves (Figure 2(b)) at the same temperatures as taken the $\rho - H$ curves. In accord with the changes in resistivity, slopes of magnetization at the transition fields slightly increase in the field-increasing runs. (See downward open-triangles in lower panel of Figure 2(b)). Namely, it turned out that the resistive drop is due to the field-induced increase of RE magnetic moments. To clarify the magnetic phase transition of RE spins, the temperature dependence of magnetization was measured in several magnetic field intensities. Figure 3 shows the $M-T$ curves (a) and the Curie-Weiss plots (b) calculated from the $M - T$ curves. As one can see from Figure 3(b), the small magnetic moment state (upper branch) is shifted to the large moment one (lower branch) with temperature. The p_{eff} of each state was estimated to be $\sim 2.80\mu_B/RE$ ion (small moment state) and $\sim 3.53\mu_B/RE$ ion (large one) by means of subtracting the temperature independent magnetization due to field-induced spin canting of manganese ions (M_{min}). The phase transition temperatures

$$(Nd_{0.9}Sm_{0.1})_{0.45}Sr_{0.55}MnO_3$$

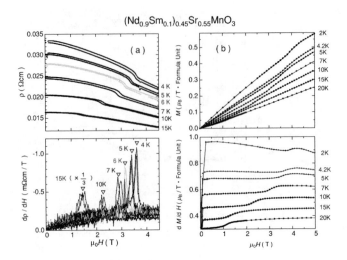

Figure 2: *The isothermal MR (left-upper panel) and M − H curves (right-upper panel) for the $(Nd_{0.9}Sm_{0.1})_{0.45}Sr_{0.55}MnO_3$ crystal at several fixed temperatures below 15 K. The curves in each lower panel are derivatives and slopes of those in each upper panel for emphasizing the inflection points, at which the phase transition fields were indicated by downward open-triangles. See also the phase diagram in Figure 4.*

indicated by downward open-triangles agree well with that obtained by the isothermal $\rho - H$ and $M − H$ measurements.

Let us consider the origin of the magnetic phase transition of the RE ions. As is well known, the ground-state J-multiplets of Nd^{3+} and Sm^{3+} are $^4I_{9/2}$ and $^6H_{5/2}$, respectively. The degeneracy of ground-state J multiplet of a RE ion in a crystal lattice is partly removed by the crystalline electric field interactions. In the case of Nd^{3+}, tenfold degeneracy of the $^4I_{9/2}$ ground state splits into five Kramers doublets. According to the recent neutron inelastic scattering measurements of the RE nickelates with the similar perovskite crystal structure [10], the lowest and second-lowest lying J multiplets of the Nd^{3+} ion are $^4I_{9/2}$ and $^4I_{11/2}$, respectively, which are split by the spin-orbit interaction and are separated by ~160 meV. The highest energy level of Kramers doublets of the ground $^4I_{9/2}$ state approaches the lowest level of that of the second-lowest lying $^4I_{11/2}$ state by the crystalline field splitting. As a result, energy level mixing between the different J multiplets (different spin states) can probably occur with increase of temperature. Application of a magnetic field, which induces the Zeeman splitting of the Kramers-conjugate state, also favors such a level mixing. In the case of Sm^{3+}, the energy separation from the highest level of the $^6H_{5/2}$ state to the lowest level of the $^6H_{7/2}$ state is ~20 meV [10], which is similar to the energy scheme in the case of Nd^{3+}. We have not performed neutron scattering experiments on the present samples yet. If the energy level of the present system is similar to the above mentioned nickelates, the origin of the phase transition from the small moment state to the large one in RE ion is thought to be a "metamagnetic" transition of the $4f$ spin by application of magnetic field and temperature. The level mixing between the different J multiplets induced by magnetic field is often observed for the heavy Fermion systems such as Pr.

$(Nd_{0.9}Sm_{0.1})_{0.45}Sr_{0.55}MnO_3$

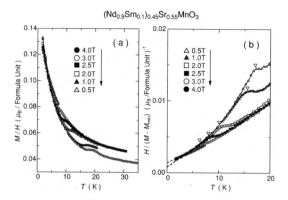

Figure 3: *The temperature dependence of magnetization (a) and inverse magnetization (b) normalized by applied magnetic fields for the $(Nd_{0.9}Sm_{0.1})_{0.45}Sr_{0.55}MnO_3$ crystal. The Curie-Weiss plot (b) is deduced by subtracting the temperature independent magnetization due to field-induced spin canting of manganese ions (M_{min}). The transition temperatures are defined as inflection points indicated by downward open-triangles in the right panel (b), in which the dashed lines in low temperatures mean the least-squares fittings of upper branch (small magnetic moment $\sim 2.80\mu_B/RE$ ion) and lower one (large one $\sim 3.53\mu_B/RE$ ion).*

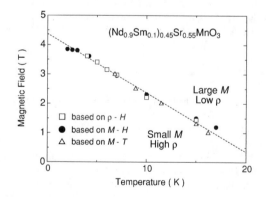

Figure 4: *The electronic and magnetic phase diagram in the temperature−magnetic-field plane for the $(Nd_{0.9}Sm_{0.1})_{0.45}Sr_{0.55}MnO_3$ crystal. The phase boundary (dashed line) was determined by the isothermal MR (open squares), by the $M - H$ measurements at fixed temperatures (solid circles), and by the temperature dependence of magnetization in magnetic fields (open triangles). The transition magnetic fields in the isothermal measurements and the transition temperatures in the $M - T$ measurements in fixed magnetic fields are defined as the inflection points of the each measurement. See also the downward open-triangles in the lower panels of Figure 2 and Figure 3(b).*

Figure 4 shows the thus-obtained electronic and magnetic phase diagram for t (Nd$_{0.9}$Sm$_{0.1}$)$_{0.45}$Sr$_{0.55}$MnO$_3$ crystal. Each transition point determined by the $\rho - H$, $M -$ and $M - T$ measurements coincides with the other. The transition magnetic fields from t small magnetic moment and high resistive state to the large moment and low resistive o linearly decrease with temperature and disappear above 25 K. We have carried out simi experiments for whole solid solution series. The obtained phase diagrams indicate that i crease of Sm content y, or equivalently decrease of W and magnetic moment of RE io induces a systematic expansion of low resistive states.

CONCLUSIONS

We have obtained the electronic and magnetic phase diagram for the RE-based mangan relevant to the RE spins in low temperatures by systematic measurements. We have fou resistive drops due to the "metamagnetic" transition of the RE ions in the layered AF ph of the manganites. These results suggest that the charge-transport of $3d$ electrons can controlled through the change of not only $3d$ spins but also $4f$ RE spins by application magnetic fields. In this study, there is no evidence of the ferromagnetic or AF ordering the RE $4f$ spins. Such orderings may occur by means of interaction caused by the coupl between $4f$ spins and $3d$ conducting electrons, for example, RKKY-like interactions.

ACKNOWLEDGMENTS

The authors would like to thank R. Kajimoto, H. Kawano, and H. Yoshizawa for neutr scattering measurements, and N. Miura and T. Hayashi for magnetization measurements i pulsed ultra-high magnetic field, and Y. Tomioka, Y. Taguchi, and Y. Tokura for permitt our using a SQUID magnetometer. The present work is partly supported by Grant-in-ℓ for Scientific Research on Priority Areas (A) No. 12046255 from the Ministry of Educati Science, Culture and Sport of Japan.

REFERENCES

1. H. Kawano, R. Kajimoto, H. Yoshizawa, Y. Tomioka, H. Kuwahara and Y. Toku Phys. Rev. Lett. **78**, 4253 (1997).
2. H. Kuwahara, T. Okuda, Y. Tomioka, A. Asamitsu and Y. Tokura, Phys. Rev. Lett. 4316 (1999).
3. T. Akimoto, Y. Maruyama, Y. Moritomo, A. Nakamura, K. Hirota, K. Ohoyama ℓ M. Ohashi, Phys. Rev. **B 57** R5594 (1998).
4. K. Noda, R. Kawasaki, and H. Kuwahara, (in preparation).
5. J.E. Gordon, R.A. Fisher, Y.X. Jia, N.E. Phillips, S.F. Reklis, D.A. Wright and A. Ze Phys. Rev. **B 59**, 127 (1999).
6. J.-G. Park, M.S. Kim, H.-C. Ri, K.H. Kim, T.W. Noh and S.-W. Cheong, Phys. Rev **60**, 14804 (1999).
7. H. Kuwahara, Y. Moritomo, Y. Tomioka, A. Asamitsu, M. Kasai, R. Kumai ϵ Y. Tokura, Phys. Rev. **B 56**, 9386 (1997).
8. R. Kajimoto, H. Yoshizawa, H. Kawano, H. Kuwahara, Y. Tokura, K. Ohoyama ϵ M. Ohashi, Phys. Rev. **B 60**, 9506 (1999).
9. T. Hayashi, K. Noda, H. Kuwahara, and N. Miura, (in preparation).
10. S. Rosenkranz, M. Medarde, F. Fauth, J. Mesot, M. Zolliker, A. Furrer, U. Sta P. Lacorre, R. Osborn, R.S. Eccleston, V. Trounov, Phys. Rev. **B 60**, 14857 (1999)

Mat. Res. Soc. Symp. Proc. Vol. 658 © 2001 Materials Research Society

Temperature Dependence of Magnetic Compton Profile in DyCo5

Hayato Miyagawa, Yasuhiro Watanabe, Akihisa Koizumi[1],
Nobuhiko Sakai[1], Masaichiro Mizumaki[2],
Yoshiharu Sakurai[2], Tetsuya Nakamura[3] and Susumu Nanao
Institute of Industrial Science, University of Tokyo
7-22-1, Roppongi, Minato-ku, Tokyo 106-8558, Japan
[1]Faculty of Science, Himeji Institute of Technology,
3-2-1, Kouto, Kamigori, Ako-gun, Hyogo 678-1297, Japan
[2]Japan Synchrotron Radiation Research Institute (JASRI),
1-1-1, Kouto, Mikazuki-cho, Sayo-gun, Hyogo 679-5198, Japan
[3]RIKEN (The Institute of Physical and Chemical Research),
2-1, Hirosawa, Wako, Saitama 351-0198, Japan

ABSTRACT

Magnetic Compton profiles (MCPs) of a $DyCo_{5.4}$ single crystal were measured at 10 K, 200 K and 300 K. The temperature dependence of the spin moment, which is deduced from the areas under the normalized MCPs, is significantly different from that of the total magnetization measured by a superconducting quantum interference device (SQUID). This difference is due to the presence of a substantial amount of the orbital moment on a Dy site that does not contribute to the magnetic Compton scattering cross section. The analysis of the MCPs reveals that the absolute value of the spin moment increases with increasing temperature and that the spin magnetic moment of the conduction electrons has an opposite sign to the total spin magnetization in the covered range of temperature.

INTRODUCTION

Magnetic Compton Scattering (MCS) is a very powerful tool for the study of various magnetic materials[1]. This technique has been demonstrated by Sakai *et al.* using a γ-ray source [2] for the first time, and then has been developed further since circularly polarized intense x-rays from synchrotron radiation sources have become available. Within the impulse approximation [3], the Compton scattering cross section is proportional to the normal Compton profile (NCP), $J(p_z)$, which is defined as the one-dimensional projection of the electron momentum density, $n(\mathbf{p})$, onto the scattering vector taken along the z-axis,

$$J(p_z) = \int\int n(\mathbf{p})\, dp_x dp_y = \int\int [\, n_\uparrow(\mathbf{p}) + n_\downarrow(\mathbf{p}) \,]\, dp_x dp_y , \qquad (1)$$

where $n_\uparrow(\mathbf{p})$ and $n_\downarrow(\mathbf{p})$ depending on the momentum are the majority and minority spin densities, respectively. If the incident x-rays are circularly polarized, a small spin-dependence arises in the scattering cross section. Reversing the magnetization of a sample changes the sign of the spin-dependent contribution, which enables us to separate the spin part of the cross section. The resultant profile, known as the magnetic Compton profile (MCP), is a projection of the momentum density of electrons with unpaired spins (Eq. 2).

$$J_{mag}(p_z) = \int\int [\, n_\uparrow(\mathbf{p}) - n_\downarrow(\mathbf{p}) \,]\, dp_x dp_y \qquad (2)$$

$J(p_z)$ and $J_{mag}(p_z)$ are normalized by the following conditions,

$$\int_{-\infty}^{+\infty} J(p_z)\, dp_z = N \qquad\qquad (3)$$

$$\int_{-\infty}^{+\infty} J_{mag}(p_z)\, dp_z = F_s \qquad\qquad (4)$$

where N and F_s are the number of total electrons and the total spin moment per formula unit, respectively. Therefore, the area of the normalized MCP is proportional to the spin moment. The MCS is solely sensitive to the spin moment; the orbital one is not measured. This feature is one of the important differences from x-ray elastic magnetic scattering, which measures both spin and orbital contributions to the scattering cross section [4]. Another difference is that the MCS easily probes the spin polarization of conduction electrons, since they are localized around $p_z = 0$ in momentum space.

The magnetic compound $DyCo_5$ has a $CaCu_5$ type structure, where the magnetic moment on the Dy site is ferrimagnetically coupled with that on the Co site through a negative exchange interaction between the Dy $5d$ electrons and Co $3d$ electrons [5]. A compensation point has been observed at 148 K in the spontaneous magnetization curve in $DyCo_{5.2}$ single crystals[5]. Local anisotropies of the Dy and Co atoms are of opposite sign, which leads to a spin reorientation between 325 K and 367 K due to the varying dependence of the anisotropies on the temperature [6]. Although macroscopic anisotropies associated with the spin reorientation in $DyCo_5$ have been studied [7, 8], reliable values of magnetic moments on the Dy and Co sites have not been reported because of the high absorption cross-section of Dy for thermal neutrons. In the present study the MCS technique has been applied to $DyCo_5$ to elucidate detailed information on the magnetic moments.

EXPERIMENTAL DETAILS

A spherical single crystal of $DyCo_5$ with a radius of 2 mm was cut from an ingot grown by the Bridgman method. The composition was found to be $DyCo_{5.4}$ by inductively coupled plasma spectroscopy. The magnetization was measured with a SQUID up to 2 T.

The MCPs for $DyCo_5$ were measured on the high-energy inelastic scattering beamline (BL08W) at SPring-8. A schematic drawing of the experimental arrangement is shown in Figure 1. The incident x-ray energy of 274 keV was selected by the (771) reflection of an asymmetric Johann monochromator. The energy spectra of Compton scattered x-rays in the sample were measured by a Ge solid state detector (SSD) with a scattering angle of 175 degrees. The overall momentum resolution was 0.66 atomic units (a.u., where 1 a.u. = 1.99 x 10^{-24} kg m/s). The sample's magnetization was reversed by a 2 T superconducting magnet [9] in the sequence of (+,-,-,+,-,+,+,-), where (+) represents the direction of the magnetic field parallel to the scattering vector and (-) vice versa. The switching time was 6 seconds and the dwell time was 60 seconds. The MCPs were measured along the [-2, 1, 0] direction at 10 K, 200 K and 300 K. At these temperatures, the magnetic field of 2 T is sufficient to saturate the magnetization. The total number of counts accumulated in each NCP was ~ 8 x 10^7, leading to ~ 3 x 10^5 in the MCPs.

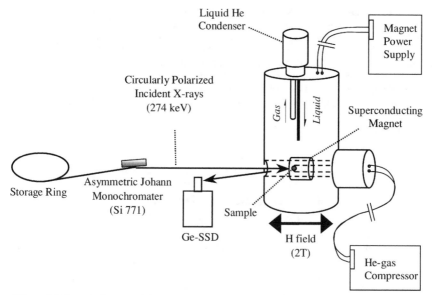

Figure 1. Schematic drawing of the experimental arrangement at BL08W of SPring-8.

RESULTS AND DISSCUSSION

The total moment on the Dy site is ferrimagnetically coupled with that on the Co site. On the Dy site, the spin moment is aligned with the large orbital moment. The spin moment on the Co site is aligned opposite to the moments on the Dy site, where the Co orbital moment is almost quenched. Below the compensation temperature at 130 K (see Figure 3), the total magnetization in DyCo$_5$ is dominated by Dy atoms. Then the spin moment on the Co site is aligned antiparallel to the external magnetic field, which brings a negative MCP at T = 10 K in Figure 2(a). Above 130 K, Co atoms become dominant in the total magnetization, then positive MCPs are observed at 200 K and 300 K as shown in Figure 2(b) and 2(c). The experimental MCPs exhibit similar shapes above and below the compensation temperature, although their signs are opposite. This fact shows that the spin magnetization produced by five Co atoms in the formula unit is much larger than that produced by one Dy atom over the temperature range of the study.

The integrated area of the MCP divided by that of the NCP gives the relative value of the spin moment. The temperature dependence of the spin moment is shown in Figure 3, together with the magnetization curve measured by SQUID with a magnetic field of 2 T. It is found that the temperature dependence of the spin moment is different from that of the magnetization curve. The difference between them is well explained by the presence of a substantial amount of the orbital moment on the Dy site, which is not detected by the MCS technique.

The line-shape analysis of MCP in terms of the characteristically different orbital profiles has been demonstrated in several cases [1].   Present results in Figure 2 clearly show that there is a spin moment associated with conduction electrons, opposed to ones associated with the localized Co 3d electrons.   The conduction electron part forms a dip at the low momenta ($|P_z| < 1.5$ a.u.), while the localized part exhibits a single peak profile extended to $|P_z| = 5$ a.u.   The former is superimposed on the latter with an opposite sign.   Thus, figure 2 shows the fact that the spin magnetic moment of the conduction electrons has an opposite sign to the total spin magnetization in DyCo$_5$ in the covered range of temperature.

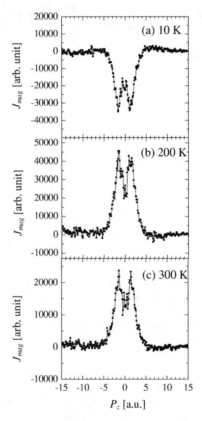

Figure 2.   *MCPs for DyCo$_5$ at (a) 10 K, (b) 200 K and (c) 300 K. The scattering vector is along the [-2, 1, 0] direction.*

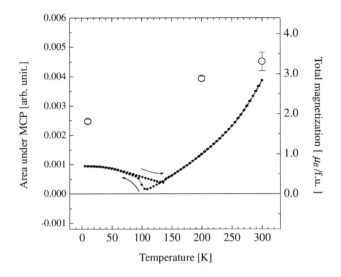

Figure 3. *Area under MCP (open circle) and total magnetization (filled circle with a solid line) in DyCo₅. The area under MCP, which is normalized by that under NCP, is proportional to the spin moment in DyCo₅.*

CONCLUSIONS

Magnetic Compton profiles for $DyCo_{5.4}$ have been measured at 10 K, 200 K and 300 K using circularly polarized synchrotron x-rays. The temperature dependence of the spin moment in $DyCo_{5.4}$ has been deduced from the area of MCP normalized by that of NCP. The results show that the absolute value of the spin moment increases with increasing temperature and that the spin magnetic moment of the conduction electrons has an opposite sign to the total spin magnetization in $DyCo_5$.

ACKNOWLEDGEMENT

This work has been performed at SPring-8 with the approval of Japan Synchrotron Radiation Research Institute (JASRI) (Proposal No. 1999B0370-ND-np). The authors are grateful to Dr. N. Hiraoka and Mr. Y. Kakutani of Faculty of Science, Himeji Institute of Technology, and Dr. M. Itou of JASRI for giving us both experimental and theoretical supports. This work was partially supported by Grant-in-Aid for Scientific Research from the Japanese Ministry of Education, Science, Sports and Culture.

REFERENCES

1. N.Sakai, *J. Appl. Cryst.*, **29**, 81, (1996).
2. N. Sakai and K. Ono, *Phys. Rev. Lett.*, **37**, 351, (1976)
3. P. M. Platzman and N. Tzoar, *Phys. Rev.*, **B2**, 3556, (1970)
4. M. Blume and D. Gibbs, *Phys. Rev.*, **B37**, 1779, (1988)
5. J. P. Liu, X. P. Zhong, and F. R. de Boer, *J. Appl. Phys.*, **69**, 5536, (1991)
6. D. Gignoux and D. Schmitt, *J. Magn. Magn. Mater.*, **100**, 99, (1991)
7. A. G. Berezin, *Sov. Phys. : JETP*, **52**(1), 135, (1980)
8. A. G. Berezin, *Sov. Phys. : JETP*, **52**(3), 561, (1980)
9. N. Sakai, *J. Synchrotron Rad.*, **5**, 937, (1998)

New Materials,
Meso/Nanoporous Materials

Mat. Res. Soc. Symp. Proc. Vol. 658 © 2001 Materials Research Society

Effect of Polymerization of Precursor Solutions on Crystallization and Morphology of $Ce_{0.9}Gd_{0.1}O_{1.95}$ Powders

Shuqiang Wang, Masanobu Awano and Kunihiro Maeda
Synergy Ceramics Laboratory, Fine Ceramics Research Association,
Shidami Human Science Park, Shimo-Shidami, Moriyama-ku, Nagoya 463-8687, Japan

ABSTRACT

Synthesis of $Ce_{0.9}Gd_{0.1}O_{1.95}$ (CGO) powders from polymeric precursor solutions formed by mixing nitrates and ethylene glycol at 60-85℃ was investigated with emphasis on the effect of the polymerization of the precursor solution on the crystallization and morphology of the derived intermediate and the resultant oxide powders. It was revealed by FTIR that the molecular structures of the polymeric precursor solution change from aldehyde or ketone groups to carboxylic acid and carboxylate groups with increasing heating time. TG-DTA analyses demonstrated the temperature shifting and the disappearance of the exothermic reactions of the derived powders with different heating times of the polymeric precursor solutions. Furthermore, it was identified by XRD, SEM and TEM that the derived powders can be changed from well crystallized organic formates of $Ce_{1-x}Gd_x(HCOO)_3$ with dendritic growth to loose agglomerated cubic CGO powders with grain sizes below 10 nm.

INTRODUCTION

Rare earth oxide-doped ceria has been investigated as a promising alternative solid oxide electrolyte to replace stabilized zirconia for use in solid oxide fuel cells (SOFC) [1] and electrochemical NOx decomposition [2] at lower operation temperatures. However, difficult sintering the doped ceria to dense ceramics has been the issue for applications requiring dense electrolyte membranes even if the powders were prepared from precursor solution or by the sol-gel process [3-5]. Polymeric precursor methods, originating from the Pechini technique [6] have been employed to synthesize various mixed-cation oxide powders for electrolyte and electrode applications [7]. It is noticed that the organic precursors working as chelating agents and a resin vehicle can also produce combustion heat, which results in increased high temperature during powder calcinations making strongly agglomerated large crystallites [7] and consequently decreasing the sintering activity of the derived powders. In this paper, the crystallization and the morphology of the powders derived from precursor solutions without using citric acid, similar to those used by Chen et al. [4] and Huang et al. [5], were investigated with emphasis on the effect of the polymerization processes.

EXPERIMENTAL DETAILS

The starting precursor solution was prepared by mixing $Ce(NO_3)_3 \cdot 6H_2O$ and $Gd(NO_3)_3 \cdot 6H_2O$ in the molar ratio Ce : Gd = 0.9 : 0.1, with 20 ml distilled water, 80 ml ethylene glycol, and 20 ml concentrated (60%) nitric acid. The as-prepared transparent solution with 0.04 mole final oxide was then heated on a hot plate at about 60-85℃ with magnetic stirring for the polymerization treatments. The changes in the viscosity of the solution were measured at room temperature by means of a viscometer (VM-1G-L, Yamaichi Electronics Co., Ltd.). The structure changes of the polymeric precursor solutions with different heating times were investigated using a Fourier transform infrared spectrometer (FTIR) (Janssen, JASCO Corporation). The solutions were dried at 130℃ in an oven, then the decomposition and reaction processes of the dried powders or gels were examined with a TG-DTA thermal analyzer (Thermo Plus, TG8120, Rigaku Denki Co., Ltd.) in ambient atmosphere with a heating rate of 5℃/min from room temperature to 1000℃. The crystallization and microstructure of the derived powders were characterized by a RINT2000 (Rigaku Corporation) X-ray diffractometer (XRD), a JSM-5600 (JEOL, Ltd.) scanning electron microscope (SEM), and a JEM-4000FX (JEOL, Ltd.) transmission electron microscope (TEM).

RESULTS AND DISCUSSION

Figure 1 shows the change in the viscosity of the polymeric precursor solution as a function of the heating time. The viscosity of the solution increases steeply with increasing heating time. The typical FTIR transmission spectra of the polymeric precursor solutions

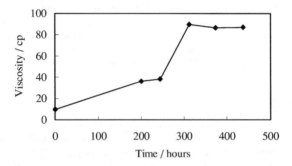

Figure 1. *Viscosity changes of the polymeric precursor solution as a function of heating time at 60-85 ℃ in air.*

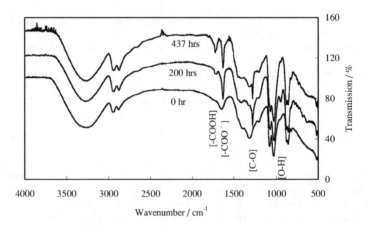

Figure 2. *FTIR transmission spectra of the polymer precursor solutions as a function of heating time.*

with different heating times are shown in Figure 2. Changes in the molecular structures of the precursor solutions during the heating process were identified. The presence of C=O and O-C=O groups at about 1600 and 1338 cm^{-1}, respectively, were indicated in the initial solution. With heating, however, carboxylate groups [-COO$^-$] exhibiting strong asymmetric stretching at about 1600 cm^{-1}, carboxylic acid groups [-COOH] exhibiting a C=O stretch at about 1700 cm^{-1} and signature of esters at about 1300 cm^{-1} [8] were identified in the solutions after heating times of 200 and 437 hours. These spectral features were not observed in the solutions with shorter heating times [4].

After drying the solutions at 130℃ for 24 hours, three kinds of powders were obtained. The powder derived from the initial solution exhibits dark brown color and sticky gel-like particles, while the powders derived from the solutions heated at 80℃ for 200 and 437 hrs are medium and light brown in color and are loosely agglomerated fine particles. Figure 3 shows the TG-DTA curves of these powders. It is found from the DTA curves that

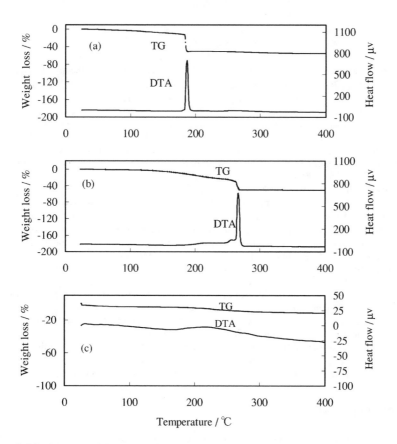

Figure 3. *TG-DTA curves of the powders derived from the precursor solutions with heating times of (a) 0 hr, (b) 200 hrs and (c) 437 hrs at about 80 ℃, followed by drying at 130 ℃ in air.*

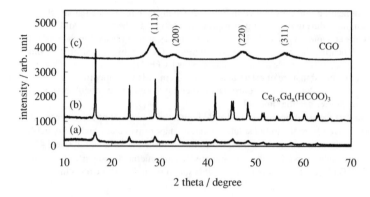

Figure 4. XRD patterns curves of the powders derived from the precursor solutions with heating times of (a) 0 hr, (b) 200 hrs and (c) 437 hrs at about 80 ℃, followed by drying at 130 ℃ in air.

the strong exothermic peak at about 185℃ for the powder derived from the initial solution shifts to about 265℃ when the heating time of the solution was increased to 200 hours. Furthermore, the exothermic peak disappears almost completely when the solution heating time was increased to more than 430 hours. Obviously, the exothermic peak shift toward higher temperature would promote the crystallization and even the phase transformation of the derived powders, which was demonstrated by the XRD analyses, as described below.

Figure 4 shows the XRD patterns of the dried powders corresponding to those shown in Figure 3. It is identified that both the powders derived from the initial solution and from the solution heated for 200 hours show the crystalline formate structures ($Ce_{1-x}Gd_x(HCOO)_3$), and the latter exhibits much higher crystallinity. Moreover, it was found that when the heating time of the solution was further increased to more than 430 hours, the derived powder at 130℃ exhibits the cubic oxide (CGO) structure. Typical SEM micrographs of these powders are shown in Figure 5. It was observed that the powder derived from the initial solution exhibits large-sized gel particle morphology (Figure 5a). When the heating time of the solution increased to 200 hrs, the derived powder shows dendritic growth (Figure 5b), which should possess the crystalline structure of formate, $Ce_{1-x}Gd_x(HCOO)_3$, as identified by XRD (Figure 4). TEM observations revealed that the formate particles are formed by nano-sized crystallites. The formation mechanism of the anisotropically grown formate particles is to be investigated together with the analysis of the changes in molecular structures of the polymeric precursor solutions during the heating process. Furthermore, it was observed that the loose agglomerated powder with cubic oxide (CGO) structure (Figure 4), derived from the precursor solution heated for 437 hours at about 80℃, exhibits uniformly fine-grained morphology (Figure 5c). The lattice substructures were observed by TEM, from which the grain sizes of the powder shown in Figure 5c are determined to be below 10 nm.

CONCLUSIONS

The changes in molecular structure of the polymeric precursor solution during the heating process at about 80℃ were revealed by FTIR spectra. TG-DTA analyses of the

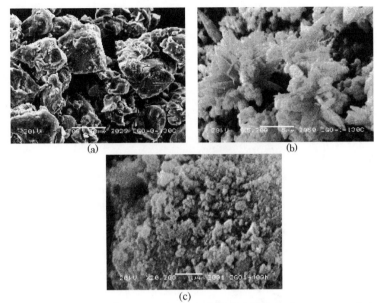

(a)   (b)

(c)

Figure 5. *SEM micrographs curves of the powders derived from the precursor solutions with heating times of (a) 0 hr, (b) 200 hrs and (c) 437 hrs at about 80 ℃, followed by drying at 130 ℃ in air.*

derived powders at 130°C demonstrated the temperature shifting and the disappearance of the exothermic reaction of the precursor solutions with increasing heating time. Furthermore, it was identified by XRD, SEM and TEM that the derived powders could be changed from well crystallized organic formates of $Ce_{1-x}Gd_x(HCOO)_3$ with dendritic growth to loose agglomerated fine grained oxide (CGO) with grain sizes below 10 nm with increasing polymerization of the precursor solution.

ACKNOWLEDGMENTS

This work has been supported by METI, Japan, as part of the Synergy Ceramics Project. The authors are members of the Joint Research Consortium of Synergy Ceramics.

REFERENCES

1. B.C.H. Steele, *Solid State Ionics*, **129**, 95-110 (2000).
2. T. Hibino, K. Ushiki, Y. Kuwahara, and M. Mizono, *J. Chem. Soc. Faraday Trans.*, **92**, 4297-4300 (1996).
3. I. Reiss, D. Braunshtein, and D. S. Tannhauser, *J. Am. Ceram. Soc.*, **64**, 479-85 (1993).
4. C.C. Chen, M.M. Nasrallah, and H. U. Anderson, *J. Electrochem. Soc.*, **140**, 3555-60 (1993).
5. K. Huang, M. Feng, and John B. Goodenough, *J. Am. Ceram. Soc.*, **64**, 479-85 (1993).
6. M. Pechini, U. S. Pat., No. 3 330 697 (11 July 1967).
7. P.A. Lessing, *Ceramic Bulletin*, **68**, 1002-7 (1989).
8. D.W. Brown, A.J. Floyd, and M. Sainsbury, *Organic Spectroscopy*, (John Wiley & Sons Publisher, New York, 1988) p. 39-49.

Mat. Res. Soc. Symp. Proc. Vol. 658 © 2001 Materials Research Society

Decoration of silicon carbide nanotubes by CoFe$_2$O$_4$ Spinel nanoparticles.

Claude Estournès[1], Cuong Pham-Huu[2], Nicolas Keller[2], Marc J. Ledoux[2],
[1] Groupe des Matériaux Inorganiques Institut de Physique et Chimie des Matériaux de Strasbourg UMR7504 CNRS-ULP-ECPM 23 rue du Loess 67037 Strasbourg Cedex France
[2] Laboratoire de Chimie des Matériaux Catalytiques, ECPM, GMI, CNRS, Université Louis Pasteur, 25 rue Becquerel, 67087 Strasbourg Cedex 2, France.

ABSTRACT

Silicon carbide nanotubes have been prepared and decorated with CoFe$_2$O$_4$ spinel. These nanocomposites were prepared by incipient wetness impregnation of the dried SiC support with a stoichiometric aqueous solution of the cobalt and iron nitrates. After drying, subsequent annealing treatments were performed under air in the temperature range 873 to 1073 K. X-ray diffraction and high resolution transmission electron microscopy reveal the presence of spinel nanoparticles at the surface of the tubes. Magnetic properties of the nanocomposites show in the magnetic field explored (up to 5 T) that saturation magnetization is not observed neither at room temperature nor at 5Kelvin. The nanometric character and the nature of the spinel particles are also confirmed by the coercive field (1.6 T) observed at low temperature and the curie temperature.

INTRODUCTION

The design and synthesis of materials of nanometer size are currently the subject of intense research because their properties differ considerably from those of the corresponding bulk materials. Many researchers have been active in preparing systems in which magnetic nanocrystalline particles are separated by means of a non-magnetic material in order to improve the magnetic properties.

Bulk cobalt ferrite CoFe$_2$O$_4$ is usually assumed to have a collinear ferrimagnetic spin structure. It is a partially inverse spinel, the ratio Fe(A)/Fe(B) depends on the prepartion process. Co-ferrite has a very high cubic magnetocrystalline anisotropy accompanied by a reasonable saturation magnetization. These properties make it a promising material, whether or not nanoparticles can be prepared, for use in the production of isotropic permanent magnets, magnetic recording and magnetic fluids. A number of techniques (chemical reactions [1-4], hydrothermal method [5], microemulsion [6,7], sonochemistry [8], mechanical alloying [9] and glass crystallisation [10]) were recently used for producing fine pure Co-ferrite powders. However, studies on supported or embedded CoFe$_2$O$_4$ nanoparticles are rare [11] compared to, say, works done in metal/oxide thin films or composites.

The matrices or supports commonly used for nanocomposites are either polymers or oxides like silica or alumina. Unluckily, these phases may degrade or react with the magnetic phase upon thermal treatment. Silicon carbide is a well known chemically inert material, which does not react with the supported phase. It has been reported that despite the chemical inertness of its surface SiC exhibits a relatively good dispersion capacity for metals or oxide phases, due to the presence of a thin layer of silicon oxycarbide on the topmost surface [12]. Some of the authors have recently succeeded in the synthesis of a medium surface area silicon carbide nanotubes with different diameters. The research described in the present article concerns the synthesis and use of these SiC nanotubes as supports for preparing $CoFe_2O_4$ decorated nanotubes and the characterization of such a nanocomposites.

EXPERIMENTAL DETAILS

X-ray diffraction data were collected on a Siemens D-500 instrument using Co-Kα radiation (0.1789 nm). Transmission electron microscopy have been performed using a Topcon EM002B UHR electron microscope operating at 200 kV. Temperature dependent (300-1000 K) magnetization measurements were made on a Pendulum magnetometer operating at fixed field of 1 Tesla. Field dependent magnetization measurements of the powder samples were carried out using a Princeton Applied Research vibrating sample magnetometer Model 155 (VSM) and a Quantum Design SQUID MPMS-XL (AC and DC modes and maximum static field of ±5T).

RESULTS and DISCUSSIONS

The shape memory synthesis

Some years ago Ledoux and co-workers have developed [13-15] a new method of preparing carbides, called the shape memory synthesis (SMS), reflecting the fact that the original macrostructural features of a template material (its macroscopic shape) is retained after synthesis of a new solid. For the preparation of SiC, this method consisted of the generation of SiO vapour by heating a Si/SiO$_2$ mixture at 1473-1623 K, using a method similar to that developed by Kennedy and North [16], this vapour then being pumped through a shaped carbon. The gas-solid reaction between SiO and C occurred at a lower temperature (1273-1523 K) and produced ß-SiC material having the same shape as the initial carbon material, with a high surface area (30-200 m^2/g). The CO formed as a co-product was rapidly pumped out of the reaction zone, pushing the equilibrium towards SiC formation.

Synthesis of 100 nm diameter SiC nanotubes

A challenge had to be tackled in order to obtain SiC nanotubes. The conditions of reaction between SiO and C had to be fine-tuned because such nanostructures, the starting carbon material and the final SiC structure, were not expected to be as stable as bigger ones. Carbon nanotubes with an average diameter that ranged between 100 and 150 nm were supplied from Applied Science Ltd (USA). Theses materials have a BET surface area of 10 m^2/g. These tubes were submitted as such to the SiO vapor at 1473 – 1523 K. The product of the reaction, about 10 g, was calcined at 873 K in order to eliminate the remaining unreacted carbon, if any, and physical characterizations were performed on this final material. The SEM picture reproduced in Figure 1 clearly show that the nanotube shape of carbon was preserved and that most of the SiC tubes had an open end. The internal diameter was measured as between 30 and 100 nm (average 90 nm), which means that the diameter of the starting carbon nanotubes has not been modified by the reaction. X-Ray diffraction unambiguously confirmed that this final material was pure ß-SiC relatively well crystallized. During the C-SiC transformation a significant increase of the surface area was observed, increased to $45m^2/g$ instead of $10m^2/g$, along with a development of mesopores.

CoFe$_2$O$_4$ decoration of the SiC nanotubes

The nanocomposites were prepared by incipient wetness impregnation of the SiC nanotubes with a stoichiometric aqueous solution of the cobalt and iron nitrates. After drying, subsequent annealing treatments were performed under air in the temperature range 673 to 1073 K. Due to the known high dispersion quality of the SiC support, only very weak diffraction lines of oxide spinel nanoparticles were observed on X-ray powder diffraction patterns of the samples annealed up to 1073 K for 2 hours.

Figure 2 shows TEM microgrph of the silicon carbide nanotube decorated with Co-ferrite. EDX confirms the presence of Co and Fe codeposited on the surface of the nanotubes. The coating appears to consist of aggregates of small particles few nm in diameter that yield difficult a determination of an average size as well as a size distribution.

Magnetic studies were carried out on samples annealed above 673 K. Figure 3 shows the thermal evolution of the magnetization measured at a magnetic field of 1 T. The magnetization decreases when temperature rises and approaches zero near 800 K. This confirms the formation of CoFe$_2$O$_4$ particles as it is know that the bulk Curie temperature is about 790 K [17].

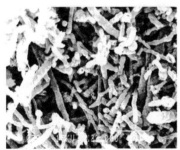

Figure 1: SEM micrograph of SiC nanotubes.

Figure 2: TEM micrograph of the silicon carbide nanotubes decorated with Co-ferrite.

The distinguishable hysteresis phenomena observed at room temperature on isothermal magnetization curves (Figure 4) suggests that the particles formed have a ferrimagnetic behavior. The increase of magnetization (after treatment above 773 K) as well as the increase in coercive field can be attributed to crystallization improvement after annealing or increase of the size of the particles. It can be also observed that, whatever annealing temperatures, saturation is never reached at room temperature in the magnetic field range explored (±1.7 T) or at 5 K up to 5 Tesla for the sample treated at 873 K (Figure 5). It is well known that the magnetic field needed to reach the saturation magnetization depends on the size of the particles. According to Noumen et al., at 10 K for particles having an average size equal to 2 and 3 nm, the saturation is not reached even for a magnetic field equal to 4 Tesla whereas it is observed for 5 nm particles [1]. This indicates the presence of small particles with super-paramagnetic behavior in our samples.

From Figure 5, it is to be noted that the coercive field measured exceeds 1.6 Tesla. This value remains lower than that reported at 5 K for single domain particles [18] but, it is exceptionally high considering that it only corresponds to an average value due to size distribution of the particles. In Figure 6 we show the Zero Field Cooling-Field Cooling (ZFC-FC) curves of the sample annealed at 873 K. In ZFC measurements, the sample is cooled from 400 K to 5 K without any external field. Then a magnetic field of 10^{-3} Tesla is applied, and the magnetization of the sample is measured following the rising temperature.

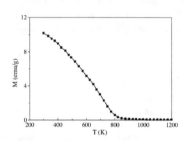

Figure 3: *Thermal evolution of the magnetization measured at 1Tesla for the sample annealed at 673 K.*

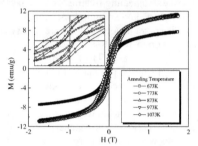

Figure 4: *Isothermal magnetization for samples annealed at different temperatures.*

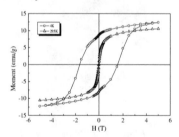

Figure 5: *Magnetization curves for the sample annealed at 873 K.*

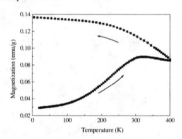

Figure 6: *Zero Field Cooling-Field Cooling (ZFC-FC, H=3 Oe) curves of the sample annealed at 873 K.*

The shape of Figure 6 evidence a single broad maximum near 320 K and then the magnetization starts to decrease. In FC measurements the sample is cooled from 400 to 5 K with 10^{-3} Tesla applied magnetic field, and the magnetization measurement is done with decreasing temperature.

When lowering the temperature, the FC curve shows a maximum at 5K and is never superimposed with the ZFC curve. Generally, when nanoparticles are cooled below their blocking temperature under a magnetic field, the magnetization direction of all the nanoparticles is frozen in the field direction. This explains the deviation of the two curves. However, when all the particles are at the same relaxation super-paramagnetic state (ie: above their T_B) the two curves should superimpose. The fact that this is not observed in Figure 7 can be the sign that some of the particles remain blocked at 400 K. Such behaviors are characteristics of super-paramagnetic particles with a wide size distribution.

CONCLUSIONS

Synthesis of silicon carbide nanotubes decorated with $CoFe_2O_4$ spinel nanoparticles and their magnetic characterization have been described. No saturation is observed at room temperature for samples annealed at any temperature. The coercive field exceeds 1.6 Tesla at 5 K, unprecedented for any ferrite. Detailed work following these directions will be published in the near future.

ACKNOWLEDGEMENTS

Authors thank Dr. M. Richard-Plouet and Mrs G. Ehret for Transmission microscopy. The technical assistance for SQUID measurements of A. Derory is much appreciated. C.E. is grateful to Dr. M. Kurmoo for fruitful discussions.

REFERENCES

1. N. Noumen, P. Bonville and M.P. Pileni, *J. Phys. Chem.* **100** (1996) 14410.
2. S. Li, V. T. John, C. O'Connor, V. Harris and E. Carpenter, *J. Appl. Phys.* **87 (9)** (2000) 6223.
3. C. Liu, B. Zou, A. J. Rondinone and Z. J. Zhang, *J. Am. Chem. Soc.* **122** (2000) 6263.
4. A. J. Rondinone, A. C.S. Samia and Z.J. Zhang, *J. Phys. Chem. B* **104** (2000) 7919.
5. K.J. Davies, S. Wells, R.V. Upadhyay, S.W. Charles, K. O'Grady, M. El Hilo, T. Meaz and S Mørup, *J. Magn. Magn. Mater.* **149** (1995) 14.
6. V. Pillai and D. O. Shah, *J. Magn. Magn. Mater.* **163** (1996) 243.
7. A.J. Rondione, A. C.S. Samia and Z.J. Zhang, *J. Phys. Chem.* **103** (1999) 6876.
8. K. V.P.M. Shafi, A. Gedanken, R. Prozorov, J. Balogh, *Chem. Mater.* **10** (1998) 3445.

9. J. Ding. P. G. McCormick, R. Street, *Solid State Comm.* **95** (1995) 31.
10. R. Müller, W. Schüppel, *J. Magn. Magn. Mater.* **155** (1996) 110.
11. S.R. Mekala, J. Ding, *J. Alloys and Compounds* **296** (2000) 152.
12. C. Pham-Huu, C. Estournès, B. Heinrich and M. J. Ledoux, *J. Chem. Soc., Farad. Trans.* **94** (1998) 435.
13. M.J. Ledoux, J. Guille, S. Hantzer and D. Dubots, *U.S. Pat.* N° **4 914 070** (1990).
14. C. Pham-Huu, N. Keller, S. Roy, M.J. Ledoux, C. Estournès and J.L.Guille, *J. of Mater. Science* **34** (1999), 3189.
15. N. Keller, C. Pham-Huu, M.J. Ledoux, C. Estournès and G. Ehret, *Appl. Cat. A : General* **187** (1999) 255.
16. P. Kenedy and B. North, *Proc. Brit. Ceram. Soc.* **33** (1983) 1.
17. Landolt-Böernstein eds. Springer.
18. M. Grigova, H.J. Blythe, V. Blaskov, V. Rusanov, V. Petkov, V. Masheva, D. Nihtianova, Ll.M. Martinez, J.S. Muñoz and M. Mikhov, *J. Magn. Magn. Mater.* **183** (1998) 163.

Mat. Res. Soc. Symp. Proc. Vol. 658 © 2001 Materials Research Society

A New Three-dimensional Lanthanide Framework Constructed by Oxalate and 3,5-pyridinedicarboxylate

Michael A. Lawandy, Long Pan, Xiaoying Huang and Jing Li*
Department of Chemistry, Rutgers University, Camden NJ 08102
Tan Yuen and C. L. Lin
Department of Physics, Temple University, Philadelphia, PA 19122

ABSTRACT

Hydrothermal reactions of rare-earth metal acetates and nitrates with oxalic acid ($H_2C_2O_4$) and 3,5-pyridinedicarboxylic acid (H_2pddc) in triethylamine/water solutions have yielded five isostructural polymers with the general formula: $[Ln(pddc)(C_2O_4)_{1/2}(H_2O)_2]\cdot H_2O$. [Ln = La(1), Pr(2), Nd(3), Eu(4) and Er(5)]. These compounds crystallize in monoclinic crystal system, space group $P2_1/n$, Z = 4, with only slight variations in their unit cell parameters: 1 a = 7.747(2), b = 9.954(2), c = 15.134(3) Å, β = 98.64(3)°, V = 1153.8(4) Å^3; 2 a = 7.707(2), b = 9.895(2), c = 15.006(3) Å, β = 98.54(3)°, V = 1131.7(4) Å^3; 3 a = 7.688(2), b = 9.897(2), c = 14.955(3) Å, β = 98.43(3)°, V = 112 5.6(4) Å^3; 4 a = 7.638(2), b = 9.842(2), c = 14.809(3) Å, β = 98.42(3)°, V = 1101.2(4) Å^3; 5 a = 7.573(2), b = 9.761(2) Å, c = 14.630(3) Å, β = 98.10(3)°, V = 1070.7(4) Å^3. The structure of these compounds is composed of 2D Ln(pddc) layers that are interconnected by chelating oxalate. Within the layer, each rare-earth metal forms a monodentate bond with each of the four pddc groups. The metal centers in the neighboring layers are bridged through μ_4-oxalate, resulting in a three-dimensional framework. The remaining two sites around the eight-coordinate Ln are occupied by water molecules. Compounds 2, 3, and 5 exhibit paramagnetic behavior and 1 is diamagnetic. They are thermally stable up to 250°C.

INTRODUCTION

The demand for new materials with practical applications has promoted research in the design and synthesis of functional coordination polymers that possess attractive properties, such as zeolite-like characteristics, catalytic activity, magnetism and non-linear optical behavior [1, 2]. Suitable choices of metals and ligands based on their coordination habits and geometric preferences often produce novel structures with interesting and specific properties. Recently, others and we have investigated a number of systems involving multi-ligands, for example, magnetically active systems containing transition metals and oxalate/4,4'-bipyrdrine [3, 4]. Rare earth metals are attractive as metal centers for construction of magnetically interesting coordination polymers that also exhibit other interesting and advantageous properties, such as high dimensionality, optical activity, and thermal stability. In this study, we report a novel three-dimensional framework built upon lanthanide metals and two ligands, oxalate and 3,5-pyridinedicarboxylate.

EXPERIMENTAL DETAILS

Chemicals and Reagents. All chemicals were used as purchased without purification. $La(O_2CCH_3)_3 \cdot 1.5H_2O$ (99.9%), $Pr(NO_3)_3 \cdot 6H_2O$ (99.9%), $Nd(NO_3)_3 \cdot 6H_2O$ (99.9%,), $Eu(NO_3)_3 \cdot 6H_2O$ (99.9%), $Er(NO_3)_3 \cdot xH_2O$ (99.9%), oxalic acid ($H_2C_2O_4 \cdot 2H_2O$, 97%), Et_3N

(99+%) were purchased from Alfa Aesar. 3,5-pyridinedicarboxylic acid (H_2pddc, 98%) w~~~
purchased from ACROS.

Synthesis of La(pddc)(C$_2$O$_4$)$_{1/2}$(H$_2$O)$_2$]·H$_2$O (1). The reactions of La(O_2CCH$_3$)$_3$·1.5H$_2$~
(0.0686 g, 0.2 mmol), and $H_2C_2O_4$·2H$_2$O (0.0252 g, 0.2 mmol), H_2pddc (0.0167 g, 0.1 mmol~
Et$_3$N (0.35 ml, 0.25 mmol) and H_2O (10 mL) in the mole ratio of 2:2:1:2.5:5555 in a 23 mL ac~
digestion bomb at 150°C for a period of six days resulted in long clear needle crystals of **1**. Th~
product was washed with water and acetone and dried in air. Compound **1** was obtained
relatively high yield (0.0549 g, 68.3%).

Synthesis of Ln(pddc)(C$_2$O$_4$)$_{1/2}$(H$_2$O)$_2$]·H$_2$O [Ln = Pr (2), Nd (3), Eu (4), Er (5)]. Simil~
reactions as described for **1** were carried out for other lanthanides compounds, yieldir
transparent yellow plate crystals for **2**; brownish-yellowish plate crystals for **3**; irregular-shape
clear crystals for **4**; and transparent flat pink crystals for **5**. The yields were calculated to b~
72.5% (0.0586 g), 85.7% (0.0698 g), 73.3%(.0609 g), and 37.9% (0.0364 g), for **2**, **3**, **4**, and~
respectively.

Crystallographic Studies. A column-like colorless transparent crystal of **1** (0.25 × 0.04 × 0.0~
mm), a plate-like yellow transparent crystal of **2** (0.20 × 0.11 × 0.005 mm), a plate-lik~
brownish-yellowish transparent crystal of **3** (0.15 × 0.15 × 0.04 mm), a column-like colorle~
transparent crystal of **4** (0.30 × 0.06 × 0.02 mm), and a plate-like pinkish crystal of **5** (0.22~
0.12 × 0.03 mm) were selected for the crystal structure analysis. Each crystal was mounted on~
tip of a glass fiber in air and placed onto the goniometer head of an Enraf-Nonius CAD
automated diffractometer. Using graphite-monochromated Mo Kα radiation, the unit cell da~
were refined by 25 well-centered reflections. Raw data were corrected for Lorentz ar~
polarization effects, and an empirical absorption correction based on ψ-scan data [5] was appli~
in each case. The structures were solved using the SHELX-97 program [6]. The non-hydrog~
atoms were located by direct phase determination and subsequent difference Fourier synthes~
and subjected to anisotropic refinement. The hydrogen atoms were located from differen~
Fourier maps but were not refined. The full-matrix least-square calculations on F^2 were appli~
on the final refinements. The unit cell parameters, along with data collection and refineme~
details, are given in Table I. Selected bond lengths are reported in Table II. Crystal drawin~
were generated by SCHAKAL 97 [7]. X-ray powder diffraction analyses were performed on~
Rigaku D/M-2200T automated diffraction system (Ultima$^+$). All measurements were ma~
between a 2θ range of 5 and 80° at the operating power of 40 kV/40 mA.

Thermal Analysis. Thermogravimetric analyses (TGA) of the compounds **2**, **3**, and **5** we~
performed on a computer-controlled TA Instrument 2050 TGA analyzer. Samples were load~
into platinum pans and heated with a ramp rate of 10°C/min from room temperature to 800°C.

Magnetic Measurements. Magnetic susceptibility $\chi(T)$ and magnetization $M(H)$ measuremen~
on polycrystalline samples of **1**, **2**, **3**, and **5** were performed using a Quantum Design SQU~
magnetometer. In $\chi(T)$ measurements, the temperature was varied from 2 K to 300 K. Magne~
fields of 500 G and 5 kG were applied in $\chi(T)$ measurements for each compound. $M(H)$ w~
measured at 2 K for all samples. In the $M(H)$ measurements, the applied magnetic field w~
increased from 0 to 50 kG and then decreased back to 0.

Table I. Crystallographic data for **1-5**.

	1	**2**	**3**	**4**	**5**
formula	$C_{16}H_{18}La_2N_2O_{18}$	$C_{16}H_{18}Pr_2N_2O_{18}$	$C_{16}H_{18}NdN_2O_{18}$	$C_{16}H_{18}Eu_2N_2O_{18}$	$C_{16}H_{18}Er_2N_2O_{18}$
fw	804.14	808.14	814.80	830.24	860.84
Space group	$P2_1/n$ (No.14)	$P2_1/n$ (No.14)	$P2_1/n$ (No.14)	$P2_1/n$ (No.14)	$P2_1/n$ (No.14)
a(Å)	7.747(2)	7.707(2)	7.688(2)	7.638(2)	7.573(2)
b(Å)	9.954(2)	9.895(2)	9.897(2)	9.842(2)	9.761(2)
c(Å)	15.134(3)	15.006(3)	14.955(3)	14.809(3)	14.630(3)
β(°)	98.64(3)	98.54(3)	98.43(3)	98.42(3)	98.10(3)
V(Å^3)	1153.8(4)	1131.7(4)	1125.6(4)	1101.2(4)	1070.7(4)
Z	2	2	2	2	2
T, K	293(2)	293(2)	293(2)	293(2)	293(2)
λ, Å	0.71073	0.71073	0.71073	0.71073	0.71073
$\rho_{calc, gcm}$$^{-3}$	2.315	2.372	2.404	2.504	2.670
μ(mm^{-1})	3.748	4.351	4.659	5.743	7.887
R^a($I>2\sigma(I)$)	0.0263	0.0392	0.0321	0.0281	0.0325
R_w^b	0.0512	0.0901	0.0867	0.0676	0.0809

$^a R = \Sigma | |F_o| - |F_c| | / \Sigma |F_o|$. $^b R_w = [\Sigma[w(|F_o^2|-|F_c^2|)^2]/\Sigma w(F_o^2)^2]^{1/2}$.
Weighting: **1**, $w = 1/\sigma^2[F_o^2+(0.02P)^2]$, where $P = (F_o^2 + 2F_c^2)/3$. **2**, $w = 1/\sigma^2(F_o^2+(0.05P)^2)$. **3**, $w = 1/\sigma^2(F_o^2+(0.06P)^2)$.
4, $w = 1/\sigma^2(F_o^2+(0.04P)^2)$. **5**, $w = 1/\sigma^2[F_o^2+(0.03P)^2+ 10.0P]$.

Table II. Selected bond lengths (Å) for **1-5**.

bond	**1**	**2**	**3**	**4**	**5**
M-O(2)^i	2.418(4)	2.369(5)	2.360(3)	2.314(4)	2.262(5)
M-O(1)	2.425(3)	2.373(5)	2.358(5)	2.326(4)	2.261(5)
M-O(3)^{ii}	2.433(3)	2.398(5)	2.388(3)	2.339(4)	2.292(5)
M-O(4)^{iii}	2.480(3)	2.437(5)	2.418(3)	2.381(4)	2.316(5)
M-O(6)^{iv}	2.536(3)	2.499(5)	2.483(4)	2.441(4)	2.391(5)
M-O(5)	2.559(3)	2.534(5)	2.516(3)	2.481(4)	2.439(5)
M-O(8)	2.567(3)	2.531(5)	2.510(4)	2.467(4)	2.407(5)
M-O(7)	2.580(3)	2.532(5)	2.522(3)	2.475(4)	2.412(6)

Symmetry codes: i -x+1,-y+1,-z   ii -x+1/2,y-1/2,-z+1/2   iii x,y-1,z   iv -x,-y+1,-z

DISCUSSION

Structures. Compounds **1-5** are isostructural and crystallize in the monoclinic system, space group $P2_1/n$, $Z = 4$. The local coordination of the metal is shown in Figure 1. Each metal (Ln) is coordinated to eight oxygen atoms, two of which are from one oxalate (O5, O6), four from four different pddc (O1, O2, O3, O4), and the remaining two from two water molecules (O7, O8). Every μ_4-pddc connects to four different metals to form a 2D network extending along the [10-1] direction and the b axis (Figure 2). The two-dimensional layers of [Ln(pddc)]$^+$ are cross-linked by oxalate ligands via chelating modes to lead to a three-dimensional structure. One pair of oxygen atoms of oxalate chelate to one metal of one layer and the other pair of oxygen atoms chelate to another metal of the adjacent layer. As shown in Figure 2, the shortest metal-metal distances within the [Ln(pddc)]$^+$ layer are 5.15 (i) and 6.34 Å (ii) for **1**; 5.11 (i) and 6.29 Å (ii) for **2**; 5.10 (i) and 6.28 Å (ii) for **3**; 5.06 (i) and 6.24 Å (ii) for **4**; and 5.01 (i) and 6.17 Å (ii) for **5**, respectively. The shortest metal-metal separations between the adjacent layers are 6.53, 6.46, 6.43, 6.35 and 6.26 Å, for **1-5**, respectively. The average bond length between Ln and the oxygen of pddc is shorter than that between Ln and the oxygen of oxalate, which in turn is shorter than the bond length between Ln and the oxygen of water. Interestingly, the nitrogen of pddc does not participate in bonding in contrast to the Co-pddc compound in which nitrogen and two chelating carboxylate groups connecting three cobalt metals in three directions to give rise to a 2D structure [8].

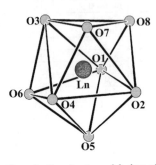

Figure 1. *Coordination polyhedron of a metal ion in [Ln(pddc)(C₂O₄)₁/₂(H₂O)₂]·H₂O (Ln = La, Pr, Nd, Eu and Er).*

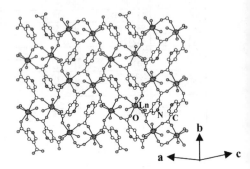

Figure 2. *View of the two-dimensional network of [Ln(pddc)(C₂O₄)₁/₂(H₂O)₂]·H₂O (Ln = La, Pr, Nd, Eu and Er). Metal atoms are shown as larger dark gray circles.*

Magnetic Properties. As expected, the $\chi(T)$ data for [La(pddc)(C$_2$O$_4$)$_{1/2}$(H$_2$O)$_2$]·H$_2$O (**1**) is diamagnetic with a magnitude of $\chi_{dia} = -2 \times 10^{-4}$ emu per mole of La^{3+}. However a tiny small paramagnetic component shows up in the low temperature region, which may come from a non detectable impurity in the sample. The $\chi(T)$ data for [Nd(pddc)(C$_2$O$_4$)$_{1/2}$(H$_2$O)$_2$]·H$_2$O (**3**) and [Er(pddc)(C$_2$O$_4$)$_{1/2}$(H$_2$O)$_2$]·H$_2$O (**5**) are essentially the same in shape, exhibiting typical Curie paramagnetic behavior. The $\chi(T)$ data for [Pr(pddc)(C$_2$O$_4$)$_{1/2}$(H$_2$O)$_2$]·H$_2$O (**2**) looks Curie paramagnetic like, except for a noticeable deviation from it for the temperatures below 20 K. In the data of dχ/dT vs. T, slope changes below 20 K are seen. Because the $\chi(T)$ values for **2** and **3** are much smaller than that of **5**, we only show $\chi(T)$ of **5** as a representative in Figure 3. The insert in Figure 3 shows the χT vs. T plots for **2**, **3**, and **5**. The $\chi(T)$ data for **5** can be fit with simple Curie law very well for all temperatures, and a paramagnetic effective moment of Er^{3+} $\mu_{eff} = 9.1$ μ_B is obtained from the Curie constant. For compounds **2** and **3**, the $\chi(T)$ data above 5 K were fit to an equation: $\chi(T) = C/(T + \theta)$. The paramagnetic effective moments, $\mu_{eff} = 3.6$ μ_B for Pr^{3+} and $\mu_{eff} = 3.4$ μ_B for Nd^{3+}, were obtained. The fitted θ values for **2** and **3** are –41 K and –30 K, respectively. The effective moments of Pr^{3+}, Nd^{3+}, and Er^{3+}, obtained from fitting, are very close to the theoretical values for trivalent lanthanide group ions (3.58 μ_B for Pr^{3+}, 3.62 μ_B for Nd^{3+}, and 9.59 μ_B for Er^{3+}). Isothermal magnetization $M(H)$ measured at 2 K for **2**, **3**, and are shown in Figure 4. At 50 kG, the M value of **5** reaches 4.9 μ_B and the $M(H)$ curve starts to be saturated at 15 kG. In contrast, the M value of **2** only reaches 0.42 μ_B at 50 kG, and an upturn in the slope around 30 kG is observed. The observations in $M(H)$ of **2**, together with the slope change observed in dχ/dT vs. T of **2**, indicate a possible long range antiferromagnetic ordering of Pr^{3+} in this compound. The $M(H)$ curve for **3** seems to begin a saturation above 40 kG and a M value of 1.1 μ_B is reached at 50 kG. Furthermore, small hysteresis effects are seen in $M(H)$

both **2** and **3**. The observed magnetic behavior of the $[Ln(pddc)(C_2O_4)_{1/2}(H_2O)_2]\cdot H_2O$ series suggests that the 4f-electrons of the rare-earth ions in these compounds are very much localized. The rare-earth ions are pretty much evenly separated from each other in a range of 5.0-6.5 Å throughout the 3D network. However, there is a connection between every pair of Ln ions by the oxalate units along the [101] direction. It is interesting to compare the effect of the oxalate bridging in this series of compounds, with its strong bridging functions to the magnetic interactions exhibited in many transition-metal molecular magnetic compounds [9]. For **5**, the simple Curie-paramagnetic behavior seems to suggest that the effect of oxalate bridging is at its minimum. Yet the antiferromagnetic coupling between the rare-earth ions, indicated by the fitted θ values for **2** and **3** (−41 K and −30 K, respectively) may be related to the oxalate bridging.

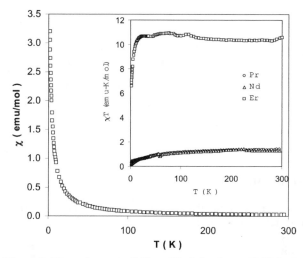

Figure 3*. Magnetic susceptibility $\chi(T)$, defined as $\chi(T)/H$ for **3**. The applied field is 5 kG. Insert: χT vs. T plots for **2**(o), **3** (Δ), and **5**(□).*

Thermal Stability. The weight loss curves show that compounds **2**, **3**, and **5** start to collapse at 450°C. Each compound underwent a three-step decomposition process. The three water molecules per formula were lost around 250°C. This was confirmed via percent loss calculations. In all compounds, the loss of pddc and oxalate were completed before the temperature reached 800°C. The residues of each of these compounds were a mixture of elemental metal and metal oxides.

CONCLUSIONS

In this study, we have successfully obtained a new three-dimensional lanthanide framework constructed by oxalate and 3,5-pyridinedicarboxylic acid via hydrothermal synthesis. Unlike the Ln-pdc systems, where the lanthanide contraction leads to several different structures [10], the Ln-pddc compounds crystallize in the same structure. Magnetic studies show that Compounds **2**,

3, and **5** are primarily paramagnetic, while Compound **1** is diamagnetic. These compounds ar thermally stable up to 250°C.

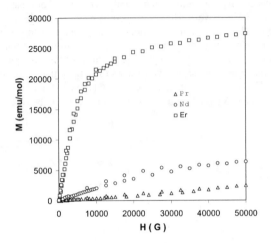

Figure 4. Isothermal magnetization M(H), measured at 2 K for 2(Δ), 3(o), and 5(□).

ACKNOWLEDGMENTS

   Financial support from the National Science Foundation (DMR-9553066 and DMI 9633018) is gratefully appreciated.

REFERENCES

1. A. K. Cheetham, G. Ferey and T. Loiseau. *Agnew. Chem. Int. Ed.* **38**, 3268 (1999).
2. O. M. Yaghi and M. O'Keefe. *J. Solid State Chem.* 152, 1 (2000).
3. S. Kitagawa, T. Okubo, S. Kawata, M. Kondo, M. Katada and H. Kobayashi. *Inorg. Chem.* **34**, 4790 (1995).
4. J. Y Lu, M. A. Lawandy and J. Li. *Inorg. Chem.* **38**, 2695 (1999).
5. G. Kopfmann and R. Hubber. *Acta Crystallogr.* **A24**, 348 (1968).
6. G. M. Sheldrick. *SHELX: program for structure refinement*; University of Goettingen: Germany, 1997.
7. E. Keller. *SCHAKAL 97: a computer program for graphical representation of crystallographic models*; University of Freiburg: Germany, 1997.
8. M. J. Platter, A. J. Roberts and R. A. Howie. *J. Chem. Res. (S)* 240 (1998).
9. O. Kahn *Molecular Magnetism*, (Wiley-VCH, 1993).
10. L. Pan, X. Huang, J. Li, Y. Wu, N. Zheng. *Angew. Chem. Int. Ed.* **39**, 527 (2000).

Mat. Res. Soc. Symp. Proc. Vol. 658 © 2001 Materials Research Society

Synthesis of Mesophase Cerium(IV) Oxides via Surfactant Templating

Masahiko Inagaki, Atsushi Hozumi, Yoshiyuki Yokogawa and Tetsuya Kameyama
National Industrial Research Institute of Nagoya,
1-1 Hirate-cho, Kita-ku, Nagoya 462-8510, Japan.

ABSTRACT

Mesostructured cerium(IV) oxides were directly synthesized via the hexamethylenetetramine (HMT) homogeneous precipitation method. Cerium(III) nitrate, a templating agent, HMT and water were mixed in a molar ratio of $1 : 0.1\text{-}2 : 1\text{-}15 : 100\text{-}1000$ and kept at 65 to 95°C for 10 to 40 hours. 1-Hexadecanesulfonic acid sodium salt was used as the templating agent. Low-angle XRD patterns showed that the as-prepared powders had a lamellar structure with $d_{100}=4.1$ nm. Wide-range ($2\theta=10\text{-}70°$) XRD patterns also exhibited reflection peaks of CeO_2 with a fluorite structure. However, the low-angle reflection peaks disappeared in calcined specimens.

INTRODUCTION

Recently the study of mesoporous materials based on the supra-molecular templating method has progressed rapidly and the achievements have attracted much attention. This templating technique has also been extended to the synthesis of non-silicate mesoporous materials. Mesoporous Al_2O_3 [2], Nb_2O_5 [3], Ta_2O_5 [3], ZrO_2 [4-8], TiO_2 [9, 10], MnO_2 [11], HfO_2 [12], SnO_2 [13] and rare-earth oxides [16] have been synthesized over the past years. Stucky and co-workers have also reported the synthesis of several mesoporous oxides and mesopoorous mixed oxides [1,14,15]. Mesoporous materials based on transition metal oxides are promising as highly functional electrical, magnetic or optical host materials, as well as catalysts and molecular sieves. Cerium(IV) oxide (ceria) and trivalent or divalent cation doped ceria have capabilities for redox chemistry, photocatalysis, UV absorption and oxygen ion conductivity. If such ceria can be synthesized as mesoporous solids, these materials might find application in gas sensors, fuel cells, dye-sensitized solar cells and high-surface-area redox catalysts. To our knowledge, no studies have been conducted on the synthesis of mesostructured cerium(IV) oxides. Here, we report on the direct synthesis of such a mesostructured ceria by a combination of the hexamethylenetetramine (HMT) homogeneous precipitation method [17] and supramolecular templating.

EXPERIMENTAL PROCEDURE

Cerium(III) nitrate (CeN), a templating agent, HMT and water were mixed in a molar ratio of $1 : 0.1\text{-}2 : 1\text{-}15 : 100\text{-}1000$. 1-Hexadecanesulfonic acid sodium salt (HDSAS) was used as the templating agent. HDSAS was dissolved in H_2O at 65°C. CeN and HMT were then added to the HDSAS solution and it was stirred for about 30 min at 65°C. The mixed solution was placed in a Teflon™ vessel, which was sealed with a cap and then heated in an oven maintained at a temperature 65-95°C. The vessel was kept at that temperature for 10-40 hours. The resulting dispersion was filtered through a 1 μm filter. The precipitates were washed with water several times and then lyophilized. In order to remove the template agent, the as-prepared powders were calcined at 350°C. These powders were evaluated by X-ray diffraction (XRD) (MPX³, MAC Science, Japan) and micro Fourier transform infrared spectroscopy (micro FT-IR) (MFT-200, JASCO, Japan) using the KBr disk technique. Themogravimetric analysis (TGA) and differential thermal analysis (DTA) were conducted at the heating rate of 5°C/min in air by a differential thermal analyzer (DPS-8151, Rigaku, Japan).

RESULTS AND DISCUSSION

Figure 1 shows typical low-angle XRD patterns of the as-prepared powders synthesized with different surfactant/cerium (S/C) ratios. When synthesized with an S/C ratio between 0.2 and 0.8, these patterns showed peaks at d = 4.16, 2.09 and 1.39 nm, which are attributable, respectively, to (001), (002) and (003) of lamellar phases. At other S/C ratios, the diffraction patterns exhibited extra peaks at d = 4.75, 2.40 and 1.61 nm, which indicates the existence of other mesostructural orders or mixed mesostructures. The wide-angle (2θ=10-70°) XRD patterns also exhibited reflection peaks of CeO_2 with a fluorite structure. This indicates that the inorganic framework may consist of nanocrystalline cerium(IV) oxide domains. Yada et al. reported the synthesis of a series of mesostructured rare-earth oxides by the homogeneous precipitation method using urea [16]. Unfortunately, the urea-based method yields trivalent cerium carbonate compounds [17]. Heat treatment above 500°C is required to obtain cerium(IV) oxides [17], but such heat treatment usually destroys the mesostructure. With our present method, mesostructured cerium(IV) oxides can be obtained without difficulty.

In the TGA scans (not shown), a weight loss of about 38% was observed between room temperature and 350°C, with the greatest weight loss at ca. 240°C for S/C = 0.2 specimens. The templates/ceria ratio of the as-prepared powders was roughly estimated at 0.34 for S/C = 0.2 specimens from the above weight loss. Such difference of composition

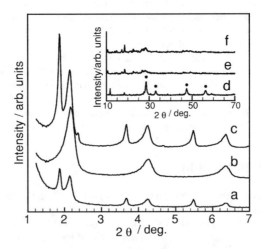

Figure 1. XRD patterns of the as-prepared specimens synthesized with surfactant/cerium ratios of 0.1 (a, d), 0.5 (b, e), and 1.2 (c, f). Circle : CeO_2.

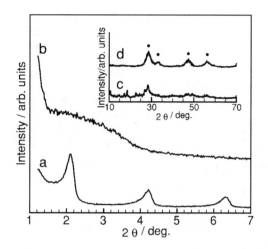

Figure 2. XRD patterns of the as-synthesized (a, c) and the calcined (b, d) specimens. Circle : CeO₂.

between initial surfactant/cerium ratio and the ratio of the as-prepared powders became smaller for initial S/C around 0.5. The DTA curve showed a large exothermic peak at 238°C and smaller ones at 276 and 319°C. These thermal events may be due to the loss of the template. This was confirmed in the FT-IR spectra for the calcined specimens. Strong absorption bands of C-H stretching vibrations at 2915 and 2850 cm^{-1}, asymmetric stretching of -SO_3 groups at 1175 cm^{-1} and symmetric stretching of -SO_3 groups at 1066 cm^{-1} all disappeared after the calcination at 350°C. The low-angle XRD patterns of the calcined specimens showed very broad or no diffraction peaks (Figure 2). This indicates that the mesostructures of our specimens disappeared during thermal treatment owing to the thermolability of lamellar structures.

SUMMARY

In this study, we have successfully synthesized mesostructured cerium(IV) oxide by a combination of the hexamethylenetetramine homogeneous precipitation method and supra-molecular templating. Low angle XRD patterns indicated that the as-prepared powders had a lamellar structure (d_{100} = 4.1 nm) when the surfactant/cerium ratio was between 0.2 and 0.8. However, the low-angle reflection peaks disappeared in calcined specimens. This indicates that the mesostructured CeO_2 prepared in the present study was thermally unstable.

REFERENCES

1. Q. Huo, D. I. Margolese, U. Ciesla, P. Feng, T. E. Gier, P. Sieger, R. Leon, P. M. Petroff, F. Sch th and G. D. Stucky, *Nature*, **368**, 317 (1994).

2. D. M. Antonelli and J. Y. Ying, *Angew. Chem. Int. Ed*, **34**, 2014 (1995).
3. D. M. Antonelli and J. Y. Ying, *Chem. Mater.*, **8**, 874 (1996).
4. U. Ciesla, S. Schacht, G. D. Stucky, K. K. Unger, F. Sch th, *Angew. Chem. Int. Ed.*, **35**, 541 (1996).
5. J. A. Knowles, M. J. Hudson, *Chem. Commun.*, 2083 (1995).
6. A. Kim, P. Bruinsma, Y. Chen. L. Wang, J. Liu, *Chem. Commun.*, 161 (1997).
7. G. Pacheco, E. Zhao, A. Garcia, A. Sklyarov, J. J. Fripiat, *Chem. Commun.*, 491 (1997)
8. S. M. Wong, J. Y. Ying, *Chem. Mater.*, **10**, 2067 (1998).
9. F. Vaudry, S. Khodabandeh and M. E. Davis, *Chem. Mater.*, **8**, 1451 (1996).
10 R. L. Putnman, N. Nakagawa, K. M. McGrath, N. Yao, I. A. Aksay, S. M. Gruner, A. Navrotsky, *Chem. Mater.*, **9** 2690 (1997).
11. Z. Tian, W. Tong, J. Wang, N. Duan, V. V. Krishnan and S. L. Suib, *Sience* **276**, 926 (1997).
12. P. Liu, J. Liu, A. Sayari, *Chem. Commun.*, 557 (1997).
13. G. K. Severin, T. M. AbdelFatah, T. J. Pinnavaia, *Chem. Commun.*, 1471 (1998).
14. P. Yang, D. Zhao, D. I. Margolese, B. F. Chmelka and G. D. Stucky, *Nature*, **396**, 152 (1998).
15. P. Yang, D. Zhao, D. I. Margolese, B. F. Chmelka and G. D. Stucky, *Chem. Mater.* , **11**, 2813 (1999).
16. M. Yada, H Kitamura, A. Ichinose, M. Machida and T Kijima, *Angew. Chem. Int. Ed.*, **38**, 3506 (1999).
17. P. L. Chen and I. W. Chen, *J. Am. Ceram. Soc.* **76** ,1577 (1993).

Mat. Res. Soc. Symp. Proc. Vol. 658 © 2001 Materials Research Society

Electronic and Optical Studies of Coexisting 5- and 6-Atom Rings in Tetrahedral a-C

R.M. Valladares♦, A.G. Calles♦, M.A. Mc Nelis◇, Ariel A. Valladares§*
♦Depto. de Física, Facultad de Ciencias-UNAM, Apartado Postal 70-542, México, D.F., 04510, MEXICO.
◇Apartado Postal 70-464, México, D.F., 04510, MEXICO.
*Instituto de Investigaciones en Materiales-UNAM, Apartado Postal 70-360, México, D.F., 04510, MEXICO.

ABSTRACT

We simulate both the amorphous cluster a-$C_{57}H_{52}$ that contains 5-atom planar rings and 6-atom boat-type rings and the crystalline cluster c-$C_{59}H_{60}$ that contains *only* chair-type 6-atom rings, as in the tetrahedral crystal (diamond structure). We carry out *ab initio* calculations using the *DMol* code and report the total energy, the electronic (density of states) structure and optical properties of the two types of clusters, with and without relaxation of the structures, in order to see the effect of the types of atom rings found in the amorphous structures compared to those found in the crystal structure.

INTRODUCTION

Recently, renewed interest in the atom topology of group IV semiconductors has appeared, in particular for silicon and carbon in themselves and also for the contrasting influences of 5-atom and other size rings and bonding in both materials ([1] and references contained therein). The silicon fourfold-coordinated crystalline clathrate structures Si(34) (fcc) and Si(46) (sc) have become relevant for their potential to generate wide band gap silicon semiconductors which makes them useful from the technological viewpoint [2]. The clathrate structures are formed with a higher proportion of 5-atom rings than that of 6-atom rings. It should be borne in mind that it is not possible to construct a crystalline structure with *only* 5-atom rings [3].

Carbon is a more versatile element and its chemistry is richer than that of silicon, nevertheless finding crystalline clathrate structures experimentally has been an unsuccesful endeavour and the rigidity of the carbon-carbon bonds has been invoked as the responsible factor in these unfruitful attempts. Amorphous carbon, however, has a variety of bondings, atom topologies and coordination numbers. In general both sp^2 and sp^3 hybridizations coexist leading to 3d- and 2d-like structures and the presence of 5-atom and 7-atom rings, in addition to 6-atom boat-type rings, is not uncommon. It follows then, that in the amorphous counterpart it may be possible to create clathrate-like structures since the non-crystallinity allows the formation of structures that contain 5- and 6-atom rings.

The influence of 5-atom ring topology on amorphous semiconductors is still not well understood since attempts to characterize its structure have led to experimental and theoretical results that are largely sample dependent [4]. There are indications that different topologies affect the properties and electronic structure of silicon and carbon, sometimes in opposite ways [5]. Adams *et al.* [6] have found that the theoretical clathrate structures of carbon manifest the

---

§ Contacting author. Email: valladar@servidor.unam.mx

opposite effect in the energy gap than the clathrate structures in silicon; *i.e.*, the band gap f
carbon in these structures is *smaller* than the band gap in diamond.

On the diamond-like lattices the density of states (DOS) for the valence band has thr
prominent areas that consist mainly of *s*-like states, *sp*-like states and *p*-like states, in order
increasing energy. That is, for carbon one has 2*s* and 2*p* states with some 2*sp* component [?
Due to the symmetry of the 6-atom rings in crystalline lattices, the *s* states can form antibondi
states having a node on each bond, whereas with the pentagonal rings this cannot happe
Therefore, it is expected that in pentagonal structures the *s* states of the fivefold rings would ha
a lower energy than the corresponding *s* states of the sixfold rings thereby possibly creating a g
in the middle of the valence band, separating *s* and *p* states, for both silicon and carbon [8].
similar argument is invoked for the lowering of the top of the valence band in pentagon
structures compared to that of crystalline structures.

The study of clusters can shed new light on these problems since it allows the analysis
particular properties, specific restrictions, or geometries that may elucidate the role of t
different factors that are relevant to the electronic structure of these materials. We ha
undertaken a systematic study of amorphous clusters that simulate the bulk by forcing t
outermost atoms to be fixed to account for the inertia of the bulk, fitting gaussian functions to t
energy levels, for a variety of bond angles, bond lengths and atom ring topologies, in covale
four-fold coordinated specimens [5, 9-11]. Here we apply these techniques to the study of t
electronic and optical properties of two carbon clusters, one amorphous and the other crystallir
in order to better comprehend the effect of the type of atom rings found in each.

THE CLUSTERS

The clusters were constructed using the builder module of the *InsightII* graphical u
interface of MSI [12]. The amorphous tetrahedral cluster of the type a-$C_{57}H_{52}$ [11], (Figure
contains 5-atom planar rings and 6-atom boat-type rings that are found in amorphous solids t
never in crystalline ones. The 52 hydrogens are used to passivate the outermost dangling bon
The cluster has one central atom, four nearest neighbors, *nn*; 12 second neighbors, *2n*; 4 th
neighbors, *3n*; 24 fourth neighbors *4n* and 12 fifth neighbors, *5n*. They have the T_d point gro
symmetry with 17 degrees of freedom. The five interior angles of a given 5-atom ring are: c
of 109.47° at the common vertex of the six pentagons; two of 106.84° closest to the previc
angle and the remaining two of 108.425° . All interior angles of the 6-atom rings are 109.4
The interatomic distances are 1.54 Å and the H atoms were placed at a distance of 1.07 Å.

A four-fold coordinated crystalline cluster of the type c-$C_{59}H_{60}$, with *only* 6-atom cha
type rings was studied as well, Figure 2. We used the interatomic distance of the bulk, *i.e.*, 1
Å and fourfold coordination. The cluster was hydrogenated placing the 60 H atoms at a distar
of 1.07Å, maintaining tetrahedral symmetry. The central atom has 4 nearest neighbors, *nn*;
second neighbors, *2n*; 12 third neighbors, *3n*; 6 fourth neighbors, *4n*; 12 fifth neighbors, *5n* a
12 sixth neighbors, *6n*. The cluster has the T_d point group symmetry with 20 degrees of freedor

THE METHOD

The theoretical DFT-LDA based method used allows the energy calculation a
relaxation of each cluster and provides the energy spectrum and the optical transitions for
structures implemented in *DMol* [13]. All calculations were carried out for unrestricted s

using the local density approximation (LDA) [14]. The solutions to the DFT equations were calculated variationally and self-consistently and these solutions provide the molecular wave function, the electron energy levels and the intensity of the dipole transitions permitting the evaluation of the optical properties of the system. Two types of calculations were carried out: unrelaxed and relaxed. In the first case we allowed the relaxation of the central, first neighbor and hydrogen atoms, maintaining the rest fixed. For the second case only the hydrogen atoms were held fixed allowing all the other atoms to move.

Figure 1. a-$C_{57}H_{52}$ cluster.

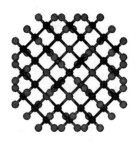

Figure 2. c-$C_{59}H_{60}$ cluster.

RESULTS

Density of states

In order to obtain information of the electronic properties of the clusters it is necessary to construct the total DOS for each one of them. To understand the DOS curves and the calculation of energy gaps we have to identify adequately the highest occupied molecular orbital (HOMO) and the lowest unoccupied molecular orbital (LUMO). In order to define them we have followed the procedure outlined in previous work [11]. The energy gap of the clusters is defined as the difference between the HOMO and the LUMO values thus obtained.

Table I. Energy values (eV) for the HOMO, LUMO, energy and optical gaps for all clusters.

Cluster	a-$C_{57}H_{52}$				c-$C_{59}H_{60}$			
	HOMO	LUMO	Energy Gap	Optical Gap	HOMO	LUMO	Energy Gap	Optical Gap
Unrelaxed	-5.71	-1.03	4.68	4.68	-5.04	-0.86	4.18	4.20
Relaxed	-5.67	-1.06	4.61	4.61	-5.06	-0.89	4.17	4.27

In Table I the values obtained from the simulations for the pertinent energy levels and energy gaps are given. Comparing Figures 3 and 4, which correspond to the unrelaxed clusters, it can be seen that the top of the valence band, the HOMO, of the amorphous cluster has a lower energy value than the corresponding crystalline one. The LUMO behaves the same way; it has a lower value for the amorphous cluster than for the crystalline one. The same tendency is observed for the relaxed amorphous and crystalline clusters (Figures 5 and 6) and is due to the 5-

atom topological structures of the rings. In both cases, relaxed and unrelaxed, the amorphous
C atom cluster has a larger energy gap than the crystalline 59 C atom cluster.

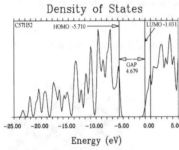

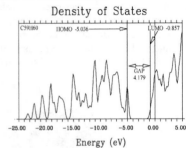

Figure 3. *Total DOS for unrelaxed a-C₅₇H₅₂.*   **Figure 4**. *Total DOS for unrelaxed c-C₅₉H*

We find as well that when comparing the relaxed and unrelaxed calculation results, the
is basically no difference for the crystalline structure but, in the amorphous case there is
reduction in the size of the energy gap due to shifting values of both the HOMO and the LUMO
in the first case to a higher energy value and in the second case to a lower one. We find
expected that the total energy for the relaxed clusters is lower than that corresponding to th
unrelaxed clusters.

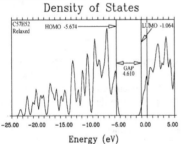

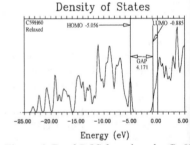

Figure 5. *Total DOS for relaxed a-C₅₇H₅₂.*   **Figure 6**. *Total DOS for relaxed c-C₅₉H*

When doing the relaxed amorphous cluster calculation we maintained tetrahedr
symmetry. We find that the carbon bond length in the five-atom rings changes from 1.54 Å i
all bonds in the unrelaxed cluster to the following situation: the two atoms closest to the fi
neighbour remain at 1.54 Å, the distance between these atoms and their nearest neighbo
changed to 1.55 Å and the distance between the remaining two atoms in the pentagon changed
1.56 Å.

Optical absorption

For the unrelaxed a-C₅₇H₅₂ cluster we find an absorption spectrum with an optical gap
4.68 eV (Figure 7). There are six transitions at this energy value and each one has an intensity
0.05. The unrelaxed c-C₅₉H₆₀ cluster has an absorption spectrum with an optical gap of 4.20 e
(Figure 8). There are six transitions at this energy value and each one has an intensity of 4 X 1

eV (not shown in the Figure). The situation for their relaxed counterparts can be seen in Figures 9 and 10. For the a-$C_{57}H_{52}$ cluster we find an absorption spectrum with an optical gap of 4.61 eV (Figure 9). There are six transitions at this energy value and each one has an intensity of 0.05. The c-$C_{59}H_{60}$ cluster has an absorption spectrum with an optical gap of 4.27 eV (Figure 10). There are six transitions at this energy value and each one has an intensity of 0.01.

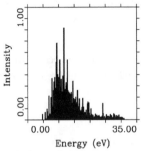

Figure 7. Optical absorption spectrum for unrelaxed a-$C_{57}H_{52}$.

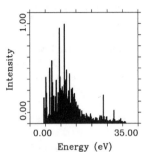

Figure 8. Optical absorption spectrum for unrelaxed c-$C_{59}H_{60}$.

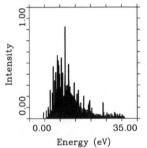

Figure 9. Optical absorption spectrum for relaxed a-$C_{57}H_{52}$.

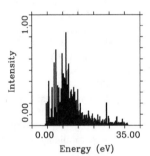

Figure 10. Optical absorption spectrum for relaxed c-$C_{59}H_{60}$.

CONCLUSIONS

We have carried out *ab initio* calculations on two types of clusters, one with only 6-atom chair-type crystalline-like rings and one with 6-atom boat-type rings and 5-atom rings in a pentagonal arrangement, with and without relaxation, in order to see the effect of the ring types found in amorphous structures compared to those found in crystalline structures.

The effect of the topology on the electronic structure and on the optical absorption spectrum of the relaxed and unrelaxed amorphous clusters is clear. The top of the valence band of the amorphous moves noticeably downward with respect to the crystalline cluster whereas the bottom of the conduction band does not change as noticeably. The bottom of the valence remains the same for all clusters, relaxed and not relaxed. Since the selection rules in the amorphous clusters are less important it is expected that the energy and optical gaps will be the same. This is indeed the case as can be seen in Table I. On the other hand in the crystalline clusters, where selection rules are more relevant, there is a difference in the values obtained for the energy a optical gaps. We find that for both the crystalline relaxed and unrelaxed clusters the optical g is larger than the energy gap.

By comparing Figure 7 with Figure 9 and Figure 8 with Figure 10 it can be seen th some well defined allowed transitions in the regular (unrelaxed) cluster disappear in the relax counterpart. This is due to the loss of regularity in the atomic structure of the cluster wh relaxation is allowed even though tetrahedral symmetry was preserved.

ACKNOWLEDGEMENTS

We thank DGAPA-UNAM for financial support through the project IN101798 *(Estructura Electrónica y Topología Atómica de Silicio Amorfo Puro y Contaminado).*

REFERENCES

1. S. Roy, K. Sim and A. Caplin, *Philos. Mag.* B**65**, 1445 (1992).
2. G. Adams, M. O'Keeffe, A. Denkov, O. Sankey and Y.Huang, *Phys. Rev.* B**49**, 80 (1994).
3. N. Mott and E. Davis, *Electronic Processes in Non-crystalline Materials*, Oxfo University Press, 1971, p. 272.
4. J. Robertson, *J. Non-Cryst. Solids* **198-200**, 615 (1996).
5. R.M. Valladares, C.C. Díaz, M. Arroyo, M.A. Mc Nelis & A. Valladares, *Phys. Rev.* B**6** 2220 (2000).
6. G.B. Adams, M.O'Keeffe, A.A. Demkov, O.F. Sankey & Y. Huang, *Phys. Rev.* B**4** 8048 (1994).
7. D. Papaconstantopoulos, *Handbook of the Band Structure of Elemental Solids*, Plenu New York, 1986.
8. S. Saito and A. Oshiyama, *Phys. Rev.* B**51**, 2628 (1995).
9. A.A. Valladares, R. Valladares, A. Valladares, M. Mc Nelis, *Synth. Met.* **103**, 25 (1999).
10. R. Valladares, A. Calles, A.A. Valladares, *Synth. Met.* **103**, 2570 (1999)
11. A.A. Valladares, A. Valladares, R. Valladares and M. Mc Nelis, *J. Non-Cryst. Solids* **23** 209 (1998).
12. *Insight II User Guide*, Release 4.00 (San Diego, Molecular Simulations, Inc., September 1996).
13. *Quantum Chemistry, DMol User Guide*, Release 960 (San Diego, Molecular Simulatior Inc., September 1996). See also B. Delley, *J. Chem. Phys.* **92**, 508 (1990), and *J. Che Phys.* **94**, 7245 (1991).
14. S. Vosko, L. Wilk and M. Nusair, *Can. J. Phys.* **58**, 1200 (1980).

Mat. Res. Soc. Symp. Proc. Vol. 658 © 2001 Materials Research Society

Modified Sol-Gel Synthesis of Vanadium Oxide Nanocomposites Containing Surfactant Ions, and Their Partial Removal

Arthur Dobley, Peter Y. Zavalij, Jürgen Schulte, and M. Stanley Whittingham
Chemistry Department and the Institute for Materials Research,
State University of New York at Binghamton,
Binghamton, New York 13902-6016, U.S.A.

ABSTRACT

Recently, there has been much interest in creating new layered transition metal oxides. Vanadium oxides may be used as sorbents, catalysts, and cathodes in lithium batteries. The modified sol-gel technique allows for some control towards the final structure of the compound. Using this technique, a new layered vanadium oxide phosphate material, containing the surfactant dodecylphosphate, has been synthesized. The compound was analyzed using powder XRD, TGA, SEM, FTIR, TEM, and solid state NMR for both ^{51}V and ^{31}P. $V_2O_3(PO_4C_{12}H_{25})_3Na_{2-x}K_x(H_2O)_{3.2}$ is the general formula of the layered product with an interlayer spacing of 36.6 Å. The initial compound is composed of a vanadium oxide phosphate layer sandwiched between two hydrocarbon layers. The synthesis, composition, and structure of the initial compound will be discussed. Interestingly, when this compound is calcined to 400°C, the structure changes and is possibly hexagonal. Preliminary results are presented on this calcined material.

INTRODUCTION

Since the discovery at Mobil [1, 2] of MCM-41, a mesoporous structured aluminum silicate, there has been great interest in extending the research to include transition metals [3, 4]. However, very little of this work has targeted vanadium oxides. This is surprising as many vanadium oxide compounds are of particular interest as they have many potential commercial applications such as molecular sieves, sorbents, catalysts, and energy storage devices. Vanadium oxides have a particularly rich structural chemistry [5] and have also been found to form a wide range of inorganic/organic materials [6]. Along with the quest of discovering new materials analogous to MCM-41, new techniques are being used to synthesize these compounds. One of the techniques is the modified sol-gel method. Sol-gel synthesis involves a sol (a fluid colloidal system) turning into a gel (jelly-like mass with 3D system). This occurs by a metal alkoxide undergoing hydrolysis and polymerization. Further drying yields a xerogel. Organics present in the product make it a 'modified' sol-gel. Templating, done by surfactants, forms micelles (shapes) in solution. Metal oxides can be fashioned to a particular shape (i.e. tunneled, layered structures) around these micelles. The surfactants can then be removed to leave behind the metal oxide with the desired nano-structure. Earlier work in our laboratory showed that vanadium oxide surfactant materials were not mesoporous, despite TEM indications of a 40 Å lattice, but rather were composed of Keggin-like vanadium oxide clusters [7, 8]. Single crystal X-ray diffraction confirmed the presence of clusters and of organic/inorganic systems swellable by a wide range of solvents [9].

EXPERIMENTAL

A modified sol-gel method [10] was used to react a vanadium alkoxide with the surfactant dodecylphosphate. The air sensitive reagents required the use of a double manifold system with vacuum and argon gas lines, together with ground joint glassware. Also a means of cooling the violent or exothermic reactions is needed, and water or slush baths were used. A reflux set up was used with a cooling bath of ice water to react 5 grams of sodium metal (Aldrich) with 300 ml of isopropanol (Aldrich) in 300 ml of diethyl ether (Aldrich). After completion, the alcohol and ether were removed by vacuum to leave the white needle like crystals of $NaOCH(CH_3)_2$. The following synthesis, of the vanadium alkoxide, was modified from that of Turevskaya [11]. About 250 ml of dry diethyl ether was added to the flask with the $NaOCH(CH_3)_2$, and a dry ice acetone bath was applied. Then 3.4 ml of VCl_4 (Aldrich) was added slowly via a syringe to the reaction vessel. The resulting NaCl was removed by a fritted filter, and the ether was vacuumed off to leave $V[OCH(CH_3)_2]_4$. Vanadium (IV) isopropoxide was then reacted with acetylacetone (also known as 2,4-pentanedione from Aldrich) with a 1:1 molar ratio in isopropanol. Acetylacetone acts as a chelating agent to slow down the hydrolysis reactions of the metal alkoxide. If hydrolysis occurs too fast, then amorphous phases will be obtained. The isopropanol (iPrOH) was removed by vacuum. A 3.4 wt % aqueous solution of the surfactant was prepared by dissolving dodecylphosphate (from Lancaster Synthesis Inc.) into pH 5 adjusted water (with concentrated HCl). The 3.4 wt % value was due to the addition of extra solvent in order to increase solubility. The reactants, in the molar ratio 1 $V[OCH(CH_3)_2]_4$: 1 acetylacetone (acac) : 1 surfactant :1 KCl were mixed in a flask for several hours. The contents of the flask were then transferred to a beaker, oven heated at 80°C for 5 days, washed with water, and then oven dried at 120°C for 1day. After drying, the solid crystalline product was black in color suggesting partial reduction of the vanadium. This synthesis scheme is shown in Figure 1.

The material, upon drying, was a black hardened gel with small variations in color. After grinding it became gray. Initial results from XRD indicated the presence of NaCl and KCl salts, byproducts of the synthesis, which were removed by washing with reverse osmosis water.

Samples were characterized via X-ray powder diffraction on a Phillips PW3040-MPD powder diffractometer using $CuK\alpha$ radiation, fitted with an Anton Parr heating stage model HTK-10 calibrated to the m.p. of lead. Thermalgravimetric analysis, to determine water and organic content as well as overall stability, was done on a Perkin Elmer TGA 7. Electron Microprobe analyses and SEM images were obtained using a JEOL 2000 Electron Microprobe. NMR analysis was performed on a Bruker AVANCE 600 MHz instrument set up for solids. Concentrated phosphoric acid (aq) was used as a reference for ^{31}P NMR, and V_2O_5 was used as reference for ^{51}V (which has a −310 ppm shift in relation to $VOCl_3$ which is set at 0 ppm) [12]. The FTIR instrument used was a Perkin Elmer 1600 with all of the samples being pressed into KBr pellets. TEM was done on a Hitachi H-7000 Electron Microscope at 125,kV with samples mounted on a lacey carbon coated copper grid after suspension in an ethanol mixture.

RESULTS

Initial Compound
The powder XRD pattern of the initial compound, shown in Figure 2(a) is consistent with a layered structure with approximate d values of 37Å, 18Å, 12Å. and 9.7Å. These observed peaks are in the 00l direction only. This lattice spacing of 37Å is consistent with a bilayer of surfactant

molecules sandwiched between two sheets of a vanadium oxide phosphate. No evidence for residual NaCl or KCl was observed after the washing.

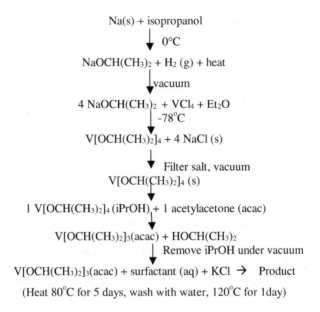

$$Na(s) + isopropanol$$

$$\downarrow 0°C$$

$$NaOCH(CH_3)_2 + H_2 (g) + heat$$

$$\downarrow vacuum$$

$$4\ NaOCH(CH_3)_2 + VCl_4 + Et_2O$$

$$\downarrow -78°C$$

$$V[OCH(CH_3)_2]_4 + 4\ NaCl\ (s)$$

$$\downarrow Filter\ salt,\ vacuum$$

$$V[OCH(CH_3)_2]_4\ (s)$$

$$\downarrow$$

$$1\ V[OCH(CH_3)_2]_4\ (iPrOH) + 1\ acetylacetone\ (acac)$$

$$\downarrow$$

$$V[OCH(CH_3)_2]_3(acac) + HOCH(CH_3)_2$$

$$\downarrow Remove\ iPrOH\ under\ vacuum$$

$$V[OCH(CH_3)_2]_3(acac) + surfactant\ (aq) + KCl \rightarrow\ Product$$

$$(Heat\ 80°C\ for\ 5\ days,\ wash\ with\ water,\ 120°C\ for\ 1day)$$

Figure 1. Synthesis scheme.

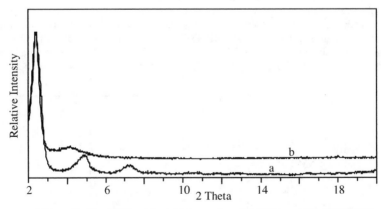

Figure 2. The XRD of the initial sample at (a) 25°C and (b) 400°C [13].

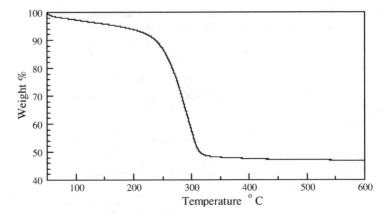

Figure 3. TGA of initial sample undergoing calcination in oxygen [13].

Thermal gravimetric analysis (TGA) in oxygen (Figure 3) or nitrogen gave a similar curve. This graph revealed a weight loss of around 10% for H_2O, and about 45% for the hydrocarbon chain of the surfactant. The TGA of the surfactant itself showed a steep weight loss around 200°C, which is also seen in the TGA of the sample.

The electron microprobe images showed a layered morphology for the material. For use as reference in the elemental analysis for the electron microprobe, $VOPO_4$ was synthesized [14] and then furnace dried at 500°C for several days. $VOPO_4$ exhibited a ratio of 1.01 P to 1.00 V. Elemental analysis, by WDS and EDS, of the initial compound showed only 4 elements with the resulting atomic ratio of $V_1P_{1.5}Na_{0.4}K_{0.3}$. The standard deviation is large for Na and K so there is probably more than what is written for the ratio given.

Solid state NMR analyses were done on the material for both [51]V and [31]P. Lapina et. al. extensively investigated [51]V solid state NMR of vanadium compounds [12]. They discovered that the coordination of vanadium oxides may be determined, together with the number of adjacent vanadium polyhedra from the chemical shift anisotropy (csa) of the static solid state NMR of [51]V

Figure 4. The [51]V static solid state NMR of the initial sample showing the chemical shift anisotropy. The vertical line is caused by the transmitter offset [13].

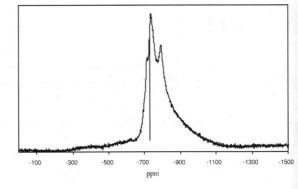

The initial compound had a peak with a range of about 350 ppm for static solid state ^{51}V NMR. This means a VO_4 tetrahedron has one of the oxygens shared with another vanadium oxide tetrahedron, forming a V_2O_7 group. A compound with a known structure containing V_2O_7 groups was also run, and it had a similar range [15]. The MAS NMR for ^{51}V gave one peak indicating 1 type of vanadium nuclei. Also performed on the sample was ^{31}P MAS NMR, which indicated 2 types of phosphate nuclei with shifts of 0.6 and 25 ppm. The surfactant dodecylphosphate was analyzed by ^{31}P MAS NMR, and showed no shift relative to phosphoric acid. Phosphates are always tetrahedral, and here they form a chain with the vanadia tetrahedra.

FTIR was also used to help elucidate the structure. Vibrations near 2900 cm^{-1} for the aliphatic C-H stretch and a P-O stretch at 1111 cm^{-1} [16] were seen. No vibrations were observed at 3000 cm^{-1} for hydroxyl bonded to a P, nor was a peak seen for the O-H stretch for a hydrogen bonded hydroxyl (which is also bonded to a P) at 2600 cm^{-1} [16]. The P=O stretch, which was seen for the dodecylphosphate at 1239 cm^{-1}, was not seen in the material. So no P=O or P-O-H bond vibrations were observed. Unfortunately some of the vanadium oxygen bonds occur in the region where the P-O exhibits a strong wide band from 900-1200 cm^{-1}, so they were not able to be observed.

TEM images showed a layered morphology with a repeat distance of 33 Å. The layers are stacked on top of one another in what appears to be a random method, like sheets of paper stacked, with the sheets at different orientations. Thus the crystals have order in one direction only, as was seen in the XRD pattern. Several crystals were analyzed and all had a similar layered appearance.

The proposed structure of the initial compound is a bilayer of surfactant molecules aligned with their hydrophobic hydrocarbon 'tails' together, and their phosphate 'heads' sticking out and bonded with a layer of vanadium oxide giving a 37Å interlayer distance. The experimental evidence is consistent with the formula $V_2O_3(PO_4C_{12}H_{25})_3Na_{2-x}K_x(H_2O)_{3.2}$. The hydrocarbon chains are covalently bonded to the phosphate tetrahedra, and most likely in the 'plane' of the tetrahedra plane (Figure 5). Sodium and potassium cations are bonded to the other oxygens of the phosphate. The few water molecules can hydrogen bond to the edges of the polyhedra.

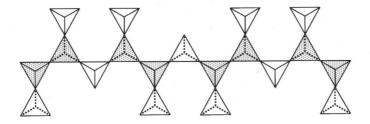

Figure 5. The structure of the initial compound with the hydrocarbon chains removed for clarity; the dark tetrahedra are V, and the light tetrahedra are P [13].

Calcined Material

Interestingly when the initial compound is heated in air to 400°C, the compound undergoes structural change. At this temperature the hydrocarbon portion of the surfactant has combusted; the phosphate remains with the vanadium oxide. This material has phosphate tetrahedra that are not bonded to hydrogens. The vanadia are tetrahedra that do not share any oxygens with other vanadium atoms. The V to P ratio is 1 to 1 from electron microprobe analyses. From the XRD pattern (Figure 2b) the two peaks can be indexed to a hexagonal unit cell (37 Å (100), 22 Å (110)) and $a = 43$ Å. More evidence is needed for the elucidation of the unit cell for this material. Further TEM analysis and other experiments will help determine the structure of the calcined material.

ACKNOWLEDGEMENTS
We gratefully acknowledge the National Science Foundation for support of this work through grant DMR-9810198.
We would like to thank David Kiemle, at the S.U.N.Y. College of Environmental Science and Forestry, for use of the NMR instrument. Also we thank at Binghamton University, Henry Eichelberger for use of the TEM, and Professor David Jenkins for use of the x-ray instrument.

REFERENCES
1. J. S. Beck, J. C. Vartuli, W. J. Roth, M. E. Leonowicz, C. T. Kresge, K. D. Schmitt, C. T.-W. Chu, D. H. Olson, E. W. Sheppard, S. B. McCullen, J. B. Higgins, and J. L. Schlenker, *Amer. Chem. Soc.*, **114**, 10834 (1992).
2. J. S. Beck, J. C. Vartuli, G. J. Kennedy, C. T. Kresge, W. J. Roth, and S. E. Schramm, *Chem. Mater.*, **6**, 1816 (1994).
3. Q. S. Huo, D. I. Margolese, U. Ciesla, P. Feng, T. E. Gier, P. Sieger, R. Leon, P. M. Petrof, F. Schuff, and G. D. Stucky, *Nature*, **368**, 317 (1994).
4. Q. Huo, D. I. Margolese, U. Ciesla, D. G. Demuth, G. D. Stucky, and et.al., *Chem. Mater.*, **6**, 1176 (1994).
5. P. Y. Zavalij and M. S. Whittingham, *Acta Cryst.*, **B55**, 627 (1999).
6. T. A. Chirayil, P. Y. Zavalij, and M. S. Whittingham, *Chem. Mater.*, **10**, 2629 (1998).
7. G. G. Janauer, A. Dobley, J. Guo, P. Zavalij, and M. S. Whittingham, *Chem. Mater.*, **8**, 209 (1996).
8. G. G. Janauer, R. Chen, A. D. Dobley, P. Y. Zavalij, and M. S. Whittingham, *Mater. Res. Soc. Proc.*, **457**, 533 (1997).
9. G. G. Janauer, P. Y. Zavalij, and M. S. Whittingham, *Chem. Mater.*, **9**, 647 (1997).
10. D. M. Antonelli and J. Y. Ying, Angew. *Chem. Int. Ed. Engl.*, **34**, 2014 (1995).
11. E. P. Turevskaya and N. Y. Turova, *Koord. Khim.*, **15**, 373 (1989).
12. Lapina O.B., Mastikhin V.M., Shubin A.A., Krasilnikov V.N., and Zamarev K.I. *Progress in NMR Spectroscopy*, **24**, 457-525, (1992).
13. A. Dobley, P.Y. Zavalij, J. Schulte, and M. S. Whittingham, *Chem. Mater.* (submitted).
14. J.W. Johnson, A.J. Jacobson, J.F. Brody and S.M. Rich *Inorg. Chem.* **21**, 3820-2825 (1982).
15. P.Y. Zavalij, F. Zhang, and M. S. Whittingham, *Acta Cryst.* **C53**, 1738 (1997).
16. R.T. Conley. *Infrared Spectroscopy.* (Allyn and Bacon, Inc. Boston 1966).

Mat. Res. Soc. Symp. Proc. Vol. 658 © 2001 Materials Research Society

Two Pillared Three-Dimensional Inorganic/Organic Hybrid Structures Composed of Bimetallic Layers Bridged by Exo-bidentate Ligands

Long Pan, Xiaoying Huang, and Jing Li*
Department of Chemistry, Rutgers University, Camden, NJ 08102, USA

ABSTRACT

The hydrothermal reactions of $Cu(NO_3)_2 \cdot 3H_2O$, K_2CrO_4 and exo-bidentate ligand bpy/bpe, (bpy = 4,4'-bipyridine, bpe = trans-1,2-bis(4-pyridyl)ethylene) result in two pillared 3D organic-inorganic hybrid compounds. Both contain bimetallic layers of copper(II) cations and chromate anions, which are directly connected by different organic ligands, leading to pillared structures with varying pillar length and thickness of layers. Crystal data for $[Cu_3(CrO_4)_2(OH)_2(bpy)_2]$ (1): triclinic, $P\bar{1}$, a = 5.358(1), b = 5.603(1), c = 13.515(3) Å, α = 81.78(3), β = 86.56(3), γ = 79.75(3)°, V = 394.91(13) Å^3, Z = 1; $[CuCrO_4(bpe)]$ (2): monoclinic, $C2/c$, a = 23.282(5), b = 11.917(2), c = 9.726(2) Å, β = 112.99(3)°, V = 2484.2(8) Å^3, Z = 8.

INTRODUCTION

Considerable efforts have recently been made to fabricate hybrid inorganic/organic materials, with a strong emphasis on their optimized functionality [1, 2]. One of the major challenges in this area is to design desired structures. The usual approach is to deliberately select suitable metal-containing inorganic fragments and organic ligands in consideration of their preferred coordination habit and possible geometry, as well as the ratio of the metal to the ligand. A reasonable strategy to build high dimensionality frameworks is realized by building pillared structures with organic pillars of changeable length and/or types to connect inorganic layers [3, 4]. The linear exo-bidentate ligands (e.g. pyrazine, 4,4'-bipyridine) as effective connectors or spacers have led to various interesting coordination structures, such as those of ladder, brick, and diamondoid [5]. While inorganic pillared layers composed of molybdenum or vanadium oxides have been found in several inorganic/organic hybrid frameworks, no 3D structures containing chromium oxides and organic pillars have been reported. To date, only one 2D structure with bimetallic layers (Cu/O/Cr) anchored with terminal pyridine ligands has appeared in the literature [6]. In this paper, we exploit the transformation of a 2D structure into a 3D one by replacing the terminal ligands with bidentate pillar molecules, which interconnect the 2D layers to give rise to a 3D network.

EXPERIMENT SECTION

Materials: $Cu(NO_3)_2 \cdot 3H_2O$, K_2CrO_4 trans-1,2-bis(4-pyridyl)ethylene (bpe), and 4,4'-bipyridine (bpy) of reagent grade quality were used as purchased without further purification.

Synthesis: $[Cu_3(CrO_4)_2(OH)_2(bpy)_2]$ (1): Hydrothermal reactions of $Cu(NO_3)_2 \cdot 3H_2O$ (0.0242 g, 0.1 mmol), K_2CrO_4 (0.0097 g, 0.05 mmol), bpy (0.0156 g, 0.1 mmol) and 10

ml deionized water with the ratio of 1:0.5:1:5555 for 3 days at 150°C followed by slow cooling to room temperature produced orange plate crystals in 65% yield (0.010 g). [CuCrO$_4$(bpe)] (2): Hydrothermal reactions of Cu(NO$_3$)$_2$·3H$_2$O (0.0242 g, 0.1 mmol), K$_2$CrO$_4$ (0.0097 g, 0.05 mmol), bpe (0.0182 g, 0.1 mmol) and 10 mL deionized water with the ratio of 1:0.5:1:5555 for 3 days at 150°C followed by slow cooling to room temperature produced orange column crystals in 77% yield (0.014 g).

X-ray analysis and structure refinement: Single crystals of **1** (0.12 × 0.12 × 0.01 mm) and **2** (0.10 × 0.06 × 0.03 mm) were selected for data collection. Each crystal was mounted on a glass fiber with epoxy resin in air. Diffraction data were collected on an Enraf-Nonius CAD4 diffractometer equipped with graphite monochromatized MoK$_\alpha$ radiation ($\lambda = 0.71069$ Å). The lattice constants were optimized from a least-squares refinement of the setting of 25 centered reflections. For each data set, three standard reflections were monitored every two hours, and raw intensities were corrected for Lorentz and polarization effects and for absorption by empirical method based on ψ-scan data [7]. The number of measured and unique reflections with R$_{int}$ were 1898, 1712 and 0.0549 in the range $3.05 \leq \theta \leq 26.97°$ for **1** and 2308, 2178 and 0.09421 in $2.70 \leq \theta \leq 24.98°$ for **2**. All non-hydrogen atoms were readily located and refined with anisotropic thermal parameters except that C3 in **2** was refined isotropically. The final full-matrix least-squares refinement on F^2 was applied to the data. Data collections were controlled by the CAD4/PC program package. The structures were solved by direct methods with SHELX-97 [8]. Details of data collection and structure refinement, as well as unit cell parameters for **1** and **2** are listed in Table I.

Table I. Crystallographic Data for **1** and **2**.

	1	2
empirical formula	C$_{10}$H$_{10}$Cr$_2$Cu$_3$N$_2$O$_{10}$	C$_{12}$H$_{10}$CrCuN$_2$O$_4$
fw	612.82	361.76
space group	P-1 (No.2)	$C2/c$ (No.15)
a (Å)	5.358(1)	23.282(5)
b (Å)	5.603(1)	11.917(2)
c (Å)	13.515(3)	9.726(2)
α (°)	81.78(3)	90
β (°)	86.56(3)	112.99(3)
γ (°)	79.75(3)	90
V (Å^3)	394.9(13)	2484.2(8)
Z	1	8
T, K	293(2)	293(2)
λ, Å	0.71073	0.71073
ρ_{calc}, $_{gcm}$$^{-3}$	2.577	1.935
μ (mm^{-1})	5.365	2.600
R^a (I>2σ(I))	0.0584	0.0664
R_w^b	0.1038	0.0843

$^aR = \Sigma| \ |F_o| - |F_c| \ |/\Sigma|F_o|$. $^b R_w = [\Sigma[w(|F_o^2| - |F_c^2|)^2]/\Sigma w(F_o^2)^2]^{1/2}$.

Weighting: **1**, $w = 1/\sigma^2[F_o^2 + 3.0P]$, where $P = (F_o^2 + 2F_c^2)/3$; **2**, $w = 1/\sigma^2(F_o^2 + (0.01P)^2)$.

RESULTS AND DISCUSSION

Both compounds were prepared by mixing $Cu(NO_3)_2 \cdot 3H_2O$, K_2CrO_4 and bpy/bpe in a molar ratio of 2:1:2 and reacting in an autoclave at 150 °C for three days. The structures of **1** and **2** were determined by single crystal X-ray diffraction. Selected bond lengths and angles are listed in Table II and III, respectively. The crystal structure analysis of **1** revealed a pillared 3D structure built upon bimetallic layers (Cu/O/Cr) that are bridged by bpy pillars (Figure 1). A single bimetallic layer is illustrated in Figure 2 (top). Two types of metals with different coordination environments are present in the layer: octahedrally coordinated copper and tetrahedrally coordinated chromium. Each layer can be regarded as composed of ribbons of edge-sharing $Cu(2)O_5N$ and $Cu(1)O_6$ octahedra interconnected via corner-sharing CrO_4 tetrahedra. Shown in Figure 2 (bottom) is a single ribbon of edge-sharing $Cu(1)O_6$ and $Cu(2)O_5N$ octahedra. There are two crystallographically independent Cu atoms within the ribbon. Each Cu2 shares one edge with another Cu2 via two oxygen atoms (O1, O1^{ii}). It also shares two edges with two neighboring Cu1 through (O1, O2), and (O1^{ii}, O3^{iii}). Each Cu1 shares four edges with four Cu2. The axial Cu-O bonds, 2.443(6) Å for Cu1 and 2.325(6), 2.378(6) Å for Cu2, are significantly longer than the average equatorial Cu-O (1.961 Å) and Cu2-O (1.983 Å) due to the Jahn-Teller distortion. The η^3-O1 attributed to oxygen of the hydroxide group is bound to two Cu2 with Cu-O distances of 1.991(6) and 2.003(6) Å, and one Cu1 with a Cu-O bond length of 1.974(6) Å. The ribbons are interconnected by corner-sharing CrO_4 tetrahedra through O2, O3 and O4, respectively, giving rise to a 2D structure motif parallel to *ab* plane (Figure 2, top). The remaining site of the CrO_4 tetrahedron is occupied by a terminal oxygen atom that points outward from the layer with a Cr-O bond length of 1.610(6) Å. The two N atoms of each bipyridine pillar form coordinated bonds with the two edge-sharing Cu2 octahedra from the neighboring bimetallic layers to result in a pillared 3D structure, as shown in Figure 1. All Cu2-N bonds are identical with a single bond length of 2.014 (7) Å. The interlayer distance is approximately 10.6 Å. Previous study has shown that use of a monofunctional pyridine (py) in the Cr-Cu-O system leads to a 2D structure $[Cu_3(CrO_4)_2(OH)_2(py)_2]$, where py serves as a terminal ligand [6]. In the present work, we have demonstrated that by replacing the monodentate ligand with a bidentate ligand (bpy), we can retain the structural configuration of the bimetallic layer $Cu_3(CrO_4)_2(OH)_2$, and at the same time, to build a 3D network by interconnecting the bimetallic layers through both pyridine N atoms. This is a neat example in which rational synthesis works.

Table II. Selected Bond Lengths (Å) for 1.

Cu(1)-O(2)^i	1.948(6)	Cu(2)-O(2)	2.325(6)
Cu(1)-O(2)	1.948(6)	Cu(2)-O(3)^{iii}	2.378(6)
Cu(1)-O(1)	1.974(6)	Cr-O(5)	1.610(6)
Cu(1)-O(1)^i	1.974(6)	Cr-O(3)	1.623(6)
Cu(1)-O(3)	2.443(6)	Cr-O(4)	1.671(6)
Cu(1)-O(3)^i	2.443(6)	Cr-O(2)^{iv}	1.691(6)
Cu(2)-O(4)	1.955(6)	Cu(1)-Cu(2)^i	3.0478(14)
Cu(2)-O(1)	1.991(5)	Cu(1)-Cu(2)	3.0478(14)
Cu(2)-O(1)^{ii}	2.003(6)	Cu(2)-Cu(2)^{ii}	3.054(2)
Cu(2)-N	2.014(7)		

Symmetry transformations used to generate equivalent atoms:
i -x+2,-y+1,-z   ii -x+2,-y+2,-z   iii x,y+1,z   iv x-1,y,z

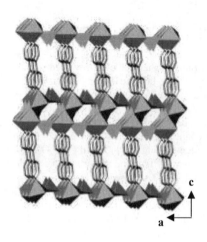

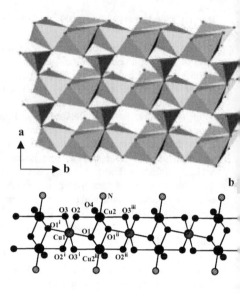

Figure 1. View of *1* along the b axis. The metal coordination polyhedra are shown (CrO₄ tetrahedra and CuO₆ and CuO₅N octahedra).

Figure 2. Illustration of bimetallic layer in *1*. Top: a single Bimetallic layer parallel to the ab plane shown corner- and edge-sharing polyhedra. Bottom: a single Cu/O/N ribbon along the b axis.

Encouraged by these results, we conducted another set of reactions under the same conditio introducing a longer organic exo-dentate ligand, bpe. These reactions have produced a n pillared compound, [Cu(CrO₄)(bpe)] (**2**). Figure 3 shows the local coordination environment metals in **2**. The copper center adopts a distorted trigonal bipyramidal (tbp) geometry throu three oxygens (O1, O2^i, O3^{ii}) from the three CrO₄ groups in the equatorial plane, and two b nitrogen atoms (N1, N2) in the axial positions with the Cu-N bond lengths of 1.985(7) a 2.004(7) Å, respectively. The Cu-O1 (1.957(5) Å), Cu-O2^i (1.998(6) Å), Cu-O3^{ii} (2.115(6) Å) the equatorial plane are essentially equivalent, falling into the range of Cu-O bond lengt reported previously for similar compounds [9]. Every CuO₃N₂ bipyramid shares corners w three adjacent tetrahedral CrO₄ groups, and vice versa, each CrO₄ shares corners with thr CuO₃N₂ bipyramids to form a 2D bimetallic layer. The layer can also be considered as consisti of two types of CrCuO rings: the 8-member (Cu₂Cr₂O₄) rings and 16-member (Cu₄Cr₄O₈) rin as illustrated in Figure 4. As in **1**, the bpe ligands bridge to the Cu atoms of the bimetallic lay to generate a 3D pillared structure with the pillared distance of *ca* 11.4 Å (Figure 5). T remaining terminal oxygen of the CrO₄ group points outward from the inorganic layer. It worth noting that the bimetallic layer in **2** has changed to a different configuration from that of It is analogous to a pillared 3D structure [CuMoO₄(bpe)] [10] with the same composition t replacing CrO₄ by MoO₄. However, only uniform 12-membered (Cu₃Mo₃O₆) rings exist in t Cu/Mo layer.

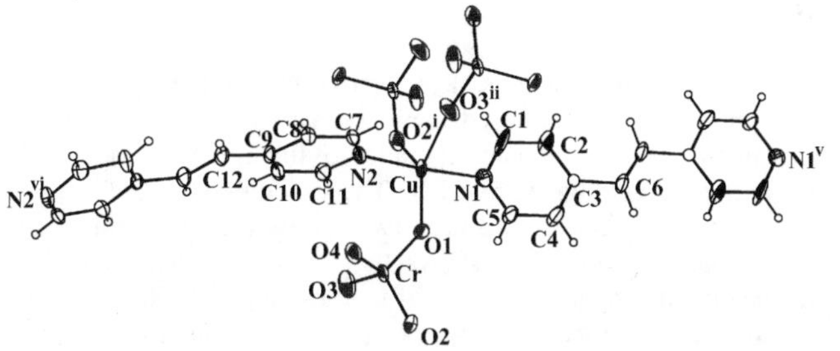

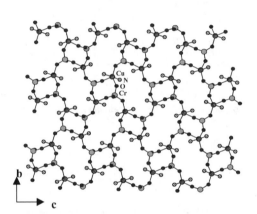

Figure 3. *Local coordination environment of metals in* **2** *with all atoms except C3 and H represented by thermal ellipsoid drawn at 50% probability level.*

Figure 4. *View of the (Cu/CrO₄) bimetallic layer parallel to the bc plane.*

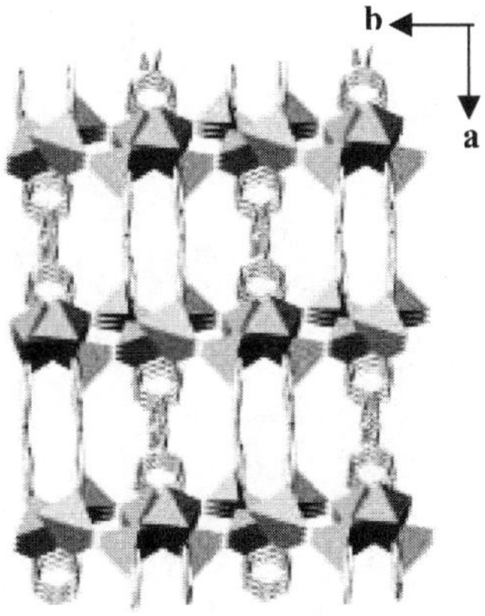

Figure 5. *Polyhedron view of* **2** *along the c-axis.*

Table III. Selected Bond Lengths (Å) and Angles (°) for **2**.

Cu-O(1)	1.957(5)	Cr-O(4)	1.611(5)
Cu-N(2)	1.985(7)	Cr-O(3)	1.615(6)
Cu-O(2)^i	1.998(6)	Cr-O(2)	1.654(6)
Cu-N(1)	2.004(7)	Cr-O(1)	1.668(5)
Cu-O(3)^{ii}	2.115(6)		
O(1)-Cu-N(2)	89.5(3)	O(4)-Cr-O(3)	108.2(3)
O(1)-Cu-O(2)^i	137.1(2)	O(4)-Cr-O(2)	108.8(3)
N(2)-Cu-O(2)^i	87.6(3)	O(3)-Cr-O(2)	109.9(3)
O(1)-Cu-N(1)	92.6(3)	O(4)-Cr-O(1)	109.4(3)
N(2)-Cu-N(1)	176.0(3)	O(3)-Cr-O(1)	110.1(3)
O(2)^i-Cu-N(1)	93.1(3)	O(2)-Cr-O(1)	110.5(3)
O(1)-Cu-O(3)^{ii}	117.8(3)	Cr-O(1)-Cu	138.5(3)
N(2)-Cu-O(3)^{ii}	89.4(3)	Cr-O(2)-Cu^{iii}	142.0(4)
O(2)^i-Cu-O(3)^{ii}	104.9(2)	Cr-O(3)-Cu^{iv}	173.2(4)
N(1)-Cu-O(3)^{ii}	86.7(3)		

Symmetry transformations used to generate equivalent atoms:
i -x+1/2,y+1/2,-z+1/2   ii x,-y,z-1/2   iii -x+1/2,y-1/2,-z+1/2   iv x,-y,z+1/2

CONCLUSION

In summary, two types of structures have been synthesized using CrO_4 as an inorganic anion. Bidentate ligands bpy and bpe were used as pillars that connect the bimetallic layers to generate pillared 3D structures. The structural configuration of the bimetallic layer in **1** is the same as that obtained by using the terminal ligand pyridine. The structural change from a 2D to a 3D network by a suitable selection of ligand is well demonstrated by the successful design and synthesis of this compound. Moreover, our work shows that pillared structures with different interlayer separations can be obtained by use of pillars of various lengths (Scheme 1). Further studies are in progress to investigate the structure changes in similar systems involving other pillared ligand and different chromate anions such as bichromate.

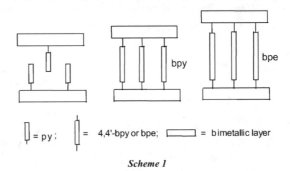

Scheme 1

ACKNOLEDGEMENT

We thank the National Science Foundation (DMR-9553066) for the financial support.

REFERENCES

1. O. M. Yaghi, H. Li, C. Davis, D. Richardson, and T. L. Groy, *Acct. Chem. Res.,* **31**, 47 (1998).
2. S. R. Batten, and R. Robson, *Angew. Chem.,* **37**, 1461 (1998).
3. A. Clearfield, *Chem. Mater.* **10**, 2801 (1998).
4. D. Li, and K. Kaneko, *J. Phys. Chem. B.,* **104**, 8940 (2000).
5. P. J. Hagrman, D. Hagrman, and J. Zubieta, *Angew. Chem.,* **38**, 2639 (1999).
6. W. Bensch, N. Seferiadis, and H. R. Oswald, *Inorg. Chim. Acta.,* **126**, 113 (1987).
7. Kopfmann, G, and Hubber R., *Acta Crystallogr,* **A24**, 348 (1968).
8. Sheldrick, G. M., SHELX-97: program for structure refinement; University of Goettingen Germany, 1997.
9. G. Smith, E. J. O'Relly, C. H. L. Kennard, and T. C. W. Mak, *Inorg. Chim. Acta,* **65**, L21 (1982)
10. D. Hagrman, R. C. Haushalter, and J. Zubieta, *Chem. Mater.,* **10**, 361 (1998).

Mat. Res. Soc. Symp. Proc. Vol. 658 © 2001 Materials Research Society

A Structural Analysis Method for Graphite Intercalation Compounds

Tatsuo Nakazawa, Kyoichi Oshida, Takashi Miyazaki, Morinobu Endo[1] and Mildred S. Dresselhaus[2]

Nagano National College of Technology, 716 Tokuma , Nagano-shi, 381-8550, JAPAN
[1]Shinshu University, 4-17-1 Wakasato, Nagano-shi, 380-8553, JAPAN
[2]Currently on leave from the Massachusetts Institute of Technology, Cambridge, MA 02139, USA

ABSTRACT

Study of the microstructure of electronic materials can be enhanced by using high resolution transmission electron microscopy (TEM) combined with the technique of digitized image analysis. We show here a practical image analysis method for the microstructures of acceptor graphite intercalation compounds (GICs) with $CuCl_2$ and $FeCl_3$ intercalates. The two dimensional fast Fourier transform (2D-FFT) was used for the frequency analysis of the TEM pictures. It is found that the lattice images of $CuCl_2$-GICs consist of different frequency images corresponding to specific frequencies. The detailed features of the stage-1 structure of the $FeCl_3$-GICs is extracted quantitatively by this method from a relatively indistinct TEM picture. The stage structure of the $CuCl_2$- and $FeCl_3$-GICs are further investigated by analyzing the reconstruction of the TEM images by means of the two dimensional inverse FFT (2D-IFFT).

INTRODUCTION

Well-staged acceptor graphite intercalation compounds (GICs) exhibit many useful electrical properties, such as a large magnitude of the conductivity with a metallic temperature-dependence. The electrical and mechanical properties of GICs are thought to strongly depend on the GIC staging structure. In order to analyze the micro-structure of materials, transmission electron microscope (TEM) observations provide one of the most useful techniques, because TEM images give very detailed information about the nature of the lattice structure and the anomalies found in a very small sample. In this study, the method of high-resolution TEM observations combined with digitized image analysis to get a more quantitative evaluation of the staging structure in GICs. Analysis of the microstructure of GICs is largely enhanced by using this technique. The GICs can be synthesized in the fiber matrix of vapor grown carbon fibers (VGCFs) [1,2] that consist of a honeycomb network of concentrically stacked cylindrical layers of graphene planes around the fiber axis. While TEM observations of well-staged acceptor GICs with $CuCl_2$ and $FeCl_3$ intercalates have revealed the staging structure of the GICs, these observations have not yielded much quantitative information. Digitized image analysis is applied here to promote the quantitative interpretation of the TEM images and to clarify the structure of GICs. TEM images of $CuCl_2$ and $FeCl_3$ GICs are evaluated by means of the image analysis method described here using a 2-dimensional (2D) fast Fourier transform (FFT).

EXPERIMENTAL DETAILS

Sample preparation

The preparation of intercalated VGCF samples in brief is as follows [3]. The precursor VGCF was obtained by the thermal decomposition of a mixture of benzene and hydrogen gas, and then heat treated at $2900°C$. The intercalation of the VGCF was performed using the two-zone growth technique. The TEM images of the lattice fringes were obtained using a transmission electron microscope (JEOL 200CX, acceleration voltage of 200keV).

Image analysis method

Two categories of image processing methods can be applied for the analysis of the microscope images. One is the measurement and the outline on the basis of the binarized microscope picture image, which is often used for the evaluation of, for example, the shape and number of pores in the materials. In the present study, another method, a spatial frequency analysis using the 2D-FFT, combined with its verification using the 2D inverse FFT (IFFT), yields an objective and quantitative analysis. Figure 1 shows the schematic flow of the spatial frequency analysis of the microscope images by the 2D-FFT and the verification by the 2D-IFFT. As shown in Figure 1, the microscope pictures are digitized by an image scanner and stored in the computer as the original digitized images. The power spectrum of the original image is obtained by a 2D-FFT and the analyzed image is reconstructed by a 2D-IFFT. It is then possible to extract the materials features that are buried in the rich information included in the original TEM image by choosing a part of the power spectrum, and then processing the 2D-IFFT in detail.

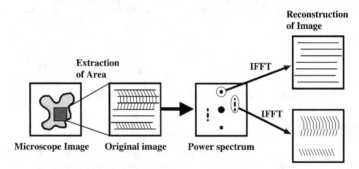

Figure 1. *The schematic flow of the digitized image analysis.*

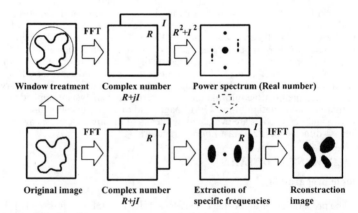

Figure 2. *Flow chart for the 2 dimensional fast Fourier transform (2D-FFT) and the 2 inverse FFT (2D-IFFT) operation.*

The schematic processes of the FFT and IFFT are shown in Figure 2. The result of the FFT is a complex number, and the power spectrum is calculated by summing each square of the real part and the imaginary part. Since the TEM image itself is thought to be a window, which cuts down the continuous data to a rectangle, the edge of the image is discontinuous and it affects the characteristics of the resulting power spectrum. In order to remove this effect, the window treatment is performed on the original image before the FFT operation (see upper flow diagram in Figure 2). In this study, we use a Hamming window for the window treatment.

The window treatment is not needed, on the other hand, when we get the reconstruction image by an IFFT operation, as shown in the lower flow chart of Figure 2. The specific frequencies are estimated by analysis of the power spectrum, and the mask pattern is made for the extraction of specific frequencies. The mask pattern is applied to the complex data calculated by the FFT operation without using a window treatment, and then the specific frequency image is obtained by the IFFT operation.

RESULTS AND DISCUSSION

CuCl$_2$ intercalated VGCF

The original TEM image of the lattice fringes of the CuCl$_2$ intercalated graphitized VGCF (CuCl$_2$-GIC) is shown in Figure 3. From this picture, it is expected that the CuCl$_2$-GICs have a complicated staging texture. Figure 4 shows the power spectrum for the TEM images shown in Figure 3 obtained by 2D-FFT processing. The power spectrum pattern observed here is very similar to the streaking, that was found in the electron diffraction pattern of the same sample. We can thus obtain the streak pattern either by image analysis of the lattice fringe image, or by electron diffraction of the material. Its characteristic pattern is distributed in a line perpendicular to the 00l lattice planes of the original image. The spectrum is very sharp for the CuCl$_2$-GICs and this is because the transverse width of the FFT power spectrum pattern is very small, indicating large lateral regions with the same stage.

Since the power spectrum is distributed only in one direction, the 2-dimensional power spectrum image can be graphed by integration perpendicular to the distribution of the spectrum [4]. Figure 5 shows the distribution of interlayer spacings obtained by integration along the horizontal-axis of the power spectrum in Figure 4. The horizontal scale of the graph is calibrated

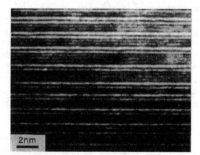

Figure 3. *The original digitized TEM image of a CuCl$_2$ graphite intercalated compound (CuCl$_2$-GIC) (above left).*

Figure 4. *The power spectrum of the TEM image of the CuCl$_2$-GIC (Figure 3) obtained by 2D-FFT. The characteristic feature of a streak pattern is observed (above right).*

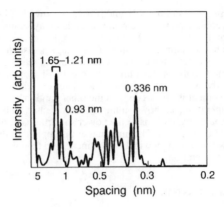

Figure 5. *The distribution of interlayer spacings derived from the power spectrum of CuCl₂-GIC (Figure 4) by integration.*

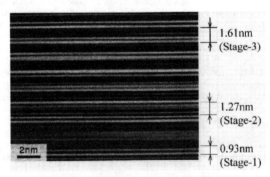

Figure 6. *The reconstructed image for the CuCl₂-GIC by 2D-IFFT.*

at the position of the large peak in the extreme right of the graph corresponding to the 002 graphite lattice planes. The spacing of the 002 planes was pre-determined as 0.336 nm by X-ray diffraction. There are many peaks in the integrated power spectrum of $CuCl_2$-GICs in the region between 0.3 and 4 nm. Stage-1, stage-2, and higher stage regions as well as graphite stacking are assumed to appear in the Figure. We observe a fairly small stage-1 peak around 0.93 nm. There another peak between 1.21 and 1.65 nm (the highest point is located at 1.396 nm). This peak is thought to consist of a stage-2 component (peak expected at 1.27 nm) and a stage-3 component (peak expected at 1.61 nm).

In order to verify the results of the analysis, the lattice fringe images were reconstructed using the data extracted from the power spectrum. The details of the reconstruction techniques were mentioned in previous papers [5-7]. Briefly, in extracting the peak at 0.336 nm, and at 1.39 nm in the power spectrum of $CuCl_2$-GICs, for example, the stripe images correspond to the 002 lattice fringes and the interlayer distance of the stage-2 or stage-3 regions can be reconstructed using the 2D-IFFT processing, respectively. These images are easily superimposed upon each other by selecting the specific frequencies and then performing the 2D-IFFT. Using the data of several peaks or of a certain interlayer spacing region, which are assumed to appear as a feature of the staging structure, the Figure of the stage structure is obtained, as shown in Figure 6. The reconstructed image is very similar to the original one (see Figures 3 and 6). The layers of the reconstructed image appear to be sharper than those in the original image, because the selection of specific frequencies act as a noise reduction operation. Some lines appear in faint form in the $CuCl_2$ intercalated layers. This appearance is caused by a lack of data when the specific frequencies were selected. If all of the frequencies present in the original image (Figure 3) could be identified and selected, a perfect reconstruction of the original image could then be achieved. The original image, however, already contains some faint lines where the intercalates enter and exit. These faint features appear as a result of the influence of the aperture setting when the TEM image was taken, which occurred likewise when the above image analysis and the specific frequency selection of the power spectrum were made. Since only part of the electron beam of the TEM is intercepted by the aperture, the aperture setting of the TEM observation may also act as a frequency selection of the electron diffraction pattern.

FeCl₃ intercalated VGCF

The original TEM image of the $FeCl_3$ intercalated graphitized VGCF ($FeCl_3$-GIC) is shown in Figure 7. Since Figure 7 is a relatively unclear picture, the quantitative analysis of the $FeCl_3$-GIC image is thought to be more difficult than that of the $CuCl_2$-GIC. The power spectrum for the TEM image of $FeCl_3$-GIC, which is shown in Figure 8, is somewhat indistinct, reflecting the unevenness of the dark and light images in the longitudinal direction of the picture.

Figure 9 shows the distribution of the interlayer spacings obtained by the method mentioned above. The horizontal scale of the graph is calibrated at the position of the peak at 0.467 nm. In the integrated power spectrum of the $FeCl_3$-GIC sample, one large peak at 0.467 nm corresponds to the spacing between the graphite plane and the intercalate plane and some small peaks were observed in the region between 0.4 and 2 nm. The small peak corresponds to the spacing of the stage-1 $FeCl_3$-GIC, which was determined as 0.934 nm and appears clearly in the power spectrum.

The reconstructed image from the characteristic features between 0.4 and 2 nm is shown in Figure 10. Again, as in the case of the $CuCl_2$-GICs, the sharper layer image corresponds to the stage-1 structure, as obtained by a digitized image analysis.

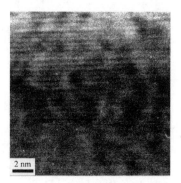 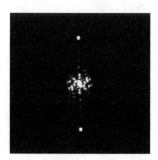

Figure 7. *The original digitized TEM image of a FeCl₃-GIC sample (above left).*

Figure 8. *The power spectrum of the TEM image of the FeCl₃-GIC (Figure 7) by 2D-FFT. The two bright spots that are symmetrically positioned about the central point correspond to the distance between hexagonal carbon layers containing one intercalate layer (above right).*

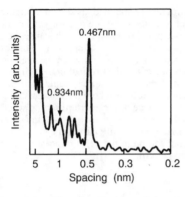 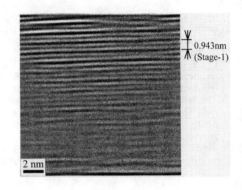

Figure 9. *The distribution of interlayer spacings derived from the power spectrum of FeCl$_3$-GIC (Figure 8) by integration (above left).*

Figure 10. *The reconstructed image for FeCl$_3$-GIC obtained by using the 2D-IFFT (above right).*

CONCLUSION

A digitized image analysis of TEM pictures by means of a 2D-FFT and a 2D-IFFT has been illustrated. The possibility of using an image analysis method, in combination with TEM picture as a quantitative evaluation tool for the material structure study was presented. The complex staging structure of CuCl$_2$-GICs was clearly revealed. A clear stage1 structure for the FeCl$_3$-GIC was also displayed based on a relatively indistinct TEM picture.

ACKNOWLEDGEMENT

The work by the author (KO) is supported by a grant-in-aid for "Research for the Future Program No. JSPS-RFTF96R11701 from the Japan Society for the Promotion of Science.

REFERENCES

1. T. Koyama, M. Endo and Y. Onuma., *Jpn. J. Appl. Phys.*, **11**, 445 (1972).
2. M. Endo, *Chemtech*, American Chemical Society, **18**, 568 (1988).
3. M. Endo, T. C. Chieu, G. Timp, M. S. Dresselhaus, *Synthetic Metals*, **8**, 251 (1983).
4. K. Oshida, M. Endo, T. Nakajima, S. L. di Vittorio, M. S. Dresselhaus, and G. Dresselhaus, *Mater. Res.*, **8**, 512 (1993).
5. M. Endo, K. Oshida, K. Kobori, K. Takeuchi, K. Takahashi, and M. S. Dresselhaus, *J. Mate Res.*, **10**, 1461 (1995).
6. K. Oshida and T. Nakazawa, *Memoirs of Nagano National College of Technology*, **34**, (2000).
7. K. Oshida, K. Kogiso, K. Matsubayashi, K. Takeuchi, S. Kobori, M. Endo, M. S. Dresselha and G. Dresselhaus, *J. Mater. Res.*, **10**, 2507 (1995).

Mat. Res. Soc. Symp. Proc. Vol. 658 © 2001 Materials Research Society

A Novel Route For The Synthesis of LiAl$_x$Co$_{1-x}$O$_2$ Battery Materials And Their Structural Properties

M. S. Tomar, A. Hidalgo, P. S. Dobal*, R. S. Katiyar*, A. Dixit*, R. E. Melgarejo, and K. A. Kuenhold[a]
Department of Physics, University of Puerto Rico, Mayaguez Campus, Mayaguez PR 00681
*Department of Physics, University of Puerto Rico, San Juan PR 00931-3343; [a]University of Tulsa, Engineering Physics Department, Tulsa, OK. 74104.

ABSTRACT

LiCoO$_2$ is an important cathode material for rechargeable batteries. Al substitution on Co sites was recently found to enhance its intercalation voltage. We have developed a reliable and less expensive method for the synthesis of such LiAl$_x$Co$_{1-x}$O$_2$ compositions using simple salts of the constituent elements and organic solvents. X-ray diffraction and Raman spectroscopy were used for structural characterization of these materials at various annealing temperatures. The phase evolutions in LiAl$_x$Co$_{1-x}$O$_2$ compositions were studied using micro-Raman spectroscopy and a phase diagram is proposed based on the observations. The phase diagram suggests that thin films of these materials can be deposited at substrate temperature at about 650°C

INTRODUCTION

Lithium cells consisting of LiMeO$_2$ (Me; 3d transition metal element) are very attractive for battery applications because of their high energy density [1-4]. Among such materials, LiCoO$_2$ became an active cathode material in commercially available Li-ion batteries due to its attractive electrochemical properties [5]. It is, however, too expensive for large-scale applications and it has relatively low capacity. It has been recently observed that the intercalation voltage increases if Co is partially replaced by some non-transition metal like Al [6]. Despite their potential for enhancing the intercalation voltage [7], the structural behavior of such non-transition metal substituted LiCoO$_2$ has not been investigated.

Solid-state reaction is very common for the synthesis of complex oxides. However, it requires high processing temperatures and is not suitable for thin film growth. The solution-based route, on the other hand, is attractive for thin film deposition by simple spin coating. We report here the synthesis of LiAl$_x$Co$_{1-x}$O$_2$ ($0.0 \leq x \leq 0.7$) by a chemical solution route at lower processing temperatures. Our motivation was to study the temperature requirements in order to achieve structural reliability for these materials. The structural evolution with increasing processing temperature in these compounds has been studied in detail by X-ray diffraction and micro-Raman scattering techniques.

EXPERIMENTAL DETAILS

For LiAl$_x$Co$_{1-x}$O$_2$ compositions x = 0.0, 0.1, 0.3, 0.5 and 0.7, a particular ratio of metals and precursors (reagent grade lithium acetate and cobalt acetate) were mixed in methoxy ethanol and 2 – ethylhexanoic acid and boiled. Al incorporation in the LiAl$_x$Co$_{1-x}$O$_2$ system was achieved using Al-hydroxide and Al-acetates. These solutions were mixed hot and dehydrated for about 50 min before refluxing. The master solutions thus obtained were slowly evaporated with constant stirring to make powders. The resulting powders were then ground and annealed in a box furnace at temperatures ranging from 450°C to 800°C. The Raman spectra were recorded in backscattering geometry using a triplemate (ISA T64000 from Jovin-Yvon) equipped with a microscope and a charge coupled device (CCD) detection system; 514.5 nm line of an Innov argon ion laser was used as an excitation source.

RESULTS AND DISCUSSION

Figure 1 shows the x-ray diffraction (Cu Kα) patterns of $LiAl_{0.5}Co_{0.5}O_2$ at different annealing temperatures. The evolution of diffraction patterns with increasing annealing temperature suggests that the materials crystallize in a layered hexagonal structure at 700°C Figure 2 shows the x-ray diffraction patterns of $LiAl_xCo_{1-x}O_2$ (x = 0.0, 0.1, 0.3, 0.5, and 0.70 annealed at 800°C in air. Samples are clearly single phase and have layered α-NaFeO$_2$-type structure (space group: $R\bar{3}m$) [8,9]. The undesired phases such as γ-LiAlO$_2$ or spinel Co$_3$O$_4$ which have been reported in these materials [7], were not detected in our sample. But, a splitting of the (003) reflection was observed for x = 0.5 and 0.7 composition samples below 600 °C. A similar splitting was observed in LiCoO$_2$ under biased condition and was attributed to the structural changes in the material by Reimers and Dahn [10]. The lattice parameters calculated from the x-ray data of $LiAl_xCo_{1-x}O_2$ (x = 0.0, 0.1, 0.3, 0.5, and 0.70) are given in Table I in the temperature range 450-800 °C. The substitution of Al for Co results in shorter a and larger c. A increases in the c parameter in Li$_x$CoO$_2$ has been understood as resulting from a decrease in the electrostatic binding energy of the Li-depleted layers [11]. The evolution of lattice constants in our samples agrees well with the recent observations [7].

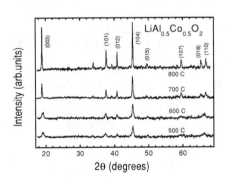

Figure 1. X-ray diffraction (Cu Kα line) patterns of $LiAl_{0.5}Co_{0.5}O_2$ at different annealing temperatures

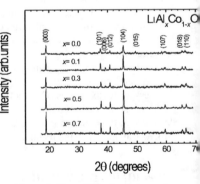

Figure 2. X-ray diffraction patterns of $LiAl_xCo_{1-x}O_2$ (x = 0.0, 0.1, 0.3, 0.5, and 0.70)

Table I. Lattice parameters (a and c) of $LiAl_xCo_{1-x}O_2$ at different annealing temperatures.

	T= 500 °C		T= 600 °C		T= 700 °C		T= 800 °C	
	a (Å)	c (Å)	a (Å)	c (Å)	a (Å)	c (Å)	a (Å)	c (Å)
$LiCoO_3$			2.43597	14.0266			2.43580	14.0274
$LiAl_{0.1}Co_{0.9}O_3$	2.43496	13.99376	2.43295	14.0069	2.43339	14.03767	2.43356	14.0582
$LiAl_{0.3}Co_{0.7}O_3$	2.43138	14.01129	2.43138	14.0399	2.42974	14.06635	2.42670	14.1606
$LiAl_{0.5}Co_{0.5}O_3$	2.42725	14.03473	2.42751	14.0443	2.42824	14.08407	2.42650	14.1644
$LiAl_{0.7}Co_{0.3}O_3$	2.42777	14.02153	2.42529	14.0223	2.42818	14.08531	2.42638	14.1659

The factor group analysis of hexagonal $R\bar{3}m$ (D_{3d}^5) $LiCoO_2$ with one formula unit (Z =1) predicts the total vibrational modes as $A_{1g} + 2 A_{2u} + E_g + 2 E_u$ of which only two (A_{1g} and E_g) modes are Raman active [12]. Inaba et al. [13] have shown the oxygen vibrations for these Raman active modes. The Raman spectra of $LiCoO_2$ prepared by the solution technique at two different temperatures are shown in Figure 3. The presence of both E_g and A_{1g} modes at about 486 and 596 cm^{-1}, respectively shows the hexagonal structure in the material. The broad features at about 1000 and 1200 cm^{-1} show the second order Raman bands. The modifications in the $LiCoO_2$ Raman spectrum with Al substitution on Co sites show an increase in the line widths of A_{1g} and E_g modes, which indicates the lattice disorder. The lattice disorder was found to be peaking at the $x = 0.5$ composition of $LiAl_xCo_{1-x}O_2$. However, the presence of only E_g and A_{1g} modes in the high temperature annealed $LiAl_xCo_{1-x}O_2$ shows that replacing Co by Al does not change the lattice structure. The lower reduced mass (μ) due to Al substitution is expected to increase the mode frequencies. But, as observed by the X-ray data, the increasing c-axis length with increasing Al content reduces the spring force constant (k). Consequently, we do not observe any significant change in the frequency $\omega = (k/\mu)^{1/2}$ of E_g or A_{1g} modes with Al substitution. Moreover, the crystallite sizes in these materials (estimated from the x-ray diffraction widths using Scherer's formula) were too large (110-184 nm) to show any quantum size effect on the phonon frequencies [14]. Unlike $LiNi_yCo_{1-y}O_2$ [15], no major reduction in peak intensities was observed in $LiAl_xCo_{1-x}O_2$ with increasing Al contents.

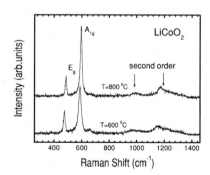

Figure 3. *Room temperature Raman spectra of $LiCoO_2$ annealed at 800 and 600 °C*

To study the phase evolutions with annealing temperature, Raman spectra were obtained from each composition annealed between 450-800°C at 50°C intervals. The room temperature Raman spectra of $LiAl_xCo_{1-x}O_2$ samples annealed at four different temperatures are displayed in Figures 4(a)-(d) for x = 0.1, 0.3, 0.5, and 0.7, respectively. The as-dried powders showed amorphous-like broad features. However, the transformation from amorphous to hexagonal structure in these materials shows interesting features. In addition to the E_g and A_{1g} modes, a third mode at about 675 cm^{-1} was observed in a 450 °C annealed $LiAl_{0.1}Co_{0.9}O_2$ sample. This frequency corresponds to the most intense Raman peak in spinel Co_3O_4 [16]. Therefore, the presence of this mode in the Raman spectra indicates the presence of residual Co_3O_4 at temperatures below 650°C. In contrast, the residual Co_3O_4 phase has been observed recently in $LiAl_{0.1}Co_{0.9}O_2$ annealed at 900° [17]. In $Li_{1-x}CoO_2$ powders prepared by electrochemical lithium deintercalation, phase transition from hexagonal to monoclinic phase has been observed for $0.47 < x < 0$ [13]. The monoclinic phase belongs to $C2/$, (C_{2h}^3) space group with Z = 2 and results in $2A_g+B_g$ Raman active modes [10]. The monoclinic distortion causes the E_g mode to split into B_g+A_g bands. We also observe an additional low frequency mode at about 44 cm^{-1} in Figure 4(a) for annealing temperature in the range 500-700 °C. Analysis of neutron diffraction data shows that the compound $LiCoO_2$ contains a small concentration of Co Li layers, which enables the structure to maintain the c/a ratio [8]. As a result, some of the Li does not go to the lattice sites and form $Li_{1-x}Al_{0.1}Co_{0.9}O_2$. This compositional variation on Li sites results in the 440 cm^{-1} peak close to the E_g mode in the monoclinic Raman spectrum up to 700°C. Above this temperature the material crystallizes in a layered hexagonal structure.

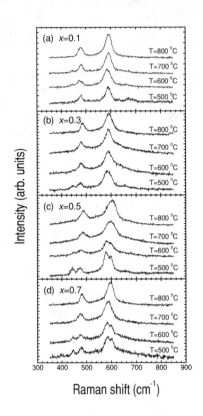

Figure 4. *Raman spectra of $LiAl_xCo_{1-x}O_2$ powders; (a) x=0.1, (b) x=0.3, (c) x=0.5, and (d) x=0.7, annealed at 500, 600, 700, and 800 °C.*

With higher Al contents the tendency of residual Co_3O_4 formation at lower temperatures reduces. Also, the monoclinic-like $Li_{1-x}Al_{0.3}Co_{0.7}O_2$ was observed only up to 600°C (Figure 4b). This shows that increasing Al up to 30 at% in $LiCoO_2$ decreases both the residual Co_3O_4 and the formation temperature of hexagonal $LiCoO_2$. Slightly different results were obtained for the most disordered $LiAl_{0.5}Co_{0.5}O_2$ sample (Figure 4c). The absence of the monoclinic structure in this composition is in line with the fact that a small concentration of Co exists in Li layers of

LiCoO$_2$. With more Al, less or negligible amounts of Co are available to reside in Li layers and one obtains LiM$_x$Co$_{1-x}$O$_2$ instead of Li$_{1-x}$M$_y$Co$_{1-y}$O$_2$ compositions. Both the E$_g$ and A$_{1g}$ modes are split at 450 °C. This splitting continues to 650 °C, above which the material shows a layered hexagonal structure with relatively broad peaks, probably due to the substitutional disorder. In the layered LiMO$_2$ (M = Co or Ni) Raman spectra corresponding to the spinel structure have been observed from low-temperature phases [18]. The irreducible representations for the spinel LiCoO$_2$ (Z = 4) allow A$_{1g}$, E$_g$, and 2T$_{2g}$ Raman active modes. Therefore, the splitting of the E$_g$ and A$_{1g}$ modes is attributed to the spinel structures formation at lower temperature. Similar spinel structure have also been recently reported in LiCoO$_2$ annealed at 400 °C [17]. The room temperature recorded Raman spectra of LiAl$_{0.7}$Co$_{0.3}$O$_2$ at different annealing temperatures in Figure 4(d) also indicate the spinel structure at lower temperature.

The Raman estimated phases or combination of phases at different temperatures for each composition are shown in Figure 5. With greater Al substituting for Co, residual Co$_3$O$_4$ phase disappears. At low Al contents, the low-temperature phase transforms to a high temperature hexagonal phase through a monoclinic-like Li$_{1-x}$M$_y$Co$_{1-y}$O$_2$ intermediate phase. With increasing Al, no monoclinic-like phase was detected. This confirms the neutron diffraction findings [8] of small concentration of Co in Li layers of LiCoO$_2$. At high Li concentration (50 at% or more), a spinel structure was detected in low-temperature annealed materials. Buta and coworkers [19], using the first principles method, have calculated a similar temperature composition phase diagram for LiAl$_x$Co$_{1-x}$O$_2$. The calculated phase diagram in general agrees with our experimental observations – no tetragonal γ-LiAlO$_2$ phase was predicted for 0.0 $\leq x \leq$ 0.7 and 0 $\leq$ T $\leq$ 800 °C. However, their calculations do not consider other crystal structures that have been identified using Raman spectroscopy in LiAl$_x$Co$_{1-x}$O$_2$.

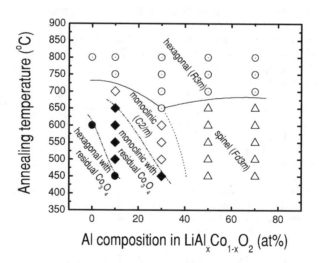

Figure 5. *Different phases observed in the Raman spectra of LiAl$_x$Co$_{1-x}$O$_2$ at different Al contents and annealing temperatures. Symbols* O, ●, ◊, ♦ *and* Δ *represent the hexagonal (Space group: R$\overline{3}$m), hexagonal with residual Co$_3$O$_4$, monoclinic, monoclinic with residual Co$_3$O$_4$, and spinel phases, respectively*

CONCLUSIONS

We have successfully prepared $LiAl_xCo_{1-x}O_2$ compositions using a chemical solutio route. This study presents the detailed Raman investigations on a superior class of Li batter materials. It is concluded that when Al is substituted for Co in $LiCoO_2$, the structure of th resulting material significantly depends on Al content as well as on the annealing temperatur Below 50 at% Al substitution the material crystallizes into a monoclinic-like structure at lowe temperature, while a low-temperature spinel phase was detected for Al contents of 50 at% c more. All these compositions crystallize in a layered hexagonal structure when annealed abov 650 °C.

ACKNOWLEDGEMENTS

This work was supported by a Department of Energy grant under Experimental Progra to Stimulate Competitive Research (DoE-EPSCoR grant).

REFERENCES

1. *for reviews*, see *Mater. Res. Soc. Bulletin*, Vol. 25 No.3 (2000).
2. R. Koksbang, J. Barker, H. Shi, and M. Y. Saidi, *Solid State Ionics*, **84**, 1 (1996).
3. C. D. W. Jones, E. Rossen, and J. R. Dahn, *Solid State Ionics*, **68**, 65 (1994).
4. R. Stoyannova, E. Zhecheva, and L. Zarkova, *Solid State Ionics*, **73**, 233 (1994).
5. K. Mitzushima, P. C. Jones, P. J. Wiseman, and J. B. Goodenough, *Mater. Res. Bull.* **15**, 497 (1980).
6. G. Ceder, Y. –M. Chiang, D. R. Sadoway, M. K. Aydinol, Y. –I. Jang, and B. Huang, *Nature* **392**, 694 (1998).
7. Y. –I. Jang, B. Huang, H. Wang, D. R. Sadoway, G. Ceder, Y. –M. Chiang, H. Liu, and H. Tamura, *J. Electrochem. Soc.* **146**, 862 (1999).
8. R. J. Gummow, M. M. Tackeray, W. I. F. David, and S. Hull, *Mater. Res. Bull.* **27**, 327 (1992); *ibid.* **28**, 1177 (1993).
9. E. Rossen, J. N. Rimers, and J. R. Dahn, *Solid State Ionics*, **62**, 53 (1993).
10. J. N. Reimers and J. R. Dahn, *J. Electrochem. Soc.* **139**, 2091 (1992).
11. A. Honders, J. M. der Kinderen, A. H. van Heerden, J. H. W. de Wit, and G. H. J. Broer *Solid State Ionics*, **14**, 205 (1984); *ibid.* **15**, 265 (1985).
12. R. K. Moore and W. B. White, *J. Am. Ceram. Soc.* **53**, 679 (1970).
13. M. Inaba, Y.Iriyama, Z. Ogumi, Y. Todzuka, and A. Tasaka, *J. Raman Spectrosc.* **28**, 6 (1997).
14. D. Bersani, P. P. Lottici, and X.-Z. Ding, *Solid State Commun.* **72**, 73 (1998).
15. M. Inaba, Y. Todzuka, H. Yoshida, Y. Grincourt, A. Tasaka, Y. Tomida, and Z. Ogumi, *Chemistry Letters* 889 (1995).
16. V. G. Hadjiev, M. N. Iliev, and I. V. Vergilov, *J. Phys. C: Solid State Phys.* **21**, L119 (1998).
17. G. Chen, W. Hao, Y. Shi, Y. Wu, and S. Perkowitz, *J. Mater. Res.* **15**, 583 (2000).
18. W. Huang and R. Frech, *Solid State Ionics* **86-88**, 395 (1996).
19. S. Buta, D. Morgan, A. van der Ven, M. K. Aydinol, and G. Ceder, *J. Electrochem. Soc* **146**, 4335 (1999).

Mat. Res. Soc. Symp. Proc. Vol. 658 © 2001 Materials Research Society

Reducibility study of the LaM$_x$Fe$_{1-x}$O$_3$ (M = Ni, Co) perovskites

C. Estournès[1], H. Provendier[2], L. Bedel[2], C. Petit[2], A.C. Roger[2], A. Kiennemann[2]
[1] IPCMS Groupe des Matériaux Inorganiques, UMR7504 CNRS-ULP-ECPM, 23 rue du Loess 67037 Strasbourg Cedex France
[2] Laboratoire de la Réactivité Catalytique, des Surfaces et Interfaces, UMR 7515, ECPM-CNRS-ULP, 25 rue Becquerel, 67087 Strasbourg Cedex 2, France.

ABSTRACT

Solid solutions of LaM$_x$Fe$_{1-x}$O$_3$ (with M = Ni and Co) have been used in the Fischer-Tropsch reaction (CO + H$_2$ → Hydrocarbons + CO$_2$) and in the partial oxidation of methane (CH$_4$ + 1/2 O$_2$ → CO + 2 H$_2$). In both catalytic reactions, the active catalyst is reported to be reduced metal particles; their size and their interactions with the support induce large differences in the product distribution.

In the nickel system, after total reduction by TPR all catalysts exhibit ferromagnetic behavior at room temperature. *In situ* magnetization in 1 Tesla on cooling the sample under reducing atmosphere shows one magnetic transition for each sample indicating one Curie temperature. These Curie temperatures are in between those known for bulk nickel and iron and decrease with the initial nickel content of the perovskite. This indicates that nickel is reduced first and induces the reduction of iron, leading to the formation of an alloy.

In the cobalt system, *in situ* magnetization on heating the sample shows a sharp increase of the magnetization only for x = 0.25, 0.40 and 1, corresponding to the formation of metallic cobalt nanoparticles. All other materials present only one increase of the magnetization for temperatures similar to those observed for the second reduction in TPR corresponding to the formation of CoFe alloys.

INTRODUCTION

In recent years, perovskites have been the subject of intense study from academics and interest from industrialists. The general stoichiometry is ABO$_3$ where A and B are generally trivalent ions, the first one usually being a rare earth while the latter is a transition metal. This class of materials displays a range of interesting and useful physical properties when the metals are substituted [1-5]. Due to their electrical properties they have been studied as cathodic materials in electrochemical devices, such as solid oxide fuel cells, as barrier electrodes [6], as well as oxidation catalysts [7,8]. In this latter domain, it is well known, particularly in the Fischer-Tropsch reaction and in partial oxidation reactions, that the active catalyst is metallic particles; the size and nature of which produce large differences in the distribution of products. Oxide structures have been used to control the material during the reduction step and the catalytic reactions with good efficiency [9-12].

This paper reports the synthesis and reducibility studies of LaM$_x$Fe$_{1-x}$O$_3$ (M = Ni, Co) perovskites using *in situ* magnetic measurements and temperature programmed reduction (TPR), in order to point out the mechanisms occurring during the catalytic reactions.

EXPERIMENTAL DETAILS

All the samples of the solid solutions of $LaM_xFe_{1-x}O_3$ (with M = Ni and Co) have been prepared by thermal decomposition of mixed propionates. The starting materials, nitrates (La, Fe, Ni) or acetates (La, Co) and Fe^0, are separately dissolved in hot propionic acid and stirred under reflux. After stoichiometric mixing of the three boiling solutions, propionic acid is evaporated until the formation of a resin occurs. The mixture is then calcined under air at 750°C for several hours.

The materials obtained were characterized by powder X-ray diffraction (XRD) on a Siemens D-5000 diffractometer using Cu Kα radiation. The temperature programmed reduction (TPR) of 50 mg of sample was carried out under 10% of H_2 (50 ml/min), heating from 300 to 1200 K at 15°C/min. This method was completed by dynamic magnetization using a homemade Faraday pendulum operating in a field of 1T, the magnetization of the sample was followed during thermal treatment under the same reducing flow as in TPR in order to point out the presence of metal or not.

RESULTS AND DISCUSSION

$LaNi_xFe_{1-x}O_3$ solid solution

Previous study by x-ray diffraction and transmission electron microscopy on all the samples of the solid solution $LaNi_xFe_{1-x}O_3$ (with $0 \leq x \leq 1$ where x varies by increments of 0.1) have shown that only one homogeneous phase with perovskite structure is formed [13]. The activity of these materials has been shown in partial oxidation of methane for syngas production. The catalysts were tested in the temperature range 673 and 1073 K in a mixture of methane with CO_2, H_2O or O_2, depending on the reaction tested, with a complement of argon in such a way that the total flow is constant. Characterization of the materials after catalytic tests showed different phases depending on the oxidant and the nickel content of the perovskite. In all cases, for $x \leq 0.1$ the low activity of the catalysts has been explained by non-reduction of this system. When methane is mixed with CO_2 a total reduction of nickel is observed for the entire solid solution. A complementary reduction of iron appears as well and the diffraction line corresponding to the metallic phase shifts with the initial composition of the perovskite. Under O_2, for $0.2 \leq x \leq 0.4$, a perovskite enriched in iron is observed by XRD together with nickel oxide, the latter corresponding to a rapid reoxidation of the metallic nickel that left the perovskite structure during the catalytic test. For this reaction, if total reduction of nickel is observed for $x \geq 0.5$, the presence of metallic nickel on XRD patterns is detected only for $x \geq 0.9$, while it appears oxidized in all other cases. This indicates the formation of larger particles of nickel when x increases, which passivate when they are exposed to air.

Since the active species for these catalytic reactions is generally admitted to be the reduced metal, the reducibility of the catalyst was studied. TPR up to 1173 K showed that $LaFeO_3$ is almost unreducible in the temperature range studied whereas two reduction zones appear for $LaNiO_3$, centered at 693 K and 823 K. The first peak can be attributed to the Ni^{III} to Ni^{II} reduction with formation of a metastable phase $La_2Ni_2O_5$ [14,15] and the second to the Ni^{II} to Ni^0 reduction, the latter being supported on La_2O_3. The temperature at which the second event occurs increases progressively with the iron content of the solid solution. One can say that iron stabilizes the $LaNiO_3$ perovskite structure in the tested conditions. After total reduction of the system two crystalline phases are observed by XRD: La_2O_3 and a second one of which the diffraction lines vary slowly with the initial composition of the perovskite.

Some of the samples recovered after TPR tests have been studied by magnetism. Ferromagnetic behavior was observed at room temperature across the entire solid solution. From saturation magnetization it is possible to say whether or not a system is totally reduced. Using a homemade Faraday pendulum operating in a field of 1T, the magnetization of the materials was followed during the reduction treatment under 5% H_2 in N_2. On heating, no significant change was detected for temperatures up to 1100 K, which suggests that the formation of the metallic phase occurs at temperatures higher than the Curie temperature of bulk Ni (627 K). On cooling, only one transition was observed on the magnetization curves indicating a single unique Curie temperature (Figure 1.). Moreover, the values of these Curie temperatures (Table I) are in between those of bulk nickel and iron and decrease with increasing nickel content of the perovskite. Consequently, at the end of reduction it is possible to say that an alloy is formed and its composition determined by energy dispersive X-ray analysis, is close to the initial metal ratio of the perovskite.

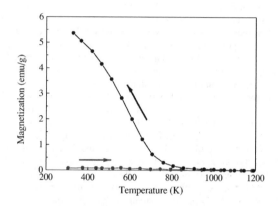

Figure 1. Magnetization (at 1T) of $LaNi_{0.3}Fe_{0.7}O_3$ under 5% H_2 in N_2 from 300 to 1200K.

Table I. Curie temperature of the alloy as a function of perovskite composition, x, in $LaNi_xFe_{1-x}O_3$.

x	0	0.3	0.7	1
Tc (K)	1073	923	700	627

In order to clarify the process occurring at the beginning of the second step of reduction observed in TPR, a partial reduction of the sample $LaNi_{0.3}Fe_{0.7}O_3$ was followed by *in situ* magnetization measurements. Therefore, while heating for two hours at 703 K, no significant increase of the magnetization was observed. Nevertheless, a reduction had occurred since ferromagnetic behavior was observed upon returning to room temperature. This implies that the metallic particles formed have a Curie temperature lower than 703 K, which is completely different from the value (923 K) observed for the alloy formed after total reduction of this system. Consequently, it is possible to say that in a first step, metallic nickel is formed, which catalyzes the reduction of iron [16] leading to the formation of an alloy.

The different characterizations performed after reducibility and catalytic tests seem to show that the active phase is either nickel metal or an alloy (Ni-Fe). However, the presence of NiO in XRD patterns can be explained by the formation of pyrophoric nanoparticles that oxidize when exposed to air. When nickel in the metallic state is observed in the XRD patterns, this indicates that the size of the particles formed is sufficient to avoid the total oxidation by air or that it is stabilized by the presence of iron.

$LaCo_xFe_{1-x}O_3$ solid solution

The $LaCo_xFe_{1-x}O_3$ series is composed of two solid solutions, one isomorphic to $LaCoO_3$ (rhombohedral) for x values greater than or equal to 0.5, the other one isomorphic to $LaFeO_3$ (orthorhombic) for x values less than 0.5. Previous works has shown good activity of these solid solutions in catalyzing Fischer-Tropsch syntheses (FTS) in order to produce light olefins from syngas. Here again, it is well known that metallic nanophases (like Co and Fe) are efficient in such reactions [17,18]. Therefore, the reducibility of these materials has been studied. Similar to the nickel case, all the samples (except $x = 0$), show two reduction areas in TPR profiles. The first reduction occurs between 593 and 793 K. The second reduction also occurs over a range of temperature, beginning at 843 K for $LaCoO_3$ and shifting to higher temperature as x decreases. To confirm that metal arises from the partial reduction of the perovskites, magnetic measurements were carried out from 300 to 1250 K under the same reducing flow as in TPR. An example is given in Figure 2 for $x = 0.4$.

The magnetization curve of $LaCoO_3$ shows two increases at 708 and 800 K corresponding to the formation of Co^0 (ferromagnetic, magnetization value at room temperature: 150 emu/g [19]). The two zones of Co^0 formation are the same as the two reduction areas observed in TPR. After the first reduction, the magnetization value of the perovskite remains stable between 713 and 793 K, accounting for the stability of the partially reduced system. From H_2 consumption and the magnetization value, the formula of the partially reduced $LaCoO_3$ can be deduced as 1.3 wt% $Co^0/LaCo^{II}_{0.90}Co^{III}_{0.05}O_{2.48}$. After total reduction no coercive field and saturation are observed in room temperature magnetization measurements. This implies the formation of small Co^0 particles since such super-paramagnetic behavior is only obtained for small particle sizes (< 50 Å) [19].

For the rhomboheral perovskites ($x = 0.75, 0.60$ and 0.50), a single increase in the magnetization is observed, which corresponds to the second reduction area observed in TPR. For these systems, no metallic cobalt is formed during the first reduction, which is due to the reduction of Co^{III} to Co^{II}.

For the orthorombic perovskites ($x = 0.25$ and 0.40), metallic cobalt is formed during the first reduction since the magnetization curves show an increase at 683 and 698 K respectively (Fig. 2). Saturation remains difficult to reach on room temperature magnetization whereas hysteresis loops are observed. For these two partially reduced materials, due to the high value obtained for the coercive field, it is possible to say that the cobalt metal particles formed remain monodomain; domain sizes are distributed around an average value, higher than 50 Å. From isothermal magnetization (Table II) it is possible to estimate the amount of metallic cobalt. Surprisingly, despite the fact that H_2 consumption during the partial reduction is less than in the case of $LaCoO_3$, the amount of cobalt metal produced by the partial reduction of $LaCo_{0.40}Fe_{0.60}O_3$ and $LaCo_{0.25}Fe_{0.75}O_3$ is higher, by almost 2 wt % in both cases. On cooling, no superposition is observed in Figure 2 suggesting that at high temperature both remaining cobalt and iron are reduced leading to the formation of a CoFe alloy.

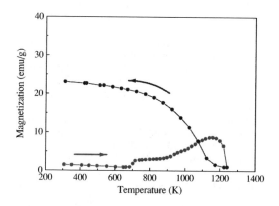

Figure 2. Magnetization (at 1T) of $LaCo_{0.40}Fe_{0.60}O_3$
under 5% H_2/ N_2 from 300 to 1250K

Table II. Magnetization at room temperature (for H = 1.7 Tesla) and Coercive
field as a function of the $LaCo_xFe_{1-x}O_3$ perovskite composition.

x	0.25	0.40	1
M at 1.7 T (in emu/g)	4.2	3.3	1.7
Coercive field (T)	0.065	0.057	-

From this study it appears that the reducibility of the $LaCo_xFe_{1-x}O_3$ perovskite is highly
dependent on the x value. Despite their lower cobalt contents the orthorombic perovskites
reduced at 673 K produce more Co^0 than the higher cobalt loaded rhombohedral ones. This
has to be correlated to the catalytic activity : Where $LaCo_{0.40}Fe_{0.60}O_3$ and $LaCo_{0.25}Fe_{0.75}O_3$
exhibit CO conversion of almost 15% and 12%, $LaCoO_3$ produces a CO conversion of 9%,
and for the rhombohedral series, the CO conversion decreases with x until $LaCo_{0.50}Fe_{0.50}O_3$.

CONCLUSIONS

The combined use of magnetic measurements and TPR demonstrates that for Co
perovskites, the ability of the perovskite to form metal particles after partial reduction, which
are catalytic active sites, depends on the crystalline structure of the starting perovskite.
Despite a lower Co content, only the orthorhombic perovskites allow the formation of Co
metal particles. These Co^0 particles remain small (monodomain), inducing a good selectivity
towards light olefins in Fischer Tropsch synthesis. For Ni perovskites, Ni^{III} is the first species
to be reduced and the effect of this reduction increseases the reducibility of iron in the
perovskite. Alloy formation is responsible for the high and stable behavior of this catalyst for
syngas production.

REFERENCES
1. Handbook on the physics, *Chemistry of rare earth, North Holland Amsterdam* (1979).
2. J.B. Goodenough, *Prog. Solid State Chem.* **40**, 185 (1971).
3. J.S. Zhou, J.B. Goodenough, B. Dabrowski, P.W. Klamut, Z. Bukowski, *Phys. Rev. B* **61**, 4401 (2000).
4. R. Von Helmot, J. Wecker, B. Holzappel, L. Schultz, K. Samwer, *Phys Rev. Let.* **71**, 2331 (1993).
5. A. Asamitsu, Y. Morimoto, Y. Tomioka, T. Arima, Y. Tokura, *Nature* **373**, 407 (1995).
6. P. Murugrave, R. Sharma, A.R. Raju, CNR Rao, *J. Phys. D* **33**, 906 (2000).
7. B. Viswanathan, *J. Sci. Indust. Res* **43**, 151 (1984).
8. A.E. Gota, G.F. Goya, R.C. Mercader, G. Punte, H. Falcon and R. Carbonio, *Hyperfine interactions* **90**, 371 (1994).
9. H. Provendier, C. Petit, C. Estournès, S. Libs, A. Kienneman, *Appl. Catal. A.* **180**, 163 (1999).
10. L. Bedel, A.C. Roger, A. Kiennemann and C. Estournès, *Am. Chem. Soc., Div. Pet. Chem.*, **45(2)**, 236 (2000).
11. L. Bedel, A.C. Roger, C. Estournès and A. Kiennemann, *Heterogeneous catalysis, L. Petrov, Ch. Bonev, G. Kadimov (Eds) Proc. of the 9th Int. Symp.* Varna Bulgaria 2000.
12. H. H. Provendier, C. Petit, C. Estournès, A. Kienneman, *Stud. Surf. Sci. Cat.* **119**, 741 (1998).
13. H. Provendier, C. Petit, J. L. Schmidt, A. Kienneman, C. Chaumont, *J. Mat. Sci.* **34**, 4121 (1999).
14. J. Choisnet, N. Abadzhieva, P. Stefanov, D. Klissurki, J.M. Bassat, V. Rives and L. Minchev, *J. Chem. Soc. Faraday Trans.* **90**, 1987 (1994).
15. M.J. Sayagues, M. Vallet-Régi, A. Caneiro, J.M. Gonzalez-Calbet, *J. Solid States Chem.* **110**, 295 (1994).
16. M.I. Nasr, A.A. Omar, M.H. Khedr, A.A. El-Glassy, *ISIJ Int.* **35-9**, 1043 (1995).
17. H. Schulz, *App. Catal. A* **186**, 3 (1999).
18. C.H. Batholomew, *Trends in CO Activation, Elsevier Amsterdam* (1991).
19. X.M. Lin, C.M. Sorensen, K.J. Klabunde, G.C. Hadjipanayis, *Langmuir* **14**, 7140 (1998).

Mat. Res. Soc. Symp. Proc. Vol. 658 © 2001 Materials Research Society

Conversion of an Aurivillius Phase $Bi_2SrNaNb_3O_{12}$ into Its Protonated Form *via* Treatment with Various Mineral Acids

Masashi Shirata, Yu Tsunoda, Wataru Sugimoto,[1] and Yoshiyuki Sugahara
Department of Applied Chemistry, School of Science and Engineering, Waseda University,
Shinjuku-ku, Tokyo 169-8555 JAPAN
[1]Department of Fine Materials Engineering, Faculty of Textile Science and Technology, Shinshu University, Ueda, Nagano, 386-8567 JAPAN

ABSTRACT

A protonated form of a layered perovskite was prepared from an Aurivillius phase $Bi_2SrNaNb_3O_{12}$ *via* acid treatments, and the effect of the type of mineral acids was investigated. The treatment with HX (X = Cl, Br, I) resulted in the formation of a protonated form $H_{1.8}[Sr_{0.8}Bi_{0.2}NaNb_3O_{10}]$, while no reactions were observed for HNO_3 and H_2SO_4 under the present experimental conditions. All the products obtained by HX-treatments exhibited layered structures and the structures of the perovskite-like slabs were preserved.

INTRODUCTION

Layered perovskites possess two-dimensional perovskite-like slabs in their structures. Among them, ion-exchangeable layered perovskites consist of perovskite-like slabs ($[A_{n-1}B_nO_{3n+1}]$; A = Sr, Ca, *etc.*, B = Ti, Nb, Ta) and monovalent exchangeable interlayer cations (M; M = Na, K, Rb, *etc.*) [1]. There have been two homologous series reported, that are called Dion-Jacobson phases ($M[A_{n-1}B_nO_{3n+1}]$) [2,3] and Ruddlesden-Popper phases ($M_2[A_{n-1}B_nO_{3n+1}]$) [4-6]. When these layered perovskites were treated with mineral acids, they were converted into corresponding protonated forms ($H[A_{n-1}B_nO_{3n+1}]$ and $H_2[A_{n-1}B_nO_{3n+1}]$) [3,6-8]. These protonated forms exhibit interesting properties including photocatalytic behavior [9], proton conduction [10], and intercalation of organic amines [11]. These protonated forms can also be utilized as precursors for oxide syntheses [12].

Aurivillius phases ($Bi_2A_{n-1}B_nO_{3n+3}$ or alternately expressed as $Bi_2O_2[A_{n-1}B_nO_{3n+1}]$) are known to exhibit excellent dielectric properties. In terms of structures, they consist of perovskite-like slabs ($[A_{n-1}B_nO_{3n+1}]$) and Bi_2O_2 layers, and it should be noted that the structures of perovskite-like slabs in the Aurivillius phases are identical with those in the aforementioned ion-exchangeable layered perovskites [13]. Suzuki *et al.* [14] reported the acid treatment of a $Bi_2SrTa_2O_9$ single crystal, and observed drastic variation in X-ray diffraction patterns; they ascribed this observation to the structural change in the perovskite-like slabs. We have recently reported that HCl-treatment of another Aurivillius phase, $Bi_2SrNaNb_3O_{12}$, led to the selective leaching of the Bi_2O_2 layers and the introduction of protons for charge compensation to form a corresponding protonated form [15]. Since no structural change in the perovskite-like slabs occurred, this reaction is considered to be the conversion of the Aurivillius phase into a protonated form of a Ruddlesden-Popper phase (though the layer charge is slightly reduced from 2- per $[A_{n-1}B_nO_{3n+1}]$, the general value for the Ruddlesden-Popper phases, to ~1.8- per $[A_{n-1}B_nO_{3n+1}]$ because of the disorder between Bi and Sr). Very recently, Gopalakrishnan *et al.* [16] reported a reverse reaction of the aforementioned reaction, namely the conversion of a

Ruddlesden-Popper phase into an Aurivillius phase through a metathesis reaction using BiOCl. Thus, the Ruddlesden-Popper phases and the Aurivillius phases are actually interconvertible.

Here, we report the effect of the type of mineral acids used on the conversion of $Bi_2SrNaNb_3O_{12}$. As acids, hydrochloric acid (HCl), hydrobromic acid (HBr), hydroiodic ac (HI), nitric acid (HNO_3), and sulfuric acid (H_2SO_4) were used. The products were analyzed X-ray diffraction (XRD), transmission electron microscopy (TEM), compositional analysis, anc scanning electron microscopy (SEM).

EXPERIMENTAL

$Bi_2SrNaNb_3O_{12}$ (BSNN) was prepared by the heat treatment of a stoichiometric mixture o $Bi_2SrNb_2O_9$ [17] and $NaNbO_3$ at 1100 °C for 6 h with intermediate grinding. The metal rati of BSNN (Bi:Sr:Na:Nb = 2.00:1.00:1.02:3) was consistent with the nominal one. The XRD pattern of BSNN can be indexed based on a tetragonal cell. The lattice parameters refined with the Rietveld program RIETAN [18,19] and in the space group $I4/mmm$ were $a = 0.39007(\textbf{\textit{l}}$ and $c = 3.2926(1)$ nm.

About 1 g of BSNN was dispersed in 200 mL of 6 M HX (X = Cl, Br, I), 6 M HNO_3, or 3 H_2SO_4 at room temperature for 72 h. The resulting product was washed with water, and drie at 120 °C.

XRD patterns were obtained using a Rigaku RINT-2500 diffractometer using monochromated CuKα radiation. Compositions of metals were determined by inductively-coupled plasma emission spectrometry (ICP; Nippon Jarrell Ash, ICAP 575 MARK II) after dissolving samples using a mixture of HNO_3, HF, and HCl. The amounts of protons were determined from mass losses above 160 °C by using thermogravimetry (TG; MacScience, TG-DTA 2000S) operated in air. Morphology of samples was studied by scanning electron microscopy (Hitachi, S-2500). Transmission electron microscopy was performed using Hitachi H-8100A.

RESULTS AND DISCUSSION

Figure 1 shows XRD patterns of the starting material BSNN and the products of BSNN treated with various acids. When BSNN is treated with HX (X = Cl, Br, I), the XRD pattern drastically change. The (002) peak, that corresponds to the thickness of one perovskite-like slab and one Bi_2O_2 layer in BSNN, shifts to higher angles after the HX-treatment. Correspondingly, its higher orders clearly shift to higher angles, indicating the presence of layered structures possessing smaller repeating distances in the HX-treated products. On the contrary, no shifts were observed for $(hk0)$ peaks (typically (110) and (200) in Figure 1), suggesting the preservation of the perovskite-like slabs after the HX-treatments. It should al be noted that these three XRD patterns were very similar, indicating similar reactivity irrespective of the type of the X^- anion. The peaks of the products treated with HX were successfully indexed based on a tetragonal cell. The lattice parameters of the product treated with HCl are $a = 0.391\pm0.002$ and $c = 1.39\pm0.02$ nm. The a parameter of the HCl-treated product is essentially identical with that of BSNN, while the c parameter is smaller than that of BSNN, consistent with the aforementioned observations.

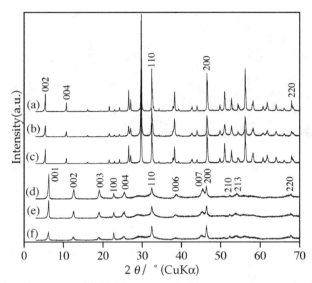

Figure 1. *XRD patterns of (a) BSNN, (b) BSNN treated with HNO₃, (c) BSNN treated with H₂SO₄, (d) BSNN treated with HCl, (e) BSNN treated with HBr, and (f) BSNN treated with HI.*

The XRD patterns of BSNN treated with HNO_3 and H_2SO_4 are completely different from those of BSNN treated with HX, and are essentially identical with that of BSNN.   Thus, under the present experimental conditions, the reactions of BSNN with acids take place only when HX (X = Cl, Br, I) is used even with the same proton concentration; halide ions seem to play an important role in this conversion reaction.

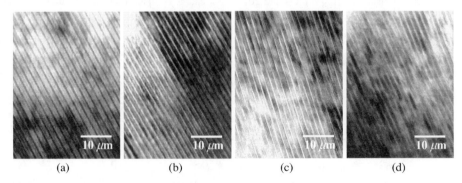

(a)   (b)   (c)   (d)

Figure 2. *TEM images of (a) BSNN, (b) BSNN treated with HCl, (c) BSNN treated with HBr, and (d) BSNN treated with HI.*

Figure 3. Scanning electron micrographs of (a) BSNN, (b) BSNN treated with HCl, (c) BSNN treated with HBr, and (d) BSNN treated with HI.

The presence of layered structures in the products treated with HX is clearly shown by TEM images. Figure 2 demonstrates the TEM images along the [010] zone. Alternating lines with dark and light contrasts are observed for BSNN and all the products treated with HX, consistent with the layered structures. The repeating distance of the dark and light rows in BSNN is about 1.6 nm ($\sim c/2$), and decreases to about 1.4 nm for all the products treated with HX, consistent with the XRD observations ($c = 1.39 \pm 0.02$ nm for the product treated with HCl).

In order to investigate the structures of the products treated with HX, their morphology was investigated with SEM (Figure 3). All the HX-treated products exhibit a morphology that is very similar to that of BSNN. Thus, dissolution of BSNN and subsequent crystallization are very unlikely as a mechanism of the present reactions with HX.

The variation in the compositions is demonstrated in Table I. The drastic decreases in Bi contents (from 2.0 to $\sim$0.2) are clearly shown for all the HX-treated products. On the other hand, slight losses of Sr ($\sim$0.2 Sr per 3 Nb) are always detected for the HX-treated products. We ascribe these observations to the selective leaching of the Bi_2O_2 layers containing Sr due to cation disorder ($Bi_{1.8}Sr_{0.2}O_2$) [15,20]. For charge compensation, 1.8 H^+ are introduced per Nb. Thus, the overall reaction can be written as follows:

$$(Bi_{1.8}Sr_{0.2}O_2)[Sr_{0.8}Bi_{0.2}NaNb_3O_{10}] \xrightarrow{\text{HX}} H_{1.8}[Sr_{0.8}Bi_{0.2}NaNb_3O_{10}]$$

Table I. Variation in compositions of BSNN after the treatment with HX ($X = Cl, Br, I$) (in molar ratio)

	BSNN	HX-treated		
		HCl	HBr	HI
Bi	2.0	0.21	0.25	0.23
Sr	1.0	0.80	0.77	0.78
Na	1.0	1.0	1.0	1.0
Nb*	3	3	3	3
H	-	1.8	1.8	1.8

*set to be three

For dissolution of metal oxides, the effects of anions have been reported.   Based on a surface complexation model [21], anions can influence the rate of dissolution *via* the formation of surface complexes, inductive effects, and thermodynamic changes in solubility.   Some of these factors may play a role in the significant effect of anions observed in the present study.

CONCLUSIONS

We have demonstrated that the type of anion in the mineral acids exhibited a significant effect on the conversion of the Aurivillius phase $Bi_2SrNaNb_3O_{12}$ into its protonated form.   HX (X = Cl, Br, I) was effective for this conversion, while HNO_3 and H_2SO_4 did not lead to successful conversion under the present experimental conditions.   These observations imply that halide ions play an important role in the conversion reaction.

ACKNOWLEDGMENTS

The authors would like to thank Prof. Kazuyuki Kuroda (Department of Applied Chemistry, Waseda University) for valuable discussion and Mr. Minekazu Fujiwara (Material Characterization Central Laboratory, Waseda University) for his assistance in TEM observation. This work was financially supported partly by the Grant-in-Aid for Scientific Research (No. 10555221) from the Ministry of Education, Science, Sports, and Culture, Japan.

REFERENCES

1.   J. Gopalakrishnan, *Rev. Solid State Sci.* **1**, 515 (1988).
2.   M. Dion, M. Ganne and M. Tournoux, *Mater. Res. Bull.* **16**, 1429 (1981).
3.   A. J. Jacobson, J. W. Johnson and J. T. Lewandowski, *Inorg. Chem.* **24**, 3727 (1985).
4.   S. N. Ruddlesden and P. Popper, *Acta Crystllogr.* **10**, 538 (1957).
5.   S. N. Ruddlesden and P. Popper, *Acta Crystllogr.* **11**, 54 (1958).
6.   J. Gopalakrishnan, V. Bhat and B. Raveau, *Mater. Res. Bull.* **22**, 413 (1987).
7.   M. Dion, M. Ganne and M. Tournoux, *Rev. Chim. Miner.* **23**, 61 (1986).
8.   A. J. Jacobson, J. T. Lewandowski and J. W. Johnson, *J. Less-Common Met.* **116**, 137 (1986).
9.   Y. Ebina, A. Tanaka, J. N. Kondo and K. Domen, *Chem. Mater.* **8**, 2534 (1996).
10.   M. A. Subramanian, J. Gopalakrishnan and A. W. Sleight, *Mater. Res. Bull.* **23**, 837 (1988).
11.   A. J. Jacobson, J. W. Johnson and J. T. Lewandowski, *Mater. Res. Bull.* **22**, 45 (1987).
12.   P. J. Olivier and T. E. Mallouk, *Chem. Mater.* **10**, 2585 (1998).
13.   C. N. R. Rao and B. Raveau, *Transition Metal Oxide, 2nd edition*, (Wiley-VCH, 1998) pp.74-75.
14.   M. Suzuki, N. Nagasawa, A. Machida and T. Ami, *Jpn. J. Appl. Phys.* **35**, L564 (1996).
15.   W. Sugimoto, M. Shirata, Y. Sugahara and K. Kuroda, *J. Am. Chem. Soc.* **121**, 11601 (1999).
16.   J. Gopalakrishnan, T. Sivakumar, K. Ramesha, V. Thangadurai and G. N. Subbanna, *J. Am. Chem. Soc.* **122**, 6237 (2000).
17.   Ismunandar and B. J. Kennedy, *J. Solid State Chem.* **126**, 135 (1996).
18.   F. Izumi, Rietveld Analysis Programs RIETAN and PREMOS and Special Applications, *Rietveld Analysis,* ed. R. A. Young,   (Oxford University Press, 1993) pp.236-253.
19.   Y. I. Kim and F. Izumi, *J. Ceram. Soc. Jpn.* **102**, 401 (1994).
20.   S. M. Blake, M. J. Falconer, M. McCreedy and P. Lightfoot, *J. Mater. Chem.* **7**, 1609 (1997).
21.   M. A. Bleza, P. J. Morando and A. E. Regazzoni, *Chemical Dissolution of Metal Oxides*, (CRC Press, 1994) pp.172-177.

Mat. Res. Soc. Symp. Proc. Vol. 658 © 2001 Materials Research Society

ESCA and Vibrational Spectroscopy of Alkali Cation Exchanged Manganic Acids with the 2x2 type Tunnel Structure Synthesized via Different Chemical Routes

Masamichi Tsuji, Hirofumi Kanoh[1], and Kenta Ooi[2]
Tokyo Institute of Technology, Research Center for Carbon Recycling and Utilization, Tokyo, JAPAN.
[1]Chiba University, Faculty of Science, Chiba, JAPAN.
[2]National Institute of Advanced Industrial Science and Technology, Shikoku Center, JAPAN.

ABSTRACT

Manganese dioxides have received much attention over the last two decades as ion exchangers. Actually these are typically mixed-valence compounds and the terminology of 'dioxide' is not appropriate. The mechanisms for their variety of chemical reactivities are still open for study. On the cation uptake mechanism there are strong claims that a redox process is involved in cation uptakes by manganic acids synthesized by substituting H^+ for alkali cations incorporated in 'hydrous manganese dioxides'. The present work was carried out to physically demonstrate the alkali cation exchange mechanism on tunnel-structured manganic acids and to study the ion exchange with lattice vibrational spectroscopy. Manganic acids were prepared through the redox process using $KMnO_4$ and $MnSO_4$, and thermal decomposition of $(CH_3)_3COK$ and $MnCO_3$ at 530°C. ESCA spectra of their alkali cation exchanged forms indicated no evidence of redox process and supported the ion exchange mechanism on these materials. Their infrared absorption spectra strongly depended on their preparation routes and are closely related to their ion-exchange selectivity of each material. Thus, the vibrational spectra of manganic acids take an important role as a synthesis index together with XRD patterns.

INTRODUCTION

From the mineralogical study, it has been believed for a long time that the cation uptake mechanism on the hollandite family of MnO_2 phases is based on a redox process [1]. Our previous studies of synthesized cryptomelane-type manganic acid (CMA) demonstrated that alkali and alkaline earth metal cations, and divalent transition metal ions were exchanged with the H^+ in the tunnel through the stoichiometric ion-exchange reaction [2,3]. These were mostly based on the chemical analysis of cations involved in the exchange reactions in the solutions. Pb^{2+} selectivity was extremely large in the CMA as well as K^+ and Rb^+. Using ESCA spectra, the CMA exchanged by Pb^{2+} was investigated for defining the chemical environment of Pb, including Mn and O ions forming the crystal structure. The results were very definitive for the ion exchange of Pb^{2+} with H^+ in the tunnel site without appreciable degree of redox processes occurring [4]. Preliminary experiments using Fourier Transform Infrared Spectroscopy were informative of the change in the vibrational spectra of the crystal lattice of ion-exchange materials. The objective of the present work was to physically demonstrate the alkali cation exchange mechanism and to find out relations of the alkali cation exchange selectivity to the spectral data on the CMA, α-MnO_2 and a hollandite-type manganese dioxide (HolMO) prepared via different chemical routes.

EXPERIMENTAL

A. Preparation of manganic acids with the 2x2 type tunnel structure

The CMA phase in the H^+ form was prepared as described previously [2,3]. It is described briefly as follows. A precipitate was allowed to form by adding 0.5 dm^3 of 0.5 M $KMnO_4$ containing 1 M H_2SO_4 to 0.5 dm^3 of 1 M $MnSO_4$ containing 1 M H_2SO_4 at 60°C. The precipitate was aged overnight in the mother liquor, followed by washing with 6 M HNO_3 (0.5 dm^3), and then with water to remove a large part of the remaining H_2SO_4 and K_2SO_4. The washed product was dried at about 70°C for 3 days, ground, and sieved to obtain 100-200 mesh size particles (74-37 μm). The sieved product was then packed in a glass column (1 cm I.D. x 20 cm long) and leached continuously with concentrated HNO_3 to exchange H^+ for K^+ that was incorporated in the tunnel site. The leaching continued until the concentration of K^+ in the effluent was less than 1×10^{-5} M. The treated material was washed thoroughly and air-dried at ambient temperature. HolMO prepared by reacting $LiMnO_4$ instead of $KMnO_4$ with $MnSO_4$ in H_2SO_4 solution was used for comparison [5]. The other manganic acid, α-MnO_2, was prepared by thermal decomposition of a mixture of $(CH_3)_3COK$ and $MnCO_3$ at 530°C (20 molar % of K) followed by K^+ leaching by HNO_3 [6]. Chemicals were supplied by Wako Pure Chemical Ind.

B. Cation uptake and selectivity study

A portion of 0.250 g of the CMA and α-MnO_2 was contacted with an aliquot of 25.0 cm^3 of 0.1 M MOH and/or MNO_3 at 30°C for the cation uptake, where M denotes an alkali metal. A similar procedure was taken for the selectivity study using a mixed solution of HNO_3 and MNO at 30°C. A period of two weeks was required to attain steady state concentration in the supernatant solution. After the equilibration, the supernatant solution was separated and analyzed for metal cations by atomic absorption spectroscopy. The adsorbed amounts were determined from the difference between the initial concentration and the equilibrated concentration.

C. XRD, IR spectroscopy, and ESCA spectra

XRD was carried out using RIGAKU model RAD2A equipment with Ni-filtered CuKα radiation. IR spectra were recorded with JASCO FT/IR-420 equipment. A powder sample was dispersed in pressed CsI pellet. ESCA spectra of powder specimens were recorded using monochromated AlKα with ULVAC PHI X-ray photoelectron spectrometer model 5500M. The pressure during the spectral recording was kept at 10^{-9} Pa.

RESULTS AND DISCUSSION

A. Lattice Parameters

The lattice parameters calculated based on the XRD pattern were a_o=0.977(1) nm and c_o=0.285(1) nm for CMA, and a_o=0.9802(4) nm and c_o=0.2857(5) nm for α-MnO_2, which agree with those reported previously [2,6,7]. The lattice parameters reported for HolMO were $a_0=0.979(1)$ nm and $c_0=0.286(1)$ nm [5]. Thus, any noticeable difference in the lattice parameters cannot be observed among these specimens. The crystallite size was 24 nm for α-MnO$_2$ and 8-13 nm for CMA. These were determined from the (110) and (121) spacings by using the Scherrer's formula [8].

B. Alkali Cation Selectivity

The cation selectivity of ion exchangers can be compared using the distribution coefficients (K_d, cm^3/g) defined by the ratio of ion concentrations in solid and solution at an infinitesimal exchange. The K_d values were taken from ref. 5 for HolMO and calculated for CMA and α-MnO$_2$ using the following equation [9].

$$(K_d)_{\overline{X}_M \to 0} = \frac{C_t}{c_t} \frac{\gamma_M}{\gamma_H} (K_H^M)_{\overline{X}_M \to 0}, \tag{1}$$

where K_H^M, C_t and c_t denote the corrected selectivity coefficient, the total cation exchange capacity and the total charge concentration. γ is the activity coefficient of ion in solution. $(K_H^M)_{\overline{X}_M \to 0}$ values of α-MnO$_2$ and CMA were taken from refs. 6 and 7, respectively.

Table I. Distribution coefficients (K_d) of alkali cations on CMA, α-MnO$_2$ and HolMO at pH 1.

Cation	CMA	α-MnO$_2$	HolMO
Li	10	0.089	1.1
Na	30	0.23	45
K	3.4×10^{14}	6.2×10^4	3.3×10^3
Rb	6.8×10^{13}	9.9×10^4	1.6×10^3
Cs	4.3×10^2	1.5	36

The selectivity for K^+ and Rb^+ is greatly different among these materials, despite that these have the same group symmetry $I_{4/m}$ with the same lattice parameters. This finding indicates that chemical environment of the ion-exchange sites cannot be probed by XRD alone.

C. ESCA and Infrared Spectra

ESCA and infrared spectroscopic studies were carried out for probing the change in the chemical state of these materials associated with cation uptake. The chemical state of Mn ions can be represented by the core binding energies of Mn 2p electrons. The results are summarized in Table II. Data of MnO were cited from ref. 10.

The manganese spinel, λ-MnO$_2$, forms a typical redox system. Mn^{4+} ion changes to Mn^{3+} ion to balance the total charge upon Li incorporation into the crystal lattice. This is the well-known electrochemical process. The binding energy of Mn 2p3/2 in λ-MnO$_2$ responds to the redox processes by changing to a lower value. This is a physical indication of the redox process. But no indication in the ESCA spectra was found in the cases of the CMA and α-MnO$_2$, as shown in Table II. In these compounds, hence, the reduction of Mn^{4+} to Mn^{3+} and/or Mn^{2+} ions never occurs upon alkali cation uptake from aqueous solutions.

The ESCA spectra of the H$^+$ and alkali cation exchanged forms for the CMA and α-MnO$_2$ showed the binding energy of O 1s electrons at ca. 530 and ca. 532 eV. The latter appeared as the small shoulder of the large former peak.   The O 1s binding energy of metal hydroxides is observed at 531 to 532 eV, while that of metal oxides appears at 528-531 eV [11].   Hence, the former peak can be assigned to the framework oxygen and the latter to the hydrogen bonded $-O^{\delta-}-H^{\delta+}\cdot\ \cdot O^{\delta-}$ group mentioned above or H_2O in the tunnel constituent.

Table II.   Binding energies of Mn 2p electrons for CMA and α-MnO_2.

Material (cation content)	Mn 2p$_{3/2}$ (eV)	Mn 2p$_{1/2}$ (eV)
CMA (H$^+$ form)	642.5	654.2
CMA (0.84 mequiv K$^+$/g)	642.7	654.3
CMA (1.0 mequiv Rb$^+$/g)	642.5	654.2
α-MnO_2 (H$^+$ form)	642.0, 642.8	654.0
α-MnO_2 (0.75 mequiv Li$^+$/g)	642.5	654.1
α-MnO_2 (1.0 mequiv K$^+$/g)	642.0, 643.1	654.0
β-MnO_2	642.0	653.9
γ-MnO_2	642.2	654.0
λ-MnO_2	642.0, 642.8	654.0
$LiMn_2O_4$	641.0, 642.5, 643.5	654.0
α-Mn_2O_3	641.7	652.5
MnO	640.6±0.2	652.2±0.2

Glemser and his coworkers extensively investigated MnO_2 modifications including α-, β-, δ-, η-, and ϵ-forms by means of XRD, thermal analyses and IR spectra from 760 to 420 cm^{-1} [12-14].   Later, Kolta, et al. studied IR spectra of α-, β-, γ-, and δ-MnO_2 heated at elevated temperatures to find that the bands at 595±20 cm^{-1} and 535±20 cm^{-1} are related to the content of an amorphous constituent [15].   A large unresolved absorption band at 500-600 cm^{-1} is characteristic of these IR spectra.   It comprises several absorption bands.   Infrared spectra of naturally occurring manganese oxide minerals have been extensively studied by Potter and Rossman [16,17].   They have pointed out that an absorption band at 310 cm^{-1} was characteristic of α-MnO_2 phase.   The same authors did not detect the absorption bands assigned to CO_2 possibly adsorbed by these minerals, though Hariya, et al. pointed out the presence of CO_2 in the tunnel of synthesized α-MnO_2 containing Na$^+$ ions and naturally occurring hollandite [18].

Recently IR spectra of synthetic MnO_2 phases were studied to better understand the ion-exchange process [19].   These showed that the absorption bands were characteristic of each crystal phase and also depended on the mode of preparation, sort and content of cations.   The present study with high-resolution infrared spectrometer more clearly revealed the vibrational spectra of the CMA in the range of 1600-250 cm^{-1} (Figure 1).   An absorption band at ca. 1530 cm^{-1} will not be due to CO_3^{2-}, because adsorption of CO_2 in the strongly acidic tunnel site of the CMA is probably not possible.   The intensity of an absorption band at 1360 cm^{-1} of the CMA increased by uptake of K$^+$ and Rb$^+$ ions.   Ion exchange occurred in acidic solution, and carbonate inclusion is negligible.   It may be ascribed to the change in the lattice vibration mode associated with elongation of the lattice parameter [20].   A new absorption band appeared at 1046 cm^{-1} by Li$^+$ and K$^+$ uptake.   The relative intensity of several absorption bands at 600-250 cm^{-1} greatly changed by alkali cation exchange.   The most striking phenomena occurred at 745 cm^{-1}.   It shifted to the lower wavenumber side upon alkali cation exchange.   Thus, the shift of this absorption band will be a useful parameter for determination of alkali cation content.   The absorption bands at 1610-1620 cm^{-1}, ca. 3400 and 3200 cm^{-1} are assigned to the bending mode and the stretching mode of water and/or OH group, and the shoulder due to the hydrogen bond, respectively.   These findings on the infrared absorption bands suggest the possibility for determination of cation uptake in the H$^+$/M^{n+} binary exchange on this material, and indicate needs of detailed reinvestigation of IR spectra of manganese oxide phases.

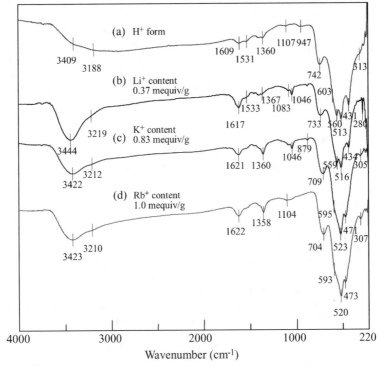

Figure 1. *Infrared spectra of CMA exchanged by Li$^+$, K$^+$, and Rb$^+$.*

Other tunnel compounds, α-MnO$_2$ and HolMO, showed similar IR spectra in the low wavenumber region.   They showed an absorption band at 705 and 720 cm^{-1}, respectively, corresponding to the absorption band of the CMA at 742 cm^{-1}.   These were never shifted by alkali cation uptake.   The absorption bands at 474-478 cm^{-1} were observed on the α-MnO$_2$ exchanged by Li$^+$ and K$^+$, corresponding to that of the CMA at 431 cm^{-1}.   This band was not shifted by alkali cation uptake.   The α-MnO$_2$ was synthesized by decomposition of starting materials at elevated temperatures.   Hence the crystal lattice can be assumed 'stiff' and the wavenumber of these absorption bands did not change upon entering of cation and extraction. The latter material (HolMO) showed unresolved and broadened spectra in the range of 800-400 cm^{-1}.   The crystal lattice of these two compounds appeared rigid and they less selectively accommodate alkali cations, even though they have the same tunnel dimension as the CMA from the viewpoint of the X-ray crystallography. Thus, the vibrational spectra of the crystal lattice the range of 1600-250 cm^{-1} were informative of the cation selectivity of manganic acids.

CONCLUSIONS

ESCA spectra clearly indicated that the CMA and α-MnO$_2$ incorporate alkali cations throu; the ion-exchange mechanism without any other side reactions.   ESCA will also serve as a goo tool for chemical environment of incorporated alkali cations.   On the other hand, very large differences in the cation-exchange selectivity of these MnO$_2$ phases with the same 2x2 type tunnel structure can be detected through observation of the lattice vibration modes with FTIR spectra.   The CMA with the most flexible lattice revealed excellent selectivity.   Thus, ESCA and FTIR spectra are important probes for cation-uptake mechanism and ion-exchange selectiv of manganese dioxide phases as well as X-ray diffractometry.

ACKNOWLEDGMENT

Part of this research was supported by the Salt Science Foundation Grant No. 9939.

REFERENCES

1. A. Byström and A. M. Byström, *Acta Crystallogr.*, **3**, 146 (1950).
2. M. Tsuji and M. Abe, *Solvent Extr. Ion Exch.*, **2**, 253 (1984).
3. M. Tsuji and S. Komarneni, *J. Mater. Res.*, **8**, 611 (1993).
4. M. Tsuji and Y. Tanaka, *J. Mater. Res.*, **16**, 108 (2001).
5. Q. Feng, H. Kanoh, Y. Miyai and K. Ooi, *Chem. Mater.*, **7**, 148 (1995).
6. Y. Tanaka, M. Tsuji and Y. Tamaura, *Phys. Chem. Chem. Phys.*, **2**, 1473 (2000).
7. M. Tsuji and Y. Tamaura, *Solvent Extr. Ion Exch.*, **18**, 187 (2000).
8. B. D. Cullity, *Elements of X-ray Diffraction*, p.102, Addison-Wesley, Reading, UK (1978).
9. M. Tsuji and S. Komarneni, *Sep. Sci. Technol.*, **26**, 647 (1991).
10. M. Oku, K. Hirokawa and S. Ideda, *J. Electron Spectrosc. Radiat. Phenom.*, **7**, 465 (1975).
11. J. F. Moulder, W. F. Stickle, P. E. Sobol, and K. D. Bomben, *Handbook of X-Ray Photoelectron Spectroscopy*, Physical Electronics, Minnesota, USA (1995).
12. O. Glemser, G. Gattow, and H. Meisier, *Z. Anorg. Allg. Chem.*, **309**, 1 (1961).
13. G. Gattow and O. Glemser, *Z. Anorg. Allg. Chem.*, **309**, 20 (1961).
14. G. Gattow and O. Glemser, *Z. Anorg. Allg. Chem.*, **309**, 121 (1961).
15. G. A. Kolta, F. M. A. Kerim and A. A. A. Azim, *Z. Anorg. Allg. Chem.*, **384**, 260 (1971).
16. R. M. Potter and G. R. Rossman, *Amer. Mineral.*, **64**, 1199 (1979).
17. R. M. Potter and G. R. Rossman, *Amer. Mineral.*, **64**, 1219 (1979).
18. H. Miura, Y. Takeda, and Y. Hariya, *Mineral. Mag.*, **16**, 437 (1984).
19. M. Tsuji and M. Abe, *Annual Report of Tokyo Institute of Technology Radioisotope Laboratory*, Tokyo, July 1985, p.46.
20. M. Tsuji and S. Komarneni, *J. Mater. Res.*, **8**, 3145 (1993).

Mat. Res. Soc. Symp. Proc. Vol. 658 © 2001 Materials Research Society

A New Microporous Silicate With 12-Ring Channels

Jacques Plévert,[1†] Yoshihiro Kubota,[2] Takahisa Honda,[2]
Tatsuya Okubo[1] and Yoshihiro Sugi[2]
[1]Dept of Chemical System Engineering, The University of Tokyo,
Tokyo 113-8656, Japan.
[2] Dept of Chemistry, Faculty of Engineering, Gifu University,
Gifu 501-1193, Japan.

ABSTRACT

A new molecular sieve GUS-1 is synthesized under hydrothermal conditions using a new organic compound acting as structure-directing agent. The structure determination reveals a framework with one-dimensional 12-membered ring channels running parallel. The structure exhibits the same channel net as zeolite ZSM-12 and shows the same projection of the framework along the pores as mordenite.

INTRODUCTION

An effective way to synthesize molecular sieves with new framework topology consists of using new organic structure-directing agents (SDA). A novel crystalline microporous silicate GUS-1 is obtained using a newly designed organic compound, 1,1'-butylenedi(4-aza-1-azonia-2,5-dimethylbicyclo[2.2.2]octane) dihydroxide as SDA [1]. The structure determination of the new microporous solid is described in this report.

EXPERIMENTAL

The synthesis of GUS-1 was performed under hydrothermal conditions in the temperature range 150-175°C for 10-20 days. The composition of the synthesis mixture is $1SiO_2$: $0.1R^{2+}(OH^-)_2$: $0.1NaOH$: $50H_2O$, where R^{2+} stands for cation **1** (**1** consists of 4 diastereomers).

GUS-1 tends to form with another molecular sieve ZSM-12, formed of channels with 12 membered-ring aperture [2]. GUS-1 in pure form has not been obtained so far.

The structure of GUS-1 has been determined from synchrotron powder diffraction data of a sample containing the two phases GUS-1 and ZSM-12, calcined at 700°C for removing the SDA. The X-ray data were collected on the beam line X7A at the National Synchrotron Light

Source at Brookhaven Laboratory. The sample was loaded in a glass capillary of diameter 1.0 mm. The powder pattern was recorded at room temperature at the wavelength $\lambda = 1.1964$ Å.

STRUCTURE DETERMINATION

The particular sample used for structure-solution contains the two phases GUS-1 and ZSM-12 in approximate proportion 40% and 60%. Its N_2 adsorption isotherm is very similar to the isotherms of pure ZSM-12 samples, suggesting that the unknown phase is a molecular sieve with 12-ring channels too.

Scanning electron micrographs reveal particles of different morphologies. The largest amount of crystals shows particles of needle shape, similar to the typical morphology of crystals of ZSM-12. Overgrowth of thin needles is found and it is tempting to assign them to the phase GUS-1 (Figure 1). In addition, particles of fan-like shape can be observed in the sample, which probably correspond to zeolite SSZ-31 as the characteristic (001) reflection is found in the powder pattern, as trace, at d = 14.4 Å [3].

The powder pattern was indexed using the program TREOR [4], giving an orthorhombic cell with systematic absences consistent with the space group Cmmm and subgroups and possibly C222$_1$. GUS-1 possesses a small axis $c = 5.0$ Å, limiting the structure-determination to a search of the 2-dimensional projection of the zeolite into the ab plane. Models have been built with Cmmm symmetry and 12-ring channels running along the c-axis. A good candidate was found showing strong similarities to mordenite. Additional searches were performed using the program FOCUS [5,6]. The structure factors of GUS-1 were extracted by full-profile fitting [7] using the GSAS software package [8] and normalized after estimation of the scale factor by a Wilson plot. The mordenite-like structure found by model building was generated with the highest frequency among a variety of similar frameworks with 12-ring channels but different connectivity.

Distance Least Squares (DLS) refinement [9] of the atomic positions gives a residual R-value R_{DLS} as low as $R_{DLS} = 0.0029$ for space group C222, but a high value, $R_{DLS} = 0.0150$, for space group Cmmm due to Si-O-Si bond angles of 180° imposed by symmetry.

The Rietveld refinement was performed in space group C222 for GUS-1 and the space group C2/c was assumed for ZSM-12. The impurity SSZ-31 was ignored during the refinement.

Figure 1. Scanning electron micrograph showing details of the GUS-1 overgrowth on ZSM-12 crystals (the bar scales 5 μm).

RESULTS

The final residuals of the Rietveld refinement are $R_{wp} = 7.4\%$, $R_p = 5.5\%$, and the individual R_F factors for the phases GUS-1 and ZSM-12 are $R_F = 9.8\%$ and $R_F = 7.9\%$ respectively. Figure 2 shows the final Rietveld refinement plot. The atomic positions of the calcined form of GUS-1 are reported in Table I. The range of angles O-Si-O varies from 136° to 169°, the distances Si-O being maintained in the range 1.58 to 1.62 Å during the refinement.

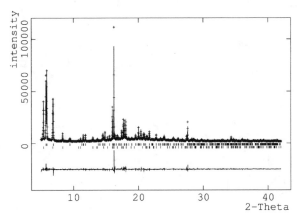

Figure 2. *Experimental and calculated powder patterns of a mixture 40% GUS-1, 60% ZSM-12. The upper vertical bars correspond to the peak positions of ZSM-12 and the lower bars to the peak positions of GUS-1 ($\lambda = 1.1964$ Å).*

Atoms	x	y	z
Si1	.3074(17)	.1879(18)	.478(8)
Si2	.1818(21)	.0767(13)	.527(11)
Si3	.0937(16)	.1222(15)	.016(13)
Si4	.0964(17)	.2819(15)	-.002(14)
O1	.3549(30)	.1874(29)	.754(15)
O2	.384(4)	.2092(25)	.307(14)
O3	.2500	.2500	.522(24)
O4	.2644(26)	.1170(21)	.477(16)
O5	.1537(30)	.0785(24)	.834(11)
O6	.212(4)	.0000	.5000
O7	.1151(31)	.1023(21)	.317(13)
O8	.0000	.1006(35)	.0000
O9	.1021(18)	.2021(24)	-.013(11)
O10	.0000	.297(5)	.0000

Table I. Fractional atomic coordinates for calcined GUS-1. Crystal data: chemical formula $Si_{32}O_{64}$, space group C222, $a = 16.4206(4)$, $b = 20.0540(4)$, $c = 5.0464(1)$ Å, $V = 1661.8(1)$ Å^3.

GUS-1 is a molecular sieve with one-dimensional 12-ring channels running along the c-axis. The channel aperture is elliptical in shape and the minor and major free diameters are 5.4 Å and 6.8 Å respectively. The projection of the structure along the pores is very familiar, as it is the same as the projection of the well-known zeolite mordenite in the ab plane (Figure 3).

On the other end, GUS-1 shows many similarities with ZSM-12 (a = 24.863, b = 5.012, c = 24.328 Å, β = 107.72° [2]). Both structures possess the same channel net composed of six-rings only and the same secondary building unit $4^2 5^4 6^2$ (Figure 4). In addition, the framework density of both structures are very close, FD = 19.3 T atoms per nm^3.

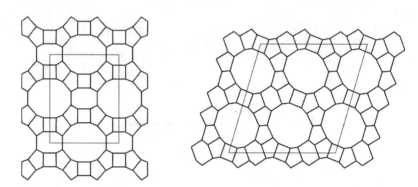

Figure 3. *The projection of the framework of GUS-1 along the pores is similar to the projection of mordenite in the ab plane (left), the framework of ZSM-12 shows some similarities with the framework of GUS-1 (right).*

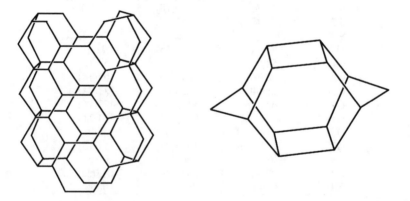

Figure 4. *Two common features to both GUS-1 and ZSM-12: the channel net (left) and the secondary building unit $4^2 5^4 6^2$ (right).*

CONCLUSION

The structure of a new high-silica microporous solid has been determined from synchrotron powder diffraction data of a mixture of two phases. The structure of GUS-1 shows similarities with both zeolite mordenite and ZSM-12. The framework topology of GUS-1 is reported in the Database of Zeolite Structures [10] under the three-letter code GON.

ACKNOWLEDGMENTS

The authors gratefully acknowledge Dr. P. Wagner for collecting the synchrotron data and Mr. T. Wakihara for taking SEM photographs. J.P. thanks Dr. T. Tatsumi for giving an opportunity to work at the University of Tokyo and Dr. M. O'Keeffe for helpful discussions. Y.K. thanks Dr. M E. Davis for useful suggestions and encouragements and Mr. M. Ogawa for supplying the precursor of SDA. Y.K. and Y.S. thanks New Energy and Industrial Technology Development Organization (NEDO) of Japan for financial support.

REFERENCES

* E-mail: JPlevert@asu.edu; kubota@apchem.gifu-u.ac.jp
† Current Address: Department of Chemistry & Biochemistry, Arizona State University, Tempe, AZ 85287-1604, USA.
1. N. Sugimoto, Y. Kubota and Y. Sugi, *Jpn. Kokai Tokkyo Koho*, No. 2000-211912 (2000) [*Chem. Abstr.*, **133**, 137465 (2000)].
2. C. A. Fyfe, H. Gies, G. T. Kokotailo, B. Marler and D. E. Cox, *J. Phys. Chem.*, **94**, 3718 (1990).
3. R. F. Lobo, M. Tsapatis, C. C. Freyhardt, I. Chan, C. Y. Chen, S. I. Zones and M. E. Davis, *J. Am. Chem. Soc.*, **119**, 3732 (1997).
4. P. E. Werner, L. Eriksson and M. Westdahl, *J. Appl. Cryst.*, **18**, 367 (1985).
5. R. W. Grosse-Kunstleve, PhD. Thesis, ETH, Zürich, 1996.
6. R. W. Grosse-Kunstleve, L. B. McCusker and Ch. Baerlocher, *J. Appl. Cryst.*, **30**, 985 (1997).
7. A. Lebail, H. Duray and J. L. Fourquet, *Mat. Res. Bull.*, **23**, 447 (1988).
8. A. C. Larson, and R. Von Dreele, Generalized Structure Analysis System, Report LAUR 86-748 (Los Alamos National Laboratory, USA, 1990).
9. C. Baerlocher, A. Hepp and W. M. Meier, DLS-76, A Program for Simulation of Crystal Structures by Geometric Refinement (ETH, Zürich, Switzerland, 1977).
10. International Zeolite Association. "Atlas of Zeolite Structure Type". http://www.iza-structure.org/databases/

Mat. Res. Soc. Symp. Proc. Vol. 658 © 2001 Materials Research Society

Deposition and Characterization of $Y_3Al_5O_{12}$ (YAG) Films and Powders by Plasma Spray Synthesis

Sujatha D. Parukuttyamma[1], Joshua Margolis[1], Haiming Liu[2], John B. Parise[1, 2, 3]
Clare P. Grey [1, 2], Sanjay Sampath[1], Perina Gouma[1] and Herbert Herman[1]
[1]Center for Thermal Spray Research, Department of Materials Science and Engineering
[2]Department of Chemistry, [3]Department of Geosciences, State University of New York at
Stony Brook, Stony Brook, New York 11794-2275, U. S. A.

ABSTRACT

YAG powders and coatings were developed for the first time by a novel precursor plasma spraying technique using the radio frequency (RF) induction plasma technique. The XRD of the as -sprayed coating confirms the presence of YAG, H-YAP or O-YAP or a mixture of the above depending on the spray conditions. ^{27}Al MAS NMR of the YAG coating corroborates the x-ray results. TEM studies on the coatings confirm that the coating consists of nano-structured particles. The successful spraying of these complex oxide coatings proves that chemistry of phase formation can be controlled in the plasma, thus opening up new avenues in material synthesis.

INTRODUCTION

Plasma or thermal spraying of metals, ceramics and composites is a well - established technology for the deposition of coatings on various substrates. In general, the traditional plasma processes utilize powder feedstocks of pre-mixed compositions to develop fine-grained deposit [1]. More recently, thermal spraying of liquid feedstocks is emerging as a viable method for the production of ceramic coatings and powders. A suspension plasma spraying (SPS) technique was reported for the production of ceramic oxides, bioceramics such as hydroxy appatite and also a series of intermetallics from suspensions of the fine powders of the corresponding materials [2].

The Y_2O_3-Al_2O_3 phase diagram contains several stable phases including $Y_3Al_5O_{12}$, YAG (Yttrium aluminum garnet), $YAlO_3$, O-YAP (Yttrium aluminum perovskite) and $Y_4Al_2O_9$, YAM (Yttrium aluminum monoclinic) [3]. In addition, a metastable yttrium aluminate phase $YAlO_3$ (H-YAP) crystallizes in the hexagonal form. Among these, the garnet phase possesses extremely high chemical stability, low electrical conductivity and high creep resistance making it useful, for example, as an insulating or refractory coating and as a promising fibre material for ceramic composites. Although synthesis of the cubic garnet phase to form powder is a very sluggish reaction, several liquid phase and soft-chemical routes have been proposed to lower the crystallization temperature of YAG [4-6]. Most of these methods produce only loose powders, which have to be processed further for making refractory coatings or sintered materials.

We have been working on developing alumina and doped alumina coatings by employing a modified version of the suspension plasma spraying where a precursor of the material is used as a liquid feed stock [7]. As an extension of this work, attempts were made to develop yttrium aluminum garnet (YAG) from sol and solution precursors considering it as a representative complex model compound. Here, we report the direct production of yttrium aluminum garnet coatings and powders by using the radio frequency induction (RF) plasma technique.

EXPERIMENTAL DETAILS

The YAG precursor with a Y:Al ratio of 3:5 and a final solid concentration of 25 g/L was sprayed using the RF plasma torch (Tafa Model 66) under a series of spray conditions. The YAG precursor sol (P1) was prepared by dispersing the required amount of boehmite powder (Catapal D. Vista Chemical Co., Houston, Texas) and yttrium nitrate hexahydrate, $Y(NO_3)_3 \cdot 6H_2O$ (Aldrich, 99.9%) in water. The pH of the slurry was adjusted to 3-4 with nitric acid and the suspension was stirred for 2-3 hours until it became a stable opaque sol. Similarly, the solution precursor (P2) was prepared by mixing yttrium nitrate and aluminum nitrate in a Y:Al ratio of 3:5 in ethanol. The spray conditions are optimized after several runs. Steel plates of area 6x6x0.2 cm^3 or larger were used as substrates. Details about spray conditions are reported elsewhere [8].

The sprayed coatings and the calcined powders were characterized by powder x-ray diffraction, XRD (SCINTAG/ PAD-V diffractometer) at a scan rate of 1 °/min using CuKα radiation. Morphological analyses were performed on a (Jeol/JSM-840A) scanning electron microscope. [27]Al MAS (Magic Angle Spinning) NMR experiments were performed with a double-tuned Chemagnetics 5-mm probe (CMX-360 spectrometer) at an operating frequency of 93.8 MHz. The TEM studies were performed on a Philips CM12 TEM running at 120kV.

RESULTS AND DISCUSSION

The diffraction pattern of the as-dried sol indicates partial crystallinity and the weak reflections correspond to AlO (OH) and $Y(NO_3)_3$ with no indication of either Y_2O_3 or Al_2O_3. The XRD pattern of the as-sprayed coating from a sol of metal stoichiometry Y: Al = 3:5 is shown in Figure 1(a). All the observed reflections in the sol sprayed coatings are identical to the reported $H-YAlO_3$ (JCPDS # 16-219), except for a low

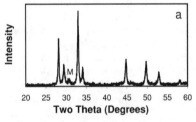
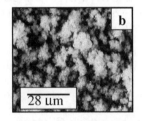

Figure 1. XRD pattern of the as sprayed coating from sol P1 (a) and the corresponding SEM picture (b) "M" on XRD pattern corresponds to YAM.

intensity one corresponding to the strongest reflection of YAM phase. The SEM micrograph of the corresponding coating is shown in Figure 1(b). It is clear from the image that the particles remain spherical in shape with very small size indicating the absence of a melting process during spraying. The absence of any Y_2O_3 or Al_2O_3 reflections in the as-sprayed coating rules out any selective decomposition of Y $(NO_3)_3$ or AlO(OH), during the atomization and spray process.

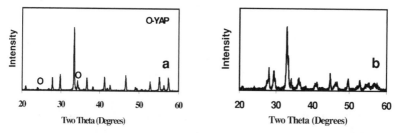

Figure 2. X-ray diffractograms of RF sprayed garnet coatings from (a) P1 and (b) P2.

In the consecutive set of experiments the as-sprayed layers from both the sol and the solution were further processed with the plasma for a couple of more cycles. A dramatic change in the phase development and a substantial growth in crystallite size were observed from P1 as evident from the sharp x-ray reflections in Figure 2(a). Almost all the reflections could be indexed based on the cubic garnet phase (JCPDS # 33-40). The reflections marked 'O' in the x-ray pattern [Figure 2(a)] correspond to the O-YAP (JCPDS # 33-41) phase. The absence of any reflections from the substrate material indicates a continuous and thick deposit of YAG. Coating from P2 [Figure 2(b)] was very difficult to index due to its complexity and phase segregation and was a mixture of YAG, H-YAP and O-YAP. The sharp and the intense reflections correspond to the YAG phase. The SEM micrographs of the above coatings from P1 and P2 are shown in Figure 3(a) & (b). There is a substantial grain growth in the sol derived coating resulting in a reasonably dense and coherent deposit [Figure 3(a)]. The coating from the solution P2, is highly porous consisting of hollow spherical particles [Figure 3 (b)]. This may be associated with the foaming of the particles during solvent evaporation.

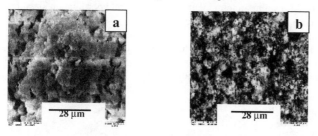

Figure 3. Surface morphologies of the plasma sprayed coatings from (a) P1 and (b) P2.

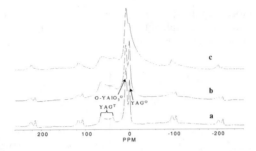

Figure 4. *NMR data on plasma sprayed coatings from (a) P1 (b) P2 and (c) as-sprayed coating from P1.*

In the garnet lattice, aluminum ions occupy both tetrahedral and octahedral coordination sites in the ratio 3:2. ^{27}Al MAS NMR is utilized for further confirmation of the garnet phase formation and different coordination states of Al centers in these sprayed materials. The spectrum of the garnet coating from P1 shows an intense narrow resonance at 0.4 ppm, with spinning side bands assigned to Al in the octahedral site and a typical line shape for Al in a distorted tetrahedral environment spreading from 70 to 30 ppm [Figure 4(b)]. In addition there is another strong sharp signal at 9.5 ppm, which is assigned to an octahedrally coordinated Al site, different from that of garnet. The x-ray pattern [Figure. 2(a)] of the same material confirms the presence of orthorhombic YAP phase, in additon to YAG which contains Al in an octahedrally coordinated environment. This phase is reported to give a single resonance at 9.4 ppm in ^{27}Al MAS NMR Therefore, the resonance at 9.5 ppm from the garnet coating is assigned unequivocally to that of octahedral Al site in orthorhombic $YAlO_3$. The NMR of the coating from precursor P2 behaved almost similarly indicating the presence of only orthorhombic $YAlO_3$ and YAG, eventhough its XRD was too complicated to index. The NMR spectrum of the as-sprayed material [Figure 4(c)] from the sol is very different from others with signs of Al residing in tetra, penta and octahedral coordination sites. Detailed NMR studies on these materials are in progress.

TEM studies on some of the sprayed materials confirm their nano structured nature, where pockets of nano particles are embedded in the matrix as shown below for the H-YAP powders collected from the as-sprayed coating and the YAG powders collected after treatment with the plasma Figure 5 (a) & (b).

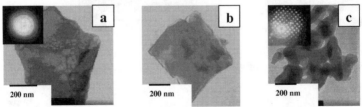

Figure 5. *TEM studies on (a) as-sprayed P1 & SAD of H-YAP phase, (b) nano-particles of P1 after plasma treatment and (c) YAG nano particles from P2 and the SAD pattern.*

The particles of the calcined powders from P2 appears to be much larger indicating the fusion of dumb-bell shaped nano partcles of YAG which grow in size on further heating. Selected area diffraction (SAD) patterns on these grains confirm that they are the cubic garnet phase.

In order to gain more insight into the phase formation to the YAG in the plasma (Ar/He), as-sprayed powders were collected, calcined and were characterized by XRD.

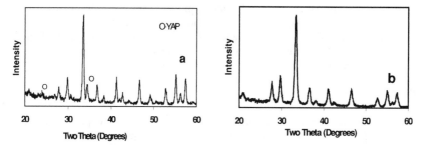

Figure 6. XRD pattern of the powder from (a) P1 calcined at 1250 °C and (b) P2 calcined at 1100 °C.

A systematic growth of the YAG phase and a disappearance of the H-YAP were observed during calcination of both the precursors. A calcination temperature of 1250°C for 6h was necessary to induce a phase change [Figure 6(a)], identical to the one produced by plasma spraying of the sol [Figure 2(a)]. The powders collected from P2 converted to single phase garnet at a much lower temperature as shown by the XRD pattern in Figure 6(b). Upon further calcination, a dramatic change in the grain size and grain growth were observed in both the cases as shown below in Figure 7 (a) and (b) wherein neck fusion of the nano grains is very clearly seen.

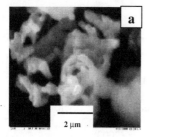 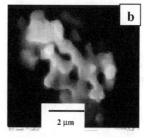

Figure 7. SEM picture of the powders from (a) P1 and (b) P2 calcined at 1350°C.

The formation of the garnet phase in the coated material from P1 indicates compositional homogeneity and integrity of the process. The precursor sol (P1) is obviously not homogeneous at an atomic level like the solution (P2) as evidenced by the presence of a minor amount of O-YAP in the former. Results on calcination of the powder from P2 further confirm the atomic scale mixing of the solution precursor.

However, the extensive phase segregation observed from P2 during spraying suggests that the rapid heating associated with the spray process did not provide enough time for the diffusion and decomposition of the metal nitrates within the droplets to produce single phase material. The more homogeneous and dense coating from the hybrid sol suggest that the Y^{3+} ions are entrapped in the boehmite sol particles, resulting in *in-situ* micro-scale reactions within the individual droplets and thereby controlling the chemistry of phase formation during spraying. Thus, by carefully controlling the chemistry of the precursor and the spray conditions it is possible to produce garnet deposits of reasonable thickness (60-100 μm), crystallinity (30-50 nm) and uniformity. Unlike normal plasma spraying, where the feed stock material melts and forms a coating upon impact with the substrate, the precursor route involves controlling the chemistry of phase formation during the spray process.

CONCLUSIONS

Films and powders of yttrium aluminum garnet ($Y_3Al_5O_{12}$) were prepared for the first time by the precursor plasma spraying through a radio frequency plasma technique. This is achieved by the injection of atomized liquid droplets of the YAG precursor sol or solution into the plasma plume, resulting in the formation of adherent and chemically controlled garnet deposits. The overall process of spraying, atomization and chemical reaction occurred within a very short time indicating the simplicity of this process. This process could further be extended to develop large area thick/thin coatings of YAG in a single step on many substrates.

ACKNOWLEDGEMENTS

This project is supported by MRSEC program of NSF under award no. **DMR-0080021**. SDP is on leave of absence from CG&CRI, Calcutta, India.

REFERENCES

1. H. Herman and S. Sampath, *Metallurgical and Ceramic Protective Coatings*, Ed (K. H. Stern) 261-289 (1996).
2. G. Schiller, M.Muller and F. Gitzhofer, *J. Therm. Spray. Tech.*, **8**[3] 389-392 (1999).
3. G. S. Corman, *Ceram. Eng. Sci. Proc.*, **12**, 1745-1766 (1991).
4. H. Wang, L. Gao and K. Niihara, *Mater. Sci. Eng. A.*, **288**, 1-4 (2000).
5. M. Veith, S. Mathur, A. Kareiva, M. Jilavi, M. Zimmer and V. Huch, *J. Mater. Chem.*, **79**[2] 385-394 (2000).
6. M. K. Cinibulk, *J. Am. Ceram. Soc.*, **83**[5] 1276-1278 (2000).
7. Sujatha Devi. P, et al., Manuscript in preparation.
8. Sujatha Devi. P, J. Margolis, H. Liu, C. P. Grey, S. Sampath, H. Herman and J. B. Parise., *J. Am. Ceram. Soc.*, (Communicated).

Mat. Res. Soc. Symp. Proc. Vol. 658 © 2001 Materials Research Society

The Synthesis and Characterization of a Novel Gallium Diphosphonate

Martin P. Attfield and Howard G. Harvey[1]
School of Crystallography, Birkbeck College, Malet Street, Bloomsbury, London, WC1E 7HX, UK.
[1]Davy-Faraday Research Laboratory, The Royal Institution of Great Britain, 21 Albemarle Street, London, W1S 4BS, UK.

ABSTRACT

The gallium diphosphonate, $Ga_2[O_3PC_2H_4PO_3](H_2O)_2F_2 \cdot H_2O$, has been prepared under solvothermal conditions from a HF/ pyridine solvent system. The structure of the material was determined by Rietveld refinement using the isostructural aluminium compound as the starting model for refinement. The structure consists of chains of corner sharing GaO_4F_2 octahedra with bridging F atoms linking the octahedra. The chains of octahedra are linked together in two directions perpendicular to the chain direction by the ethylenediphosphonate groups. The resulting structure contains a one-dimensional channel system in which extra-framework water molecules reside. Thermal gravimetric analysis of the material shows that 40 % of the extra-framework water can be reversibly absorbed. Thermodiffraction studies show that the material retains its structural integrity until all the extra-framework water is lost and then undergoes a considerable loss in crystallinity as the framework water molecules are lost. The crystallinity of the material then progressively decreases as the temperature is increased.

INTRODUCTION

The chemistry of metal phosphonates has received much interest in recent years because of the wide variety of possible structures accessible and their potential for use as sorbents, catalysts and optic materials [1]. The main attraction of such materials stems from the inclusion of an organic moiety into the compound that may be modified to impart particular chemical functionality and structure on the material and introduce a high degree of hydrophobicity. The latter property can lead to improvements in catalytic performances in particular systems [2, 3]. Initially, synthetic work focussed primarily on using phosphonic acids for the synthesis of new materials, this has been extended to include the utilization of diphosphonic acids ($(HO)_2OPRPO(OH)_2$, where R is an organic group). The attraction of the incorporation of the diphosphonic acid is that through careful selection of the R group the interlamellar distance, or pore size, and potentially the shape selective properties within the resulting structure can be controlled. Metal diphosphonates containing various metal cations have been reported [1]. As far as the authors are aware, no reports of gallium diphosphonates have hitherto been reported, which is surprising considering the possible chemical similarities between this type of material and the important gallophosphate and gallofluorophosphate family of compounds. Three gallium methylphosphonates have been reported [4, 5], one, $Ga(OH)_{0.28}F_{0.72}(CH_3PO_3) \cdot H_2O$ [4], of which has an analogous structure to AlMepO-ζ [6] and another, $Ga_3(OH)_3F_3(CH_3PO_3)_2(H_2NC_3H_6NH_3)$ [4], that contains an organic structural directing agent. Two gallium carboxyethylphosphonates have also been reported [7]. We report here the synthesis, structural characterisation and thermal behaviour of the gallium diphosphonate, $Ga_2[O_3PC_2H_4PO_3](H_2O)_2F_2 \cdot H_2O$.

EXPERIMENTAL DETAILS

The synthesis of the $Ga_2[O_3PC_2H_4PO_3](H_2O)_2F_2 \cdot H_2O$ was performed by mixing $Ga_2(SO_4)_3 \cdot 18H_2O$ (Sigma), ethylenediphosphonic acid (Lancaster), HF/ pyridine (70 wt % HF, Aldrich), pyridine (Aldrich), and de-ionised water in the molar ratio 1: 2.16: 8.7: 54: 144. The reaction mixture (initial pH = 6) was sealed in a Teflon-lined steel autoclave and heated under autogenous conditions at 170°C for 5 days. The product mixture (final pH = 6) was filtered and dried in air to yield a white crystalline powder. Microprobe analysis of the materials gave a Ga: ratio of 1 and showed the presence of fluorine in the sample.

Laboratory powder diffraction data was collected on a sample of $Ga_2[O_3PC_2H_4PO_3](H_2O)_2F_2 \cdot H_2O$ and used to determine the monoclinic unit cell of the material from the first 20 low-angle Bragg reflections using the auto-indexing program TREOR-90 [8].

A 0.5 mm diameter Lindemann glass capillary of sample was mounted on the powder diffractometer at station 2.3 at CCLRC Daresbury Laboratory, Synchrotron Radiation Source and room temperature synchrotron X-ray data were collected. The mean wavelength used was 1.2999 Å and data were collected from 4 to 80° 2θ. The diffractometer operated with a Si(111) monochromator, parallel foils prior to the detector and a scintillation detector. The sample was spun during data collection to minimise preferred orientation and sampling effects.

The starting model for the Rietveld refinement was the structure of $Al_2[O_3PC_2H_4PO_3](H_2O)_2F_2 \cdot H_2O$ [9] using the space group P 2_1/m. The background was fitted with a Cosine Fourier series running through fixed points in the profile. The profile peak shape was described using a pseudo - Voigt function with an additional term to allow for anisotropic particle size broadening effects. The final cycle of least squares refinement included the scale factor, detector zero point, background coefficients, unit cell parameters, peak shape width variation terms, a preferred orientation parameter, and positional and isotropic thermal parameters for all atoms. The isotropic thermal parameters of the Ga atoms were constrained to have the same shift, as were those of the O, F and C atoms. Final observed, calculated and difference profile plots are shown in Figure 1, and the final atomic coordinates and isotropic thermal parameters are given in Table I. The Rietveld refinement was performed using the GSA package [10].

X-ray thermodiffractometric data were collected on a pressed pellet of $Ga_2[O_3PC_2H_4PO_3](H_2O)_2F_2 \cdot H_2O$ at station 9.3 at the CCLRC Daresbury Laboratory, Synchrotron Radiation Source. The mean wavelength used was 1.8335 Å selected by a Si(111) monochromator and data were collected using position sensitive detector in the 2θ range 5 – 63° at intervals of 20°C from 40 to 300°C.

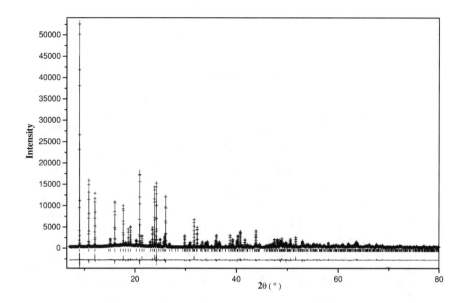

Figure 1. *The final observed (crosses), calculated (line) and difference plot for the Rietveld refinement of* $Ga_2[O_3PC_2H_4PO_3](H_2O)_2F_2 \cdot H_2O$.

Table I. Final atomic coordinates and isotropic thermal parameters for $Ga_2[O_3PC_2H_4PO_3](H_2O)_2F_2 \cdot H_2O^a$.

Atom	X	Y	Z	U_{iso} (Å^2)
Ga(1)	0.5000	0.5000	0.0000	0.0078(5)
Ga(2)	0.2293(4)	0.2500	0.1453(3)	0.0078(5)
P(1)	0.0081(5)	0.4860(2)	0.2333(4)	0.003(1)
O(1)	-0.002(2)	0.2500	-0.067(1)	0.0101(8)
O(2)	-0.279(1)	0.5185(5)	0.2055(8)	0.0101(8)
O(3)	0.504(2)	0.2500	0.341(1)	0.0101(8)
O(4)	0.202(1)	0.5396(4)	0.127(1)	0.0101(8)
O(5)	0.014(1)	0.3631(5)	0.2307(8)	0.0101(8)
O(6)	0.433(2)	0.2500	-0.337(1)	0.0101(8)
F(1)	0.458(1)	0.3489(4)	0.0435(7)	0.0101(8)
C(1)	0.086(2)	0.5264(7)	0.442(1)	0.0101(8)

[a] $a = 5.04837(3)$ Å $b = 12.36987(7)$ Å $c = 8.25470(6)$ Å $\beta = 92.064°$: $R_{wp} = 10.05$ %, $R_p = 7.82$ %, $R_I = 6.38$ % $\chi^2 = 3.15$

RESULTS AND DISCUSSION

The crystal structure of $Ga_2[O_3PC_2H_4PO_3](H_2O)_2F_2 \cdot H_2O$ is shown in Figure 2. The structure consists of chains of corner sharing GaO_4F_2 octahedra running parallel to the b axis with bridging fluorine atoms linking the octahedra. These chains contain two types of GaO_4F_2 octahedra. The first type contains gallium, Ga(1), coordinated to two fluorine atoms (2x Ga(1)-F(1) 1.916(5) Å) in a *trans* configuration with the other corners of the octahedron being occupied by four of the oxygen atoms of the diphosphonate groups, (2x Ga(1)-O(2) 1.924(6) Å; 2x Ga(1)-O(3) 2.009(6) Å). The second octahedron consists of a central gallium atom, Ga(2), bound to two fluorine atoms in a *cis* fashion (2x Ga(2)-F(1) 1.899(5) Å) with the other two equatorial positions being occupied by two diphosphonate oxygen atoms (2x Ga(2)-O(1) 1.922(6) Å). The remaining axial positions are filled by two water molecules (Ga(2)-O(4) 2.09(1) Å; Ga(2)-O(5) 2.07(1) Å). The Ga-O, Ga-F and Ga-OH$_2$ distances observed are similar to those reported in the literature [11, 12]. The internal bond angles of the GaO_4F_2 octahedra all fall within the range 80.2 – 94.9° and 172.8 – 180° as expected for octahedral units. The ethylenediphosphonate groups link the octahedral chains in the [100] and [001] directions to form a structure containing small channels running along the [100] direction. Within these channels are found extra-framework water molecules, O(6). The bond lengths (P(1)-O(1) 1.514(6) Å, P(1)-O(2) 1.496(6) Å, P(1)-O(3) 1.521(6) Å and P(1)-C(1) 1.82(1) Å) and internal tetrahedral bond angles (range 104.1 – 115.9°) of the phosphonate group all lie within the range expected for the geometry of such moieties [13]. The presence of one crystallographically independent fluorine site was confirmed by the observation of a singlet at -131.3 ppm in the ^{19}F MAS solid state NMR of the material, a chemical shift value that is similar to that reported for other bridging F atoms in Ga centred octahedra [12].

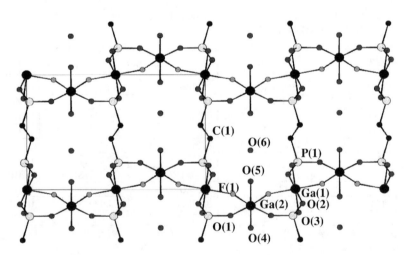

Figure 2. *The structure of $Ga_2[O_3PC_2H_4PO_3](H_2O)_2F_2 \cdot H_2O$ viewed down the a axis.*

The thermal gravimetric analysis of $Ga_2[O_3PC_2H_4PO_3](H_2O)_2F_2 \cdot H_2O$ shows two mass losses between 140 and 400°C totalling 15.9 % of the initial mass of the sample. These correspond to the removal of the extra-framework water O(6) (140 to 207°C) followed by the framework water molecules (O(4) and O(5)) (calculated weight loss is 13.5%). The slightly larger value of the observed weight compared to the calculated loss could be due to some additional fluoride loss as HF. Between 400 - 700°C weight losses associated with degradation of the organic components of the ethylenediphosphonate groups occur.

The thermodiffraction data are shown in Figure 3. The data show that on heating the material from 40 to 140°C little structural change occurs. However, between 160 and 200°C the material is rapidly transformed to a poorly crystalline phase the crystallinity of which decreases further upon heating. The temperature range at which this transformation occurs coincides with the temperature at which all the extra-framework water is removed and the framework water begins to be lost, indicating that the framework becomes unstable at this point. Separate X-ray thermodiffractometric studies with $Ga_2[O_3PC_2H_4PO_3](H_2O)_2F_2 \cdot H_2O$ using a laboratory powder diffractometer show the material to be completely amorphous by 380°C.

Heating the sample to 170°C, corresponding to the maximum rate of weight loss associated with the extra-framework water, followed by cooling to room temperature and storing in a humidifier for 22 hours, resulted in readsorption of approximately 40 % of the extra-framework water lost at 170°C.

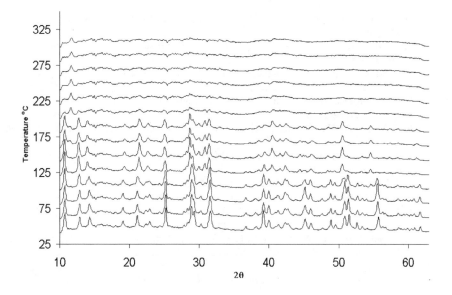

Figure 3. *The X-ray thermodiffraction data for $Ga_2[O_3PC_2H_4PO_3](H_2O)_2F_2 \cdot H_2O$.*

CONCLUSIONS

The gallium diphosphonate, $Ga_2[O_3PC_2H_4PO_3](H_2O)_2F_2\cdot H_2O$, has been prepared under solvothermal conditions and its structure determined by Rietveld refinement. The structure is found to be isostructural to the recently reported compound, $Al_2[O_3PC_2H_4PO_3](H_2O)_2F_2\cdot H_2O$ [The structure contains a one-dimensional channel system in which extra-framework water molecules reside. The material is structurally unstable after complete loss of the extra-framewo water, but can reversibly absorbed a proportion of the extra-framework water prior to this collapse, thus allowing the material to be classified as microporous.

ACKNOWLEDGMENTS

The authors would like to thank Dr. G. Sankar, Dr. A. Aliev, Dr. M. Odahaya, N. Cohen a A. Beard for help in collecting the thermodiffraction, NMR, TGA and microprobe data, respectively. MPA would like to thank the Royal Society for provision of a University Researc Fellowship and HGH would like to thank EPSRC for provision of a quota award and for fundir

REFERENCES

1. A. Clearfield in *Progress in Inorganic Chemistry*, edited by K. D. Karlin (Wiley, 1998) pp 371-510.
2. W. M. V. Rhijn, D. E. De Vos, B. F. Sels, W. D. Bossaert and P. A. Jacobs, *Chem. Commu* 317 (1998).
3. A. Bhaumik, P. Mukherjee and R. Kumar, *J. Catal.* **178**, 101 (1998).
4. C. Paulet, C. Serre, T. Loiseau, D. Riou and G. Ferey, *C. R. Acad. Sci. II C* **2**, 631 (1999).
5. M. Bujoli-Doeuff, M. Evain, F. Fayon, B. Alonso, D. Massiot and B. Bujoli, *Eur. J. Inorg. Chem.* **12**, 2497 (2000).
6. K. Maeda, Y. Hashiguchi, Y. Kiyozumi, and F. Mizukami, *Bull. Chem. Soc. Jpn.* **70**, 345 (1997).
7. F. Fredoueil, D. Massiot, D. Poojary, M. Bujoli-Doeuff, A. Clearfield A and B. Bujoli, *Chem. Commun.* 175 (1998).
8. P. E. Werner, L. Eriksson, and J. Westdahl, *J. Appl. Crystallogr.* **18**, 367 (1985).
9. H. G. Harvey, S. J. Teat and M. P. Attfield, *J. Mater. Chem.* **10**, 2632 (2000).
10. A. Larsen and R. B. von Dreele, *GSAS – General Structure Analysis System*, Tech. Rept. LA-UR-86-748, Los Alamos National Laboratory, USA (1987).
11. S. J. Weigel, S. C. Weston, A. K. Cheetham and G. D. Stucky, *Chem. Mater.* **9**, 1293 (1997
12. T. Loiseau, C. Paulet, N. Simon, V. Munch, F. Taulelle and G. Ferey, *Chem. Mater.* **12**, 13 (2000).
13. F. Serpaggi and G. Ferey, *J. Mater. Chem.* **8**, 2749 (1998).

Mat. Res. Soc. Symp. Proc. Vol. 658 © 2001 Materials Research Society

Synthesis of CoFe$_2$O$_4$ Nanoparticles via the Ferrihydrite Route

A. Manivannan, A. M. Constantinescu and M. S. Seehra

Physics Department, West Virginia University, Morgantown, WV 26506-6315

ABSTRACT

Nanoparticles of CoFe$_2$O$_4$ in the size range of 5 nm to 36 nm have been synthesized by first producing two-line Co-doped ferrihydrite and annealing it at temperatures between 573 and 1073 K. Co-doped ferrihydrite is produced by reacting appropriate amounts of FeCl$_3$/CoCl$_2$ solutions with NH$_4$OH at pH = 7. Thermogravimetric measurements provide evidence for the ferrihydrite to CoFe$_2$O$_4$ conversion between 500 and 700 K and x-ray diffraction is used to confirm the transformation to CoFeO$_4$ and determine the particle size. The lattice constant increases initially with the increase in particle size approaching a constant value above 20 nm. The measured T$_c$ equals 788 K for the 36 nm particles. Measurements of magnetization as a function of temperature and magnetic field confirm that the precursor Co-ferrihydrite is a superparamagnet with T$_b$ $\simeq$ 40 K, whereas for the 30 nm CoFe$_2$O$_4$, the coercivity H$_c$ $\simeq$ 7 kOe at 20 K, decreasing to H$_c$ $\simeq$ 800 Oe with M$_s$ $\simeq$ 60 emu/g at 300 K.

INTRODUCTION

In recent years, studies of nanoscale magnetic particles have attracted considerable interest, both from a fundamental point of view, and in view of their applications in magnetic fluids and magnetic recording [1-2]. Bulk CoFe$_2$O$_4$ is a magnetic ferrite with the inverse spinel structure and Curie temperature T$_c$ $\simeq$ 793 K. Nanoparticles of CoFe$_2$O$_4$ have been synthesized in thin film form [3], as microemulsions [4-7], and by chemical synthesis [8-10]. According to a recent paper by Grigorova et al [10], the magnetic parameters such as coercivity H$_c$, remanence M$_r$ and saturation magnetization M$_s$ vary considerably with the preparation technique, perhaps due to the relative distributions of Co^{2+}, Fe^{2+} and Fe^{3+} ions on the A (tetrahedral) and B (octahedral) sites of the inverse spinel structure. Consequently, alternative methods of synthesis are desirable.

In this work, we report the synthesis of CoFe$_2$O$_4$ nanoparticles in the size range of 5.5 nm to 36 nm using a wet chemical precipitation method, similar to the one we have used for nanoparticles of two-line ferrihydrites (FeOOH · 4H$_2$O) [11,12]. Thermogravimetric analysis (TGA) shows that at temperatures above about 500 K, the cobalt-ferrihydrite produced converts to nanoscale CoFe$_2$O$_4$, the particle size increasing with the increase in the annealing temperature. We also report some initial magnetic measurements on these nanoparticles. Details of the procedures and experimental results are presented below.

EXPERIMENTAL DETAILS

To prepare nanoscale CoFe$_2$O$_4$, the water solutions of FeCl$_3$ · 6H$_2$O and CoCl$_2$ · 6H$_2$O were mixed in appropriate amounts to yield the atomic ratio of Co:Fe = 1:2. This was followed by the addition of ammonium hydroxide to form a precipitate. This precipitate was filtered and washed until pH = 7 was achieved in the solution. The precipitate was dried overnight at 323 K. The x-ray diffraction (XRD) pattern of the dried powder measured at room temperature with a Rigaku Diffractometer and CuK_α radiation (λ = 1.5418 Å), yielded the familiar two-line pattern of ferihydrites [12,13], with no hint of $CoFe_2O_4$. Thermogravimetric measurements were carried out with a Mettler TA3000 system to determine the temperatures at which the precipitate converts to other phases. Finally, the magnetic measurements reported here were done using a SQUID magnetometer (Quantum Design Model MPMS) and a home-built vibrating sample magnetometer (VSM) with VSM head, controller and pickup coil made by EG&G.

RESULTS AND DISCUSSION

The room temperature XRD patterns of the prepared precipitate and the precipitates heated to temperatures of 323, 473, 573, 673, 773, 873, 973 and 1073 K for 5 hours each are shown in Figure 1. The precipitate heated to 323 and 473 K show only the two broad lines at $2\theta \sim 35°$ and $63°$ characteristic of nanoscale (~4 nm) ferihydrites [12-14]. Since ferrihydrite is known to convert to magnetite or hematite upon heating depending on the oxygen partial pressure [13,14], we carried out TGA measurements to determine any phase changes in the Co-doped ferrihydrite. Figure 2 shows the TGA data of mass changes versus temperature in air, clearly showing a significant loss of about 27% by 673 K. This mass loss is consistent with the transformation of $Co/FeOOH \cdot 4H_2O$ to $CoFe_2O_4$. The definite proof for this conversion comes from the XRD patterns of samples annealed at temperatures of 573 to 1073 K (Figure 1). All the observed Bragg peaks can be labeled by the known XRD pattern of $CoFe_2O_4$.

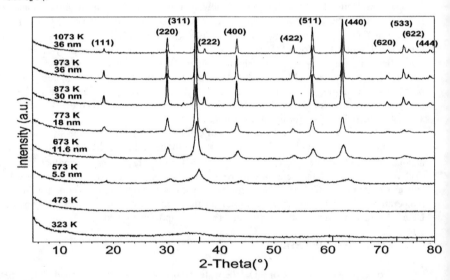

Figure 1. *Room temperature x-ray diffraction pattern of the Co-ferrihydrite precursor (323 K) and the precursor heated for 5 hrs at the temperatures indicated. The Miller indices are for the $CoFe_2O_4$ structure and particle size (nm) is determined from the (311) line.*

The data presented in Figure 1 show that the width of the Bragg peaks decreases as the annealing temperature increases. Using the Scherrer relation [15,16], the width of the strong line, the 311 peak, is used to determine the average particle size (see Figure 1). Particles of size range 5.5 to 36 nm have been obtained at the annealing temperatures of 573 to 973 K, respectively. Annealing at 1073 K did not produce any additional measurable change in the width of the (311) peak.

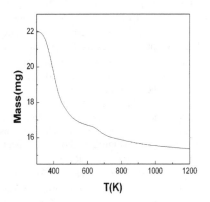
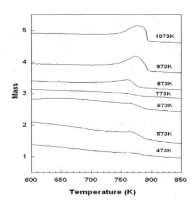

(left) **Figure 2.** *TGA plot of the change in the mass of the precursor with temperature at a heating rate of 10K/minute in air.*

(right) **Figure 3.** *TGA measurements of the change in mass with temperature in a magnetic field to determine T_c. The temperatures indicated are the annealing temperatures of the precursor.*

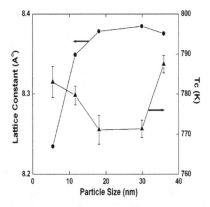

Figure 4. *Measured changes in T_c and lattice constant with particle size of $CoFe_2O_4$.*

TGA measurements in an applied field H ≃ 2 kOe were used to determine the Curie temperature T_c of the samples (Figure 3). Rapid drop of the mass just above 770 K is indicative of loss of ferromagnetism and hence T_c. T_c's so measured are plotted against the particle size in Figure 4, where the lattice constant a_o of the cubic cell determined from the analysis of the XRD data is also plotted. For size <18 nm, a_o decreases whereas T_c increases slightly, perhaps reflecting the expected slight increase in exchange coupling with reduced a_o. T_c measured here for the 36 nm is close to the expected bulk value.

The temperature variations of the measured magnetization M in H = 50 Oe are shown in Figure 5 both for the 323 K precipitate and the 36 nm $CoFe_2O_4$ particles prepared by annealing the precipitate at 973 K for 5 hrs. The data for the 323 K precipitate is similar to the observation for the 2-line ferrihydrite [12,13], except the blocking temperature T_b is about 40 K for the Co-doped precipitate as compared to T_b ≃ 60 K for ferrihydrite. The continued rise in M below T_b for the field-cooled case is commly observed in superparamagnets [17]. The data in Figure 5 for $CoFe_2O_4$ obtained by annealing the precipitate at 973 K shows two significant changes. First, the magnetization is larger for $CoFe_2O_4$ by over an order of magnitude and second T_b, determined by the separation of FC (field-cooled) and ZFC (zero-field-cooled) curves is increased to about 300 K. The hysteresis loop curves of M vs. H for the Co-doped ferrihydrite precipitate (Figure 6) show the expected superparamagnetism at 300 K and a small H_c ≃ 380 Oe below T_b (at 8 K). In contrast, the hysteresis loop curves for the 30 nm $CoFe_2O_4$ at 20, 100, 200 and 300 K (Figure 7) show the standard loops. At 20 K, the coercivity H_c ≃ 7000 Oe and M_s = 30 emu/g. With increase in temperature, H_c decreases and M_s increases approaching the magnitudes of 800 Oe and about 60 emu/g respectively at 300 K. These magnitudes are quite comparable to those reported by others [10].

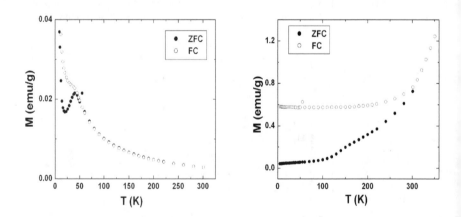

Figure 5. *Magnetization measured in H = 50 Oe plotted against temperature for the 323 K Co-ferrihydrite (left) and the 36 nm $CoFe_2O_4$ particles (right). ZFC(FC) corresponds to zero-field-cooled (field-cooled) case.*

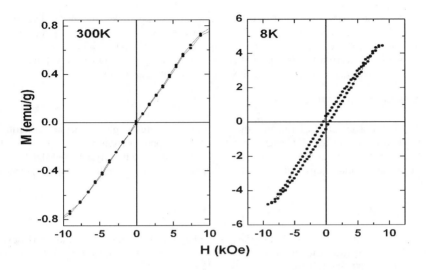

Figure 6. *The hysteresis loop data for the 323 K Co-ferrihydrite at 300 K and 8 K.*

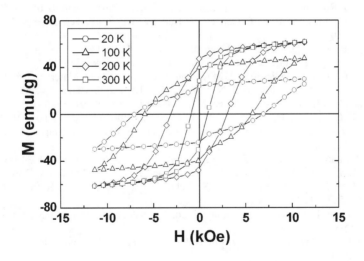

Figure 7. *The hysteresis loop measurements for the 30 nm CoFe₂O₄ particles at 20, 100, 200 and 300 K*

In summary, the synthesis of nanosized $CoFe_2O_4$ particles via the ferrihydrite route and the initial characterization of these particles by x-ray diffraction and magnetic measurements have been described. Investigations of the morphology of these particles with transmission electron microscopy and changes in the magnetic parameters with particle size are now underway. These results will be reported in the near future.

ACKNOWLEDGMENTS

We thank Dr. Lederman for the use of the vibrating sample magnetometer during the breakdown of the SQUID magnetometer. Partial support of this research from the U.S. Department of Energy (contract #DE-FC26-99FT40540) is gratefully acknowledged. Listing of the commercial equipment does not imply endorsement by the authors or their sponsors.

REFERENCES

1. *Magnetic Properties of Fine Particles* edited by J. L. Dormann and D. Fiorani (Elsevier Science, Amsterdam 1992); See also the recent review by R. H. Kodama, J. Magn. Magn. Mater. **200**, 359 (1999).
2. *Nanophase Materials: Synthesis, Properties and Applications*, edited by G. C. Hadjipayanis and R. W. Siegel (Kluwer, 1994).
3. F. Cheng, Z. Peng, C. Liao, Z. Xu, S. Gao, C. Yan, D. Wang and J. Wang, Solid St. Commun. **107**, 471 (1998).
4. A. J. Rondinone, A. C. S. Samia and Z. J. Zhang, J. Phys. Chem. B **103**, 6876 (1999).
5. S. Li, V. T. John, C. O'Conner, V. Harris and E. Carpenter, J. Appl. Phys. **87**, 6223 (2000).
6. V. Pillai and D. O. Shah, J. Magn. Magn. Mater. **163**, 243 (1996).
7. A. T. Ngo, P. Bonville and M. P. Pileni, Eur. Phys. J. B **9**, 583 (1999).
8. E. Uzunova, D. Klissurski, I. Mitov and P. Stefanov, Chem. Mater. **5**, 576 (1993).
9. V. Masheva, M. Grigorova, N. Volkov, H. J. Blythe, T. Midlarz, V. Blaskov, J. Geshev and M. Mikhov, J. Magn. Magn. Mater. **196-197**, 128 (1999).
10. M. Grigorova, H. J. Blythe, V. Blaskov, V. Rusanov, V. Petkov, V. Masheva, D. Nihtianova, L. M. Martinez, J. S. Munoz, and M. Mikhov, J. Magn. Magn. Mater. **183**, 163 (1998).
11. Z. Zhao, F. E. Huggins, Z. Feng and G. P. Huffman, Phys. Rev. B **54**, 3403 (1996).
12. M. S. Seehra, V. S. Babu, A. Manivannan and J. W. Lynn, Phys. Rev. B **61**, 3513 (2000).
13. M. M. Ibrahim, G. Edwards, M. S. Seehra, B. Ganguly and G. P. Huffman, J. Appl. Phys. **75**, 5873 (1994).
14. J. Zhao, F. E. Huggins, Z. Feng, F. Lu, N. Shah and G. P. Huffman, J. Catal. **143**, 499 (1993).
15. *X-Ray Diffraction Procedures* by H. P. Klug and L. E. Alexander (John Wiley, NY 1974) page 687.
16. M. M. Ibrahim, J. Zhao and M. S. Seehra, J. Mater. Res. **7**, 1856 (1992).
17. S. A. Makhlouf, F. T. Parker and A. E. Berkowitz, Phy. Rev. B, **55**, R14717 (1997).

Mat. Res. Soc. Symp. Proc. Vol. 658 © 2001 Materials Research Society

Study of Calcium/Lead Apatite Structure Type for Stabilising Heavy Metals

Z.L.Dong, B.Wei and T.J. White
Dept. Solid State Characterisation & Materials Processing, Environmental Technology Institute, 18 Nanyang Drive, Singapore 637723, SINGAPORE

ABSTRACT

$(Ca_xPb_{10-x})(VO_4)_6F_2$ apatites were synthesised and their microstructures were studied using powder X-ray diffraction, scanning electron microscopy and transmission electron microscopy before and after leach testing. X-ray diffraction showed that the apatites were hexagonal with $a \approx 10$Å and $c \approx 7$Å. During the leach test, Pb was released into the solution more slowly than Ca, which is desirable as the immobilisation of Pb is of importance. The experimental results also showed that V was almost undetectable in the leaching test solutions. In the $(Ca_7Pb_3)(VO_4)_6F_2$ pellet, Ca and Pb distributions were not homogenous from one grain to another. Microstructural evidence from scanning electron microscopy revealed that the dissolution via development of etch pits began at grain boundaries and inside grains, and progressed faster in Ca rich regions. These results suggest that apatites of high Pb to Ca ratio are more durable.

INTRODUCTION

Solid Waste Incineration (SWI) processes generate bottom ash and fly ash, as well as flue gas. The fly ash normally contains heavy metal oxides, such as PbO and V_2O_5, and other hazardous compounds [1,2]. One way to reduce the impact of fly ash on the environment is to stabilise the toxic species before disposal. The tailored addition of extra materials can promote the formation of more durable structure-types, eg., olivine, apatite, spinel and anhydrite structures [2-5].

This investigation is focused on the formation of apatite structures containing the heavy metals Pb and V, and the leaching characteristics of the ceramic pellets. The results of the investigation can be applied not only to incinerator ash stabilisation, but also to other industrial solid wastes.

EXPERIMENTAL DETAILS

Apatites of composition $(Ca_xPb_{10-x})(VO_4)_6F_2$ were synthesised from AR grade CaO, CaF_2, V_2O_5 and PbO powder mixtures by firing at 800°C for more than 10 hours. The apatites so obtained were pressed into pellets and heat treated again at 800°C.

Leach tests were performed to determine the mechanism of dissolution and the leach rate of each element in the pellets. The leach solution used was TCLP prescribed acetic acid of pH 2.88. The concentrations of Pb, V and Ca released into solution were measured using wavelength dispersive X-ray fluorescence spectrometer (XRF) after periods of 3, 5 and 7 days. A Siemens D5005 X-ray diffractometer (XRD), JSM-5310LV scanning electron microscope (SEM) and

JEM-3010 transmission electron microscope (TEM) were used for structural characterisation prior to and after leach tests.

RESULTS AND DISCUSSION

Formation of apatite structure

Apatite is a general term for crystalline materials with the composition of $A_{10}(BO_4)_6X_2$ this study, $(Ca_9Pb_1)(VO_4)_6F_2$, $(Ca_7Pb_3)(VO_4)_6F_2$ and $(Ca_5Pb_5)(VO_4)_6F_2$ pellets were obtained b firing the oxide mixtures at 800°C. SEM results revealed that higher Ca containing pellets tend to form two phases, whereas lower Ca content pellets yielded single-phase apatite. The backscattered electron images of the slightly polished pellet surfaces are shown in Figure 1.

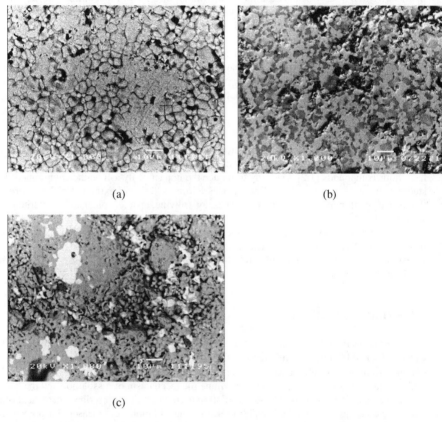

(a)

(b)

(c)

Figure 1. Backscattered electron images from three different pellets: (a) $(Ca_5Pb_5)(VO_4)_6F_2$, (b) $(Ca_7Pb_3)(VO_4)_6F_2$, (c) $(Ca_9Pb_1)(VO_4)_6F_2$. The bright regions in (b and (c) have higher Pb concentrations than the dark regions.

X-ray diffraction analysis revealed that the apatite structure formed using these heat treatment conditions. For the two-phase apatite sample, no split diffraction peaks could be detected as the compositional variations of these apatites were not large enough to give this effect. The XRD pattern and HRTEM image from apatite $(Ca_7Pb_3)(VO_4)_6F_2$ with lattice parameters $a \approx 10Å$ and $c \approx 7Å$ are given in Figures 2 and 3 respectively. Rietveld simulation and refinement of the XRD pattern has been completed using a two-phase assemblage with each component having slightly different Ca/Pb ratios and lattice parameters [6].

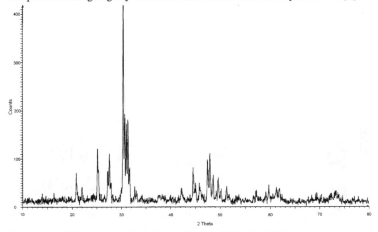

Figure 2. *Diffraction patterns from $(Ca_7Pb_3)(VO_4)_6F_2$ pellet showing apatite structure.*

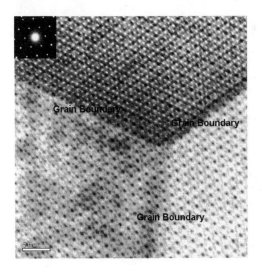

Figure 3. *HRTEM image taken from zone axis [0001] of hexagonal $(Ca_7Pb_3)(VO_4)_6F_2$. The diffraction pattern was taken from the upper part of the image.*

Leaching characteristics of apatite pellet

A leach test was performed on the $(Ca_7Pb_3)(VO_4)_6F_2$ pellet (see Figure 4). The data indicated that leach losses of V and Pb were much slower than that of Ca. A possible explanatic is that some solid phase(s) containing Pb and V might precipitate and the concentrations of the corresponding elements in the leach solution reduced, although it should be noted that precipitates were not observed by SEM. More accurate leach liquid analysis will be conducted for the other apatite pellets and the results will be presented in subsequent paper.

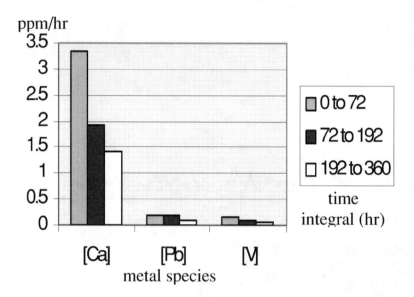

Figure 4. *Concentrations of Ca, Pb and V in the solutions after different leaching intervals.*

The microstructure of the same pellets after leaching was examined again using SEM (see Figure 5). For single-phase apatite $(Ca_5Pb_5)(VO_4)_6F_2$, etch pits were observed on the polished surface. Grain boundaries also dissolved preferentially. It was likely that the etch pits penetrated along particular crystallographic direction(s)/plane(s).

For the two-phase apatite sample $(Ca_7Pb_3)(VO_4)_6F_2$, the etch pits in the dark regions were already connected, so pull-out morphology was observed. The bright regions in the same apatite pellet (Figure 5 (b)) demonstrated similar etch pit morphology as the single-phase pellet (Figure 5(a)). The experimental evidence suggests that the Pb rich areas (bright regions) dissolved much more slowly than the Ca rich areas (black regions). Pb rich areas were more durable. Further work will be carried out with the aid of TEM to better understand the stabilisation mechanism for different toxic species.

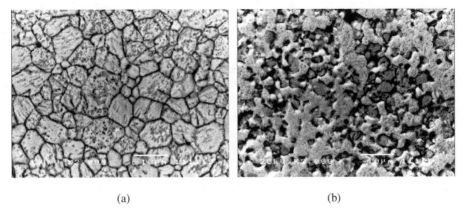

| (a) | (b) |

Figure 5. SEM observation of pellet surfaces after leach test:
(a) $(Ca_5Pb_5)(VO_4)_6F_2$, (b) $(Ca_7Pb_3)(VO_4)_6F_2$.

CONCLUSIONS

$Ca_xPb_{10-x}(VO_4)_6F_2$ apatite structure was formed by firing oxide mixtures at 800°C. During leach testing in acetic acid, Pb and V were extracted more slowly than Ca. Precipitation of Pb- and V-rich phases was not observed with SEM.

The dissolution rate is faster in Ca rich regions than in Pb rich regions. Apatites of high Pb to Ca ratio are more stable. The potential to accommodate more Pb in this type of apatite minimises the addition of CaO, which in turn reduces the volume of waste for final disposal.

ACKNOWLEDGEMENTS

The authors would like to express their thanks to Dr. Karin Laursen for her assistance in XRF analysis and to ETI for supporting this project.

REFERENCES

1. S.Auer, H.J.Kuzel, H.Pollmann and F.Sorrentino, "Investigation on MSW fly ash treatment by reactive calcium aluminates and phases formed", *Cement and Concrete Res.*, **25** (6), 1347 (1995).
2. T.J.White and I.A.Toor, "Synthetic mineral immobilization technology for the stabilization of incinerator ashes", *The 1995 International Incineration Conference*, Seattle, May 1995.
3. J.V.Bothe Jr. and P.W.Brown, "Apatite formation in the $CaO-PbO-P_2O_5-H_2O$ system at 23° ± 1°C", *J. Am. Ceram. Soc.*, **83** (3), 612 (2000).

4. P.Zhang and J.A.Ryan, "Transformation of Pb(II) from cerrusite to chloropyromorphite in the presence of hydroxyapatite under varying conditions of pH", *Environ. Sci. Technol.*, **33** (4), 625 (1999).

5. P.Zhang and J.A.Ryan, "Formation of chloropyromorphite from galena (PbS) in the presence of hydroxyapatite", *Environ. Sci. Technol.*, **33** (4), 618 (1999).

6. T.J.White, Z.L.Dong and B.Wei, "X-ray Rietveld simulation and refinement of calcium lead apatite system used for stabilization of toxic metals", to be published.

Micro/Meso/Nanoporous
Materials: Inorganic-Organic Hybrids

Mat. Res. Soc. Symp. Proc. Vol. 658 © 2001 Materials Research Society

Transition Metal Based Zeotypes: Inorganic Materials at the Complex Oxide – Zeolite Border

Paul F. Henry, Robert W. Hughes and Mark T. Weller
University of Southampton, Highfield, Southampton, SO17 1BJ, UK.

ABSTRACT

The synthesis of compounds of the type $CsA^{III}M^{IV}O_4$ and $(Cs,Rb)A^{II}PO_4$ (A^{III} = Al, Ga, Fe; A^{II} = Ni, Co, Cu; M = Si, Ge, Ti,), using gel decomposition techniques at high temperatures is described. Several new materials in this family have been characterised using powder X-ray diffraction and are shown to adopt the zeolite ABW framework topology. Evidence for the large monovalent cation templating the formation of the ABW framework is presented. Extension of the method to other A^{III} and M^{IV} (A = Mn, Co, B, In; M = Sn, Zr) framework cations, which are known to adopt tetrahedral geometry in other structures, has proved unsuccessful.

INTRODUCTION

The existence of zeolites in the Li_2O - Al_2O_3 - SiO_2 - H_2O system, including the ABW structure type, was first reported almost 50 years ago by Barrer [1]. However, the first structure determination of Li-ABW ($LiAlSiO_4.H_2O$), by Kerr, using powder X-ray diffraction techniques, was not carried out until the early seventies [2]. The structure was later confirmed by single crystal X-ray diffraction studies [3] and the hydrogen (deuterium) positions determined from powder neutron diffraction studies [4]. The structure can be described as being made up of 4-, 6- and 8-membered rings, based on AlO_4 and SiO_4 tetrahedra arranged in accordance with Löwenstein's rule, with the largest channels lying along a two fold screw axis. In Li-ABW this main channel is represented by the c axis and the Li^+ counter cations occupy sites within these channels along with any zeolitic water.

The zeolite ABW topology, Figure 1, is adopted by a wide range of materials, demonstrating a considerable range of chemical compositions, some of which have been tabulated by Stucky et al [5]. Examples of systems with monovalent (T^+) / hexavalent (T^{6+}) [6,7], divalent (T^{2+}) / pentavalent (T^{5+}) [8,9] and trivalent (T^{3+}) / tetravalent (T^{4+}) [10,11] tetrahedral centres have been characterised each with a monovalent counter cation (e.g. group I, Tl^+ and NH_4^+ ions). Two fluoroberyllates have also been found to adopt this topology further expanding the compositional limits of the structure type [12,13].

The highest symmetry that is found for ABW type framework topology is Imma, however, to date only one example has been characterised, $CsAlTiO_4$ [14]. In this case the framework cations Al and Ti are statistically disordered, probably a result of the similar ionic radii of tetrahedral Al^{3+} and Ti^{4+} (0.39 Å and 0.42 Å respectively) [15]. The majority of orthorhombic ABW materials adopt the space group $Pna2_1$, due to framework cation ordering, and monoclinic materials adopt the space group $P2_1$. Several other space groups have been found but only when the counter cation is caesium. In the cases of monoclinic symmetry the angle β deviates only slightly from orthogonal (examples range from 90 ° to 91.03 °).

The non-centrosymmetric nature of the ABW framework, coupled with the wide range of chemical compositions available, indicates that ABWs have potential uses as ferroelectric and

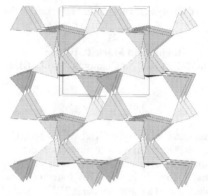

Figure 1. *Polyhedral representation of the ABW framework viewed along the main 8-ring channels.*

non-linear optical materials. Here, we present results on the high temperature synthesis and characterisation, using powder X-ray diffraction techniques of several new members of this family.

EXPERIMENTAL

Polycrystalline samples of $CsAMO_4$ (A = Al, Ga, Fe; M = Si, Ge, Ti) were prepared as follows. Stoichiometric quantities of the required starting materials from LUDOX (40% by wt. SiO_2 in water, Aldrich), $Al(NO_3)_6.9H_2O$ (98+%, Aldrich), Ga_2O_3 (99.999%, Strem), GeO_2 (99.98%, Avocado), TiO_2 (99.9%, Aldrich) and $Fe(NO_3)_6.9H_2O$ (99.9%, Aldrich) were added to 50 ml warm 2M HNO_3. A 10 % stoichiometric excess of CsOH (50% by weight solution in water, Aldrich) was then added to the solution, in order to compensate for the high volatility of caesium salts at the intermediate temperatures used in the experimental procedure. Ethanol (100ml) was added under constant stirring followed by concentrated (.880 specific gravity) ammonia solution, added dropwise, to induce gel formation. The mixture was heated to dryness, with constant stirring, over a period of 12 hrs. The resultant solid was then partially decomposed in an alumina crucible at a temperature of 250 ºC for 12 hrs. The powder obtained was thoroughly ground and heated for a further 16 hours at 600 ºC, then 850 ºC and finally 1000 ºC, except in the case of gallogermanate, $CsGaGeO_4$, where this caused melting and so the maximum temperature used was 850 °C. After each heat treatment a powder X-ray diffraction (PXD) pattern was collected. The final annealing was repeated until there were no observable changes in the PXD pattern.

$(Cs,Rb)A^{II}PO4$ (A = Ni, Co, Cu) materials were prepared using a similar route. Stoichiometric quantities of $A(NO_3)_2.6H_2O$ (Aldrich, 99.9%), $(NH_4)_2HPO_4$ (Aldrich, 99%) and CsOH or RbOH (Aldrich, 50% wt. solution in water) were dissolved in 2M nitric acid. Ethanol (100ml) was added to the solution and the pH raised by addition of .880 ammonia solution until precipitation occurred. The gel formed was then heated to dryness on a hotplate. Further decomposition of the ethanolic precursor was achieved by heating at 225 °C for 12 hours. The resulting powder was ground thoroughly and annealed at 500 °C and then 750 °C, each for 12 hours. Unreacted soluble starting material was then removed by washing with deionised water. A final 12 hour anneal at 750 °C produced the end product.

CHARACTERISATION

All powder X-ray diffraction was carried out using a D5000 Siemens diffractometer ($Cu_{K\alpha 1}$ radiation, λ = 1.54056 Å) operating in reflection geometry. Initial characterisation involved determination of unit cell parameters, using data collected over the 2θ range 10 - 60° with a step size of 0.02°, by indexing using the PC program TREOR90 [16]. Once the unit cell type, derived from systematic absences, and lattice parameters had been calculated, PXD pattern simulations were performed using model ABW atomic positions [6], which gave reasonable agreement in all cases. Data were then collected on each compound for Rietveld analysis, using the GSAS suite of programs [17], over a period of 16 hrs over the 2θ range 10 – 110º using a step size of 0.02º.

In view of the potentially porous nature of the ABW structure, water content was measured by thermogravimetric analysis (TGA) using an STA 1500 TGA/DSC between 20 °C and 1000 °C. Several samples were chosen for elemental analysis to determine the stoichiometric ratios of the framework and extra-framework cations. Measurements were carried out on a JEOL JSM-6400 analytical scanning microscope equipped with a TRACOR series II energy dispersive X-ray analysis system.

RESULTS AND DISCUSSION

$CsA^{[III]}MO_4$ systems

The initial PXD patterns, collected after treatment of the gels at 600 ºC, showed no discernible Bragg reflection. The PXD patterns collected after annealing at 850 ºC indicated a crystalline phase to be present, although the Bragg reflections were weak and broad. Repeated annealing at 1000 ºC gave highly crystalline materials with sharp Bragg reflections in all cases except the gallogermanate, $CsGaGeO_4$. Table I shows the results obtained from the series of reactions studied; lattice parameters, unit cell volume, space group and colour are summarised. References show the materials that have been previously reported in the literature with the zeolitic ABW topology and in these cases the results obtained in this work compare well to previous studies. Several new ABW type materials were synthesised including the first examples of the ABW topology containing iron, previously communicated [18]. Detailed account of the structural determinations can be found in the next section.

It proved impossible to synthesise the caesium gallogermanate ABW analogue by high temperature methods, with melting to form a glassy material the most common outcome. This result is consistent with previous work on the synthesis of gallogermanate framework materials by Stucky et al. [19] using hydrothermal techniques. When a powdered sample was prepared, by annealing carefully at 850 °C, the PXD pattern could not be assigned to any known structure type and autoindexing programs failed to yield consistent results.

Attempts to extend the series, $CsAMO_4$, to other A^{3+}/B^{4+} (A = Mn, Co, B, In; M = Sn, Zr) cations failed to produce ABW materials, using the gel decomposition technique. Hydrothermal techniques were also investigated in an attempt to synthesise framework substituted ABWs. However, these were unsuccessful in all cases with analcime type materials ($CsAM_2O_6$ where A is the 3+ cation and B the 4+ cation) the most usual product. In the gallogermanate case, single crystals were produced and characterised by single crystal X-ray diffraction. The product was found to be identical to $CsGaGe_2O_6$, previously published by Stucky et al [19].

Ion exchange reactions (with the other group I cations) were also attempted on the new frameworks using both solution and melt routes. It was found that ion exchange reactions in solution led to no observable changes in the PXD pattern and no change in mass of the material indicating that no reaction had taken place. The same result was found with nitrate melt routes (temperatures ranged from 200 °C to 300 °C) and after washing out the excess nitrate no change in the framework was found from PXD data. High temperature melt reactions (700 °C) using excess NaCl caused the frameworks to decompose to the constituent oxides as confirmed from PXD data. To date it has proved impossible to successfully ion exchange any ABW frameworks containing Cs^+ as the counter cation. Therefore, we believe that the Cs^+ acts as a templating agent in the synthesis of ABW frameworks in the same way that amines template other framework types. However, in these cases the templating agent is not carbon based and so cann be removed by pyrolysis. We are currently synthesising rubidium analogues in the hope that rubidium is large enough to template a wide range of ABW frameworks but small enough to allow ion exchange reactions to be performed while keeping the framework intact.

Rietveld Powder X-ray Diffraction Profile Refinements

For each successful synthesis Rietveld refinements were carried out in the orthorhombic space groups $Pc2_1n$ and Imma. $Pc2_1n$ (a non-standard setting of $Pna2_1$) is the common space group found in orthorhombic ABW materials that have ordered frameworks, whereas Imma occurs when the framework atoms are disordered. The initial models used were those of $LiAlSiO_4$ [2] and $CsAlTiO_4$ [14] respectively for the two possibilities. In each case the thermal parameters of the oxygen sites (4 in $Pc2_1n$ and 3 in Imma) were constrained to be equivalent. Soft constraints were also entered for the T - O bond lengths (standard deviation of 0.005 Å) using the ionic radii [15]. These constraints were necessary due to the domination of the X-ray diffraction intensities by caesium. Refinements converged in all cases and by comparison of the fits for the two possibilities and inspection of the profile the final space groups were assigned, given in Table I together with the final profile fit factors. The derived atomic positions for $CsGaSiO_4$ ($Pc2_1n$) and $CsGaTiO_4$ (Imma) are given in Tables II and III respectively.

As stated in the introduction, only one example of an ABW with the space group Imma has been discovered to date, $CsAlTiO_4$. The reason given for the disorder was the relatively small difference in ionic radii of the two framework cations (approximately 0.03 Å). From Table I it can be seen that $CsGaTiO_4$ also adopts the disordered framework structure (Imma). The ionic radii of Ga and Ti are 0.47 Å and 0.42 Å respectively and so this result is somewhat unexpected as there are several examples in Table I of materials that have a closer match of framework ionic radii that adopt the ordered framework space group $Pc2_1n$. In order to further probe the order / disorder in the frameworks of these materials powder neutron diffraction investigations are currently in progress.

$CsA^{II}PO_4$ (A^{II} = Ni, Co, Cu) systems.

Pure materials of the compositions $CsCoPO_4$ and $CsNiPO_4$ were obtained and the characterisation of the former has been published [20]. The latter has an intense purple blue colour and the powder X-ray diffraction pattern indicates that it also adopts an ABW type framework structure though structure refinement has thus far proved impossible. $RbCoPO_4$ is already noted in the literature [19] and the structure of the rubidium $RbNiPO_4$ has recently been

Table I. Cell data and colour of framework substituted ABWs.

Compound	a / Å	b / Å	c / Å	Volume / Å^3	Space Group	R_{wp} / %	Colour
CsAlSiO$_4$[11]	9.4397(4)	5.4372(2)	8.8957(3)	456.58(4)	Pc2$_1$n	11.01	White
CsAlTiO$_4$[14]	8.9811(4)	5.7351(3)	9.9548(4)	512.75(4)	Imma	10.83	White
CsAlGeO$_4$	9.4717(5)	5.4997(3)	9.2610(5)	482.42(5)	Pc2$_1$n	12.25	White
CsGaSiO$_4$	9.3294(2)	5.4293(1)	9.2450(2)	468.28(2)	Pc2$_1$n	12.20	White
CsGaTiO$_4$	9.1470(3)	5.7761(2)	9.9344(4)	524.87(3)	Imma	11.75	White
CsFeSiO$_4$[18]	9.5858(4)	5.5538(3)	9.0476(4)	481.67(4)	Pc2$_1$n	6.00	Yellow
CsFeTiO$_4$	9.9799(4)	5.8050(3)	9.1497(4)	530.07(4)	Pc2$_1$n	7.10	Brown

Table II. Atomic positions and thermal parameters from CsGaSiO$_4$.

Atom	Site	X	Y	Z	Occupancy	$U_i/U_e \times 100$ / Å^2
Cs	4a	0.1969(2)	0.443(2)	0.4989(3)	1.0	2.69(6)
Ga	4a	0.0856(6)	-0.0602(28)	0.1781(5)	1.0	1.98(10)
Si	4a	0.4148(13)	-0.0626(29)	0.3126(9)	1.0	1.98(10)
O1	4a	0.1025(15)	-0.026(5)	-0.0147(7)	1.0	2.91(29)
O2	4a	-0.0381(20)	-0.2907(25)	0.242(4)	1.0	2.91(29)
O3	4a	0.0277(19)	0.2341(23)	0.246(4)	1.0	2.91(29)
O4	4a	0.2626(12)	-0.1370(24)	0.2409(24)	1.0	2.91(29)

Table III. Atomic positions and thermal parameters from CsGaTiO$_4$.

Atom	Site	X	Y	Z	Occupancy	$U_i/U_e \times 100$ / Å^2
Cs	4e	0	¼	0.20545(29)	1.0	2.82(13)
Ga	8I	0.1941(4)	¼	0.5797(4)	0.5	2.40(18)
Ti	8I	0.1941(4)	¼	0.5797(4)	0.5	2.40(18)
O1	4e	0	¼	0.5763(27)	1.0	4.9(5)
O2	8f	0.2279(11)	0	0	1.0	4.9(5)
O3	4c	¼	¼	¾	1.0	4.9(5)

published [21]. With copper a pure phase of composition RbCuPO$_4$ was obtained and this material adopts a highly distorted ABW type framework [22]. CsCuPO$_4$ does not form an ABW type structure.

CONCLUSIONS

Gel decomposition techniques have been used to synthesise several new frameworks with the ABW topology including the first examples containing iron, which have been communicated previously. All of the new frameworks were found to adopt the orthorhombic space group Pc2$_1$n (i.e. ordered frameworks) except CsGaTiO$_4$ that was found to be only the second example of an ABW with a disordered framework in the space group Imma. However, the latter results have yet to be confirmed using powder neutron diffraction studies. All of the frameworks were found to be anhydrous, as expected for ABWs with Cs$^+$ as the extra-framework cation. Attempts to extend the ABW topology to frameworks containing Sn(IV), Zr(IV), Co(III), Mn(III), In(III) and B(III) using this method were unsuccessful. Rubidium/caesium late-first-row-transition-metal phosphates are found with ABW or ABW related structures for $CsCoPO_4$, $CsNiPO_4$, $RbCoPO_4$, $RbNiPO_4$ and $RbCuPO_4$.

The theory that the ABW framework is templated by large monovalent cations is supported by the inability to ion exchange the caesium cations out of the framework by either solution experiments, where no reaction is found to occur, or melt reactions, which result in decomposition to the component oxides. These results indicate that the design of zeolite structures with a much broader range of chemical compositions should be possible, with the correct choice of experimental conditions, which may lead to new applications.

ACKNOWLEDGEMENTS

We would like to thank the EPSRC for funding this work under grant number GR/L20955 and a studentship for RWH.

REFERENCES

1. R.M. Barrer, E.A.D. White, *J. Chem. Soc.*, 1267 (1951).
2. I.S. Kerr, *Z. Kristallogr.*, **139**, 186 (1974).
3. E. Krogh Andersen, G. Ploug-Sørensen, *Z. Kristallogr.*, 1986, **176**, 67.
4. P. Norby, A.N. Christensen, I.G. Krogh Andersen, *Acta Chem. Scand.*, **A40**, 500 (1986).
5. P. Y. Feng, X. H. Bu, S. H. Tolbert, G. D. Stucky, *J. Am. Chem. Soc.*, **119**, 2497 (1997).
6. A.I. Kruglik, M.A. Simonov, K.S. Aleksandrov, *Kristallografiya*, 23, 494 (1978).
7. A.I. Kruglik, M.A. Simonov, E.P. Zhelezin, N.V. Belov, *Dokl. Akad. Nauk. SSSR*, **247**, 138 (1979).
8. B.M. Gatehouse, *Acta Crystallogr.*, **C45**, 1674 (1989).
9. G. Nitsch, H. Schaefer, *Z. Anorg. Allg. Chem.*, **417**, 11 (1975).
10. J. M. Newsam, *J. Phys. Chem.*, **92**, 445 (1988).
11. R. Klaska and O. Jarchow, *Naturwiss.*, **60**, 299 (1973) and *Z. Kristallogr.*, **142**, 225 (1975).
12. M.R. Anderson, I.D. Brown, S. Vilminot, *Acta Crystallogr.*, **B29**, 2625 (1973).
13. S.J. Chung, T. Hahn, *Mat. Res. Bull.*, **7**, 1209 (1972).
14. B.M. Gatehouse, *Acta Crystallogr.*, **C45**, 1674 (1989).
15. R.D. Shannon, *Acta Crystallogr.*, **A32**, 751 (1976).
16. P.E. Werner, L. Eriksson, M. Westdahl, *J. Appl. Crystallogr.*, **18**, 367 (1985).
17. A.C. Larson, R.B. Von Dreele, *General Structure Analysis System*, Los Alamos National Laboratory, LAUR 86-748 (1994).
18. P.F. Henry, M.T. Weller, *Chem. Commun.*, 2723 (1998).
19. X. Bu, P. Feng, T.E. Gier, G.D. Stucky, *J. Am. Chem. Soc.*, **120**, 13389 (1998).
20. E.M.Hughes and M.T.Weller, *Dalton Transactions*, **4**, 555 (2000).
21. P. F. Henry, M. T. Weller and R. W. Hughes, *Inorg. Chem.*, **39**, 5420 (2000).
22. P.F.Henry, R.W.Hughes, S.C.Ward and M.T.Weller, *Chem. Comm.*,1959 (2000).

Mat. Res. Soc. Symp. Proc. Vol. 658 © 2001 Materials Research Society

A Computational Study of the Translational Motion of Protons in Zeolite H-ZSM-5

M.E. Franke[1], M. Sierka[2], J. Sauer[2], U. Simon[1]
[1] Aachen University of Technology, Institute of Inorganic Chemistry, Professor Pirlet Str. 1, Aachen, Germany
[2] Humboldt University of Berlin, Institute of Chemistry/Quantum Chemistry, Jägerstr. 10/11, Berlin, Germany

ABSTRACT

The potential energy profile for proton translational motions between two neighboring Al-sites in zeolite H-ZSM-5 is calculated by a combined quantum mechanics-interatomic potential function approach. The potential energies of the six stable intermediate proton positions and the five transition structures along this path show an almost symmetrical trend reaching the maximum in the middle between these sites. Therefore, the maximum barrier will decrease with decreasing SiO_2/Al_2O_3 ratio of the zeolite. For the SiO_2/Al_2O_3 ratio examined (190) an activation energy of ~ 210 kJ/mol is calculated. This is much lower than the energy of deprotonation, about 1300 kJ/mol. The deprotonation energy of an Al-OH-Si bridge is obviously partially compensated by the proton affinity of the Si-O-Si bridges.

INTRODUCTION

The interest in the proton mobility in zeolites is due to the enormous application potential of their protonic form, e.g. in heterogenous catalysis and chemical sensing. The proton motion and the corresponding activation barrier may have a significant influence on the properties and the activity, respectively, of the protons in the catalyst, which are again given by the structural environment. However, up to now no unifying concept has been developed to relate the overall activity of a zeolite catalyst to the framework structure and composition. In this paper an attempt has been made to enhance the understanding of the structure property relation in zeolites by means of theoretical calculations.

In H-zeolites the protons are strongly bound to the anionic, negatively charged zeolite lattice forming a Si-OH-Al bridge, called Brønsted site (in figure 1 the Al atom of the corresponding Brønsted site is illustrated in black color). Proton mobility arises either from on-site jumps between the four oxygen atoms surrounding one aluminum site or from inter-site translational motions between two neighboring Brønsted sites. The on-site mobility has been analyzed in detail by theoretical [1-3] and experimental [4-8] methods. Translational motions have been studied by impedance spectroscopy on H-ZSM-5. For SiO_2/Al_2O_3 ratios between 30 and 1000 activation energies between 89 and 126 kJ/mol have been found [9-11]. This is much less than the energy required to completely remove the proton from the zeolite, about 1300 kJ/mol [12]. Obviously, on its translational path the proton is stabilized by interactions with Si-O-Si framework groups. To obtain specific theoretical estimates of the activation barriers, the structures and energies of intermediate stable states and of transition structures of the proton moving from one Al-site to another are determined by quantum chemical ab initio techniques. To make the calculations on zeolites with their large unit cells feasible, we apply the combined quantum mechanics interatomic potential function method (QM-Pot) [13,14].

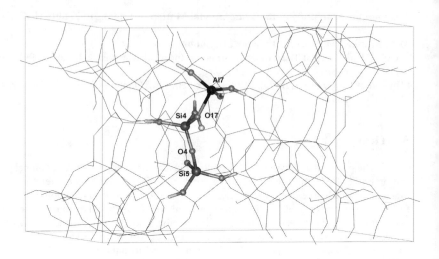

Figure 1. *The 3T cluster model embedded in the ZSM-5 lattice with one Al-atom per unit cell. The QM-Pot converged transition structure is shown.*

COMPUTATIONAL DETAILS

The embedding scheme used [12-14] partitions the entire system (S) into two parts: the inner part (I) usually containing the site in question and the outer part (O). The QM-Pot energy is defined as the sum of the energy of the inner part calculated at quantum mechanical (QM) level and the outer part calculated at the interatomic potential (Pot) level.

The dangling bonds created by partitioning are saturated by link hydrogen atoms (I) which are kept at a fixed distance. The inner part together with the link atoms form the cluster (C). The QM-Pot energy is defined as follows

$$E(S)_{QM-Pot} = E(C)_{QM} + E(S)_{Pot} - E(C)_{Pot} \qquad (1)$$

The QM-Pot calculations are performed using the QMPOT program [14]. As interatomic potential functions a shell-model ion pair-potentials [15] parametrized on density functional results on protonated form of zeolites [16] are used. All potential function calculations employ the GULP [17] program. For the calculation of the transition structures the EVB method is used, for which the necessary coupling parameters have been derived previously [14].

The QM calculations employ the TURBOMOLE [18,19] program and apply the density functional method. We adopt the B3-LYP [20] exchange correlation functional, which is known to yield superior results for hydrogen bonded systems. The basis set is triple-ζ on oxygen and double-ζ on all other atoms. We apply the fully optimized basis sets from Ahlrich's group [21].

In this work for the QM-Pot calculations clusters including three tetrahedra, so called 3T clusters, are embedded into the ZSM-5 lattice. It was shown previously [14] that this cluster size is suitable for calculating the energy barriers for the local (on-site) motion in zeolites and provides a good compromise with respect to computational time. In figure 1 this cluster embedded into the ZSM-5 lattice is shown. It is used for the calculation of the two minima Al7-O17H-Si4 and Si4-O4H-Si5 and the respective transition structure Al7-O17-Si4-O4-Si5 (numbering according to ref. 22).

RESULTS AND DISCUSSION

For the calculation of the stationary points on the potential energy surface relevant for the translational motion of protons we started with H-ZSM-5 having one Al-atom per unit cell, which is equivalent to a SiO_2/Al_2O_3 ratio of 190. The proton starts at the crystallographic Al7-O17H-Si4 position, which is the most stable one in the orthorhombic modification containing one aluminum atom [23]. The proton moves along a T-O-T – chain (T = Si, Al), which connects the two neighboring and crystallographic identical Al-atoms (T7) almost linearly. The two Brønsted sites have a spatial distance of 13.45 Å in this direction. Figure 2 shows this chain of T-O-T sites.

Figure 3 shows the calculated relative energies of the equilibrium and transition structures taking into account the zero point vibrational energy corrections as a function of their crystallographic position. The calculated energies of the six equilibrium proton positions, T-OH-Si connecting the two neighboring Brønsted sites, show an almost symmetrical trend between the neighboring Al-sites reaching its maximum for a proton position in the middle between these sites. The energies of the transition structures show a similar trend, whereas their absolute value depends on the local structure of the respective T-O-Si-O-T unit.

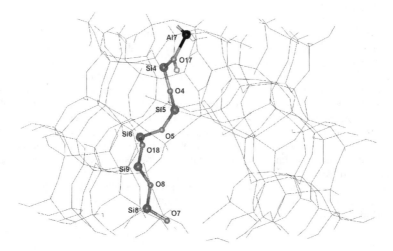

Figure 2. Chain of T-O-T sites along which the proton moves between two crystallographic identical Brønsted sites. The proton is in the initial position

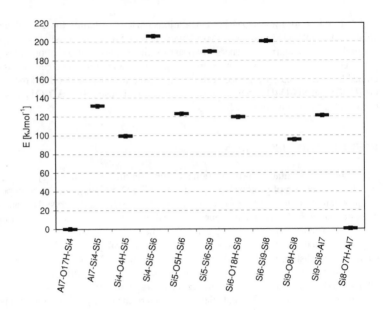

Figure 3. *Energies of minimum and transition structures as a function of their crystallographic position for a proton moving along a Si-O-Si chain*

CONCLUSION

We have calculated the potential energy profile for proton translational motions between two neighboring Al-sites in zeolite H-ZSM-5 by a combined quantum mechanics-interatomic potential function approach. The trend of the calculated energies can be explained by the influence of the Coulomb potential of the two neighboring aluminum sites, which decays with $1/r$. Thus, a decrease of the SiO_2/Al_2O_3 ratio may lead to a decrease of the distance dependent activation energy, as soon as their Coulomb potential starts to overlap, which has already been shown qualitatively in previous work [9,11]. Taking the highest energy of the transition structure as the energy to create mobile charge carriers, the translational motion of the proton may appear with an activation energy of about 210 kJ/mol for a SiO_2/Al_2O_3 ratio of 190. As a consequence, the theoretically determined activation energy for translational proton motion is significantly lower than the predicted deprotonation energy of about 1300 kJ/mol, which is in accordance with the experimental results of the impedance measurements. This indicates that the deprotonation energy of a Brønsted sites is largely compensated by the energy gained by binding the proton to the bridging Si-O-Si groups.

ACKNOWLEDGMENTS

Financial support by the Fond der Chemischen Industrie (FCI) is gratefully acknowledged.

REFERENCES

1. J. Sauer, C.M. Kölmel, J.-R. Hill, and R. Ahlrichs, *Chem. Phys. Letters*, **164**, 193 (1989).
2. J. Sauer, M. Sierka, and F. Haase, ACS Symp., Transition State in Catalysis, Dallas (1998), Series 721, American Chemical Society, Washington, 1999, pp. 358-367.
3. M. Sierka, J. Sauer, *J. Comput. Chem.*, **21**, 1470 (2000).
4. D. Freude, W. Oehme, H. Schmiedel, and B. Staudte, *J. Catal.*, **32**, 137 (1974).
5. H. Ernst, D. Freude, T. Mildner, and H. Pfeifer, in: *Proceedings of the 12th international Zeolite Conference: July 5-10, 1998, Baltimore, Maryland, USA*, ed. M.M.J. Treacy, B.K. Marcus, M.E. Bisher, and J.B. Higgins; Materials Research Society, Warrendale, 1999, Vol.4, pp. 2955-2962.
6. D. Sarv, T. Tuherm, E. Lippma, K. Keskinen, and A. Root, *J. Phys. Chem.*, **99**, 13763 (1995).
7. T. Baba, Y. Inoue, H. Shoji, T. Uematsu, and Y. Ono, *Microporous Mater.*, **3**, 647 (1995).
8. T. Baba, N. Komatsu, Y. Ono, and H. Sugisawa, *J. Phys. Chem. B*, **102**, 804 (1998).
9. M.E. Franke, U. Simon, *Solid State Ionics*, **118**, 311 (1999).
10. M.E. Franke, U. Simon, *Phys. Stat. Solidi (b)*, **218**, 287 (2000).
11. U. Simon, M.E. Franke, *Micropor. and Mesopor. Mat.*, **41**, 1 (2000).
12. U. Eichler, M. Brändle, and J. Sauer, *J. Phys. Chem. B*, **101(48)**, 10035 (1997).
13. U. Eichler, C.M. Kölmel, and J. Sauer, *J. Comp. Chem.*, **18**, 463 (1997).
14. M. Sierka, J. Sauer, *J. Chem. Phys.*, **112(16)**, 6983 (2000).
15. C.R.A. Catlow, M. Dixon, and W.C. Machrodt, in Computer simulations of solids; Lecture Notes in Physics166; C.R.A. Catlow, W.C. Machrodt (Eds), Springer-Verlag, Berlin, q982, p. 130-161.
16. M. Sierka, J. Sauer, *J. Faraday Discuss.*, **106**, 41 (1997).
17. J. Gale, *J. Chem. Soc., Faraday Trans.*, **93**, 629 (1997).
18. R. Ahlrichs, M. Bär, M. Häser, H. Horn, and C. Kölmel, *Chem. Phys. Lett.*, **162**, 165 (1989).
19. O. Treutler, R. Ahlrichs, *J. Chem. Phys.*, **102**, 346 (1995).
20. A.D. Becke, *J. Chem. Phys.*, **98**, 5648 (1993).
21. A. Schäfer, H. Horn, and R. Ahlrichs, *J. Chem. Phys.*, **97**, 2571 (1992).
22. H. Van Koningsveld, J.C. Jansen, and H. van Bekkum, *Zeolites*, **10**, 235 (1990).
23. K.–P. Schröder, J. Sauer, M. Leslie, and C.R.A. Catlow, *Zeolites*, **12**, 23 (1992).

Mat. Res. Soc. Symp. Proc. Vol. 658 © 2001 Materials Research Society

Disordered mesoporous silicates formed by templation of a liquid crystal (L₃)

Abds-Sami Malik, Daniel M. Dabbs, Ilhan A. Aksay, Howard E. Katz[1]
Princeton University, Dept. of Chemical Engineering and Princeton Materials Institute,
Princeton, NJ 08540
[1]Bell Laboratories, Lucent Technologies, Murray Hill, NJ 07974

ABSTRACT

For a wide range of technological applications the need for optically transparent, monolithic, mesoporous silicates is readily apparent. Potential areas of utility include filtration, catalysis, and optoelectronics among many others. This laboratory has previously reported on the synthesis of such materials that are formed through the addition of tetramethoxysilane to a liquid crystal solution of hexanol, cetylpyridinium chloride, and 0.2 M hydrochloric acid, and our investigation into the properties of these materials is a continuing process. We have achieved defect and fracture free material of suitable size (0.5 cm x 3 cm diameter disks) via supercritical drying of the silicate under ethanol or CO_2. The dried materials are remarkably similar to ordinary glass in strength, texture, and clarity. They possess pore volumes of *ca.* 1.0 cm^3/g, with BET surface areas >1000 m^2/g. We can re-infiltrate the dried monolith with hydroxyethylacrylate, a photo-polymerizable monomer, to create an inorganic/organic nanocomposite. There is fracturing upon re-infiltration, but preliminary tests show that the polymerization proceeds despite the mechanical failure. These findings suggest many possible applications for these unique nanocomposites.

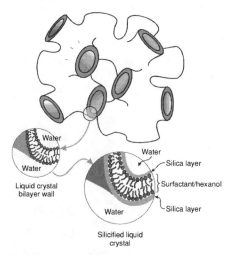

Figure 1: A sketch of the L₃ phase liquid crystal structure, derived from electron microscopy done on freeze fractured samples, as well as theoretical considerations. The pore diameters are uniform, but are randomly interconnected in all three dimensions resulting in short range order with long range disorder. The bilayer forming the liquid crystal can then be templated by a silica layer to capture the original liquid crystal structure as solid silica gel. (Source: McGrath *et al.*, *Langmuir*, vol. 16, **2000,** pp. 398-406)

INTRODUCTION

In recent years, there has been an explosion of interest in mesostructured and mesoporous materials, partly because they are seen to have high potential for applications over a wide range of technological needs. Possible areas of use include filtration, biological separation, catalysis, and patterned thin film formation for use in optoelectronics. The need for optically isotropic, transparent, monolithic material is apparent, especially for optoelectronic applications.

A liquid crystalline phase, termed the L_3 phase, has been explored for a quasiternary system formed from hexanol, cetylpyridinium chloride, and 1 wt % aqueous NaCl solution.[1] This phase, which is optically isotropic and transparent, can be described as a surfactant bilayer that is convoluted to form a sponge-like structure with randomly distributed mesopores interconnected in all three dimensions (Figure 1). This liquid crystal has also been referred to as the 'sponge phase.'[2] The addition of tetramethoxysilane to this template leads to the formation of optically isotropic, transparent silica, which possesses the original L_3 structure. These mesoporous silicas are large monoliths (several cubic centimeters) that conform to the shape of the reaction vessel.

Because of these properties, we propose that the silica L_3 could be used as a basis for constructing nanocomposite materials. It is possible to impregnate the mesopores with a photo-active monomer. Selective polymerization (with a laser) could then be used to 'write' patterns within the silica/monomer composite. Unique optoelectronic properties may be obtained.

In this paper we report on the synthesis, characterization, and processing of mesoporous silica formed by templation of an L_3 phase liquid crystal. We use supercritical ethanol or supercritical carbon dioxide to dry the silica. We then soak the dry silica in photopolymerizable monomers. Although some fracturing is observed in the test material, we can show that photopolymerization does take place within the pores of this material.

EXPERIMENTAL DETAILS

The synthesis procedure for templated silica has been reported before [3,4], and is briefly summarized here for completeness. An L_3 liquid crystal solution was made by first dissolving cetylpyridinium chloride and hexanol in 0.2 M hydrochloric acid (pH 0.7). The liquid crystal

Figure 2: Samples of dried silica, (top) super-critical ethanol, (bottom) supercritical carbon dioxide.

was then silicified by adding tetramethoxysilane (TMOS), cast into a 1 inch diameter petri dish, and allowed to age one to two days to form a gel body.

After forming a gel, the silica disk was placed directly into a 50%/50% methanol/water solution (by volume). After several washings (one day each) with progressively higher concentrations of methanol, the gel disk underwent slight syneresis, which allowed it to be removed from the petri dish. Then the washings were continued with progressively higher concentration of ethanol until the disk was in (nominally) 100% ethanol.

Supercritical drying in ethanol was accomplished by heating (260°C) the silica disk in a stainless steel, sealed pressure vessel partially filled with ethanol. The supercritical ethanol was extracted from the chamber by slowly opening a valve and allowing the hot gas to escape. (**Caution:** Work with supercritical fluids is extremely dangerous, and adequate precautions must be taken to ensure worker safety.) The pressure was eventually brought to ambient, and then the chamber was allowed to cool slowly to room temperature. Supercritical carbon dioxide was also used to dry the silica disk, using an automatic, computer-controlled extraction apparatus. In both cases, dried silica disks were obtained (Figure 2).

Disks were re-infiltrated with solvent, such as ethanol, simply by placing into the infiltrating liquid. This uncontrolled reinfiltration usually caused the monolith to granulate, forming pieces less than a millimeter in dimension. But, sometimes, larger portions measuring as much as 1 cm^2 would survive reinfiltration with liquid. These pieces were recovered and placed into liquid monomer for further studies.

Pieces that had been soaked with the acrylate (hydroxyethyl acrylate) monomer were then placed into more monomer solution with photo-initiator. The piece was taken out of the solution, placed in a Schlenk flask under flowing nitrogen, and exposed to visible light from a strong reading lamp. Infrared spectroscopy done on these pieces and suggested that polymerization had indeed occurred (Figure 3).

Surface area and bulk porosity measurements were done on the dried silica. For materials dried with supercritical ethanol, results indicated that surface areas ranged from 1200

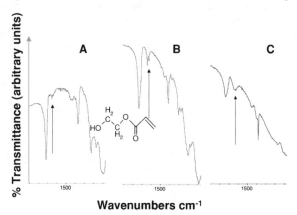

Figure 3: Infrared spectra of dried silica (A), the photomonomer (hydroxyethyl acrylate) in the porous silica (B), and the polymerized acrylate within porous silica. The change in the IR resonance around 1600 cm^{-1} indicates that polymerization has occurred. The silica did not appear to fracture during the polymerization.

cm^3/g to 1400 cm^3/g, and that approximately two thirds of the total volume was composed of pores. Silica dried using supercritical carbon dioxide had less surface area and less pore volume
Thermogravimetric analysis (TGA) also showed that silica dried in supercritical ethanol did not exhibit weight loss until after 250°C, whereas samples dried in supercritical CO$_2$ exhibited weight loss starting from 50°C.

RESULTS AND DISCUSSION

Pores of uniform dimension formed by surfactant bilayers characterize the L$_3$ liquid crystal. The TMOS undergoes a series of acid catalyzed hydrolysis and condensation steps after being added to the liquid crystal. It is then templated by the surfactant to capture the liquid crystal structure in solid form (Figure 4). After gelation, the organic surfactant is trapped in the silica matrix. Therefore, supercritical ethanol is required not only to dry the silica, but also to extract the surfactant. We found that supercritical carbon dioxide treatment did not succeed in removing the surfactant. This was apparent in the TGA experiments as well as surface area measurements and could be attributed to the poor solvating power of supercritical carbon dioxide as compared to supercritical ethanol.

The dried silica monoliths could be impregnated with solvent and monomers, but they frequently cracked or pulverized. The loss of the monolithic structure is thought to be due to the capillary pressures that must exist within the nanopores of the silica when liquid infiltrates the sample. However, we were able to obtain some pieces that did not crack upon reinfiltration. Specifically, porous silica that had been soaked in monomers were subjected to polymerization. Upon polymerization, further cracking was observed in the monoliths but infrared spectroscopy revealed that polymerization did occur in these composites.

CONCLUSIONS

The porous silicates that have been impregnated with photopolymerizable monomers are a novel nanocomposite material. We believe that they are unique and could eventually find applications as novel opto-electronic materials. Furthermore, the demonstration that the porous silica can be impregnated with solvent opens up possibilities for constructing other nano-composite materials. The pores could be impregnated with metal-alkoxides, such as vanadium

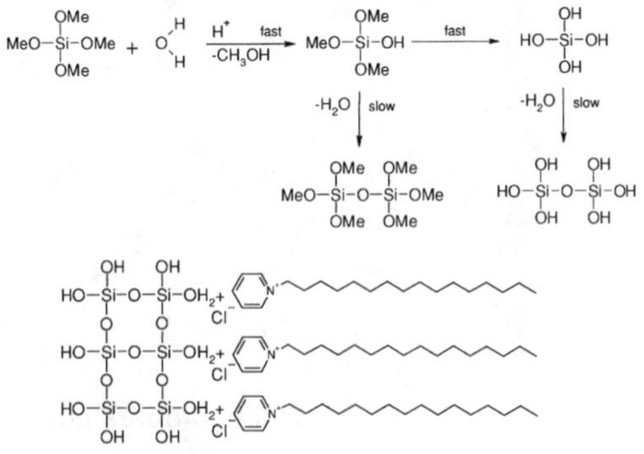

Figure 4: TMOS undergoes a series of acid catalyzed hydrolysis and condensations steps. Then, it is templated by the cetylpyridinium chloride as shown in the lower half of the figure. Only a small portion of the bilayer formed by the surfactant is shown.

alkoxide, to form a silica supported vanadium oxide for catalysis. Such a material has been reported before [5], but only in mesoporous silica powders. Our material offers the promise of fabricating larger monoliths. Metals could also be introduced through electroless deposition, which may have application as supercapacitors. Similarly the pores could be functionalized with proteins and other biomaterials for use as sensors. These and other applications for L_3 templated silica are being actively pursued in our laboratories.

ACKNOWLEDGMENTS

The authors would like to thank Prof. George W. Scherer for helpful discussions. We would also like to thank Prof. Norbert Mulders for suggesting supercritical alcohol extraction and his guidance in fabricating our supercritical extraction apparatus used in this study.

REFERENCES

1.   McGrath, K. M. *Langmuir* **1997**, *13*, 1987-1995.
2.   Schwarz, B.; Mönch, G.; Ilgenfritz, G.; Strey, R. *Langmuir* **2000**, *16*, 8643-8652.
3.   McGrath, K. M.; Dabbs, D. M.; Yao, N.; Aksay, I. A.; Gruner, S. M. *Science* **1997**, *277*, 552-556.
4.   McGrath, K. M.; Dabbs, D. M.; Yao, N.; Edler, K. J.; Aksay, I. A.; Gruner, S. M. *Langmuir* **2000**, *16*, 398-406.
5.   Morey, M.; Davidson, A.; Eckert, H.; Stucky, G. *Chem. Mater.* **1996**, *8*, 486-492.

Mat. Res. Soc. Symp. Proc. Vol. 658 © 2001 Materials Research Society

Frameworks of Transition Metals and Linkers with Two or More Functional Groups

Slavi C. Sevov
Department of Chemistry and Biochemistry
University of Notre Dame
Notre Dame, IN 46556, USA

ABSTRACT

Hybrid inorganic/organic materials with open-framework or layered structures are known for many transition metals linked by functionalized organic molecules such as organic diphosphonates, polycarboxylates, polynitriles, etc., species with more than one equivalent functional groups. We have studied the effect of pH on such a system of cobalt-methylenediphosphonate and report three new compounds, $Na_3Co[(O_3PCH_2PO_3)(OH)]$, $Na_2Co(O_3PCH_2PO_3)\cdot H_2O$, and $Co_2[(O_3PCH_2PO_3)(H_2O)]$, that form at very basic, moderately basic, and acidic conditions, respectively. More interesting structural chemistry should be expected from linkers with two or more different functionalities. Both the carboxylic and phosphonic groups in carboxyethylphosphonic acid are used to coordinate to cobalt or calcium atoms in the new compounds $Co_3(O_3PCH_2CH_2COO)_2\cdot 6H_2O$ and $Ca(O_3PCH_2CH_2COOH)\cdot H_2O$. Taking one more step further in complexity we have also studied linkers with three different functional groups, phosphonated amino acids. The structures of two new compounds, $Zn(O_3PCH_2CH(NH_3)COO)$ and $Zn(O_3PCH_2CH_2CH(NH_3)COO)$, are three-dimensional frameworks made of zinc coordinated by both the carboxylic and phosphonic ends of the organic molecules. The amino groups are protonated and terminal in the voids of the frameworks.

INTRODUCTION

Our interest in multifunctional linkers developed from research in transition metal borophosphates where tetrahedral borate and phosphate groups are connected via an oxygen corner [1], and the idea to replace that oxygen atom with another atom or group. One possibility is an organic group and to use a molecule with two functional groups, boronic and phosphonic. Unfortunately, there seem to be no such linkers synthesized yet, and our own efforts to synthesize such species have been unsuccessful so far. Thus, instead we looked at what other functional groups can "replace" the boronic end of the linker in order to study the potential frameworks built of such linkers and transition metals. We studied linkers with one phosphonic end and either phosphonic or carboxylic groups, or a double functionality of carboxylic- and amino-groups at the second end. In general, multifunctional ligands with two equivalent groups have attracted much interest in recent years mainly due to the possibility to build infinite frameworks by coordination to metal centers [2,3]. The resulting hybrid inorganic-organic compounds have the potential for use in shape recognition [4], and perhaps in stereo selectivity when chiral linkers are used [5,6]. Linkers with two or more different functional groups are even more interesting since they can provide different modes of coordination, different stability, and different structural dimensionality and motifs. Extended structures of such organic moieties with two different end groups that are both coordinated to the metal centers are very rare. Our ultimate goal is to achieve layered compounds with noncoordinated carboxylic and amino groups that are potential candidates for stereo-selective intercalation due to the chiral α-carbon. In the pursuit of this goal, layered inorganic compounds such as zirconium phosphate have been intercalated with amino acids whe the latter interact electrostatically or van der Waals-like with the host [7].

EXPERIMENTAL DETAILS

All compounds were synthesized hydrothermally in autoclaves at temperatures of 120 to 16 °C from the carbonate or nitrate of the metal, the corresponding organic bi- or tri-functional acid and eventually NaOH for adjusting the pH. The resulting solids were filtered, washed with wate ethanol and acetone, and were dried at room temperature. The structures were determined from single crystal X-ray diffraction experiments.

DISCUSSION

The effect of the pH of the systems on the structure of the resulting products was studied in detail for the cobalt-methylenediphosphonate system [8,9]. Cobalt was usually the metal of choice mainly because of its coordination versatility, i.e. capability to coordinate in tetrahedral, square-pyramidal, trigonal-bipyramidal, and octahedral geometries. At pH > 13, the system produces the 1-dimensional structure of $Na_3Co(O_3P-CH_2-PO_3)(OH)$ made of chains of corner-sharing Co-centered octahedra where the diphosphonate is coordinated via four of its six oxygen atoms [9]. The chains are oriented towards each other with their hydrophobic sides, while the sodium cations separate them at the hydrophilic sides. Lowering the pH to between 12 and 12.5 produces the 2-dimensional $Na_2Co(O_3P-CH_2-PO_3) \cdot H_2O$ made of wavy layers of cobalt-centere square pyramids connected via the diphosphonate groups [9]. The latter use five of their oxygen atoms for coordination. The inter-layer space is intercalated by water molecules and sodium cations. There is good hydrophobic-philic segregation. At even more acidic conditions, a compound with a 3-dimensional structure is formed, $Co_2(O_3P-CH_2-PO_3)(H_2O)$ [8]. It contains both tetrahedrally- and octahedrally-coordinated cobalt, and the diphosphonate is coordinated via all six oxygen atoms. The segregation of the hydrophobic and hydrophilic parts of the linkers results in the formation of channel-like openings in the structure. The very important role of the pH of such systems, unfortunately often ignored, is quite evident from this detailed study.

We also studied the effect of the length of the linker's chain, i.e. the distance between the two phosphonate groups, and found that for two and more methylene groups the structures are alway of the 2-dimensional layered type [9]. For example, the structure of $NaCo_2(O_3P-CH_2CH_2CH_2-PO_3)(OH)$ is made of inorganic layers of cobalt-centered trigonal-bipyramids and octahedra and the phosphonic groups that are covalently connected by the linkers' chains. The segregation of hydrophobic parts, the methylene groups of the linkers, and hydrophilic parts, the metal-phosphonate layers, is even more apparent, i.e. the structure is an inorganic-organic hybrid of alternating hydrophobic and hydrophilic layers.

The second type of linker studied was with a carboxylic group in addition to the phosphonic group, i.e. transition metal phosphonate-carboxylates. At the time we looked at such systems, only two compounds were known, both zinc-based and both with phosphono-propionic acid as a linker [10,11]. We studied instead cobalt and calcium with the same linker. The resulting compounds are both three-dimensional and with openings that are collapsed due to the extensive hydrogen-bonding (Figure 1). The cobalt compound, $Co_3(O_3P-CH_2CH_2-COO)_2\cdot6H_2O$, contains linear trimers of edge-sharing cobalt-centered octahedra, and the trimers are connected in a framework by the linkers (Figure 1a) [12]. The compound readily loses water at 150 °C and becomes blue, which indicates tetrahedral cobalt. The calcium analog, $Ca(O_3P-CH_2CH_2-COOH)\cdot H_2O$, has dimers instead of trimers of edge-sharing polyhedra, and these polyhedra are pentagonal bipyramids instead of octahedra due to the larger calcium that centers them (Figure 1b).

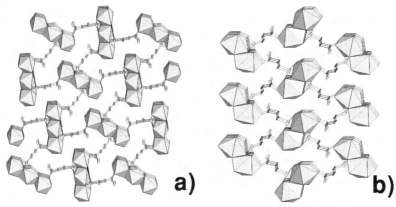

Figure 1. *The structures of: a) $Co_3(O_3P-CH_2CH_2-COO)_2\cdot6H_2O$ with trimers of edge-sharing Co-centered octahedra, and b) $Ca(O_3P-CH_2CH_2-COOH)\cdot H_2O$ with dimers of edge-sharing Ca-centered pentagonal bipyramids. The polyhedra are connected by the phosphono-propionate linkers.*

In addition to the two reported zinc-phosphonocarboxylates, there was also a compound with a linker with phosphonic and amino ends reported in the literature [13]. This indicated that amino groups can also be used as functionalities in compounds with transition metal centers. Thus, we took one more step in complexity and studied phosphono-amino-carboxylic acids as linkers, i.e. groups with three different functionalities. The two acids that were studied were 2-amino-3-phosphono-propionic and 2-amino-4-phosphono-butyric [14], and both formed compounds with zinc where all three phosphonic oxygens and one carboxylic oxygen coordinate to zinc atoms (Figure 2). Also, the amino-groups in both compounds are protonated while the carboxylic ends are deprotonated.

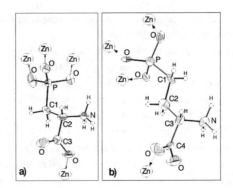

Figure 2. *Shown are the two different amino-acids: a) 2-amino-3-phosphono-propionic and b) 2-amino-4-phosphono-butyric, and the way they are found to coordinate to zinc atoms.*

The structure of the first compound, $Zn(O_3P\text{-}CH_2\text{-}CH(NH_3)\text{-}COO)$ (**1**), is a hybrid of alternating inorganic and organic layers (Figure 3). The inorganic layers, made of oxygen-connected tetrahedra centered by phosphorus and zinc, are linked by the amino acid molecules. The connectivity is achieved via the carbon–phosphorus bond of the phosphonated amino acid and with the coordination of one of the carboxylic oxygens to Zn (Figures 2 and 3). The other oxygen of the latter is noncoordinated as is the protonated amino group of the amino acid. Thus unlike other examples of bi-coordinated organic moieties such as diphosphonates, carboxylate-phosphonates, or poly-carboxylates, the organic species here are highly functionalized with hydrophilic groups, i.e. with the $-C=O$ and $-NH_3^+$ functionalities. These functional groups point toward the one-dimensional inter-linker apertures (4.36 Å O–O diameter) that exist along the *b* direction (Figure 3), and are hydrogen bonded to each other.

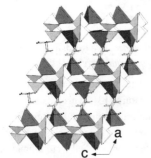

Figure 3. *A general view of the structure of compound **1**, $Zn(O_3PCH_2CH(NH_3)COO)$, along the b axis. The nitrogen atom of NH_3^+ and the non-coordinated oxygen atom of the carboxylic group are shown as filled and open circles, respectively. The darker- and lighter-shaded tetrahedra are centered by Zn and P, respectively.*

The structure of the second compound, $Zn(O_3P\text{-}CH_2CH_2\text{-}CH(NH_3)\text{-}COO)$ (**2**), presents very similar features. Its carboxylic groups are also coordinated with only one oxygen atom, and therefore, the second one is terminal, and the amino groups are again protonated (Figures 2 and 4). Crystallographically, there are two different phosphono-amino ac-as well as zinc atoms in this structure, but they are all nearly identical in coordination, bond distances and angles. Similar to **1**, the structure of **2** (Figure 4) contains phosphonate-zinc inorganic layers that are grafted by the amino acid moieties. Nevertheless, there is an additional feature in this structure, inorganic chains of $-PO_3/ZnO_4$ running along the *a* axis (the viewing direction of Figure 4). The chains are positioned between the inorganic layers and are linked to the two neighboring layers by the amino acids. This leads to increased separation between the layers, 17.5 Å (half the *b* axis), since they are formally connected by the aminoacid-PO_3-ZnO_4-aminoacid composite linker. These linkers are not perpendicular to the layers but rather are tilted, most likely due to the rigidity of the carbon angles. This gives the impression that the layers are offset or shifted along the *c* direction with respect to each other. Long narrow gaps of about 19.3 Å (measured in O–O distance) are formed between the composite linkers. All pendant amino groups and carboxyl oxygen atoms point into these gaps and are extensively hydrogen bonded through that space (3 bonds per amino group and one bond per carboxylic oxygen). Both structures are centrosymmetric and contain racemic mixtures of the corresponding amino acids.

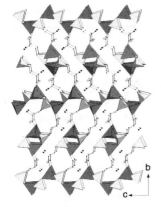

Figure 4.   *A general view of the structure of compound 2, $Zn(O_3PCH_2CH_2CH(NH_3)COO)$, along the a axis. The nitrogen atom of $-NH_3{}^+$ and the non-coordinated oxygen atom of the carboxylic group are shown as filled and open circles, respectively. The darker- and lighter-shaded tetrahedra are centered by Zn and P, respectively.*

Compounds **1** and **2** are thermally stable up to about 350 °C in a flow of air, according to TGA measurements. At this temperature, in a single sharp step, they lose the organic fraction in the form of CO_2, NH_3 and H_2O, and end up as $ZnHPO_4$. The IR spectra of the compounds (in KBr pressed pellets) show the antisymmetric and symmetric vibrations of the carboxylic groups. For compound **1** they are observed at 1627 and 1414 cm^{-1}, respectively, while doublets at 1636 and 1621 cm^{-1} for the antisymmetric and at 1446 and 1429 cm^{-1} for the symmetric vibrations are observed for compound **2**. The splitting is due to the two crystallographically different amino acids, i.e. the two slightly differently bonded carboxylic groups in the compound. The vibrations of the phosphonic groups are observed in the region 900-1200 cm^{-1}, as expected.

CONCLUSIONS

The existence of these framework compounds proves that it is possible to use many differen functionalities to coordinate to transition-metal centers and to link them in a variety of different ways. The next step is to find a way to anchor only the phosphonic group of the aminoacids in ; layer- or channel-type compound, and use the two functionalities, $-COOH$ and $-NH_3$, and the chiral α-carbon for molecular recognition purposes. Such supported amino acids can be used fo a number of organic reactions including formation of peptide bonds and longer chains of differen residues. Compounds 1 and 2 presented here are not layered or microporous, and guest molecules can not access the inside of the structures, but they clearly have the right structural motifs and are on the right track towards more open frameworks.

ACKNOWLEDGMENTS

The author thanks the National Science Foundation (DMR-9701550) for the financial support and for the Summer Program in Solid State Chemistry, and the undergraduate students Doug Lohse, Anne Distler, Stacy Hartman, and Carlos Cruz for much appreciated technical help

REFERENCES

1. S. C. Sevov, *Angew. Chem. Int. Ed. Engl.*, **35**, 2630 (1996).
2. *Comprehensive Supramolecular Chemistry*, ed. Jerry L. Atwood et al. (Pergamon, 1996).
3. A. Clearfield, In *Progress in Inorganic Chemistry*, ed. K. D. Karlin (Wiley, 1998), vol. 47, 371-510.
4. G. Cao and T. E. Mallouk, *Inorg. Chem.*, **30**, 1434 (1991).
5. G. Cao, M. E. Garcia, M. Alcala, L. F. Burgess and T. E. Mallouk, *J. Am. Chem. Soc.*, **114** 7574 (1992).
6. T. E. Mallouk and J. A. Gavin, *Acc. Chem. Res.*, **31**, 209 (1998).
7. Y. Ding, D. J. Jones, P. Maireles-Torres and J. Roziere, *Chem. Mater.*, 1995, **7**, 562.
8. D. Lohse and S. C. Sevov, *Angew. Chem. Int. Ed. Engl.*, **36**, 1619 (1997).
9. A. Distler, D. L. Lohse and S. C. Sevov, *Dalton Transact.*, 1805 (1999).
10. S. Drumel, P. Janvier, P. Barboux, M. Bujoli-Doeuff and B. Bujoli, *Inorg. Chem.*, **34**, 148 (1995).
11. S. Drumel, P. Janvier, M. Bujoli-Doeuff and B. Bujoli, *New J. Chem.*, **19**, 239 (1995)
12. A. Distler and S. C. Sevov, *Chem. Commun.*, 959 (1998).
13. S. Drumel, P. Janvier, D. Deniaud and B. Bujoli, *J. Chem. Soc., Chem. Commun.*, 1051 (1995).
14. S. J. Hartman, E. Todorov, C. Cruz and S. C. Sevov, *Chem. Commun.*, 1213 (2000).

Synthesis, New Methods,
New Materials

Mat. Res. Soc. Symp. Proc. Vol. 658 © 2001 Materials Research Society

Synthesis of Large ZSM-5 Crystals under High Pressure

Xiqu Wang and Allan J. Jacobson
Department of Chemistry and Materials Research Science and Engineering Center,
University of Houston, Houston, Texas 77204-5641

ABSTRACT

The synthesis temperature for silicalite-I (MFI) can be raised to 300 °C by applying high pressure to stabilize the structure-directing organic template. The elevated temperature and pressure favor the formation of crystals with improved quality. Prismatic silicalite-I crystals with a uniform size of about 0.7 x 0.2 x 0.2 mm have been obtained by heating a gel prepared from TMA-silicate solution, TPABr and sodium hexafluorosilicate at 250 °C under a pressure of 80 MPa. The influence of synthesis conditions on the crystal sizes has been studied by systematically changing temperature, pressure and gel compositions. Under the specific conditions of 250 °C and 80 MPa, a strong correlation was found between the crystal size and the F/Si mole ratio of the starting gel, which enables the preparation of uniform crystals of silicalite-I with preset dimensions.

INTRODUCTION

Large uniform zeolite crystals are highly desired for many uses that range from crystal structure analysis, adsorption and diffusion studies to zeolite functional materials. Synthetic zeolites obtained from hydrothermal reactions are usually in the form of microcrystalline powders. Considerable progress has been made in the growth of large zeolite crystals by carefully adjusting the solvents and starting materials. [1-4] We have been investigating the possibility of improving the size and crystal quality of microporous materials by using synthesis temperatures and pressures that are higher than those usually employed (220-400 °C, 50-200 MPa). By using this approach, we obtained crystals of the microporous titanosilicate ETS-10 suitable for single crystal X-ray structure analysis. [5] Similar experimental conditions were used to mimic the natural growth conditions of zeolite minerals. [6]

It is a general rule that high pressures and high temperatures favor the formation of dense phases rather than microporous metastable phases such as zeolites. Many synthetic zeolites are stabilized by organic templates that act as space fillers in the crystal structure or as structure-directing agents. Decomposition of the templates often determines the upper temperature limit of the zeolite synthesis. However, most organic templates can be stabilized to high temperatures by applying a high pressure. Therefore, under high pressure the synthesis temperatures of many zeolites may be substantially raised, which favors formation of large crystals because of higher crystal growth rates.

We have chosen the silica form of the synthetic zeolite ZSM-5 (silicalite-I, MFI) to test this synthesis strategy because the synthesis of silicalite-I below 200 °C has been thoroughly investigated, and it is known that the presence of organic templates is crucial for its formation. [7,8] In addition, silicalite-I is one of the most important synthetic molecular sieves and is the only one having a hydrophobic framework with 10-ring channels. The natural counterpart of ZSM-5 called mutinaite was recently discovered in cavities of ancient magmatic rocks [9], indicating that it might be formed at quite high temperatures.

EXPERIMENTAL

Synthesis

An aqueous solution of tetramethylammonium silicate (TMAOH·2SiO$_2$) (Aldrich, 99.99+%) was used as the silica source, and tetrapropylammonium bromide (TPABr) (Aldrich, 98%) as the templating agent. Other materials used were: Na$_2$SiF$_6$ (Aldrich, 98%), (NH$_4$)$_2$SiF$_6$ (Aldrich, 98%), and deionized water.

In a typical synthesis experiment, TMA-silicate solution and TPABr solution were added to a flexible Teflon tube (~ 5 x 50 mm). After adding the proper amount of hexafluorosilicate the tube was sealed in air and then inserted into a high pressure vessel. The tube was heated using a LECO HR-1B-2 high pressure - high temperature system for 4 days. The reaction vessel was then removed from the oven and cooled to room temperature in air. The product was washed with water, filtered and dried in air.

Characterizations

The products were examined with an optical microscope and with X-ray powder diffraction on a Scintag-2000 diffractometer using CuK$_\alpha$ radiation. Infrared spectra were measured using a Galaxy FTIR 5000 spectrometer with the KBr pellet method. Scanning electron microscope images were recorded on a JEOL JSM 6400 microscope. Single crystal X-ray diffraction data were measured on a Siemens SMART diffractometer equipped with a CCD area detector using MoK$_\alpha$ radiation at 293 K. The crystal structure refinements were performed using the program SHELXTL. [10]

RESULTS AND DISCUSSION

The effects of temperature and pressure

Starting with reaction mixtures of the following mole ratio
1.0 TMAOH·2SiO$_2$: 0.25 Na$_2$SiF$_6$: 0.39 TPABr : ~100H$_2$O,
single phase silicalite-I crystals were obtained in the temperature range 230-300 °C and pressure range 50-100 MPa (Fig.1). A reaction at 330 °C and 100 MPa resulted in only quartz. The best crystals of silicalite-I with a maximum dimension of ca. 0.45 mm were obtained from reactions at 250 °C. As the synthesis temperature deviates from 250 °C the crystal sizes decrease.

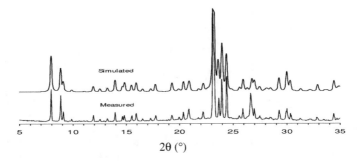

Fig.1 X-ray powder diagram of a silicalite-I sample synthesized at 300 °C and 100 MPa (bottom). The top diagram is simulated with structure data from single crystal X-ray diffraction.

Fig.2 Optical micrograph of silicalite-I crystals synthesized at 300 °C and 100 MPa.

Reactions at 300 °C and 100 MPa produced crystals with sizes of ca. 0.1 mm (Fig.2). According to Fleming and Crerar the solubility of quartz at high pH has a maximum around 230 °C. [11] This may also be true for other forms of silica and may account for our observation that the largest crystals were obtained near this temperature. The high solubility might lead to a low oversaturation and formation of fewer nucleation centers, thus favoring growth of large crystals.

Pressure has no substantial influence on the products in the range mentioned above. Reactions performed at 230-250 °C under autogenous pressures (< 4 MPa) lead to the formation of the dense phase cristobalite instead of silicalite-I, which is in agreement with previous observations [7], and probably is because of the decomposition of the template.

The infrared spectrum of a sample synthesized at 300 °C and 100 MPa is shown in Fig.3. In addition to the characteristic bands of the silicalite-I framework between 1300 - 400 cm^{-1}, typical bands for the TPA cations are clearly observed (1384, 1475, 2984cm^{-1}). Therefore, the applied pressure does stabilize the organic template at synthesis temperatures as high as 300 °C.

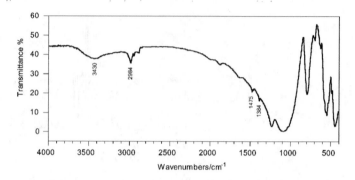

Fig.3 IR spectrum of silicalite-1 synthesized at 300 °C and 100 MPa.

Fig.4 The structure of silicalite-I refined from single crystal X-ray data, showing TPA cations located inside the channels.

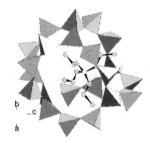

The encapsulation of the TPA template is also confirmed by crystal structure refinement with X-ray diffraction data measured on a single crystal synthesized at 250 °C and 80 MPa (Fig.4). The structure has a space group symmetry Pnma with cell dimensions a = 19.999(2) Å, = 20.007(2) Å, c = 13.398(1) Å. TPA molecules are located at intersections between the [010] and [100] 10-ring channels with each of the N-C-C-C arms pointing to a 10-ring, similar to that previously reported in the literature. [12]

The effect of fluorine

The effect of Na_2SiF_6 was investigated by varying its concentration in the reaction mixtures. At the specific conditions of 250 °C and 80 MPa, a strong correlation between the F/ ratio of the starting mixture and the size of the synthesized silicalite-I crystals was observed (Fig.5). In the absence of Na_2SiF_6, elongated prisms of silicalite-I with uniform sizes of about 0 x 0.03 x 0.03 mm were formed. As the content of Na_2SiF_6 increases the prisms became thicker addition to increasing in overall size. The largest crystals of ca. 0.7 x 0.2 x 0.2 mm were synthesized with a F/Si ratio of 1.5 (Fig.6). In reactions with F/Si ratios higher than 1.8, Na_2Si recrystallized and no silicalite-I was formed. With other conditions unchanged, Na_2SiF_6 was replaced by $(NH_4)_2SiF_6$ which has a much higher solubility. However, the sizes of silicalite-I crystals (ca. 0.3 x 0.06 x 0.06 mm) obtained did not change as the $(NH_4)_2SiF_6$ content varied. Therefore, F and Na are probably both responsible for the crystal size changes.

The majority of the silicalite-I crystals obtained in our experiments are well separated crystals with a good uniformity in size. However, when carefully examined it is found that most of them are twinned by a 90° rotation around the c axis. Such twinning is often observed in ZSM-5 crystals synthesized by other methods. A small amount of cross-intergrowing crystals were also observed in our samples. The crystal surfaces show large variations even between different crystals from the same synthesis experiment. In particular, small cavities and sometimes satellite crystallites are observed on the surfaces of crystals prepared at high F/Si ratios (Fig.7). In the final stage of crystal growth the F content in the solution increases thus increasing the solubility of silica. Fluctuations of the reaction conditions might cause dissolution of the growing crystals and formation of secondary nucleation centers.

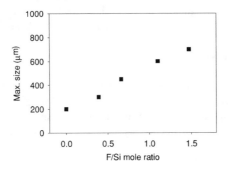

Fig.5   Correlation between silicalite-I crystal size and the reagent F/Si ratio adjusted by using Na_2SiF_6 for syntheses at 250 °C and 80 MPa.

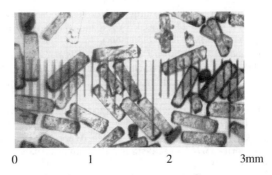

Fig.6   Optical micrograph of silicalite-I crystals synthesized at 250 °C and 80 MPa.

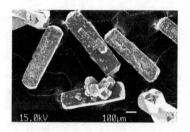

Fig.7 SEM image of silicalite-I crystals synthesized at 250 °C and 80 MPa.

CONCLUSIONS

Large silicalite-I crystals can be prepared by using temperatures and pressures that are substantially higher than usually employed. Organic templates can be stabilized at high temperatures by applying high pressures. We anticipate that the synthesis temperatures of many other zeolite materials can be substantially raised under high pressures.

ACKNOWLEDGMENTS

We thank the National Science Foundation (DMR9805881) and the R.A. Wel Foundation for financial support. This work made use of MRSEC/TCSUH Shared Experimen Facilities supported by the National Science Foundation under Award Number DMR-96326 and the Texas Center for Superconductivity at the University of Houston.

REFERENCES

1. J. F. Charnell, *J. Crystal Growth*, **8**, 291 (1971).
2. A. Kuperman, S. Nadimi, S. Oliver, G. A. Ozin, J. Garces and M. M. Olken, *Nature*, **365**, 239 (1993).
3. W. G. Klemperer and T. A. Marquart, *Mat. Res. Soc. Symp. Proc.*, **346**, 819 (1994).
4. S. Qiu, J. Yu, G. Zhu, O. Tarasaki, Y. Nozue, W. Pang and R. Xu, *Microporous and Mesoporous Mater.*, **21**, 245 (1998).
5. X. Wang and A. J. Jacobson, *J. Chem. Soc., Chem. Commun.*, 1999, 973.
6. H. Ghobarkar, O. Schaf and U. Guth, *Prog. Solid St. Chem.*, **27**, 29 (1999).
7. J. L. Guth, H. Kessler and R. Wey, *in Proc. 7th Intern. Conference on Zeolite*, Elsevier, Tokyo, 121 (1986).
8. J. C. Jansen, C. W. R. Engelen and H. van Bekkum, *in Amer. Chem. Soc. Symp. Series* **398**, 257 (1989).
9. E. Galli, G. Vezzalini, S. Quartieri, A. Alberti and M. Franzini, *Zeolite*, **19**, 318 (1997).
10. G. M. Sheldrick, *SHELXTL, Version 5.03*, Siemens Analytical X-ray Instruments, Madison, WI, (1995).
11. B. A. Fleming and D. A. Crerar, *Geothermics*, **11**, 15 (1986).
12. H. van Koningsveld, H. van Bekkum and J. Jansen, *Acta Cryst.* B, **43**, 127 (1987).

Mat. Res. Soc. Symp. Proc. Vol. 658 © 2001 Materials Research Society

Synthesis of Tetramethylammonium Polyoxovanadates

Nathalie Steunou, Laure Bouhedja, Jocelyne Maquet, Jacques Livage
Chimie de la Matière Condensée, UMR CNRS 7574
Université Pierre et Marie Curie, Paris, France

ABSTRACT

Tetramethylammonium (TMA) polyoxovanadates have been precipitated from aqueous solutions around pH 7. Decavanadate clusters $(TMA)_4[H_2V_{10}O_{28}] \cdot 4H_2O$ are formed at room temperature whereas a layered $(TMA)[V_4O_{10}]$ mixed valence compound is formed under hydrothermal conditions. ^{51}V NMR spectra recorded on the solution at different temperatures show that upon heating decavanadate clusters are progressively transformed into cyclic $[V_4O_{12}]^{4-}$ metavanadates. This suggests that the mixed valence polyoxovanadates are formed via the ring opening polymerization of metavanadate precursors.

INTRODUCTION

The synthesis of polyoxovanadates from aqueous solutions under mild conditions leads to a large variety of compounds. The decavanadate cluster $[H_nV_{10}O_{28}]^{(6-n)-}$ has already been characterised for a long time, both in the solution and in the solid state [1]. New layered polyoxovanadates such as $(TMA)[V_4O_{10}]$, $(TMA)[V_3O_7]$ and $(TMA)[V_8O_{20}]$ have been recently synthesized under hydrothermal conditions by Whittingham et al [2]. A large family of hollow clusters encapsulating anionic or neutral species such as $(TMA)_6[V_{15}O_{36}Cl] \cdot 4H_2O$ or $(Cs)_9[H_4V_{18}O_{42}I] \cdot 12H_2O$ have been reported [3]. Most of these polyoxovanadates are formed at pH values close to neutrality where V^V coordination changes from five to four. Therefore a more detailed analysis has been undertaken in order to get a better understanding of the chemical processes leading from molecular precursors in the solution to solid compounds.

EXPERIMENTAL

Synthesis of tetramethylammonium polyoxovanadates

The vanadate salts of the organic tetramethylammonium (TMA) cation were synthesized via the dissolution of a V_2O_5 powder (1g) into an aqueous solution of TMAOH (10%, pH=13, 5 mL). The V_2O_5/TMAOH molar ratio was close to 1. At room temperature the dissolution of the oxide was rather slow. An orange suspension was obtained after about one day. The vanadium pentoxide in excess was removed by filtration and the resulting clear solution (pH = 6.7) was slowly evaporated at room temperature for one week giving rise to orange crystals of compound 1. Black shiny flat crystals corresponding to compound (2) precipitated from a yellow solution (pH≈7.2) when the mixture was heated under hydrothermal conditions (48 h at 200°C) in a Parr Teflon-lined digestion bomb of 23 mL capacity.

X-ray structures of $[H_2V_{10}O_{28}][N(CH_3)]_4 \cdot 4 H_2O$ (1) and $(TMA)[V_4O_{10}]$ (2)

The structures of compounds (1) and (2) were determined by single crystal X-ray diffraction. Measurements were collected at room temperature on a CAD4 diffractometer using MoKα radiation ($\lambda = 0.71069$ Å) and graphite monochromator.

The asymmetric unit of compound (1) is composed of one $[H_2V_{10}O_{28}]^{4-}$ anion, four tetramethylammonium cations and four water molecules of crystallization [4]. A polyhedral view of the $[H_2V_{10}O_{28}]^{4-}$ anion is represented in Figure 1. The decavanadate anion which has already been described in the literature is a highly condensed polyanion made of ten VO$_6$ octahedra sharing corners and edges. All vanadium have distorted octahedral geometry and taking into account the ideal symmetry of the decavanadate anion D$_{2h}$, three types of VO$_6$ octahedra can be distinguished. The main crystallographic data for compounds (1) are given in Table I [4].

Figure 1 : Polyhedral view of $[H_2V_{10}O_{28}]^{4-}$.

Table I. X-ray data of
$[H_2V_{10}O_{28}]$ $[N(CH_3)]_4 \cdot 4H_2O$[4]

Formula	$C_{16}H_{58}N_4O_{32}V_{10}$
Crystal System	Triclinic
a (Å)	12.626(3)
b (Å)	12.639(4)
c (Å)	15.733(2)
α (°)	68.07(2)
β (°)	83.09(2)
γ (°)	78.71(2)
V (Å^3)	2281.1(9)
Space group	P -1

The X-ray structure of compound (2) was resolved and found to be identical to the TMA[V_4O_{10}] compound already described by Whittingham et al [5]. Compound (2) is made up of double chains of edge-sharing [VO$_5$] tetragonal pyramids with TMA$^+$ cations distributed between the [V_4O_{10}]$^-$ layers (Figure 2). The neutrality of this compound suggests that one out of four vanadium atoms has been reduced to V^{4+}.

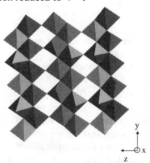

Figure 2. X-ray structure of $N(CH_3)_4[V_4O_{10}]$ [5].

Room temperature ^{51}V NMR of the solutions

^{51}V NMR solution spectra were recorded on a Brucker AC 300 (or MSL 400) spectrometer operating at 78 MHz (or 105.25MHz), using a 90° pulse width of 11 μs, and a relaxation delay of 1 s. Neat $VOCl_3$ was used as an external reference for chemical shifts (δ = 0 ppm).

The ^{51}V NMR spectrum of the orange solution obtained at room temperature exhibits three peaks at δ = -427 ppm, -502 ppm and -518 ppm (Figure 3(a)) due to the decavanadate polyanion $[H_nV_{10}O_{28}]^{(6-n)-}$ that contains three different vanadium sites in the ratio 2:4:4 [6]. The two smaller peaks at δ = -581 ppm and δ=-591 ppm can be assigned to the cyclic metavanadates, $[V_4O_{12}]^{4-}$ and $[V_5O_{15}]^{5-}$ respectively [7]. Integration of these peaks shows that decavanadates correspond to 91% of the whole V^{5+} and metavanadates only 9%. The ^{51}V NMR spectrum of the supernatant yellow solution obtained after hydrothermal treatment is quite close to that obtained at room temperature. It displays the same resonances assigned to the decavanadates and metavanadates species. Only the intensity of the signals changes. Surface measurements of the ^{51}V NMR peaks shows that 44% of the vanadium is as decavanadate species and 56% as metavanadates (30.5 and 25.5 % for the tetramer and pentamer respectively).

Temperature dependence of the ^{51}V NMR spectra of decavanadate-metavanadate solutions (pH = 6.7).

These experiments have been performed on the supernatant orange solution (pH = 6.7) from which decavanadate crystals were precipitated. The solution was heated and a ^{51}V NMR spectrum was recorded at different temperatures within the NMR apparatus from room temperature to 100°C. Finally, ^{51}V NMR spectra were recorded on the solution heated to 200°C for 30 min or for 3 h before cooling to room temperature. The ^{51}V NMR spectra recorded at different temperatures are presented in Figure 3.

Figure 3 shows that the small peak corresponding to the cyclic $[V_5O_{15}]^{5-}$ species disappears rapidly around 40°C. The relative intensity of the three decavanadate peaks decreases compared to the peak corresponding to the cyclic metavanadate $[V_4O_{12}]^{4-}$ (Figure 3). The molecular ratio [decavanadate]/[metavanadate] decreases progressively from 91/9 at room temperature to 76/24 at 100°C. At a given temperature, it also decreases as a function of the heating time. Upon heating at 200°C for example, the ratio decreases from 24/76 after half an hour to 3/97 after 3 hours. It has to be pointed out that this decavanadate-metavanadate transformation is quite reversible. A ratio of 82/18 is again observed when the former solution heated at 150°C for 12 hours is cooled down to room temperature. The overall intensity of ^{51}V NMR peaks decreases with the temperature as some black solid phase forms above 150°C while the solution turns green.

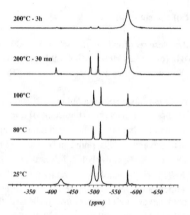

Figure 3. ^{51}V *NMR spectra of decavanadate solutions heated at different temperatures (a) T = 25°C; (b) T = 80°C; (c) T = 100°C; (d) T = 200°C for 30 min, (e) 200°C for 3 h (Larmor frequency = 78 MHz; Number of scans, NS = 400 for (a), (b), (c); Larmor frequency = 105.25 MHz; NS = 44 for (d) and (e)).*

DISCUSSION

These experiments show that different polyoxovanadates can be obtained depending on the experimental conditions. The nature of solute species in aqueous solutions mainly depends on pH. In the pH range 5-9, both decavanadate and metavanadate species are in equilibrium in aqueous solutions.

$$2[H_2V_{10}O_{28}]^{4-} + 4H_2O \rightarrow 5[V_4O_{12}]^{4-} + 12H^+ \text{ (eq.1)}$$

Around pH $\approx$ 6 the main species is the well known decavanadate polyanion that can be evidenced clearly by ^{51}V NMR (Figure 3). At room temperature, these solutions give rise to decavanadate crystals. There is a straightforward correlation between V^{5+} species in the solution and in the solid state.

This is no longer the case under hydrothermal conditions. There is no obvious structural relationship between the decavanadate polyanion observed in the supernatant solution at room temperature and the layered compound (TMA)[V_4O_{10}]. However ^{51}V NMR experiments show that the equilibrium between decavanadate and metavanadate species is progressively displaced towards the formation of cyclic metavanadates [V_4O_{12}]$^{4-}$ when the temperature increases.

It may then be assumed that at 200°C, under hydrothermal conditions, the layered (TMA)[V_4O_{10}] vanadate is formed via the ring opening polymerization (ROP) of cyclic metavanadates [V_4O_{12}]$^{4-}$ precursors. Crystallization from metavanadate aqueous solutions is always accompanied by polymerization of the anion. Actually metavanadates are based on [VO_4] tetrahedra and precipitation at room temperature currently leads to the formation of chain compounds made of corner sharing [VO_4] tetrahedra as in KVO_3 or edge sharing [VO_5] trigonal bipyramids as in $KVO_3 \cdot H_2O$ [8].

Hydrothermal conditions may favour the coordination expansion of vanadium V^{5+} from four to five or even six. According to eq.1 protons are formed when decavanadate species are transformed into metavanadate. The "measured" pH of the solution decreases upon heating and protonation is known to lead to coordination expansion in this pH range. Moreover some reduction, due to the partial decomposition of organic species, occurs during the hydrothermal synthesis. V^{4+} species are then formed that are too large for tetrahedral coordination. Therefore coordination expansion may occur via the nucleophilic addition of $[VO_4]$ tetrahedra giving rise to double chains in which both V^{5+} and V^{4+} are in $[VO_5]$ square pyramids. These double chains are further linked together via corners in order to form $[V_2O_5]$-like layers. Such a condensation process may occur via oxolation between protonated V-OH bonds along the vanadate chains. Without protonation only chain compounds are formed. A possible mechanism for the formation of $(TMA)[V_4O_{10}]$ from chain metavanadates is proposed in Figure 4.

The pH of the solution appears to be a major parameter for the formation of layered phases. Hydrothermal syntheses performed by M.S. Whittingham et al [9] have shown that layered trivanadate $(TMA)[V_3O_7]$ are formed at higher pH than $(TMA)[V_4O_{10}]$. Both $[VO_5]$ and $[VO_4]$ coordinations are then observed in $[V_3O_7]$ layers whereas only $[VO_5]$ square pyramids are formed in $[V_4O_{10}]$ [9].

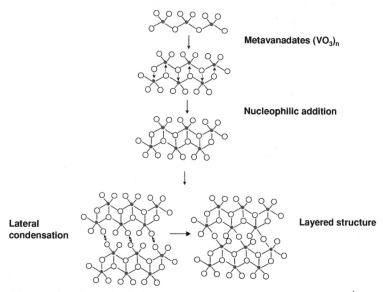

Metavanadates $(VO_3)_n$

Nucleophilic addition

Lateral condensation

Layered structure

Figure 4. Suggested mechanism for the formation of $[V_4O_{10}]^-$ layers from cyclic $[V_4O_{12}]^{4-}$ metavanadates.

REFERENCES

1. (a) M.T. Pope, A. Müller, *Angew. Chem. Int. Ed. Engl.*, **30**, 34 (1991); (b) A. Müller, H. Reuter, S. Dilionger, *Angew. Chem. Int. Ed. Engl.*, **34**, 2328 (1995).

2. T. Chirayil, P.Y. Zavalij, M.S. Whittingham, *Chem. Mater.*, **10**, 2629 (1998).

3. (a) A. Müller, M. Penk, R. Rohlfing, E. Krickemeyer, J. Döring, *Angew. Chem. Int. Ed. Engl.*, **29**, N°8 926 (1990). (b) A. Müller, R. Sessoli, E. Krickemeyer, H. Bögge, J. Meyer, D. Gatteschi, L. Pardi, J. Westphal, K. Hovemeier, R. Rohlfing, J. Döring, F. Hellweg, C. Beugholt, M. Schmidtmann, *Inorg. Chem.*, **36**, 5239 (1997).

4. N. Steunou, J. Vaissermann, J. Livage, to be published

5. P.Y.Zavalij, M.S. Whittingham, E.A. Boylan; V.K. Pecharsky, R.A. Jacobson, Z. *Kristallogr*, **211**, 464 (1996).

6. (a) S.E. O'Donnell, M.T. Pope, *J. Chem. Soc. Dalton Trans.*, 2290 (1976). (b) O.W. Howarth, M. Jarrold, *J. Chem. Soc. Dalton Trans.*, 503 (1978).

7. (a) E. Heath, O.W. Howarth, *J. Chem. Soc. Dalton Trans.*, 1105 (1981). (b) V.W. Day, W.G. Klemperer, D.J. Maltbie, *J. Am. Chem. Soc.*, **109**, 2991 (1987).

8. H.T. Evans, Z. *Kristallogr.*, **114**, 257 (1960).

9. (a) T.G. Chirayil, E.A. Boylan, M. Mamak, P.Y. Zavalij, M.S. Whittingham, *Chem. Comm.*, 33 (1997). (b) P.Y. Zavalij, T. Chirayil, M.S. Whittingham, *Acta Cryst.*, **C53**, 879 (1997).

Mat. Res. Soc. Symp. Proc. Vol. 658 © 2001 Materials Research Society

Assembly of Metal-Anion Arrays within Dion-Jacobson-Type Perovskite Hosts

Thomas A. Kodenkandath,[†] Marilena L. Viciu, Xiao Zhang, Jessica A. Sims, Erin W. Gilbert, François-Xavier Augrain, Jean-Noël Chotard, Gabriel A. Caruntu, Leonard Spinu, Weilie L. Zhou, and John B. Wiley[*]
Department of Chemistry and the Advanced Materials Research Institute, University of New Orleans, New Orleans, LA 70148-2820 jwiley@uno.edu

ABSTRACT

The construction of metal-anion arrays within Dion-Jacobson-type (DJ) perovskites was studied. A variety of hosts readily react with $CuCl_2$ to produce compounds of the form $(CuCl)[A'_{n-1}(M,M')_nO_{3n+1}]$ (A' = alkaline earth, rare earth or Bi; M/M' = Nb, Ta, and/orTi; n = 2, 3). In contrast, $CuBr_2$ only reacts with some of these hosts; this difference relative to the chloride may simply be a size effect. The formation of metal-anion arrays with other transition metals has also been examined. $(FeCl)LaNb_2O_7$ can be readily prepared from $FeCl_2$, while reactions with $FeBr_2$ and $NiCl_2$ under similar reaction conditions result in compounds of the form $M_{0.5}LaNb_2O_7$ (M = Fe, Ni). The reductive intercalation of lithium and sodium into $(CuCl)LaNb_2O_7$ was also examined and was found to produce new compounds that have a large layer expansion.

INTRODUCTION

Low-temperature ($\leq 500°C$) strategies based on topotactic methods are currently being explored for the directed synthesis of new materials. Exploitation of methods involving ion exchange and/or (de)intercalation offers an important approach to the control of valence and structure in receptive hosts. Recent efforts, in perovskite-related compounds for example, have resulted in the preparation of new mixed-valence titanates, niobates and tantalates [1-4], alternating layered 3-D perovskites [5], and mixed-metal compounds [6,7]. Similarly, our group has been examining the construction of metal-anion arrays within Dion-Jacobson-type perovskite hosts [8,9]. We recently reported the synthesis of $(CuX)LaNb_2O_7$ (X = Cl, Br). Herein we describe the extension of this chemistry to other host systems and other transition-metal halides as well as the behavior of $(CuCl)LaNb_2O_7$ with respect to reductive intercalation.

EXPERIMENTAL

Synthesis. Compounds belonging to the Dion-Jacobson (DJ) series of layered perovskites, $A[A'_{n-1}(M,M')_nO_{3n+1}]$ (A = alkali metal, H, NH_4, A' = alkaline earth, rare earth or Bi, M/M' = Nb, Ta, Ti, n = 2, 3) were synthesized by standard high-temperature solid-state or low-temperature ion exchange reaction methods as described in the literature [10-15]. Starting mixtures for compounds where A = K, Rb were made with the stoichiometric amounts of the appropriate oxides and a 20 - 25% molar excess of A_2CO_3 (to compensate for the loss of the oxide due to

---

[†] Current address American Superconductor, Westborough, MA 01581
[*] To whom correspondence should be addressed

volatilization). Reagents were ground together and heated in alumina reaction vessels at 850 - 900°C for 12 h and then 1050 - 1100°C for 24 - 48 h with one intermittent grinding. The products were washed thoroughly with distilled water and dried overnight at 120°C. $HLaNb_2O_7$ was prepared by the ion exchange of $KLaNb_2O_7$ with 6 M HNO_3 at 60°C for 24 h [10] and $LiLaNb_2O_7$, $NaLaNb_2O_7$, and $NH_4LaNb_2O_7$ were synthesized by ion exchange of $RbLaNb_2O_7$ with the corresponding nitrates at 350°C for 3 days [10]. Unit cell parameters of the $A[A'_{n-1}(M,M')_nO_{3n+1}]$ host materials were in good agreement with the values reported in the literature [10-15].

The synthesis of the series $(CuX)[A'_{n-1}(M,M')_nO_{3n+1}]$ (X = Cl, Br) was carried out by a single-step ion exchange reaction between the parent DJ phases and the cupric halides. The host compounds were mixed thoroughly with a two-fold molar excess of the anhydrous CuX_2 (dried overnight at 120°C), sealed in evacuated (< 10^{-3} torr) Pyrex tubes and heated (325 - 350°C) for - 12 days. The samples were then washed with copious amounts of cold and hot distilled water to remove the excess cupric halides and AX byproduct. Exchanges with other transition-metal halides were carried out under similar conditions, but often at slightly higher reaction temperatures (350°C - 450°C). Reductive intercalation studies of $(CuCl)LaNb_2O_7$ were carried out over a 24 hr period at room temperature under inert atmosphere in hexane solution with n-butyllithium and $NaB(C_2H_3)_3H$. Intercalated samples contained an amorphous elemental copper byproduct, which starts to crystallize on heating by 100°C.

Characterization. Chemical analyses were carried out with energy dispersive spectroscopy (EDS) on a JEOL (Model JSM-5410) scanning electron microscope (SEM) equipped with an EDAX (DX-PRIME) microanalytical system. Parent compounds and anhydrous metal halides were used as standards. Powder X-ray diffraction data were collected between 5 - 95°2θ (step = 0.02° with a 2 s count time) on a Philips X'Pert-MPD diffractometer equipped with a graphite monochromator and Cu-K_α (λ = 1.5418 Å) radiation.

RESULTS

Other host systems. The assembly of metal-anion arrays within perovskite hosts represents an important route to new perovskite-related metal oxyhalides. As part of the interest in this chemistry it is important to determine the utility of this approach to a variety of Dion-Jacobson-type perovskite hosts. We have carried out a systematic study of these reactions for the ion exchange of both copper chloride and copper bromide [9]. Table I compares the exchange behavior of a series of double and triple-layered perovskites. While all of the hosts are found to exchange with $CuCl_2$ to form the corresponding $(CuCl)[A'_{n-1}(M,M')_nO_{3n+1}]$ compounds, the bromide reactions show some degree of selectivity.

Other transition metals. The ability to form metal-anion arrays becomes even more significant if we can extend this chemistry beyond copper cations. With respect to other transition metals, we have looked at several first row transition metals. Interestingly, under the same reaction conditions used to prepare the copper halides (325°C), other transition metal chloride arrays are not readily obtained. The only clear success that we have observed so far has been for the synthesis of $(FeCl)LaNb_2O_7$ from $RbLaNb_2O_7$ and $FeCl_2$ at higher temperatures (425°C). Figure 1 shows a comparison of the diffraction pattern for $(FeCl)LaNb_2O_7$ versus those of the host and $(CuCl)LaNb_2O_7$. The shift of the first reflection, the 001, to lower angles relative to $RbLaNb_2O_7$ reflects the increase in the c parameter on insertion of the metal-chloride array. The tetragonal unit cell parameters for the host, the iron oxychloride, and the copper oxychloride

Table I. Exchange behavior of copper halides in Dion-Jacobson-type perovskites.

Compd	chloride	bromide	layer orientation[a]	layer spacing[b] (Å)
$LiLaNb_2O_7$	✓	✗	S	10.16^{10}
$NaLaNb_2O_7$	✓	✗	S	10.45^{10}
$HLaNb_2O_7$	✓	✗	E	10.459^{10}
$RbNdNb_2O_7$	✓	✗	E	10.733^{11}
$KLaNb_2O_7$	✓	✗	P	10.77^{10}
$NH_4LaNb_2O_7$	✓	✓	E	10.95^{10}
$RbLaNb_2O_7$	✓	✓	E	10.989^{10}
$RbLaTa_2O_7$	✓	✓	E	11.1053^{12}
$RbBiNb_2O_7$	✓ [c]	✗	E	11.22^{13}
$RbCa_2Nb_3O_{10}$	✓	✗	E	14.909^{14}
$RbSm_2Ti_2NbO_{10}$	✓	✓	E	$15.0572(6)^{d}$
$RbPr_2Ti_2NbO_{10}$	✓	✓	E	$15.1517(5)^{d}$
$RbLa_2Ti_2NbO_{10}$	✓	✓	E	15.24^{15}

[a]E = layers are eclipsed, S = layers are staggered, and P = layers are partially eclipsed
[b]Layer spacing corresponds to distance between A' cations in adjacent perovskite layers
[c]Reaction incomplete even after three weeks and always contains some amount of BiOCl
[d]This work.

are $a = 3.896(9)$ Å and $c = 11.027(2)$ Å, $a = 3.8736(9)$ Å and $c = 11.835(3)$Å, and $a = 3.8792(1)$ Å and $c = 11.7282(3)$ Å, respectively. We have had some preliminary results that indicate that other transition-metal oxyhalides exist, but the exact conditions under which these compounds can be prepared have not yet been established.

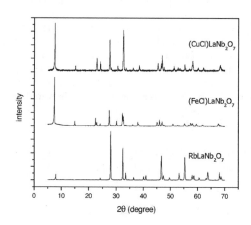

Figure 1. *X-ray powder diffraction pattern of (FeCl)LaNb₂O₇ versus RbLaNb₂O₇ and (CuCl)LaNb₂O₇.*

We have also found that it is possible to carry out reactions that result in compounds of the form $M_{0.5}LaNb_2O_7$. $Cu_{0.5}LaNb_2O_7$ can be prepared by the controlled decomposition of $(CuCl)LaNb_2O_7$ under dynamic vacuum at 450°C [16]. This compound has not been observed, however, in any of our exchange reactions involving copper halides. This is in contrast to other systems where $M_{0.5}LaNb_2O_7$ (M = Fe, Ni) phases have been prepared by direct exchange with $FeBr_2$ and $NiCl_2$.

$$RbLaNb_2O_7 + \tfrac{1}{2} MX_2 \rightarrow M_{0.5}LaNb_2O_7 + RbX$$
(M = Fe and Ni for X = Br and Cl, respectively)

Reductive intercalation of (CuCl)LaNb$_2$O$_7$. We have investigated the reductive intercalation chemistry of $(CuCl)LaNb_2O_7$ with n-butyllithium and $NaB(C_2H_3)_3H$. As would be expected, Cu(II) is reduced to the metal, but what is unexpected is that the rest of the structure remains intact to give products of the form $(A_xCl)LaNb_2O_7$ (A = Li, Na; x $\geq$ 2). The changes in the unit cell parameters are shown in Table II. While the structures of these compounds are still under investigation, preliminary modeling studies strongly support the formation of a double alkali-chloride layer within the perovskite host.

DISCUSSION

Copper halides are found to exchange with a number of Dion-Jacobson-type perovskites to result in a series of new copper oxyhalides. Copper chloride is found to react with all the hosts listed in Table I, and the reaction does not appear to be influenced by the orientation (eclipsed, partially eclipsed or staggered) or the spacing between the perovskite layers. In contrast copper bromide demonstrates some selectivity. This difference, with the exception of $RbBiNb_2O_7$, correlates directly with the spacings of the perovskite layers, and may therefore simply be due to the size difference between chloride (1.67Å) and bromide (1.82 Å) [17].

The construction of metal-anion arrays within receptive hosts will be more interesting and of greater utility if it can be applied to transition metals halide systems beyond those of copper. The ability to synthesize $(FeCl)LaNb_2O_7$ indicates that there may be some flexibility in this chemistry and it may just be a question of identifying suitable conditions for these reactions to take place. It would appear that in some systems there might be a competition with the formation of the $M_{0.5}LaNb_2O_7$ compounds. In both iron bromide and nickel chloride, these products are found exclusively. If one can determine how to direct these reactions to the corresponding $(MX)LaNb_2O_7$ compounds, *i.e.*, $(FeBr)LaNb_2O_7$ and $(NiCl)LaNb_2O_7$, then this should open the door to the formation of other isostructural divalent transition metal complexes.

Table II. Unit cell parameters and cell volumes for intercalation products.

Compound	Tetragonal unit cell (Å)		Cell volume (Å^3)
$RbLaNb_2O_7$	a = 3.896(9)	c = 11.027(2)	167.4
$(CuCl)LaNb_2O_7$	a = 3.8792(1)	c = 11.7282(3)	176.49
$(Li_xCl)LaNb_2O_7$	a = 3.8957(1)	c = 12.564(1)	190.68
$(Na_xCl)LaNb_2O_7$	a = 3.915(2)	c = 13.39(1)	205.2

Our interests in low-temperature topotactic methods involve the development of multistep reactions as well as the search for new mixed-valence compounds. With this goal in mind, we have examined the reactivity of $(CuCl)LaNb_2O_7$ with respect to reductive intercalation. Interestingly, this appears to produce a new set of compounds where the copper is substituted by at least two alkali metal cations (Li or Na). This would represent an important change relative to the iron and copper halide systems in that these compounds, as indicated by the large increase in the layer spacing, must contain a more extensive interlayer assembly, possibly a double metal-halide array.

CONCLUSIONS

The assembly of metal-anion arrays within receptive host compounds offers an important topotactic route to new materials. While we have started to establish this method in terms of various hosts and metals beyond copper, much more work is needed to truly exploit it for the routine fabrication of new layered compounds. Of particular interest will be the formation of compounds that contain layer features with an established correlation to important properties, such as superconductivity and colossal magnetoresistance. It is expected that as the subtleties of various reaction details are determined, we will be able to apply this approach to a diverse set of compounds.

ACKNOWLEDGMENTS

The authors gratefully acknowledge the support provided by the National Science Foundation (DMR-9983591) and the Department of Defense (DARPA MDA972-97-1-0003).

REFERENCES

1. R. A. McIntyre, A. U. Falster, S. Li, W. B. Simmons, Jr., C. J. O'Connor and J. B. Wiley *J. Am. Chem. Soc.* **120**, 217 (1998).
2. J. N. Lalena, B. L. Cushing, A. U. Falster, W. B. Simmons, Jr., C. T. Seip, E. E. Carpenter, C. J. O'Connor, and J. B. Wiley, *Inorg. Chem.* **37**, 4484 (1998).
3. A. R. Armstrong and P. A. Anderson *Inorg. Chem.* **33**, 4366 (1994)
4. K. Toda, T. Teranishi, M. Takahashi, Z-.G. Ye and M. Sato *Solid State Ionics* **113-115**, 501 (1998).
5. R. E. Schaak and T. E. Mallouk *J. Am. Chem. Soc.* **122**, 2798 (2000).
6. K.-A. Hyeon and S.-H. Byeon *Chem. Mater.* **11**, 352 (1999).
7. J. Gopalakrishnan, T. Sivakumar, K. Ramesha, V. Thangadurai and G. D. Subbanna *J. Am. Chem. Soc.* **122**, 6237 (2000).
8. K. A. Thomas, J. N. Lalena, W. L. Zhou, E. E. Carpenter, C. Sangregorio, A. U. Falster, W. B. Simmons, Jr., C. J. O'Connor, and J. B. Wiley *J. Am. Chem. Soc.* **121**, 10743 (1999).
9. T. A. Kodenkandath, A. S. Kumbhar, W. L. Zhou, and J. B. Wiley *Inorg. Chem.* (in press).
10. J. Gopalakrishnan, V. Bhat, and B. Raveau, *Mat. Res. Bull.* **22**, 413 (1987).
11. M. Dion, M. Ganne, and M. Tournoux *Rev. Chim. Miner.* **23**, 61 (1986).
12. K. Toda and M. Sato *J. Mater. Chem.* **6**, 1067 (1996).

13. M. A. Subramanian, J. Gopalakrishnan, and A. W. Sleight *Mat. Res. Bull.* **23**, 837 (1988).
14. M. Dion, M. Ganne, and M. Tournoux *Mat. Res. Bull.* **16**, 1429 (1981).
15. J. Gopalakrishnan, S. Uma, and V. Bhat, *Chem. Mater.* **5**, 132 (1993).
16. T. A. Kodenkandath and J. B. Wiley (manuscript in preparation).
17. Radii are from R. D. Shannon *Acta Crystallogr.* **A32**, 751 (1976).

Mat. Res. Soc. Symp. Proc. Vol. 658 © 2001 Materials Research Society

Preparation of New Bismuth Oxides by Hydrothermal Reaction

N. Kumada, T.Takei, N. Kinomura and A. W. Sleight[1]
Faculty of Engineering, Yamanashi University, Miyamae-cho 7, Kofu 400-8511 Japan,
[1]Department of Chemistry, Oregon State University, Corvallis, OR 97331, USA

ABSTRACT

Hydrothermal reactions using $NaBiO_3 \cdot nH_2O$ produced a variety of new bismuth oxides that were not prepared by high temperature reaction. Some of them have pentavalent bismuth such as Bi_2O_4, $LiBiO_3$, ABi_2O_6 (A = Mg, Zn) and $AgBiO_3$. When using transition metal or rare-earth metal solutions, new bismuth oxides with trivalent bismuth were obtained. In the case of $Cr(NO_3)_3$ solution a chromium bismuth oxyhydroxide, $HBi_3(CrO_4)_2O_3$, was prepared at 180°C, while in A_2CrO_4 (A = Li, Na, K) solution a new bismuth chromium oxide, $Bi_8(CrO_4)O_{11}$, was prepared above 180°C. When A_2MoO_4 (A = Li, Na, K) solution was used, a new phase appeared above 220°C. By using A_2WO_4 (A = Li, Na, K) solution β-Bi_2O_3 and Bi_2WO_6 were obtained above 220°C.

INTRODUCTION

We have reported several new bismuth oxides prepared by hydrothermal reaction using hydrated sodium bismuth oxide, $NaBiO_3 \cdot nH_2O$ as a starting material [1-12]. Table 1 summarizes new bismuth oxides prepared by hydrothermal reaction with $NaBiO_3 \cdot nH_2O$. Some of them have pentavalent bismuth such as Bi_2O_4 with the β-Sb_2O_4-type structure [6], $LiBiO_3$ with the $LiSbO_3$-related structure [7], ABi_2O_6 (A = Mg, Zn) with the trirutile-type structure [8] and $AgBiO_3$ with the ilmenite-type structure [10]. One of them, Bi_2O_4 is the first example of a mixed valent bismuth oxide with distinct crystallographic sites of Bi^{3+} and Bi^{5+}. These bismuth oxides with Bi^{5+} cannot be prepared by usual high temperature reaction.

In the course of this investigation we attempted hydrothermal reaction with $NaBiO_3 \cdot nH_2O$ in A_2MO_4 (A = Li, Na, K; M = Mo, W) solution. In this paper, we will review the crystal structures of bismuth oxides prepared by hydrothermal reaction with $NaBiO_3 \cdot nH_2O$ and report the latest results from these reactions.

Table I. Products of hydrothermal reaction with $NaBiO_3 \cdot nH_2O$.

Solution	Reaction Temp. (°C)	Product	Ref.
LiOH	120	$LiBiO_3$	[7]
AOH (A = Li, Na, K)	120~160	Pyrochlore Type	[2]
AOH (A = Li, Na, K)	160~190	Bi_4O_7	[5]
KOH	200	New allotropic Bi_2O_3	[10]
ANO_3 (A = Li, Na, K)	120~160	Bi_2O_4	[6]
$A(NO_3)_2$ (A = Ca, Sr, Ba)	180	Pyrochlore Type	[1]
$AgNO_3$	70	$AgBiO_3$	[11]
$A(NO_3)_2$ (A = Mg, Zn)	90~130	ABi_2O_6	[8]
$Ln(NO_3)_2$ (Ln = La, Nd)	180	(Ln,Bi)OOH	[3]
$Ln(NO_3)_2$ (Ln = Y, Sm, Eu, Gd, Tb, Dy, Er, Yb)	140~200	$(Ln,Bi)_3O_4NO_3$	[9]
$Cr(NO_3)_3$	180	$HBi_3(CrO_4)_2O_3$	[4]
A_2CrO_4 (A = Li, Na, K)	180~220	$Bi_8(CrO_4)O_{11}$	[12]

EXPERIMENTAL

The starting material (1 g) of $NaBiO_3 \cdot nH_2O$ (Nacalai Tesque Inc.) was put into an autoclave (70 mL) with a teflon lining with various solutions (30 mL). The reaction temperature was 70~260°C and the reaction duration was 4~7 days. The solid products were separated by centrifuging, washed with distilled water and dried at 50°C. The products were identified using an X-ray powder diffractometer with CuKα radiation. The thermal stability of products was investigated with TG-DTA. When single crystals were obtained, the crystal structure was determined by single crystal X-ray diffraction analysis. In the case of powder samples the crystal structure refinement was carried out by the Rietveld method [13] using neutron powder diffraction data.

RESULTS and DISCUSSION

Bismuth oxides with Bi^{5+} were obtained at reaction temperatures below 190°C for 4 days. At higher temperature the bismuth oxides with only Bi^{3+} were prepared. In the case of KNO_3 solution at 120~160°C, Bi_2O_4 was obtained and Bi_4O_7 appeared above 160°C [5]. While in KOH solution at 120~160°C, a pyrochlore-type compound was obtained instead of Bi_2O_4 for 4 days. Under the condition of higher molar ratio (K/Bi~200) above 180°C a new allotropic form of Bi_2O_3 as well as β- and γ-Bi_2O_3 were prepared [10]. The crystal structures of Bi_2O_4 and new form of Bi_2O_3, which were prepared for the first time by hydrothermal reaction, are different from each other as shown in Figure 1, however both structures can be derived from a fluorite-type structure.

On the other hand in LiOH solution at lower temperature (~120°C) $LiBiO_3$ is obtained. The crystal structure is related to that of $LiSbO_3$ (Figure 2) [7]. At low temperature below 130°C the ilmenite-type $AgBiO_3$ [11] and the trirutile-type ABi_2O_6 (A=Mg,Zn) [8] appeared. These complex bismuth oxides with only Bi^{5+} have the crystal structural types found in complex antimony oxides with Sb^{5+}. These new bismuth oxides prepared at low temperature are considered to be formed from a topotactic reaction by ion-exchange. Ion-exchange reactions of its Sr^{2+} and Ba^{2+} ions do occur, and in the case of Sr^{2+}, the Na^+ ion is almost completely replaced [14]. For alkaline-earth metal ions, pyrochlore-type compounds with mixed valent Bi^{3+} and Bi^{5+} were prepared from alkaline-earth nitrate solutions at 180°C [1].

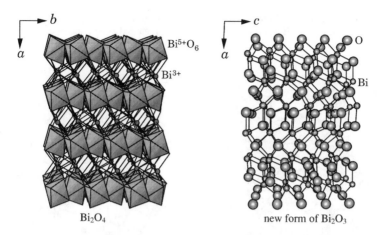

Figure 1. *Crystal structure of Bi₂O₄ and new Bi₂O₃.*

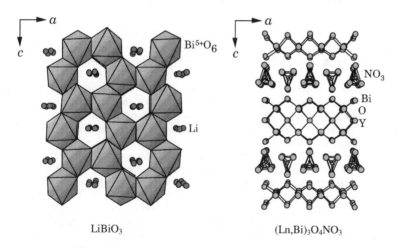

Figure 2. *Crystal structure of LiBiO₃ and (Ln,Bi)₃O₄NO₃.*

When the rare-earth metal solutions were used, the bismuth oxides with only Bi^{3+} were prepared at a lower temperature than in the case of alkali metal or alkaline-earth metal solutions. Two types of compounds were prepared in rare-earth metal nitrate solutions; one is $(Ln,Bi)OOH$ for Ln=La, Nd [3] and another is $(Ln,Bi)_3O_4NO_3$ for Ln = Y, Sm, Eu, Gd, Tb, Dy, Er, Yb [9]. For Ln = Pr the X-ray powder pattern of the product was similar to that of $(Ln,Bi)_3O_4NO_3$, however, several weak unindexed peaks were observed, and for Ln = Ce the products were not well crystallized. The crystal structure of $(Ln,Bi)_3O_4NO_3$ is shown in Figure 2. Both $(Ln,Bi)OOH$ and $(Ln,Bi)_3O_4NO_3$ may be regarded as derivatives of the Sillen structure.

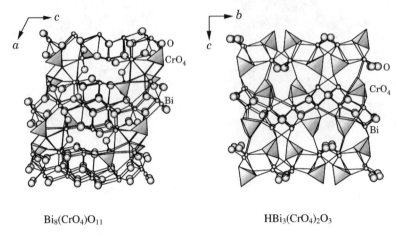

$Bi_8(CrO_4)O_{11}$ $HBi_3(CrO_4)_2O_3$

Figure 3. *Crystal structure of $Bi_8(CrO_4)O_{11}$ and $HBi_3(CrO_4)_2O_3$.*

There are also two types of compounds prepared using chromium solutions; one is $HBi_3(CrO_4)_2O_3$ prepared in $Cr(NO_3)_3$ solution [4] and another is $Bi_8(CrO_4)O_{11}$ prepared in A_2CrO_4 (A = Li, Na, K) solution [12]. Both compounds have similar layered structures with CrO_4 tetrahedron as shown in Figure 3.

In the case of A_2MO_4 (A = Li, Na, K; M = Mo, W) solution Bi_2O_4 appeared at 140ºC, regardless of solution, but above 180°C Bi_4O_7, β-Bi_2O_3, Bi_2WO_6 and a new phase were obtained, depending on reaction temperature and solution as summarized in Table II. From the chemical analysis of the new phase the Bi/Mo ratio was 6.3 and no alkali metal was detected. No weight change was observed in the TG-curve as shown in Figure 4, suggesting that neither water molecules nor Bi^{5+} was contained. From these results the chemical composition of the new phase was found to be $Bi_{6.3}MoO_{12.45}$. The X-ray powder pattern of this new phase is similar to that of Bi_4O_7, which is thought to have the triclinic pyrochlore-type structure with the lattice parameters of $a = 10.412(8)$Å, $b = 11.110(9)$Å, $c = 10.718(8)$Å, $\alpha = 90.38(5)º$, $\beta = 90.08(7)º$ and $\gamma = 92.38(6)º$ [5,15] as shown in Figure 5. All diffraction peaks can be indexed with a monoclinic cell, the dimension of which is approximately half of that of Bi_4O_7. Lattice parameters were calculated to be $a = 5.769(1)$Å, $b = 5.675(1)$Å, $c = 5$ .542(2)Å and $\beta = 90.82(2)º$.   This new compound is thought to have a crystal structure related to pyrochlore. The weak exothermic peak at 648ºC in the DTA curve corresponds to the transformation to a stable phase $3Bi_2O_3 \cdot MoO_3$ [16], because the X-ray powder pattern of the sample heated up to 900ºC is very similar to that of the sample prepared by heating a mixture of Bi_2O_3 and MoO_3 with molar ratio of 3:1 at 850ºC. According to the Bi_2O_3-MoO_3 phase diagram, a broad range of composition in the vicinity of $3Bi_2O_3 \cdot MoO_3$ was observed [16], and the crystal structure in this composition range was based on the fluorite [17]. The endothermic reaction at 864ºC is reversible and the corresponding peak is also observed for the sample prepared by high temperature reaction. This transformation is now under investigation.

Table II. Products of hydrothermal reaction in A_2MO_4 (A=Li,Na,K:M=Mo,W) solution.

| Solution | Reaction temperature | | | |
	140°C	180°C	220°C	260°C
Li_2MoO_4	Bi_2O_4	Bi_2O_4+new phase	new phase	new phase
Na_2MoO_4	Bi_2O_4+β-Bi_2O_3	β-Bi_2O_3	new phase	new phase*
K_2MoO_4	Bi_2O_4	Bi_4O_7	new phase	new phase*
Li_2WO_4	Bi_2O_4+β-Bi_2O_3	β-Bi_2O_3	β-Bi_2O_3	Bi_2WO_6
Na_2WO_4	Bi_2O_4	β-Bi_2O_3	β-Bi_2O_3	β-Bi_2O_3
K_2WO_4	Bi_2O_4	Bi_4O_7	Bi_4O_7	Bi_4O_7

* indicates coexistence of a small amount of γ-Bi_2O_3 single crystals.

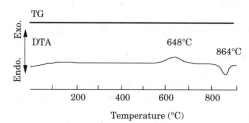

Figure 4. *TG-DTA curves of a new phase .*

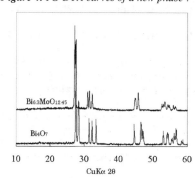

Figure 5. *X-ray powder patterns of a new phase and Bi_4O_7.*

CONCLUSIONS

A variety of new bismuth oxides were prepared by hydrothermal reaction using hydrated sodium bismuth oxide, $NaBiO_3 \cdot nH_2O$, as a starting material. Generally at higher temperatures trivalent bismuth oxides appeared and at lower temperatures pentavalent bismuth oxides were obtained. No pentavalent bismuth oxides containing rare-earth metals were prepared and only $AgBiO_3$ and $ZnBi_2O_6$ were obtained for pentavalent bismuth oxides containing transition-metals. A new bismuth molybdenum oxide, $Bi_{6.3}MoO_{12.45}$ with a pyrochlore-related structure was prepared by hydrothermal reaction in A_2MoO_4 (A = Li, Na, K) solution above 220ºC.

ACKNOWLEDGMENTS

This work was supported by Grant-in-Aid from The Ministry of Education, Science, Sports and Culture of Japan (No.12650668).

REFERENCES

1. N. Kumada, M. Hosoda and N. Kinomura, *J. Solid State Chem.*, **106**, 476 (1993).
2. N. Kinomura , M. Hosoda, N. Kumada and H. Kojima, *J. Ceram. Soc. Japan,.* **101**, 966 (1993).
3. N. Kumada, N. Kinomura, S. Kodialam and A. W. Sleight, *Mater. Res. Bull.*, **29**, 497 (1994).
4. S. Kodialam, N. Kumada, R. Mackey and A. W. Sleight, *European J. Solid State and Inorg. Chem.*, **31**, 739 (1994).
5. N. Kinomura and N. Kumada, *Mater. Res. Bull.*, **30**, 129 (1995).
6. N. Kumada, N. Kinomura, P. M. Woodward and A. W. Sleight, *J. Solid State Chem.*, **116**, 281 (1995).
7. N. Kumada, N. Kinomura, N. Takahashi and A. W. Sleight, *J. Solid State Chem.*, **126**, 121 (1996).
8. N. Kumada, N. Kinomura, N. Takahashi and A. W. Sleight, *Mater. Res. Bull.*, **32**, 1003 (1997).
9. N. Kumada, N. Kinomura, N. Takahashi and A. W. Sleight, *J. Solid State Chem.*, **139**, 321 (1998).
10. N. Kumada and N. Kinomura, *Mat. Res. Soc. Symp. Proc.*, **547**, 227 (1999).
11. N. Kumada, N. Kinomura, and A. W. Sleight, *Mater. Res. Bull.*, in press.
12. N. Kumada, ,T. Takei and N. Kinomura, in preparation.
13. F. Izumi, *Kobutsugaku Zasshi,* **17,** 37 (1985).
14. N. Kumada, N. Kinomura and A. W. Sleight, *Solid State Ionics*, **122**, 183 (1999).
15. B. Begemann and M. Jansen, *J. Less-Common Metals*, **156**, 123 (1980).
16. M. Egashira, K. Matsuo, S. Kagawa and T. Seiyama, *J. Catal.*, **58**, 409 (1979).
17. D. J. Buttrey, D. A. Jefferson and J. M. Thomas, *Mater. Res. Bull.*, **21**, 739 (1986).

Mat. Res. Soc. Symp. Proc. Vol. 658 © 2001 Materials Research Society

Chemical Synthesis and Properties of Layered $Co_{1-y}Ni_yO_{2-\delta}$ Oxides ($0 \le y \le 1$)

A. Manthiram, R. V. Chebiam and F. Prado
Texas Materials Institute, ETC 9.104,
University of Texas at Austin,
Austin, TX 78712

ABSTRACT

Layered $Co_{1-y}Ni_yO_{2-\delta}$ oxides with $0 \le y \le 1$ have been synthesized by chemically extracting lithium from $LiNi_{1-y}Co_yO_2$ with NO_2PF_6 at ambient temperature. The samples have been characterized by X-ray diffraction, wet-chemical analyses, infrared spectroscopy, and magnetic susceptibility measurements. While $NiO_{2-\delta}$ retains the initial O3 ($CdCl_2$ structure) layer structure of $LiNiO_2$, $CoO_{2-\delta}$ consists of a mixture of P3 and O1 (CdI_2 structure) phases that are formed by a sliding of the oxide ions in the initial O3 structure. $CoO_{2-\delta}$ and $NiO_{2-\delta}$ have oxygen contents of, respectively, 1.67 and 1.95 and the oxygen content increases with increasing Ni content, y, in $Co_{1-y}Ni_yO_{2-\delta}$. While $CoO_{2-\delta}$ exhibits metallic conductivity as revealed by the absence of absorption bands in the infrared spectrum, $NiO_{2-\delta}$ exhibits semiconducting behavior due to a completely filled t_{2g} band. Magnetic data reveal a transition from antiferromagnetic to ferromagnetic correlations as the Ni content in $Co_{1-y}Ni_yO_{2-\delta}$ increases.

INTRODUCTION

Transition metal oxides having a near-equivalence of metal:d and oxygen:2p energies are characterized by a small charge transfer gap [1] and they consist of metal ions in high oxidation states. Some examples are oxides containing $Mn^{3+/4+}$, $Fe^{3+/4+}$, $Co^{3+/4+}$, $Ni^{3+/4+}$, and $Cu^{2+/3+}$ couples. Such oxides have drawn considerable attention in recent years due to their interesting electronic properties such as high temperature superconductivity, colossal magnetoresistance, and metal-insulator transitions [2-6]. However, most of the previous studies have focused mainly on oxides having perovskite structures with 180° M-O-M bonds. Relatively, less information is available on oxides having 90° M-O-M and direct M-M bonds (for example, rock salt and spinel structures) in which the transition metal ions exist in a high oxidation state. The lack of a detailed investigation of their electronic properties is mainly due to the difficulties in accessing these oxides having unusually high oxidation states such as Fe^{4+}, Co^{4+} and Ni^{4+}. The objective of this investigation is to access such oxides by soft chemistry procedures and investigate their properties.

Oxides with the general formula $LiMO_2$ (M = Co and Ni) crystallize in a layer structure in which the Li^+ and M^{3+} ions occupy the alternate (111) planes of the rock salt structure. With an edge-shared octahedral arrangement for the M^{3+} cations, this structure provides 90° M-O-M and direct M-M interactions. These $LiMO_2$ (M = Co and Ni) oxides have also drawn much attention in recent years due to their use as positive electrodes in lithium-ion batteries [7]. These layered oxides offer a convenient route to synthesize MO_2 phases having highly oxidized M^{4+} ions by extracting the lithium ions from the $LiMO_2$ phases. However, previous studies have focused mainly on the electrochemical extraction of lithium from these oxides [8-11]. The electrochemically synthesized samples are generally contaminated with carbon, binder, and electrolyte and therefore, it is difficult to characterize their physical properties and oxygen content. Synthesis of bulk samples by chemically extracting lithium from $LiMO_2$ is needed to characterize their physical properties. However, the most commonly used oxidizing agents such as I_2 and Br_2 are not strong enough to extract all the lithium from $LiCoO_2$ and $LiNiO_2$. Additionally, lithium has been extracted by treating the $LiMO_2$ oxides with dilute acids, but such a procedure leads to the incorporation of protons into the lattice. We present in this paper the use of the oxidizing agent NO_2PF_6 [12] in acetonitrile medium to extract all the lithium from $LiCo_{1-y}Ni_yO_2$ and obtain $Co_{1-y}Ni_yO_2$ for $0 \leq y \leq 1$. Characterizations of the $Co_{1-y}Ni_yO_2$ phases by X-ray diffraction, redox titrations to obtain the oxygen contents, infrared spectroscopy, and magnetic measurements are presented in this paper.

EXPERIMENTAL

Layered $LiCo_{1-y}Ni_yO_2$ ($0.5 \leq y \leq 1$) samples were synthesized by a sol-gel procedure [13] with a final firing temperature of $750°C$ in oxygen atmosphere for 24 hours. Layered $LiCoO_2$ was synthesized by solid state reaction between Li_2CO_3 and Co_3O_4 at $900°C$ in air for 24 hours. Chemical extraction of lithium from $LiCo_{1-y}Ni_yO_2$ was carried out by stirring the oxides for 2 days under argon atmosphere in acetonitrile medium containing excess NO_2PF_6. The molar ratio of $LiCo_{1-y}Ni_yO_2 : NO_2PF_6$ in the reaction mixture was increased from 1 : 1.5 to 1 : 2 as the nickel content, y, increases from 0 to 1. During this process, the following reaction occurs:

$$LiCo_{1-y}Ni_yO_2 + NO_2PF_6 \rightarrow Co_{1-y}Ni_yO_2 + NO_2 + LiPF_6 \qquad (1)$$

The products formed after the reaction were filtered under argon atmosphere, washed with acetonitrile to remove the byproducts and any unreacted oxidizing agent, and dried under vacuum at ambient temperature. Crystal chemistry of the samples was characterized by X-ray diffraction. Lithium content was determined by atomic absorption spectroscopy and oxygen content was determined by iodometric titration [14]. Fourier transform infrared (FTIR) spectra were recorded with KBr pellets.

RESULTS AND DISCUSSION

Figure 1 shows the X-ray diffraction patterns of $Co_{1-y}Ni_yO_2$ ($0 \leq y \leq 1$) that were obtained by chemically extracting lithium from $LiCo_{1-y}Ni_yO_2$. The X-ray pattern of the end member NiO_2 could be fit by the Rietveld program on the basis of the O3 layer structure ($CdCl_2$ structure) similar to that of the parent $LiNiO_2$. It has been suggested in the literature [11] that the presence of some nickel in the lithium plane in the initial material stabilizes the O3 structure for the end member NiO_2 while the absence of nickel in the lithium layer stabilizes the O1 structure. In view of this, the observation of the O3 structure for the chemically prepared NiO_2 may suggest the presence of a small amount of nickel in the lithium plane in our initial $LiNiO_2$ sample. However Rietveld analysis of our initial $LiNiO_2$ could not locate any nickel in the lithium plane. On the other hand, the CoO_2 sample was found to consist of a mixture of P3 and O1 (CdI_2 structure) phases. This result is in contrast to the single O1 phase [9] or a mixture of two O1 phases with different oxygen contents [10] reported for the electrochemically prepared CoO_2. The P3 and O1 phases are formed by a sliding of the oxide ions in the initial O3 structure. For $0 \leq (1-y) \leq 0.3$ in $Co_{1-y}Ni_yO_2$, the O3 structure is maintained. For Co content $(1-y) > 0.3$, the crystallinity of the O3 structure is slowly destroyed to give ultimately a mixture of P3 and O1 phases for $CoO_{2-\delta}$.

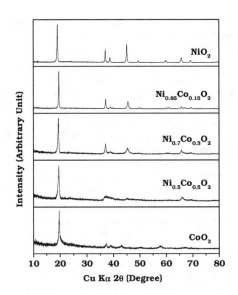

Figure 1. *X-ray diffraction patterns of* $Co_{1-y}Ni_yO_2$.

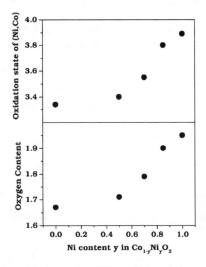

Figure 2. *Variations of the oxidation states of Co and Ni and the oxygen content with Ni content in* $Co_{1-y}Ni_yO_{2-\delta}$.

Figure 2 shows the variations of the oxidation state of the transition metal ions (average oxidation state of cobalt and nickel ions) and the oxygen content with Ni content in $Co_{1-y}Ni_yO_2$ $NiO_{2-\delta}$ has an oxygen content close to 2 with a $\delta = 0.05$ and an oxidation state of 3.9+ for Ni. T oxidation state of the transition metal ions in $Co_{1-y}Ni_yO_{2-\delta}$ decreases with increasing Co conten for $0 \leq (1-y) \leq 0.5$ and thereafter remains constant around 3.4+. $CoO_{2-\delta}$ has an oxygen content c 1.67 with considerable amount of oxygen vacancies. We believe that the presence of a large amount of oxygen vacancies in $CoO_{2-\delta}$ makes the initial O3 structure unstable and facilitates the sliding of the oxide ions to form the P3 and O1 structures.

The variations in the oxygen content with Ni content can be understood by considering the qualitative energy diagram shown in Figure 3. In the case of $LiCoO_2$ with a $Co^{3+}:t_{2g}^{6}$ configuration, the electrons are removed initially from the t_{2g} band on extracting lithium. Since the t_{2g} band overlaps with the top of the O:2p band, electrons will be removed from the O:2p band as well at low lithium contents. Removal of electrons from the O:2p band leads to an oxidation of the O^{2-} ions and a loss of oxygen from the lattice resulting in a significant amount oxygen vacancies in $CoO_{2-\delta}$. On the other hand, in the case of $LiNiO_2$ with a $Ni^{3+}:t_{2g}^{6}e_g^{1}$ configuration, the electrons are removed from the e_g band on extracting lithium. Since the e_g band lies well above the top of the O:2p band, $NiO_{2-\delta}$ does not suffer from much oxygen loss. A a result, the oxygen content decreases with increasing Co content 1-y in $Co_{1-y}Ni_yO_{2-\delta}$ (Figure 3)

Since the $Co_{1-y}Ni_yO_{2-\delta}$ samples are metastable and tend to disproportionate at T > 100 °C, it difficult to obtain sintered pellets and measure the electrical resistivity. However, the nature of electrical conduction can be ascertained by examining the infrared spectra. Figure 4 shows the FTIR spectra of $Co_{1-y}Ni_yO_{2-\delta}$. $CoO_{2-\delta}$ does not show any absorption bands suggesting metallic behavior, which is consistent with that found by Menetrier et al [15]. On the other hand, absorption bands develop with increasing Ni content y in $Co_{1-y}Ni_yO_{2-\delta}$ suggesting semiconducting behavior. The differences in electrical behavior can be understood by considering the energy diagram in Figure 3. A direct Co-Co interaction via the partially filled t_2

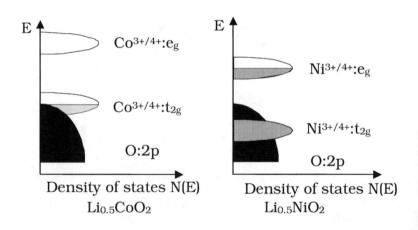

Figure 3. *Comparison of the qualitative energy diagrams of $Li_{1-x}CoO_2$ and $Li_{1-x}NiO_2$.*

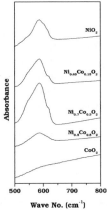

Figure 4 (left panel): FTIR spectra with labels NiO₂, Ni₀.₈₅Co₀.₁₅O₂, Ni₀.₇Co₀.₃O₂, Ni₀.₅Co₀.₅O₂, CoO₂; x-axis Wave No. (cm⁻¹) 500 600 700 800; y-axis Absorbance.

Figure 5 (right panel): plot with legend CoO₁.₆₇, Co₀.₅Ni₀.₅O₁.₇₁, Co₀.₃Ni₀.₇O₁.₇₉, Co₀.₁₅Ni₀.₈₅O₁.₉, NiO₁.₉₅; y-axis 1/χₘ (mol/emu); x-axis Temperature (K).

Figure 4. *FTIR spectra of $Co_{1-y}Ni_yO_{2-\delta}$.*

Figure 5. *Variation of inverse molar magnetic susceptibility with temperature for $Co_{1-y}Ni_yO_{2-\delta}$. The inset shows a magnification of the low temperature region.*

band in $CoO_{2-\delta}$ leads to metallic conduction. In contrast, a completely filled t_{2g} band in $NiO_{2-\delta}$ leads to semiconducting behavior. Additionally, the Ni^{3+} ions would be oxidized first before the Co^{3+} ions are oxidized since $Ni^{3+/4+}$:e_g band lies above the $Co^{3+/4+}$:t_{2g} band. This factor together with a lower oxidation state of the transition metal ions with oxygen vacancies (Figure 2) leads to semiconducting behavior (Figure 4) in $Co_{1-y}Ni_yO_{2-\delta}$ for $0.5 \leq y \leq 1$.

Figure 5 shows the variation of the reciprocal molar magnetic susceptibility with temperature for $Co_{1-y}Ni_yO_{2-\delta}$. The data suggest a gradual transition from antiferromagnetic to ferromagnetic correlations with increasing nickel content, y, as indicated by a change in the sign of the Weiss constant (Table I). Table I also gives the observed effective magnetic moment in the region 150-320 K along with the calculated effective moment based on low spin configurations for $Co^{3+/4+}$ and $Ni^{3+/4+}$ ions. While the moments of the y = 0.5 and 0.7 samples are close to the calculated values, those of the y = 0, 0.85 and 1 samples are higher than the calculated values. The discrepancy could be due to some intermediate spin configuration or a partial disproportionation of the metastable $Co_{1-y}Ni_yO_{2-\delta}$ phases during handling and measurements.

Table I. *Magnetic data of $Co_{1-y}Ni_yO_{2-\delta}$.*

Composition	Oxidation state of (Co, Ni)	θ (K)	Experimental magnetic moment (μ_B)	Calculated magnetic moment (μ_B)
$CoO_{1.67}$	3.32	-260	1.74	0.95
$Co_{0.5}Ni_{0.5}O_{1.71}$	3.42	-60	1.10	1.22
$Co_{0.3}Ni_{0.7}O_{1.79}$	3.58	-25	1.09	1.19
$Co_{0.15}Ni_{0.85}O_{1.90}$	3.8	2	1.26	0.93
$NiO_{1.95}$	3.9	85	0.93	0.55

CONCLUSIONS

Metastable layered $Co_{1-y}Ni_yO_{2-\delta}$ oxides have been synthesized successfully for $0 \leq y \leq 1$ by chemically extracting lithium from $LiCo_{1-y}Ni_yO_{2-\delta}$. The nickel-rich $Co_{1-y}Ni_yO_{2-\delta}$ phases maintain the initial O3 layer structure (CdCl$_2$ structure) of $LiNiO_2$ with oxygen contents close to 2. In contrast, the cobalt-rich phases consist of a mixture of P3 and O1 (CdI$_2$ structure) phases with significant amount of oxygen vacancies. While $CoO_{2-\delta}$ is metallic, the $Co_{1-y}Ni_yO_{2-\delta}$ phases with $0.5 \leq y \leq 1$ are semiconducting. The differences in structure, oxygen content, and electrical properties could be rationalized on the basis of qualitative energy diagrams. The $Co_{1-y}Ni_yO_{2-\delta}$ phases exhibit a gradual transition from antiferromagnetic to ferromagnetic correlations with increasing Ni content.

ACKNOWLEDGEMENTS

Financial support by the Center for Space Power at the Texas A&M University (A NASA Commercial Space Center) and Welch Foundation Grant F-1254 is gratefully acknowledged.

REFERENCES

1. J. Zaanen, G. A. Zawatzky, and J. W. Allen, *Phys. Rev. Lett.*, **55**, 418 (1985).
2. J. G. Bednorz and K. A. Muller, *Z. Phys.*, **B 64**, 189 (1986).
3. R. Von Helmolt, J. Wecker, B. Holzapfelh, L. Schultz, and K. Samwer, *Phys. Rev. Lett.*, **71**, 2331 (1993).
4. J. B. Torrance, P. Lacorre, C. Asavaroengchai, and R. M. Metzger, *J. Solid State Chem.*, **90**, 168 (1991).
5. J. B. Torrance, P. Lacorre, C. Asavaroengchai, and R. M. Metzger, *Physica C.*, **182**, 351 (1991).
6. J. B. Torrance, P. Lacorre, A. I. Nazzal, E. J. Ansaldo, and Ch. Niedermayer, *Phys. Rev.*, **B 45**, 8209 (1992).
7. D. Linden, *Handbook of Batteries*, ed. D. Linden (McGraw-Hill, 1995) p. 14.1.
8. T. Ohzuku, A. Ueda, and M. Nagayama, *J. Electrochem. Soc.*, **140**, 1862 (1993).
9. G. G. Amatucci, J. M. Tarascon, and L. C. Klein, *J. Electrochem. Soc.*, **143**, 1114 (1996).
10. J. M. Tarascon, G. Vaughan, Y. Chabre, L. Seguin, M. Anne, P. Strobel, and G. Amatucci, *J. Solid State Chem.*, **147**, 410 (1999).
11. L. Croguennec, C. Pouillerie, and C. Delmas, *J. Electrochem. Soc.*, **147**, 1314 (2000).
12. A. R. Wizansky, P. E. Rauch, and F. J. DiSalvo, *J. Solid State Chem.*, **81**, 203(1989).
13. T. Armstrong, F. Prado, Y. Xia, and A. Manthiram, *J. Electrochem. Soc.*, **147**, 435 (2000).
14. A. Manthiram, J. S. Swinnea, Z. T. Sui, H. Steinfink, and J. B. Goodenough, *J. Amer. Chem Soc.*, **109**, 6667 (1987).
15. M. Menetrier, I. Saadoune, S. Levasseur, and C. Delmas, *J. Mater. Chem.*, **9**, 1135 (1999).

Solid-State Ionics, Battery Materials, Thermo Power, Optical Materials

Mat. Res. Soc. Symp. Proc. Vol. 658 © 2001 Materials Research Society

Preparation, Electrical Properties, and Water Adsorption Behavior of $(1-x)Sb_2O_5 \cdot xM_2O_3 \cdot nH_2O$ (M = Al, Bi, and Y; $0 \leq x \leq 1$)

Kiyoshi Ozawa, Yoshio Sakka, and Muneyuki Amano
National Institute for Materials Science
1-2-1 Sengen, Tsukuba, Ibaraki 305-0047, Japan

ABSTRACT

Bi, Y, and Al-doped antimonic acids with the empirical formula $(1-x)Sb_2O_5 \cdot xM_2O_3 \cdot nH_2O$ (M = Al, Bi and Y; $0 \leq x \leq 1$) have been prepared by the direct reaction of an aqueous H_2O_2 solution with metal alkoxides. The electrical properties have been investigated by ac and dc conductivity measurements using a polycrystalline compact disc as a sample, in connection with the water adsorption behavior. The $(1-x)Sb_2O_5 \cdot xM_2O_3 \cdot nH_2O$ (M = Bi and Y) materials exhibit a solid solution of cubic antimonic acid ($Sb_2O_5 \cdot nH_2O$) with Bi_2O_3 and Y_2O_3 in the range of $x = 0 - 0.1$, and the ionic transference numbers of the materials are all over 0.98. The proton conductivity of $(1-x)Sb_2O_5 \cdot xM_2O_3 \cdot nH_2O$ (M = Bi and Y; $x = 0.1$) at room temperature increases to over 10^{-3} Scm^{-1} with an increase in the relative humidity. The conductivity is closely affected by the water adsorption behavior, which seems to be related to the introduction of oxygen vacancies.

INTRODUCTION

Cubic antimonic acid ($Sb_2O_5 \cdot nH_2O$) is known to be a room-temperature proton conductor [1]. The crystal structure of cubic $Sb_2O_5 \cdot nH_2O$ is represented by a cubic pyrochlore structure and has a three-dimensional framework built up of vertex-linked $Sb_2O_6^{2-}$ octahedra [2]. In this framework there are interconnected channels in which a large number of water molecules are located [2,3]. The proton conductivity of cubic $Sb_2O_5 \cdot nH_2O$ is considered to occur by a Grotthuss-type mechanism through the hydrogen bond networks of the water molecules [2,4].

Thus, a variety of conductivity can be expected for cubic $Sb_2O_5 \cdot nH_2O$ by changing the amount of water molecules, which may be affected by variation in the channel size and/or by the introduction of oxygen vacancies besides the ambient humidity. From this point of view we have prepared Bi, Y, and Al-doped antimonic acids with the empirical formula $(1-x)Sb_2O_5 \cdot xM_2O_3 \cdot nH_2O$ (M = Bi, Y and Al) for which Sb^{5+} in the cubic $Sb_2O_5 \cdot nH_2O$ is partially substituted by M^{3+} with a different ion radius from Sb^{5+} ($r(Sb^{5+}) = 0.60$ Å; $r(Bi^{3+}) = 1.06$ Å; $r(Y^{3+}) = 0.90$ Å; $r(Al^{3+}) = 0.54$ Å) [5].

The preparation and proton conductivity of the Bi-doped antimonic acids were previously reported elsewhere [6]. In this study, the preparation of the Y and Al-doped antimonic acids have been briefly demonstrated, and then the electrical properties are discussed in connection with the water adsorption behavior.

EXPERIMENTAL

The Y and Al-doped antimonic acids were prepared by the reaction of aqueous H_2O_2 solution with metal alkoxides ($Sb(O\text{-}iso\text{-}C_3H_7)_3$, $Y(O\text{-}iso\text{-}C_3H_7)_3$ and $Al(O\text{-}iso\text{-}C_3H_7)_3$), according to the previously reported method [6]. In short, $Y(O\text{-}iso\text{-}C_3H_7)_3$ and $Al(O\text{-}iso\text{-}C_3H_7)_3$ were proportionally weighed for values of $x = 0$, 0.05, 0.1, 0.2, 0.4, 0.6, 0.8 and 1 in

$(1-x)Sb_2O_5 \cdot xM_2O_3 \cdot nH_2O$ (M = Y and Al). Each weight quantity of $Y(O\text{-}iso\text{-}C_3H_7)_3$ an Al$(O\text{-}iso\text{-}C_3H_7)_3$ was dissolved in a small amount of 2-ethoxyethanol, and mixed wit Sb$(O\text{-}iso\text{-}C_3H_7)_3$. The mixtures were added in limited amounts to a 30% aqueous H_2O_2 solutio then refluxed at $\sim100°C$ for 3 h. Subsequently, the excess H_2O_2 in the solutions was decompose with several platinum foils, and then the organic residue was removed by extraction with dieth ether. Finally, the solutions were dried by evaporation at 120°C to produce th $(1-x)Sb_2O_5 \cdot xM_2O_3 \cdot nH_2O$ (M = Y and Al) powders. These powders were used for the x-ra diffraction (XRD) measurements, thermogravimetric (TG) and differential thermal analysi (DTA), water adsorption isotherms, and electrical properties.

With regard to the electrical properties, the ionic transference numbers and proton conductivit at room temperature were investigated. The ionic transference numbers were determined by simplified polarization method using dc conductivity measurements. The proton conductivity wa evaluated under various controlled levels of relative humidity employing an ac impedanc method in the frequency range of 100 Hz – 10 MHz using a Hewlett-Packard 4194 analyzer. Fc the electrical properties measurements, compact polycrystalline discs of 13 mm diameter, ~1 mi thickness, and $\sim55\%$ relative density were prepared by pressing the sample powders at 147 MP, Furthermore, nickel sponges of ~11 mm diameter were attached to both sides of the discs a electrodes, and platinum wires were connected to the nickel sponges using a silver paste. Th water adsorption isotherms were measured at 25°C using a computer-controlled automati adsorption machine (Belsorp 18: Bell Japan, Inc.).

RESULTS AND DISCUSSION

It is confirmed that all the Bragg reflections of $(1-x)Sb_2O_5 \cdot xM_2O_3 \cdot nH_2O$ (M = Y; $0 \le x \le 0.2$ can be assigned to the cubic $Sb_2O_5 \cdot nH_2O$. In contrast, reflections from the tetragonal AlSbO$_4$ ar observed in the XRD profiles of $(1-x)Sb_2O_5 \cdot xM_2O_3 \cdot nH_2O$ (M = Al; $0.05 \le x \le 0.2$) besides th cubic $Sb_2O_5 \cdot nH_2O$.

In general, the lattice parameters of many room-temperature proton conductors are influence by their water content, which usually varies according to the ambient humidity [7]. Thus, in orde to estimate the lattice parameters, the anhydrous composites prepared by heat-treatment at 480°C were used in this study. The lattice parameters of cubic Sb_2O_5 for the anhydrous composites ar plotted as a function of the M content x in Figure 1. The values of the Bi-doped anhydrou composites are also shown in Figure 1 [6]. It is obvious that variations in the lattice parameter fc $(1-x)Sb_2O_5 \cdot xM_2O_3$ (M = Bi and Y) follow Vegard's law in the x range of $0-0.1$, whereas variatic in $(1-x)Sb_2O_5 \cdot xM_2O_3$ (M = Al) is nearly unchanged. These results indicate that th $(1-x)Sb_2O_5 \cdot xM_2O_3 \cdot nH_2O$ (M = Bi and Y) materials are solid solutions of cubic $Sb_2O_5 \cdot nH_2O$ wit Bi$_2O_3$ and Y_2O_3 in the x range of $0-0.1$ and the $(1-x)Sb_2O_5 \cdot xM_2O_3 \cdot nH_2O$ (M = Al) materials are mixture of cubic $Sb_2O_5 \cdot nH_2O$ and aluminium oxides.

Figure 2 shows the water adsorption isotherms of $(1-x)Sb_2O_5 \cdot xM_2O_3 \cdot nH_2O$ (M = Y and Al; 0 $x \le 0.2$) after evacuation at room temperature for 10 h. The vertical line V indicates the mol; quantities of adsorbed water per one mole of $(1-x)Sb_2O_5 \cdot xM_2O_3 \cdot nH_2O$ with the individual startin water content n. The n values were found to be $2.0 - 2.3$ from the TG measurements. For th Y-doped antimonic acids, it is obvious that the V values are affected not only by the relativ pressure but also by the yttrium content x as shown in Figure 2 (A). For example, the V values c the Y-doped antimonic acid for $x = 0.1$ and 0.2 steeply increase as the relative pressure become greater than ~0.6. There is a difference in the V value of 1.2 moles from that for $x = 0$ when th

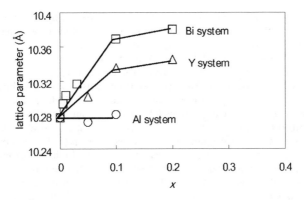

Figure 1. Plots of the lattice parameters of cubic Sb_2O_5 for $(1-x)Sb_2O_5 \cdot xM_2O_3$ (M = Al, Bi, and Y) as a function of the M content.

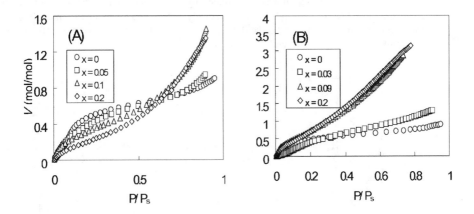

Figure 2. Water adsorption isotherms at 25°C. (A) $(1-x)Sb_2O_5 \cdot xY_2O_3 \cdot nH_2O$ (x = 0, 0.05, 0.1 and 0.2) and (B) $(1-x)Sb_2O_5 \cdot xAl_2O_3 \cdot nH_2O$ (x = 0, 0.03, 0.09 and 0.2).

relative pressure reaches 0.9. Such behavior is in excellent agreement with that of the Bi-doped antimonic acids [6]. It is found that there is no significant difference in the BET specific area measured by nitrogen molecules for $(1-x)Sb_2O_5 \cdot xM_2O_3 \cdot nH_2O$ (M = Bi and Y; $0 \leq x \leq 1$); the specific area is 66.3 and 76.2 m^2/g for M = Bi and Y, respectively. The introduction of oxygen vacancies is likely to occur for the Bi and Y-doped antimonic acids, as the result of the quantitative analyses of Sb^{3+} [8]. Such great increase in the quantities of adsorbed water measured

for the Bi and Y-doped antimonic acids is most likely to be caused by the introduction of oxyg
vacancies. Certainly, there is a lattice expansion for Bi and Y-doped antimonic acids. Howev
the lattice expansion is only 1%, if so much, as shown in Figure 1.

On the other hand, it is found for the Al-doped antimonic acids that the V values
$(1-x)Sb_2O_5 \cdot xM_2O_3 \cdot nH_2O$ (M = Al; $x = 0.09$ and 0.2) increased more steeply with an increase in t
relative humidity than those of the Y-doped antimonic acids as shown in Figure 2 (B). The
values become approximately three times as large as those of the Y-doped antimonic acids for x
0.1 and 0.2 at a relative pressure of ~0.7. Such large V values are likely to be caused by the wat
molecules adsorbed on surface of the particles. As investigated below, the large quantities
adsorbed water for the Al-doped antimonic acids scarcely take part in the proton conductivity.

Figure 3 is a typical polarization-depolarization curve for a Bi-doped antimonic acid at 20
with 1 V dc bias. A clear polarization-depolarization curve can be observed, and such a cle
curve is also observed for the remaining specimens. The ionic transference number (t_i) is given
the following equation:

$$t_i = \sigma(ion)/\sigma(total) = 1 - \sigma(electron)/\sigma(total) = 1 - \sigma(\infty)/\sigma(0) = 1 - I(\infty)/I(0) \qquad (1)$$

where $I(0)$ and $I(\infty)$ are the dc current at initial and infinite times, respectively. In practic
however, the exact $I(0)$ can not be determined by ac measurement because of a time lag. It
considered that $I(0)$, i.e., $\sigma(total)$, is more precisely determined by ac measurement [9]. The ior
transference number was found to be 0.98, and those of all the other specimens (Bi and Y-dop
antimonic acids) were also found to be over 0.98.

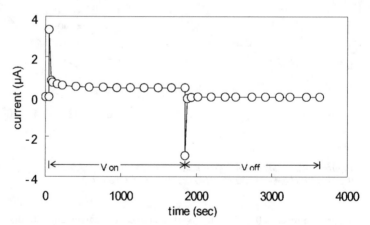

Figure 3. Polarization-depolarization curve of $(1-x)Sb_2O_5 \cdot xBi_2O_3 \cdot nH_2O$ ($x =$
0.1) at 20°C.

Figure 4 indicates the proton conductivity of $(1-x)Sb_2O_5 \cdot xM_2O_3 \cdot nH_2O$ (M = Y and Al; $0 \le x \le 0.$
as a function of the relative humidity. It is found as shown in Figure 4 (A) that t

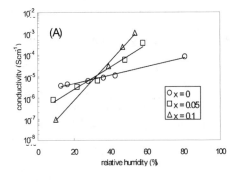 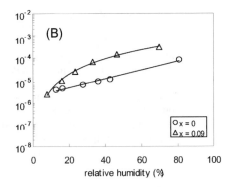

Figure 4. The proton conductivity at 20°C as a function of relative humidity. (A) $(1-x)Sb_2O_5 \cdot xY_2O_3 \cdot nH_2O$ ($x = 0, 0.05$ and 0.1) and (B) $(1-x)Sb_2O_5 \cdot xAl_2O_3 \cdot nH_2O$ ($x = 0$ and 0.09).

conductivities of the Y-doped antimonic acids increase with an increase in the relative humidity, and the increase rate becomes greater with the Y content x. Moreover, it should be noted that the conductivity for $x = 0.1$ shows a high conductivity of over 10^{-3} Scm^{-1} even at a relative humidity of 53%. Such a conductivity behavior is considered to be associated with the water adsorption behavior, as is the case with the Bi-doped antimonic acids [6].

On the other hand, variation in the conductivity for the $(1-x)Sb_2O_5 \cdot xM_2O_3 \cdot nH_2O$ (M = Al) is not as significant as expected from a large quantity of adsorbed water molecules. For example, the conductivity for $x = 0.09$ varies within two orders magnitude with the relative humidity in the range of 7.8 – 70%. These results suggest that indicate that a large quantity of adsorbed water molecules for $(1-x)Sb_2O_5 \cdot xM_2O_3 \cdot nH_2O$ (M = Al) hardly takes part in the conductivity.

CONCLUSIONS

We have demonstrated the preparation, water adsorption behavior, and electrical properties of $(1-x)Sb_2O_5 \cdot xM_2O_3 \cdot nH_2O$ (M = Bi, Y and Al). The materials of $(1-x)Sb_2O_5 \cdot xM_2O_3 \cdot nH_2O$ (M = Bi and Y) exhibit a solid solution of cubic $Sb_2O_5 \cdot nH_2O$ with Bi_2O_3 and Y_2O_3 in the x range of $0 - 0.1$. The conductivity of $(1-x)Sb_2O_5 \cdot xM_2O_3 \cdot nH_2O$ (M = Bi and Y; $0 \le x \le 0.1$) is influenced by the water adsorption behavior and increases to over 10^{-3} Scm^{-1} with the relative humidity. In contrast, the conductivity of the $(1-x)Sb_2O_5 \cdot xM_2O_3 \cdot nH_2O$ (M = Al) acids does not increase as significantly as compared to that of the Bi and Y-doped antimonic acids, though a large quantity of adsorbed water molecules is observed. The water adsorption behavior for Bi and Y-doped antimonic acids is likely to be caused by the introduction of oxygen vacancies.

REFERENCES

1. W. A. England and R. C. T. Slade, *Solid State Commun.*, **33**, 997 (1980).
2. W. A. England, M. G. Cross, A. Hamnett, P. J. Wiseman, and J. B. Goodenough, *Solid State Ionics*, **1**, 231 (1980).

3. Y. Sakka, K. Sodeyama, T. Uchikoshi, K. Ozawa, and M. Amano, *J. Am. Ceram. Soc.*, **79**, 16 (1996).
4. Ph. Colomban, C. Doremieux-Mori, Y. Piffard, M. H. Limage, and A. Novak, *J. Mol. Stru* **213**, 83 (1989).
5. CRC Handbook of Chemistry and Physics, *Ionic Radii in Crystals*, ed. D. R. Lide (CRC Pres 1999) pp.**12**-14-**12**-15.
6. K. Ozawa, Y. Sakka, and M Amano, in *Solid-State Ionics V*, ed. G. A. Nazri, C. Julien, and A Rougier (Mater. Res. Soc. Symp. Proc., **548**, Warrendale, PA, 1999) pp.599-604.
7. S. Courant, Y. Piffard, P. Barboux, and L. Livage, *Solid State Ionics*, **27**, 189 (1988).
8. K. Ozawa, unpublished data.
9. S. Y. Bae, M. Miyayama, and H. Yanagida, *J. Am. Ceram. Soc.*, **77**, 891 (1994).

Mat. Res. Soc. Symp. Proc. Vol. 658 © 2001 Materials Research Society

High Oxidation State Alkali-Metal Late-Transition-Metal Oxides

David B. Currie, Andrew L. Hector, Emmanuelle A. Raekelboom, John R. Owen and Mark T. Weller*
Department of Chemistry, University of Southampton, Highfield, Southampton SO17 1BJ, UK

ABSTRACT

$Li_2NaCu_2O_4$ has been prepared by solid state reaction under high-pressure (250 Atm) oxygen. A structural study, using time-of-flight powder neutron diffraction on a sample made with ^7Li, shows a material isostructural with $Li_3Cu_2O_4$, with sodium occupying the octahedral and lithium the tetrahedral A-cation sites. A ^7Li MAS-NMR study of $Li_3Cu_2O_4$, $Li_2NaCu_2O_4$ and Li_2CuO_2 has been used to confirm the Li/Na site ordering.

INTRODUCTION

Portability is one of the primary concerns in the design of many new electronic devices. Thus the reduction in size and weight of primary and secondary (rechargeable) battery systems is a key concern. The main focus of current interest is the lithium battery technology based on a lithium (or intercalation host) anode and a cathode, which is usually a layered or spinel oxide of manganese or cobalt. Our aim is to synthesise interesting high oxidation state transition metal oxide systems, to investigate their structures and link these with their utility as cathode materials. High oxidation states are particularly of interest as a means of maximising cell potential. To this end our synthetic strategy involves reactions in high pressure oxygen and sealed systems with highly oxidised reagents.

The number of copper based cathode systems that have been studied previously is quite small. As a primary battery system, Li-CuO cells [1] have a low but convenient working voltage of 1.5 V and are able to operate at relatively high temperatures (135°C). Use of a copper oxide phosphate ($Cu_4O(PO_4)_2$) cathode increases the working voltage to 2.5 V [1], but the recharging of cells is still not possible. These systems use the Cu^+/Cu^{2+} redox couple. Li-CuFeO_2 cells have a working voltage of 1 V [2]. The cells have been shown to cycle but lose capacity rapidly. Copper K-edge X-ray absorption and ^{57}Fe Mössbauer spectroscopy have shown [2] that copper is reduced to Cu^0 and iron remains Fe^{3+}. The mechanism by which these cells operate appears to be the replacement of copper with lithium, yielding $LiFeO_2$, with deposition of elemental copper.

In order to achieve higher cell potentials with a copper based cathode, the Cu^{2+}/Cu^{3+} redox couple is a promising candidate. In the spinel $Li_{2.02}Cu_{0.64}Mn_{3.34}O_8$ the copper(III)-copper(II) reduction yields a mid-discharge potential of 4.9 V (a manganese(III)-manganese(IV) feature is observed at 4 V) [3]. A further Cu^{2+}/Cu^{3+} system is derived from Li_2CuO_2. This system was investigated by Arai et al [4], who showed that on charging the voltage remains almost constant at around 3.1 V until the (calculated) composition $Li_{1.2}CuO_2$, after which it rose to about 3.5 V, reaching 3.7 V at a calculated composition $LiCuO_2$.

Our interest in the Arai system was heightened by the report of a $Li_{1.5}CuO_2$ phase, since the structure of $Li_3Cu_2O_4$ (Figure 1) was first reported by ourselves [5]. This material is difficult to synthesise at ambient pressure, requiring several weeks of heating cycles, but under oxygen at around 200 Atm pressure high quality samples can be prepared in a few hours. $Li_3Cu_2O_4$ contains

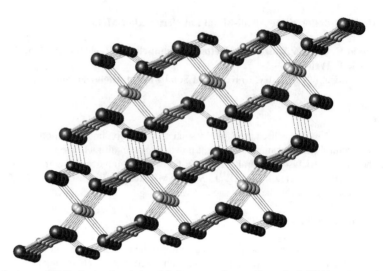

Figure 1. *Structure of $Li_3Cu_2O_4$ showing chains of CuO_4 square planes, split octahedral lithium sites and tetrahedral lithium sites.*

a stoichiometric, though disordered, mixture of copper(II) and copper(III). Its structure consists of chains of edge-sharing CuO_4 square planes, separated by alternate layers of octahedral and tetrahedral lithium ions. Powder neutron diffraction was used [6] to show that the lithium ions in the octahedral sites disorder, with equal numbers of lithium ions displaced toward the top and bottom of the octahedron and effective square pyramidal coordination of lithium. We have also re-examined the electrochemistry and found that we can effectively cycle $Li_3Cu_2O_4$/Li coin cells at around 3.3 V [7].

EXPERIMENTAL

$Li_2ACu_2O_4$ (A=Li,Na) were prepared from the binary oxides by heating at 700°C under ~250Atm oxygen for 4 hours in a gold crucible. Excess Li_2O (20%) and Na_2O (30%) were added to overcome losses through volatilisation. $^7Li_2O$ was made from $^7LiOH.H_2O$ (Aldrich) by dissolving in ethanol, adding hydrogen peroxide and firing the resultant precipitate at 700°C. The products, which were well-sintered grey-black powders, were slightly hygroscopic, possibly due to traces of unreacted alkali-metal oxide, so were handled in air only for short periods. Li_2CuO_2 was made from copper carbonate and lithium carbonate (20% excess), heated in air for 16 hours at 700°C.

Powder X-ray diffraction was performed with a Siemens D5000 diffractometer using Cu-$K_{\alpha1}$ radiation. Patterns used for structure refinement were collected for 16 hours over a 2θ range of 10-110°. Powder neutron data were collected on samples made with ^7Li for 2 hours on the Polaris medium resolution diffractometer at the Rutherford Appleton Laboratory. The backscattered detector data, with a time-of-flight range of 2000-19580 μs, were used for refinement of the structure of $Li_2NaCu_2O_4$. Lattice parameters were refined using the program *Cell* and full profile Rietveld analysis was carried out using the *GSAS* program suite [8].

^{7}Li MAS-NMR (I=$^3/_2$, 92.6% abundance) were collected with a MAS speed of 14.8 kHz using a Bruker MSL-400 solid state NMR spectrometer. A 1M aqueous LiCl solution was used as external chemical shift reference.

The conditions used in the initial electrochemical characterisation experiments are the same as those used for $Li_3Cu_2O_4$ [7].

RESULTS AND DISCUSSION

Since the octahedral lithium site in $Li_3Cu_2O_4$ was clearly underbonded, as evidenced from the split site [6], the stability of the material was considered likely to be enhanced by introducing sodium into the structure. Sodium was thought likely to localise on the octahedral site and remove the split site disorder. Interestingly it was found that $Li_2NaCu_2O_4$ exists as a discreet compound and not as a solid solution with $Li_3Cu_2O_4$. Attempts to prepare a sample of intermediate sodium content yielded mixtures of both phases. Empirical though this evidence is, there is an implication as to the significance of the stoichiometric Cu^{2+}/Cu^{3+} composition found in $Li_3Cu_2O_4$ and $Li_2NaCu_2O_4$. It is possible that there is some localised charge ordering and work is underway to investigate whether there is long range order at low temperature. It has also thus far proved impossible to prepare compounds such as $Li_2CaCu_2O_4$, $Li_2MgCu_2O_4$ or $Li_3Ni_2O_4$.

Structure refinement of $Li_2NaCu_2O_4$

The similarity of the PXD pattern to that of $Li_3Cu_2O_4$ [5] was immediately apparent, with a large shift in the reflection positions to higher d-spacings. The lattice parameters were refined and entered into GSAS with the PND data, using the atom positions and space group of $Li_3Cu_2O_4$. Initially the lattice parameters, profile coefficients and background were refined. Weak, broad reflections due to the vanadium sample can were excluded from the data. The octahedral lithium site in $Li_3Cu_2O_4$ was found to be split to two disordered sites displaced from a mean octahedral position [6]. Sodium was initially placed on this mean position (½,½,½) since it is considerably larger than lithium and could be expected to better fill the site. Refinement of all the atom positions and temperature factors led to a good fit, the introduction of anisotropic temperature factors further improved this fit.

Refinement of the sodium occupation, either by allowing vacancies or by substituting lithium, resulted in no change from full sodium occupation within 3x e.s.d. Lithium could also not be placed close to the mean position (with the sodium occupancy reduced accordingly), it refined to a distant site with very small fractional occupation and no improvement in fit statistics. The positive neutron scattering length of sodium and negative scattering length of lithium would make neutron diffraction highly sensitive to any disorder of this nature.

Finally a joint refinement of the X-ray and neutron data was carried out. It was necessary to cut the number of neutron data points so that the two data sets had approximately equal significance. No significant movement was observed in the atom positions and, crucially, the reflection intensities matched well, indicating the lithium/sodium site occupations derived from the neutron refinement were probably correct.

The final fit statistics, obtained using the full neutron data set only, are listed in Table I.

Table I. *Refined PND parameters for Li₂NaCu₂O₄*

Atom	x	y	z
Cu	0.1617(2)	0	0.2530(2)
O1	0.0680(2)	½	0.3127(2)
O2	0.2528(2)	½	0.1809(3)
Li	0.1357(4)	½	0.8774(6)
Na	½	½	½

C2/m, a=10.2733(2)Å, b=2.80324(3)Å, c=7.58532(9)Å, β=119.6903(8)°

Atom	U11	U22	U33	U12	U13	U23
Cu	0.57(4)	0.30(3)	0.76(4)	0	0.52(3)	0
O1	0.95(5)	0.64(4)	1.54(6)	0	1.04(4)	0
O2	0.93(6)	0.86(5)	1.03(6)	0	0.62(5)	0
Li	1.39(13)	1.79(13)	1.47(14)	0	0.65(11)	0
Na	1.59(10)	0.86(7)	1.39(9)	0	1.18(8)	0

R_{wp}=1.56%, R_p=3.78%, χ^2=0.948

^{7}Li MAS-NMR

In order to test the lithium site populations, ^{7}Li NMR spectra were collected for Li_2CuO_2 (all tetrahedral Li), $Li_3Cu_2O_4$ (2:1 tetrahedral:octahedral Li) and $Li_2NaCu_2O_4$ (expected all tetrahedral). Spectra are shown in Figure 2. Li_2CuO_2 has a single peak at around 340ppm, with multiple spinning sidebands. The spectrum of $Li_3Cu_2O_4$ has two major features, a strong peak at close to 0ppm and a smaller peak at around 400ppm. The spinning sidebands do not allow accurate measurements of the relative intensity but these main features may be assigned as the octahedral lithium and the tetrahedral lithium. The peak height ratio of about 5:1 (Oh:Td) compares with a known compositional ratio of the two sites of 1:2. This reflects the relative response of the two lithium sites. The ^{7}Li NMR spectrum of $Li_2NaCu_2O_4$ shows a very different octahedral:tetrahedral peak height ratio of approx 3:4, i.e. there is a marked reduction in the proportion of octahedral lithium ions. Combining the two data sets a ratio of octahedral:tetrahedral lithium of about 1:7 is obtained. However there will be a large contribution to the tetrahedral peak in $Li_3Cu_2O_4$ from the strong spinning sidebands of the octahedral peak so the real ratio will be much higher.

Structure of Li₂NaCu₂O₄

The structure of $Li_2NaCu_2O_4$ is found to be very similar to that of $Li_3Cu_2O_4$, with small increases in the lattice parameters, as would be expected to result from the substitution of lithium with larger sodium ions. The average Cu-O bond length of 1.895Å is slightly shorter than the distance in $Li_3Cu_2O_4$ of 1.910Å, possibly reflecting the underbonding of the octahedral lithium site in this material. The tetrahedral lithium site has an average Li-O distance of 1.964Å, comparable with the 1.931Å distance in the all-lithium material. The average Na-O distance in the octahedral site is 2.385Å which agrees remarkably well with the prediction from ionic radii [9] of 2.40Å.

The open structure of this material, combined with the demonstrated ability of $Li_3Cu_2O_4$, give promise regarding the application of $Li_2NaCu_2O_4$ as a lithium battery cathode material.

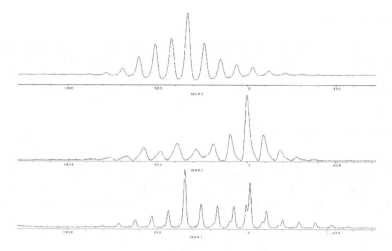

Figure 2. ^7Li MAS-NMR spectra of Li_2CuO_2 (top), $Li_3Cu_2O_4$ (middle) and $Li_2NaCu_2O_4$ (bottom).

Electrochemistry

Since the abstraction of lithium from $Li_3Cu_2O_4$ is from the tetrahedral site [7], it was expected that the behaviour of $Li_2NaCu_2O_4$ would be very similar. The lithium site which is occupied by sodium in $Li_2NaCu_2O_4$ undergoes no lithium abstraction in experiments on $Li_3Cu_2O_4$.

The open-circuit voltage of a freshly assembled $Li/Li_2NaCu_2O_4$ cell was 3.10 V. The behaviour on the first charge is similar to that previously described for $Li_3Cu_2O_4$ [7]. A gradual sloping increase in potential is observed to above 4 V with no significant plateau to indicate lithium deintercalation. In $Li_3Cu_2O_4$ this was attributed to the formation of a passivating layer, through reaction with the electrolyte, and a plateau is observed in subsequent cycles. However $Li_2NaCu_2O_4$ shows little sign of de-/re-intercalation after several cycles and the capacity was found to be less than 20mAhg^{-1}.

These are preliminary results and it is possible that further work will yield more success. In particular we have not yet explored high cell potentials. However it appears that, whilst $Li_3Cu_2O_4$ is a fairly good cathode material, the isostructural $Li_2NaCu_2O_4$ is virtually inactive. These results may indicate that the octahedral lithium site is part of the pathway for lithium ion abstraction, but it seems more likely that the presence of sodium interferes with the formation of the passivating layer on the cathode and the reaction with the electrolyte continues to occur over many cycles.

CONCLUSIONS

$Li_2NaCu_2O_4$ is isostructural with $Li_3Cu_2O_4$, with sodium preferentially occupying the octahedral sites. Both these materials demonstrate the applicability of high pressure oxygen in th synthesis of potential lithium battery cathode materials. An initial study shows $Li_2NaCu_2O_4$ to have very little electrochemical activity, whereas $Li_3Cu_2O_4$ is a promising cathode material.

ACKNOWLEDGEMENTS

This work was supported by a grant from the Engineering and Physical Sciences Research Council GR/M03610. Thanks to Dr. B. Gore and Prof. M. W. Anderson of UMIST Centre for Microporous Materials for the NMR data and to Dr. R. I. Smith of ISIS at the Rutherford Appleton Laboratory for assistance with the powder neutron diffraction data collection.

REFERENCES

1. M. Grimm, *IEEE Transactions on Consumer Electronics*. **32**, 700 (1986).
2. A. M. Sukeshini, H. Kobayashi, M. Tabuchi and H. Kageyama, *Sol. St. Ionics*. **128**, 33 (2000).
3. H. Kawai, M. Nagata, H. Tukamoto and A. R. West, *J. Power Sources*. **81-82**, 67 (1999).
4. H. Arai, S. Okada, Y. Sakurai and J. Yakami, *Sol. St. Ionics*. **106**, 45 (1998).
5. M. T. Weller, D. R. Lines and M. T. Weller, *J. Chem. Soc., Dalton Trans*. 3137 (1991).
6. D. B. Currie and M. T. Weller, *J. Mater. Chem*. **3**, 229 (1993).
7. E. A. Raekelboom, A. L. Hector, M. T. Weller and J. R. Owen, *J. Power Sources*. **97-98**, 465 (2001).
8. R. B. Von Dreele and A. C. Larson, Generalised Structure Analysis System, Los Alamos, NM (1998).
9. R. D. Shannon, *Acta Cryst., Sect. A*. **32**, 751 (1976).

Mat. Res. Soc. Symp. Proc. Vol. 658 © 2001 Materials Research Society

In situ XAFS study on cathode materials for lithium-ion batteries

Takamasa Nonaka, Chikaaki Okuda, Yoshio Ukyo and Tokuhiko Okamoto
TOYOTA Central Research & Development Laboratories., Inc.,
Nagakute, Aichi, 480-1192, Japan

ABSTRACT

Ni and Co K-edge X-ray absorption spectra of $LiNi_{0.8}Co_{0.2}O_2$ have been collected using *in situ* coin cells. To investigate the electronic and structural changes accompanied by the capacity fading during electrochemical cycling and keeping batteries at high temperatures, the cells with different cycling states and keeping conditions (temperature, time) were prepared. Upon charging the cell, the Ni and Co K absorption edge shifted towards higher energy, and the good correlation between the range of chemical shifts upon charging and the capacity of the cell was observed. From quantitative analysis of EXAFS data, it was revealed that the capacity fading is closely related to the Jahn-Teller distortion of the NiO_6 octahedron.

INTRODUCTION

$LiNi_{0.8}Co_{0.2}O_2$ is one of the current candidates for a cathode material of advanced rechargeable batteries with high capacity. It is known that the stability of $LiNi_{0.8}Co_{0.2}O_2$ is superior to that of $LiNiO_2$ because of exhibiting a single-phase region upon oxidation from 3.0V to 4.1V [1]. However, the capacity fading occurs not only during charge/discharge cycling but also when batteries are kept at high temperatures. And this capacity fading is most important problem for practical use. From the standpoint of overcoming the capacity fading, it is essential to understand the electronic and structural changes accompanied by the capacity fading. For this purpose, *in situ* XAFS analysis is very useful, because it will give the information on the local structure around an absorber atom and its electronic structure without destructing the battery for a measurement. Some studies applying *in situ* XAFS analysis for $LiNiO_2$ or $LiCoO_2$ have been already reported [2-3]. But an in situ XAFS study on the capacity faded battery has not been reported yet. So we prepared the batteries with various capacities and have measured *in situ* Ni and Co K-edge absorption spectra of $LiNi_{0.8}Co_{0.2}O_2$.

EXPERIMENTAL DETAILS

Figure 1 shows a drawing of the coin cell newly developed for *in situ* XAFS measurements in a transmission mode. By using 0.4 mm Beryllium windows, the X-rays can penetrate throw the cell. XAFS data can be obtained at various voltages without taking out the cathode material from the cell. To investigate the changes resulting from cycling and keeping at high temperatures, the cells with different cycling states and keeping conditions (temperature, time) were prepared. The cells used in this study are listed in Table I.

Ni and Co K-edge XAFS data were collected using beamline BL16B2 in SPring-8 (Hyogo, Japan). The incident X-rays were monochromatized using Si (111) double-crystal monochromator, and the harmonic content of the beam was minimized by Rh-coated Si mirror inclined to 5 mrad. The X-ray intensities were monitored using ionization chambers filled with nitrogen gas for the incident beam and a mixture of argon (25%) and nitrogen (75%) for the transmitted beam.

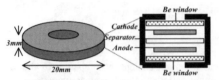

Figure 1. *Schematic drawing of the in situ coin cell.*

Table I. The conditions of cells.

Cell no.	Condition	Capacity (relative value)
1	Initial state (no treatment)	NA
2	After one charge/discharge cycle*	100
3	After 515 charge/discharge cycles*	9.5
4	After keeping at 80 ℃ for 3 days**	64.3
5	After keeping at 60 ℃ for 25 days**	14.7

* The charge/discharge cycling have been done at rate of 1mA/cm^2 in the range of voltage from 3.0V to 4.1V. ** Kept at charged state which corresponds to a voltage of 4.1V.

RESULTS AND DISCUSSION

XANES

Figures 2 (*a*) and (*b*) show the Ni K-edge and Co K-edge XANES spectra of LiNi$_{0.8}$Co$_{0.2}$O$_2$ for several samples. In both edges, chemical shifts of the edge peak energy were found. And it should be noticed that some structures were observed in the energy region 8335-8340 eV (indicated by arrows in the figures), which are discussed later.

The graphical comparisons of the edge peak energies as a function of the cell voltages are shown in Figures 3 (*a*) and (*b*). The edge peak energy E$_p$ is defined here as the energy at maximum height of the edge jump. Continuous shifts toward higher energies are thought to indicate the increases in the average oxidation states of Ni upon lithium removal [3]. In capacity faded samples, the ranges of chemical shifts upon charging are less than that in "After one cycle". A good correlation between the range of chemical shift and the capacity of the cell was found. It is also worth noticing that the chemical shifts of Co and Ni K-edge occur in the same way in spite of difference of the electronic structure between Ni and Co. The reason for this phenomenon is not clear now.

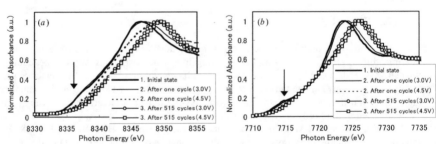

Figure 2. *(a) Ni K-edge and (b) Co K-edge XANES spectra of LiNi$_{0.8}$Co$_{0.2}$O$_2$.*

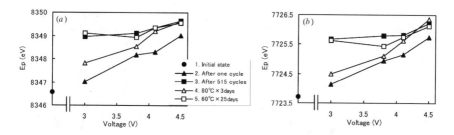

Figure 3. *Graphical comparisons of the edge peak energies (E_p) as a function of the voltages for (a) Ni K-edge and (b) Co K-edge.*

EXAFS

Fourier-transforms of the Ni K-edge EXAFS spectra for several samples are shown in Figure 4. The first peak at around 1.5Å to Ni-O interactions and the second one at around 2.5Å corresponds to Ni-Ni interactions. The Ni-O peak height of the sample in its initial state is lower than that of capacity faded samples. This phenomenon is explained by the local Jahn-Teller distortion of the NiO_6 octahedron due to the low spin Ni^{3+}. Distorted NiO_6 octahedral coordination such as 4(shorter)+2(longer) Ni-O bonds causes the apparent decrease in the height of the Ni-O peak due to the interference of imaginary and real part of the FT [3]. Figure 5 shows the heights of the Ni-O peak as a function of the cell voltage. Upon charging, the extent of the local distortion is getting lower, and the distorted NiO_6 octahedron turns into a regular octahedron. This phenomenon is resulting from gradual changes of the average valence of Ni from 3+ to 4+; therefore, the shape of Figure 5 is very similar to that of Figure 3 (a).

The averages of the Ni-O distances deduced from quantitative analysis of EXAFS data are shown in Figure 6. The single-shell curve-fitting were performed with the coordination numbers of oxygen fixed to 6. The phase shifts and backscattering amplitudes were obtained from the tabulated functions calculated by McKale *et al.* [4]. The estimated errors on distances (~ 0.02) are also shown in the figure.

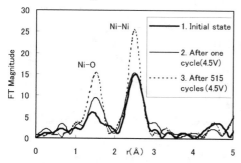

Figure 4. *Fourier-transforms of Ni K-EXAFS spectra for $LiNi_{0.8}Co_{0.2}O_2$.*

In "After one cycle", the Ni-O distance obviously decreases upon charging, while in "After 515 cycles", the distance does not change. It is supposed that the change of the Ni-O distance is originating mainly from the change of the ion-radii accompanied by the oxidation of Ni^{3+} to Ni^{4+}.

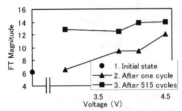

Figure 5. *A comparison of the heights of Ni-O peaks in FT spectra as a function of the voltages.*

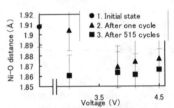

Figure 6. *A comparison of the averages of Ni-O distances as a function of the voltages.*

Molecular orbital calculations for XANES spectra

To assign the structures that were observed in the energy region 8335-8340 eV of XANES spectra, first-principles molecular orbital calculations were performed using the discrete variational (DV) Xα cluster method [5]. Figure 7 (*a*) shows a model cluster used in the present calculations. To investigate the effect of the local distortion of NiO_6 octahedron, the model cluster with a regular octahedron and the one with a distorted octahedron (4 shorter Ni-O and 2 longer Ni-O) were used for comparison. In this work, a program code SCAT [6] was used to calculate the electronic states of the model clusters. The atomic orbitals used in the calculations are $1s$-$4p$ for Ni, and $1s$-$3p$ for O.

XANES spectra reproduced by the calculations are shown in Figure 7 (*b*). The absorption intensity at around 8337 eV in the calculated spectrum with distorted octahedron was larger than that with regular octahedron. From these calculations, it is suggested that the observed structures are originating from the local Jahn-Teller distortion of the NiO_6 octahedron.

As shown in Figure 3 (*b*), the similar structures were found in Co K-edge XANES spectra. But, at present, we have no idea about these structures.

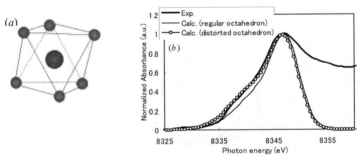

Figure 7. *(a) The model cluster used for calculations, (b) A comparison of experimental and calculated Ni K-edge XANES spectra.*

REFERENCES

1. E. Levi, M. D. Levi, G. Salitra, D. Aurbach, R. Oesten, U. Heider and L. Heider, *Solid State Ionics* **126**, 97-108 (1999).
2. A. N. Mansour, J. McBreen, C. A. Melendres, *J. Electrochemical Society* **146**(8), 2799-2809 (1999).
3. I. Nakai, K. Takahashi, Y. Shiraishi, T. Nakagome, F. Izumi, Y. Ishii, F. Nishikawa and T. Konishi, *J. Power Sources* **68**, 536-539 (1997).
4. A. G. McKale, B. W. Veal, A. P. Paulikas, S. K. Chan and G. S. Knapp, *J. American Chemical Society* **110**, 3763 (1988).
5. E. Ellis, H. Adachi and F. W. Averill, *Surface Science* **58**, 497 (1976).
6. H. Adachi, M. Tsukada and C. Satoko, *J. Physical Society of Japan* **45**, 497 (1978).

Mat. Res. Soc. Symp. Proc. Vol. 658 © 2001 Materials Research Society

Chemical delithiation, thermal transformations and electrochemical behaviour of iron-substituted lithium nickelate.

Pedro Lavela, Carlos Pérez-Vicente, and José L. Tirado
Laboratorio de Química Inorgánica, Universidad de Córdoba,
Campus de Rabanales. Edificio C3, Planta 1. 14071 Cordoba, Spain

ABSTRACT

Chemical deintercalation in Fe-substituted lithium nickelate and its effects on the thermal stability and electrochemical behaviour are studied. A sample with Fe:Ni ratio of 1:9 was used as the starting material. Chemical deintercalation of the ceramic product was achieved by acid treatment with 0.6 M aqueous hydrochloric acid solutions at room temperature. The atomic Fe:Ni ratio remained unaffected while the Li:(Fe+Ni) ratio decreased significantly down to ca. 0.5 after acid treatment. Infrared spectroscopy was used to discard a proton exchange side reaction. The initial open circuit voltage (OCV) of non-aqueous electrolyte lithium cells using the chemically deintercalated solids was ca. 3.7 V, while 3.0 V were obtained with the pristine oxide. Heat treating the deintercalated solids lead to oxygen evolution at 230ºC with the formation of spinel rock-salt structure solids at 600ºC. The improved thermal stability as compared with iron-free lithium nickelate is an interesting factor for battery safety.

INTRODUCTION

The solid state chemistry of layered lithium nickelate is extremely attractive from an academic point of view, but also due to its aptitude towards relevant applications in advanced battery technology. Although its structure and intercalation properties were evaluated more than forty years ago [1, 2], recent works are still devoted to improve its synthesis [3, 4], structural characterization [5, 6] and electrochemical performance [7]. Within the latter objective, partial replacement of nickel by Mn [8], Fe [9], Co [10, 11], Al [12] and/or Mg [10, 11, 13] has allowed the formation of new materials with more or less success in improving the capacity and cycling properties. Recently, the electrochemical behavior of iron-substituted lithium nickelate was studied [13]. ^{57}Fe Mössbauer spectroscopy revealed the simultaneous oxidation of nickel and iron in the electrochemical cell.

On the other hand, the possible oxidation of the transition metal together with lithium extraction by chemical procedures was early reported [14, 15]. The chemical delithiation with hydrochloric acid of lithium nickelate takes place by disproportionation of nickel(III), leading to a partial dissolution of the solid with a simultaneous oxidation of nickel in the remaining solid. The chemical deintercalation process was shown to induce several transformations in the solid, including the loss of oxygen with a conversion to the spinel structure at ca. 270ºC and the final conversion to a rock-salt related solid [15].

The aim of this communication is to extend the study of the chemical deintercalation reactions to Fe-substituted lithium nickelate, and to evaluate the effects of this substitution on the thermal stability and electrochemical behavior of the resulting solids.

EXPERIMENTAL DETAILS

The pristine sample of ceramic oxide with $LiFe_{0.1}Ni_{0.9}O_2$ nominal stoichiometry was prepared from stoichiometric amounts of Fe_2O_3 and NiO. Prolonged grinding with an amount of Li_2CO_3 in slight excess to the stoichiometric proportion intimately mixed the powdered solids. The mixture was heat treated at 700ºC for 20 h under a dynamic oxygen atmosphere and then slowly cooled (furnace off) to room temperature. The solid was ground in an agate mortar and the heat treatment was repeated three times.

Chemical deintercalation of the ceramic products was achieved by acid treatment, according to [14]. 0.6 M aqueous hydrochloric acid solutions were used in the treatment at 20ºC for periods of 4 and 6 h. These samples will be henceforth referred to as Fe1B and Fe1C, respectively. The chemical composition of pristine and acid treated solids was confirmed by Energy Dispersive X-ray (EDX) and Atomic Absorption (AA) spectroscopies.

X-ray Powder Diffraction data (XPD) were obtained with a Siemens D5000 apparatus provided with CuK_α radiation and graphite monochromator. Step-scan recordings for structure refinement by the Rietveld method were carried out using 0.04°2θ steps of 12 s duration. The computer program DBWS9000 was used in the calculations [16]. FTIR transmittance spectra were obtained with a BOMEM apparatus, using KBr pellets.

The electrochemical properties of $Li/LiPF_6(DEC:EC;1:1)/mixed$-oxide test cells were evaluated by using a multichannel *MacPile* system. Cell cathodes were prepared by pressing at ca. 4 tons, a mixture of 82 % of active material (ca. 4 mg), 12 % 4N carbon black (Strem) and 6% Ethylene propylene diene monomer (EPDM) terpolymer binder. Step Potential Electrochemical Spectroscopy (SPES) measurements were performed by using voltage steps of 10 mV with 6 min duration.

RESULTS AND DISCUSSION

The chemical composition of the pristine sample of lithium nickelate was determined by AA and EDX analysis, which revealed an atomic Fe/Ni ratio of 0.1. This ratio remained unaffected by treating the sample with hydrochloric acid at 4 and 6 h. In contrast, the atomic Li/Ni ratio decreased significantly down to ca. 0.5 after acid treatment, which may be taken as indicative of a chemical deintercalation process. No difference in Li/Ni ratio was found between the two treatments, which can be interpreted by assuming that the maximum deintercalation is reached after 4 h. Nevertheless, as recent works on spinel manganese oxides [17-19] have unequivocally proven that lithium can be partly exchanged by acid treatment, several spectroscopic techniques were applied to our samples to discard this possibility. First, the FTIR spectra did not agree with the presence of hydrogen in the acid treated solids, as no band at ca. 3340 cm^{-1} ascribable to stretching vibration of lattice –OH groups was present. Moreover, the band at ca. 910 cm^{-1} which has been ascribed to lattice coupling vibration of the H^+-form in spinel oxides [17-19] was also absent for chemically delithiated samples.

From the above result concerning the constancy in the Fe/Ni ratio in the solids and the presence of iron nickel qualitatively detected in the solution, the reaction taking place in aqueous medium can be expressed as:

$$(0.9+x)LiNi_{0.9}Fe_{0.1}O_2 + 4xH^+ \rightarrow 0.9Li_{1-x}Ni_{0.9}Fe_{0.1}O_2 + 0.9xNi^{2+} + 0.1xFe^{3+} + 1.9xLi^+ + 2xH_2O$$

This stoichiometry is valid in the $0 \le x \le 0.9$ interval. This approach is in perfect agreement with the distribution of oxidation states found by Delmas group in electrochemically deintercalated solids [7, 9].

On the other hand, an electrochemical proof that the transition metal ions were oxidized by the chemical deintercalation process, similarly to what was previously described in $LiNiO_2$ [14], was obtained by measuring cell potentials of $Li/LiClO_4(PC:EC::1:1)/Li_{0.46}Fe_{0.1}Ni_{0.9}O_2$

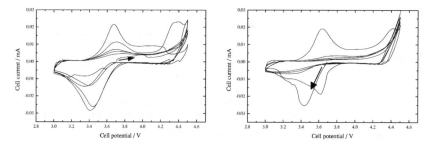

Figure 1. *Step potential electrochemical spectroscopy (SPES) plots of cell current vs. cell potential recorded by beginning with the charge (left) and the discharge (right) of the cell.*

cells. The initial voltage value after relaxation was ca. 3.7 V (Figure 1), while 3.0 V was obtained with the fully intercalated oxide (pristine sample).

Once the chemical deintercalation process was demonstrated, the possible changes in crystal structural parameters induced by this process were evaluated by using diffraction techniques. The partially deintercalated materials obtained by chemical procedures showed no significant transformations while stored at room temperature under an air atmosphere during the period in which the different experiments were carried out. Thus, the X-ray diffraction pattern could be recorded in air. Data for three samples (pristine and both acid treated samples) were used in a refined using the R-3m space group with Ni and Fe mainly in 3a sites, Li in 3b and O in 6d sites. In addition, some mixing of 3a and 3b sites was needed to improve the fit. Figure 2 shows the experimental, calculated and difference profiles, which evidence a good agreement with the model used in the analysis. No significant differences were found between the XRD data of 4 h and 6 h acid treated samples. Table I shows the results of the Rietveld refinement of X-ray diffraction data for pristine and 4 h samples. The refined occupancy of 3b sites by Ni atoms was ca. 0.015 in the pristine sample and this value was fixed in the analysis of the chemically deintercalated samples. From the unit cell parameters in Table I, a strong increase in the c/a ratio (ca. 2.6 %) is shown which clearly evidences the enhanced trigonal distortion caused by chemical deintercalation in these layered solids.

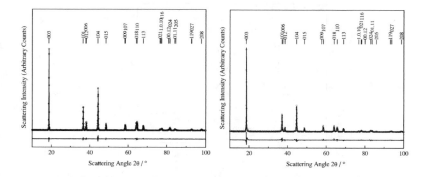

Figure 2. *Experimental (circles), calculated (full line) and difference (below) X-ray diffraction profiles of pristine $LiFe_{0.1}Ni_{0.9}O_2$ (left) and chemically deintercalated $Li_{0.46}Fe_{0.1}Ni_{0.9}O_2$ (right) samples.*

Table I. Composition, and results of Rietveld refinement of powder X-ray diffraction data of pristine and chemically delithiated samples.

SAMPLE Nominal composition	Unit cell Parameters (Å)	z_{oxygen}	Occupancy for sites 3a	Occupancy for sites 3b	R_{wp}/R_{exp} R_{BRAGG}
$LiFe_{0.1}Ni_{0.9}O_2$	$a = 2.8830(7)$ $c = 14.243(2)$	0.2582(4)	Ni = 0.898(4) Fe = 0.102(4)	Ni = 0.015(4) Li = 0.985(4)	1.21 2.93
$Li_{0.4}Fe_{0.1}Ni_{0.9}O_2$	$a = 2.8380(10)$ $c = 14.381(3)$	0.2643(5)	Ni = 0.898 Fe = 0.102	Ni = 0.015 Li = 0.400	1.46 6.81

A detailed knowledge about the thermal stability of electrode materials based on layered lithium nickelate at different depths of charge (deintercalation) is crucial from the viewpoint of their potential applications in the lithium battery industry, particularly if work under abusive conditions is evaluated. Thus, Li_xNiO_2 with $x < 1$ is known to transform into a spinel-related solid above 200ºC, followed by a conversion into a rock-salt related material with oxygen evolution at ca. 600ºC. The use of chemically deintercalated samples to study these thermal transformations allows us to examine larger amounts of additive-free materials as compared with samples prepared by electrochemical procedures.

The thermal analysis of chemically deintercalated Fe-containing lithium nickelate was carried out by using thermogravimetric (TGA) and differential thermal (DTA) analyses. These results are shown in Figure 3. The TGA is characterized by two weight loss processes. The first one takes place between 220 °C (noted T1 in Figure 3) and 238 °C (T3). The differential TGA (DTGA) curve also included in Figure 3 shows a peak centered at 229 °C (T2). The endothermic nature of this process is evidenced by a peak centered at 230 °C in the DTA curve, which is included as an inset in Figure 3. As described previously for pure $LiNiO_2$ [15], this effect could be ascribed to a structural transformation from hexagonal (R-3m) to spinel (Fd-3m) structure, accompanied by oxygen evolution and a partial reduction of Ni ions. Taking into account the percentage of weight loss, the final M:O ratio (M=Li+Fe+Ni) was determined to be 3:4, in agreement with the stoichiometry of a classical spinel structure without extra cation

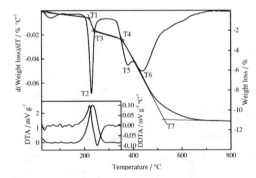

Figure 3. *Thermogravimetric (TG) and differential thermogravimetric (DTG) curves. Inset: differential thermal analysis (DTA) and its derivative (DDTA) for an acid treated sample with $Li_{0.46}Fe_{0.1}Ni_{0.9}O_2$ composition.*

vacancies. Moreover, an X-ray diffraction pattern indexable according to the spinel structure was obtained for a sample prepared by interrupting the TG experiment at 294 °C. This pattern was recorded at RT and showed a strong variation in the relative intensity of the Bragg reflections from the 4 h-deintercalated layered product. A complete Rietveld refinement was not possible, because of the low crystallinity of the sample. However, the unit cubic cell parameter could be calculated, $a = 8.31(2)$ Å. The equivalent rhombohedral unit cell parameters included in Table II evidence that this structural modification is accompanied by a considerable increase of the size of the unit cell, from $a_R = 5.81$ Å to 5.88 Å (ca. 1.3 %).

In addition, the presence of Fe partially substituting Ni was shown to result in an increase of the temperature of the rhombohedral into spinel transformation (230°C) as compared with that observed for $LiNiO_2$: 217.3 °C [11]. This effect is particularly interesting from the point of view of the safety of batteries.

Further heating results in a second and complex weight loss, which starts at 354°C (T4). The DTGA (Figure 3) shows a double peak profile, with maxima at 380°C (T5) and 441°C (T6), indicating that the oxygen evolution takes place in two steps, also accompanied by a Ni and Fe reduction to reach the oxidation state 2+. The final product has a MO stoichiometry, M=Li+Fe+Ni. The XRD pattern recorded at RT of the solid heated at 800 °C corresponded to a NaCl type structure, the unit cell parameter being $a = 4.13770(8)$ Å. The equivalent rhomboedral parameter $a_R = 5.852(1)$ Å (Table II) indicates that the spinel into NaCl transformation is accompanied by a decrease in the size of the unit cell.

Finally, the correspondence between chemical and electrochemically deintercalated solids was evaluated by recording cyclic voltammograms of lithium cells using the chemically deintercalated solids as the starting active cathode material. The plots obtained by the SPES technique are shown in Figure 1. Due to the presence of significant amounts of Ni^{4+} and Fe^{4+} in the samples and the incomplete deintercalation obtained by acid treatment, the electrochemical data could be obtained alternatively starting by charge or discharge of the cells. The profiles are partly affected by the sequence used in the study. Nevertheless, the basic characteristics are similar and both curves agree well with the results reported by the Delmas group [9].

The main redox signals are observed at ca. 3.4 V and 3.7 V for the reduction and oxidation processes, respectively. As it has been recently reported by Delmas et al. [9], these peaks were ascribed to the competitive oxidation of the 3d transition metals during charging from Ni^{3+} and Fe^{3+} to Ni^{4+} and Fe^{4+}, respectively, with the aid of Mössbauer spectroscopy.

Table II. Hexagonal (a_H and c_H), rhombohedral (a_R and α_R) and cubic (a) unit cell parameters of pristine, chemically deintercalated samples and heat-treated chemically-deintercalated samples.

Sample Nominal composition	Hexagonal and cubic unit cell parameters (Å)	Equivalent rhombohedral unit cell parameters (Å,º)
$LiFe_{0.1}Ni_{0.9}O_2$ R-3m	$a_H = 2.8830(7)$ $c_H = 14.243(2)$	$a_R = 5.799(1)$ $\alpha_R = 59.63(2)$
$Li_{0.4}Fe_{0.1}Ni_{0.9}O_2$ RT, R-3m	$a_H = 2.8882(9)$ $c_H = 14.381(4)$	$a_R = 5.807(2)$ $\alpha_R = 58.52(4)$
294ºC Fd-3m	$a = 8.31(2)$	$a_R = 5.88(1)$ $\alpha_R = 60.0$
800ºC Fm-3m	$a = 4.13770(8)$	$a_R = 5.852(1)$ $\alpha_R = 60.0$

Nevertheless, recent works on lithium, manganese spinels doped with other transition metals like Co, Cu, Cr and even Fe show an extended oxidation process of the latter elements at voltages as high as 5.0 V.
Our potential window was limited to 4.5 V. The occurrence of a not well defined oxidation process at ca. 4.4 V as a consequence of a not very stable electrolyte solution allows us to envisage additional redox couples that may be induced by the presence of iron in the structure. Further characterization studies in order to make evident the degree of reversibility of the new oxidation peak and the redox mechanism must be carried out.

ACKNOWLEDGMENT

The authors are grateful to CICYT (contract MAT1999-0741) for financial support.

REFERENCES

1. L.D. Dyer, B.S. Borie, and G.P. Smith, *J. Am. Chem. Soc.*, **76**, 1499 (1954).
2. J.B. Goodenough, D.G. Wickham, and W.J. Croft, *J. Phys. Chem. Solids*, **5**, 107 (1958).
3. J. Maruta, H. Yasuda, and M. Yamachi, *J. Power Sources*, **90**, 89 (2000).
4. R. Alcantara, P. Lavela, J.L. Tirado R. Stoyanova, E. Kuzmanova, and E. Zhecheva, *Chem. Mater.*, **9**, 2145 (1997) .
5. M.A. Monge, E. Gutiérrez Puebla, J.L. Martínez, I. Rasines, and J.A. Campa, *Chem. Mater.*, **12**, 2001 (2000).
6. L. Seguin, G. Amatucci, M. Anne, Y. Chabre, P. Strobel, J.M. Tarascon, and G. Vaugham, *J. Power. Sources*, **81-82**, 604 (1999).
7. L. Croguennec, C. Pouillerie, and C. Delmas, *J. Electrochem. Soc.*, **147**, 1314 (2000).
8. E. Rossen, C.W.D. Jones, and J.R. Dahn, *Solid State Ionics*, **57**, 311 (1992).
9. G. Prado, A. Rougier, L. Fournes, and C. Delmas, *J. Electrochem. Soc.* **147**, 2880 (2000).
10. J. Cho, *Chem Mater.* **12**, 3089 (2000).
11. C.C. Chang, J.Y. Kim, and P.N. Kumta, *J. Electrochem. Soc.*, **147**, 1722 (2000).
12. T. Ohzuku, A. Ueda, and M. Kouguchi, *J. Electrochem. Soc.* **142**, 4033, (1995).
13. C. Pouillerie, L. Croguennec, and C. Delmas, *Solid State Ionics* **132**, 15 (2000).
14. J. Morales, C. Pérez-Vicente, and J. L. Tirado, *Mater. Res. Bull.* **25**, 623 (1990).
15. J. Morales, C. Pérez-Vicente, and J. L. Tirado, *J. Thermal. Anal.*, **38**, 295 (1992).
16. R.A. Young, A. Sakthivel, T.S. Moss, and C.O. Paiva-Santos, *J. Appl. Crystallogr.*, **75**, 336 (1995).
17. Q. Feng, Y. Miyai, H. Kanoh, and K. Ooi, *Chem. Mater.*, **5**, 311 (1993).
18. Q. Feng, H. Kanoh, Y. Miyai, and K. Ooi, *Chem. Mater.*, **7**, 379 (1995).
19. P. Lavela, L. Sánchez, J.L. Tirado, S. Bach, and J.P. Pereira-Ramos, *J. Solid State Chem.*, **150**, 196 (2000).

Mat. Res. Soc. Symp. Proc. Vol. 658 © 2001 Materials Research Society

Lithium deintercalation in $LiNi_{0.30}Co_{0.70}O_2$:
Redox processes, electronic and ionic mobility as characterized by ^7Li MAS NMR and electrical properties

D. Carlier, M. Ménétrier and C. Delmas

Institut de Chimie de la Matière Condensée de Bordeaux-CNRS et Ecole Nationale Supérieure de Chimie et Physique de Bordeaux, 87 Av. Dr A. Schweitzer, 33608 Pessac cedex (France).

ABSTRACT

$LiNi_{1-y}Co_yO_2$ materials are particularly interesting as positive electrode in lithium-ion batteries. Their ^7Li MAS NMR spectra are sensitive to the presence of paramagnetic Ni^{3+} cations as nearest and next-nearest neighbors. These contact shift (or Fermi contact) interactions are used to yield information on the lithium environment in terms of paramagnetic cations. Electrochemically deintercalated $Li_xNi_{0.30}Co_{0.70}O_2$ phases have been characterized by XRD, ^7Li MAS NMR, electronic conductivity and thermoelectronic power. Ni^{3+} oxidation to Ni^{4+} occurs at the beginning of deintercalation, leading to a Ni^{3+}/Ni^{4+} hopping which causes an exchange of the ^7Li NMR signals of the Li ions interacting with nickel. For higher deintercalation amounts, the signal due to Li ions with only cobalt as first and second neighbor is also involved in the exchange upon heating, showing the onset of ionic hopping. For x close to 0.70, an increase of the NMR shift with temperature is observed, which is assigned to a hopping between Ni^{4+} and Co^{3+}. Therefore, the question as to which ion is actually oxidized during deintercalation is somewhat irrelevant around this composition. Finally, for $x < 0.70$, the presence of Ni hinders a true long-range electronic delocalisation between Co^{4+}/Co^{3+} (LS) as it happens in Li_xCoO_2 ($x < 0.70$). However, electrons are clearly delocalized in restricted zones, as seen from NMR and thermoelectronic power coefficient.

INTRODUCTION

The $LiNi_{1-y}Co_yO_2$ materials have been extensively studied in the past ten years for application as positive electrodes for lithium-ion batteries. These materials exhibit a strictly layered α-$NaFeO_2$ type structure for $y \geq 0.3$ [1], built from alternate sheets of $(Ni,Co)O_6$ or LiO_6 octahedra sharing edges. Nickel and cobalt ions are in a trivalent low-spin state forming paramagnetic Ni^{3+} ($t_{2g}^6 e_g^1$) and diamagnetic Co^{3+} (t_{2g}^6) [2].

The knowledge of the local structure in positive electrode materials is crucial for the understanding of their electrochemical properties. We therefore used ^7Li NMR, which has proved to be a good probe to characterize lithium local environment in these systems, making use of the hyperfine interactions due to their paramagnetic or metallic character [3-6]. In $LiNi_{0.30}Co_{0.70}O_2$, the localized paramagnetic Ni^{3+} ions lead to a well-resolved fine structure in the starting material: each Ni^{3+} ion influences (differently) its 1st and 2nd neighbor Li^+ ions [3].

^6Li / ^7Li NMR spectroscopy has not been extensively applied to study electrochemically deintercalated materials, in which mixed valence and Li^+ vacancies may allow electronic and ionic hopping. We performed a detailed study of the $Li_xNi_{0.30}Co_{0.70}O_2$ system for $0.40 \leq x \leq 1$ by X-ray diffraction, ^7Li MAS NMR and electrical measurements, in order to elucidate the effect of electronic and ionic mobility on the NMR spectra and to follow the oxidation of paramagnetic Ni^{3+} ions into diamagnetic Ni^{4+} and of diamagnetic Co^{3+} ions into paramagnetic Co^{4+} upon lithium deintercalation.

EXPRIMENTAL

The $LiNi_{0.30}Co_{0.70}O_2$ starting material was prepared by direct reaction from Li_2CO_3, N and Co_3O_4 in stoichiometric proportions. The mixture was heated under O_2 at 600 °C for 10 h, 800 °C for 48 h and three times at 900°C for 20 h with intermediate grinding.

The deintercalated phases were prepared by electrochemical deintercalation using sintere pellets (4 tons for an 8 mm diameter, followed by a thermal treatment of 12 h at 900 °C under oxygen) of the starting material as positive electrode without additive. The Li/liquid electrolyte/$LiNi_{0.30}Co_{0.70}O_2$ cells were assembled in an argon-filled dry box and charged at ver low current density (C/600 i.e. 600 h are needed to remove 1mole of lithium ions). $LiClO_4$ in propylene carbonate was used as electrolyte for $0.60 \leq x < 1$ and $LiPF_6$ in ethylene carbonate - dimethyl carbonate - propylene carbonate for $0.40 \leq x < 0.60$ (Merck, ZV1011 Selectipur).

^7Li MAS NMR spectra were recorded on a Bruker MSL200 spectrometer at 77.7 MHz, with a standard 4 mm Bruker MAS probe. The samples were mixed with dry silica (typically 50% in weight), in order to facilitate the spinning and improve the field homogeneity, since the may exhibit metallic or paramagnetic properties. The mixture was placed into a 4 mm diameter zirconia rotor in the dry box. Spinning speeds (v_r) of 15 kHz, single pulse and rotor-synchroniz echo sequences with $t_{\pi/2} = 3.05$ µs and a 200 kHz spectral width were used. The recycle time $D_0 = 1$ s, was long enough to avoid T_1 saturation effects. The isotropic shifts reported in parts p million are relative to an external sample of 1M LiCl solution in water.

For electronic conductivity measurements, a four-probe method was used with direct current in the 100 - 300 K range. The thermoelectronic measurements were carried out by usin equipment described elsewhere [7].

RESULTS AND DISCUSSION

The ^7Li MAS NMR spectrum of the starting $LiNi_{0.30}Co_{0.70}O_2$ material exhibits several signals (see Fig.1 for $x = 1$) due to lithium ions in different environments. One signal is situate around 0 ppm, a position typical of lithium ions situated in a diamagnetic environment. This signal is therefore assigned to lithium ions surrounded only by diamagnetic Co^{3+} ions as 1st an 2nd neighbors. Other signals recorded with larger shifts are assigned to lithium interacting with paramagnetic Ni^{3+} as 1st and/or 2nd neighbor. The paramagnetic shifts depends on the sign and amount of paramagnetic electron density transferred at the site of the lithium nucleus. In order confirm the assignment of the different signals proposed earlier in reference 3, we consider the Goodenough -Kanamori rules that we transpose to the $Ni^{3+}(e_g)$ - $O(2p)$ - $Li(2s)$ system. The spectrometer magnetic field gives the orientation of the 3d paramagnetic electrons and the sig of the spin density transferred on the Li 2s orbital is predicted by applying the Goodenough - Kanamori superexchange mechanisms [8]. As discussed in more detail in reference 9, the sign situated at +110, +95, +80, +65 ppm (see arrows in Fig. 1) are thus respectively assigned to lithium ions interacting with one Ni^{3+} as 1st neighbor and 0, 1, 2, 3 Ni^{3+} as second neighbor, an

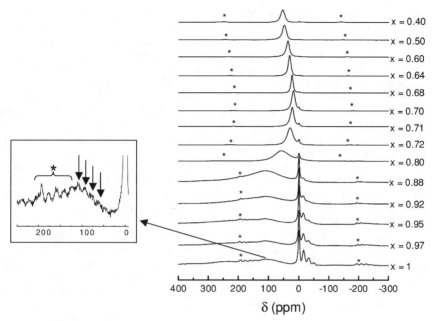

Figure 1 : ^7Li MAS NMR spectra (v =15 kHz) of various $Li_xNi_{0.30}Co_{0.70}O_2$ phases (0.40 ≤ x ≤ 1), recorded with a single pulse sequence. A close inspection of the signals situated around +110 ppm is shown on the left. (* = spinning sidebands)

the signals situated at -15, -30, -45 ppm are respectively assigned to lithium ions interacting with 1, 2, 3 Ni^{3+} as second neighbor without any Ni^{3+} as 1st neighbor, confirming the previous assignment [3].

Lithium deintercalation from $LiNi_{0.30}Co_{0.70}O_2$ occurs without phase separation. It involves 3d metal oxidation and Li^+ vacancy formation, which could lead to both electronic and ionic mobilities. The voltage curve profile suggested that the nickel ions were oxidized first during lithium deintercalation [2].

Fig. 1 shows the ^7Li MAS NMR spectra for various lithium deintercalation amounts. At the beginning of the deintercalation, the signals assigned to lithium ions interacting with Ni^{3+} as 1st and 2nd neighbor progressively disappear as x decreases, while a new broad signal appears. The fine structure disappearance shows the oxidation of paramagnetic Ni^{3+} into diamagnetic Ni^{4+} at the beginning of deintercalation. Nevertheless, this cannot be only due to Ni^{3+} oxidation, because, for x = 0.80, no signal assigned to Ni^{3+} as 1st or 2nd neighbor is observed, whereas 0.10 Ni^{3+} remain. Comparison of spectra recorded with a Hahn echo and a single pulse sequence show the existence of mobility in the material, which leads for the Hahn echo sequence to the non-refocusing of the signals concerned by this mobility.

Since mobility is involved in the emergence of the new signal, we tried to promote it in a slightly deintercalated compound by heating. Fig. 2(a) shows the ^7Li MAS NMR spectra as a function of temperature for the $Li_{0.95}Ni_{0.30}Co_{0.70}O_2$ phase. The "Ni^{3+} 1st and 2nd neighbors" fine structure remains observable at room temperature, as few Ni^{3+} ions are oxidized in this phase. As temperature increases, the "Ni^{3+} 1st and 2nd neighbors" signals progressively disappear and a new broad signal appears. Mobility is thus evidenced in the deintercalated phases and leads to an averaged NMR signal due to an exchange phenomenon. If the exchange phenomenon observed was due to ionic mobility, every lithium ion would be involved, even those with only diamagnetic Co^{3+} as 1st and 2nd neighbor, but the intensity of their signal (at 0 ppm) remains approximately the same as temperature increases. The exchange phenomenon is therefore due to an electronic hopping, occurring between Ni^{3+} and Ni^{4+} ions, so that lithium ions with nickel ion as 1st or 2nd neighbors now interact with nickel ions rapidly oscillating between the 3+ and 4+ oxidation states on the NMR time scale ($\approx 10^{-6}$s).

On Fig. 1, we observe that the averaged signal discussed above is less and less shifted as the lithium amount decreases to $x = 0.70$. Indeed, as the number of diamagnetic Ni^{4+} ions involved in the Ni^{3+}/Ni^{4+} electronic hopping phenomenon increases, the average density of unpaired electron spins seen by the lithium ions concerned decreases.

In order to increase the ionic mobility, we recorded at different temperatures the NMR spectra of $Li_{0.80}Ni_{0.30}Co_{0.70}O_2$ (not shown), which presents at room temperature, the averaged signal with no trace of the "Ni^{3+} 1st and 2nd neighbors" fine structure, and the signal at 0 ppm assigned to lithium ions interacting only with Co^{3+} as 1st and 2nd neighbors. As temperature increases, the signal at 0 ppm decreases in intensity, indicating that the time scale of the ionic mobility becomes now close to that of NMR (10^{-6}s).

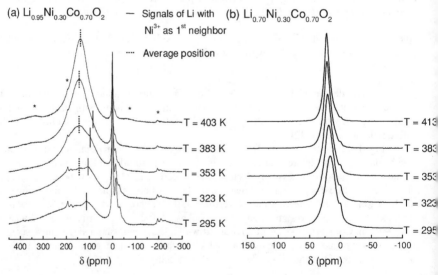

Figure 2 : ^7Li MAS NMR spectra ($\nu = 15$ kHz) recorded at various temperature with a single pulse sequence of (a) $Li_{0.95}Ni_{0.30}Co_{0.70}O_2$ and (b) $Li_{0.70}Ni_{0.30}Co_{0.70}O_2$. (* = spinning sidebands)

Fig. 1 shows that the shift of the averaged signal decreases toward 0 ppm with lithium deintercalation for $0.70 \le x \le 0.80$ and then increases for $0.40 \le x \le 0.70$, but was never observed at 0 ppm. If the reduction process associated with lithium deintercalation concerned the Ni^{3+}/Ni^{4+} and Co^{3+}/Co^{4+} redox couples successively, the $Li_{0.70}Ni_{0.30}Co_{0.70}O_2$ phase would be diamagnetic as all paramagnetic Ni^{3+} ions would be oxidized into diamagnetic Ni^{4+} ions and no diamagnetic Co^{3+} ions would be oxidized yet. The fact that the average signal is not found at 0 ppm suggests that the Ni^{3+} and Co^{3+} ions oxidation is competitive around $x = 0.70$, so that no diamagnetic phase is formed during lithium deintercalation.

The evolution of the ^7Li MAS NMR signal of $Li_{0.70}Ni_{0.30}Co_{0.70}O_2$ upon heating shows a shift of the averaged signal toward higher values (Fig. 2(b)). This behavior is unexpected, since, for paramagnetic compounds, the shift is known to decrease upon heating as the magnetic susceptibility decreases. The increase of the shift observed is not due to the presence of some Co^{4+} in the sample (as localized Co^{4+} ions lead to a loss of NMR observation of the neighboring lithium ions [5]) but to the formation of more and more paramagnetic Ni^{3+} ions upon heating. An electronic hopping between cobalt and nickel ions, which can be written as $Ni^{4+} + Co^{3+} \leftrightarrow Ni^{3+} + Co^{4+}$ is thus assumed.

In Fig. 1, the shift of the averaged signal increases for $0.40 \le x \le 0.70$. As the lithium amount decreases, oxidation of Co^{3+} to the 4+ state proceeds, so that one expects an electronic delocalisation to occur as it happens in the Li_xCoO_2 system for $x < 0.70$. For these materials, a Knight-shifted signal is then recorded, which shifts further upon deintercalation similarly to what is observed in Fig. 1 for $x < 0.70$ [5]. Indeed, electronic conductivity and thermoelectric power coefficient measurements show a clear change in behavior for $x = 0.71$ (Fig. 3), indicating that

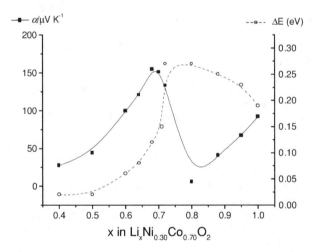

Figure 3 : *Room temperature Seebeck coefficient variation(squares and solid line) and electronic conductivity activation energy (circles and dashed line) calculated for $125 \le T \le 290$ K, versus lithium rate for $Li_xNi_{0.30}Co_{0.70}O_2$ deintercalated phases $(0.40 \le x \le 1)$.*

the Co^{3+} ions begin to be oxidized, and that the system tends to metallic behavior. However, even for $x = 0.40$, the sample is not truly metallic from the d.c. conductivity point of view as it exhibits a thermally activated electronic conductivity (Fig. 3). Obviously, the presence of nickel impedes a real long range electronic delocalisation in these materials, but electrons are delocalized within restricted zones between the cobalt ions for $x < 0.70$, leading to the observation of a "pseudo Knight-shift" in NMR.

CONCLUSIONS

^{7}Li MAS NMR spectra of partially deintercalated phases are less straightforward to interpret than the spectrum of the starting material as lithium deintercalation induces electronic and ionic hopping, which lead to exchange phenomena of the NMR lines. Electronic conductivity and thermoelectronic power coefficient measurements correlated to the NMR analysis lead to a picture of the deintercalation process in $Li_xNi_{0.30}Co_{0.70}O_2$. Oxidation of Ni an Co during lithium deintercalation is not clear-cut: nickel oxidation occurs first, but, for x close 0.70, a Ni/Co hopping occurs. The Ni^{4+} / Ni^{3+} redox couple must thus overlap with the Co t_{2g} band. For larger deintercalation amounts, cobalt ions are oxidized, but the presence of Ni^{4+} ions prevents a true long-range electronic delocalisation between Co^{3+} and Co^{4+} ions, whereas ^{7}Li NMR and thermoelectric data give evidence of some extent of delocalization.

ACKNOWLEDGMENTS

The authors wish to thank E. Marquestaut and C. Denage for technical assistance and Région Aquitaine for financial support.

REFERENCES

1. C. Delmas and I. Saadoune, *Solid State Ionics* **53-56**, 370 (1992).
2. I. Saadoune and C. Delmas, *J. Mater. Chem.* **6**, 193 (1996).
3. C. Marichal, J. Hirschinger, P. Granger, M. Ménétrier, A. Rougier and C. Delmas, *Inorg. Chem.* **34**, 1773 (1995).
4. M. P. J. Peeters, M. J. Van Bommel, P. M. C. Neilen-ten Wolde, H. A. M. Van Hal, W. C Keur and A. P. M. Kentgens, *Solid State Ionics* **112**, 41 (1998).
5. M. Ménétrier, I. Saadoune, S. Levasseur and C. Delmas, *J. Mater. Chem.* **9**, 1135 (1999).
6. Y. J. Lee, F. Wang, S. Mukerjee, J. McBreen and C. P. Grey, *J. Electrochem. Soc.* **147**, 803 (2000).
7. P. Dordor, E. Marquestaut and G. Villeneuve, *Revue Phys. Appl.* **15**, 1607 (1980).
8. J. B. Goodenough, *Magnetism and the Chemical Bond*, John Wiley & Sons, New-York London, 1963.
9. D. Carlier, M. Ménétrier and C. Delmas, *J. Mater. Chem.* **11**, 594 (2001).

Mat. Res. Soc. Symp. Proc. Vol. 658 © 2001 Materials Research Society

^{7}Li MAS NMR Studies of Lithiated Manganese Dioxide Tunnel Structures:
Pyrolusite and Ramsdellite

Younkee Paik, Young J. Lee, Francis Wang,[1] **William Bowden,**[2] **and Clare P. Grey**
Department of Chemistry, SUNY at Stony Brook, Stony Brook, NY11794-3400, U.S.A.
[1]Duracell Global Science Center, Danbury CT 06801
[2]Gillette Advanced Technology Center, U.S.A.

ABSTRACT

The one-dimensional 1×1 and 1×2 tunnel structures of manganese dioxides, pyrolusite (β-MnO_2) and ramsdellite (R-MnO_2), respectively, were chemically intercalated with LiI. Two ^{7}Li resonances were observed in lithiated pyrolusite. One isotropic resonance arising at 110 ppm shows a short spin-lattice relaxation time ($T_1 \sim 3$ ms) and was assigned to Li$^+$ ions in the 1×1 tunnel structure. The other isotropic resonance arising at 4 ppm shows a long spin-lattice relaxation time ($T_1 \sim 100$ ms) and was assigned to Li$^+$ ions in diamagnetic local environments in the form of impurities such as Li_2O or on the surface of the MnO_2 particles. Three ^{7}Li resonances were observed in lithiated ramsdellite at very different frequencies (600, 110 and 0 ppm). The resonance at 600 ppm, which is observed at low lithium intercalation levels, is assigned to Li$^+$ ions coordinated to both Mn(III) and Mn(IV) ions in the 1×2 tunnels, while the resonance at 110 ppm is due to Li$^+$ ions coordinated to Mn(III) ions and appears at higher Li levels. The resonance at 0 ppm is associated with a long spin-lattice relaxation time ($T_1 \sim 100$ ms) and is also assigned to Li$^+$ ions in diamagnetic impurities.

INTRODUCTION

Manganese dioxides are used worldwide as cathodes in lithium and zinc primary batteries [1]. These cathodes contain the highly disordered manganese dioxide EMD (electrolytic manganese dioxide), which consists of an intergrowth of the pyrolusite and ramsdellite structures [2]. In order to understand the mechanism of intercalation in this complex material, our work has focussed on understanding the intercalation processes that occur in the model systems pyrolusite (β-MnO_2) and ramsdellite (R-MnO_2). ^{7}Li MAS NMR spectroscopy has been used to study these systems since it represents an ideal method for probing the local atomic and electronic environments of the lithium ions [3].

EXPERIMENTAL DETAILS

Pyrolusite (β-MnO_2) was purchased from Fisher Scientific and ramsdellite (R-MnO_2) was purchased from Erachem. The host compounds were well ground in a mortar and dried under vacuum at 60 °C for 30 hours (β-MnO_2) or at 200 °C for 10 hours (R-MnO_2) prior to intercalation. Chemical lithiation was carried out by reacting the host compounds with LiI in acetonitrile at ~ 60 - 70 °C under a dry N_2 atmosphere for 3 days (β-MnO_2) or for 5 hours (R-MnO_2) [4]. The products were filtered and washed several times with acetonitrile and finally dried at 60 °C in air for 2 days. For comparison with previous work [5], chemical lithiation of pyrolusite (β-MnO$_2$) with n-BuLi in hexane was also carried out. The lithium and manganese contents in the lithiated samples were determined by ICP experiments (for details see table I).

Table I. Reaction conditions for lithiation and the extent of lithium incorporation into the final products.

Starting materials	Li:Mn ratio of the starting materials	Temperature and reaction time	Chemical formula of product
LiI, pyrolusite	0.2 : 1	60 °C, 3 days	Li$_{0.2}$MnO$_2$
LiI, ramsdellite	0.1 : 1	70 °C, 5 hours	Li$_{0.1}$MnO$_2$
LiI, ramsdellite	0.3 : 1	70 °C, 5 hours	Li$_{0.3}$MnO$_2$
LiI, ramsdellite	0.7 : 1	70 °C, 5 hours	Li$_{0.5}$MnO$_2$
LiI, ramsdellite	1.5 : 1	70 °C, 5 hours	Li$_{0.7}$MnO$_2$
LiI, ramsdellite	2.7 : 1	70 °C, 5 hours	Li$_{0.8}$MnO$_2$
n-BuLi, pyrolusite	0.2 : 1	60 °C, 7 days	Li$_{0.2}$MnO$_2$

The x-ray diffraction data were collected on a Scintag powder x-ray diffractometer (Cu K radiation). No significant change was observed in the diffraction patterns of the lithiated pyrolusite (β-MnO$_2$) derivatives compared with that of the host compound. In contrast, the diffraction patterns of ramsdellite (R-MnO$_2$) and its lithiated samples show an expansion of the unit cell as the lithium loading level is increased, which was also observed in a previous report [4].

^{7}Li MAS NMR experiments were performed at 77.53 MHz on a CMX-200 spectrometer with a Chemagnetics probe equipped with 4 mm rotors for MAS. Spectra were recorded with rotor synchronized Hahn-echo pulse sequence. Spin-lattice relaxation times (T$_1$) were measured with an inversion recovery pulse program. All spectra were referenced to 1.0 M LiCl solution 0 ppm as an external reference.

RESULTS & DISCUSSION

The ^{7}Li NMR spectra of lithiated derivatives of pyrolusite (β-MnO$_2$) and ramsdellite (R-MnO$_2$) are shown in figure 1 and 2, respectively. Both spectra show an isotropic resonance close to 0 ppm. Several studies have reported an isotropic resonance of ^{7}Li around 0 ppm from Li ion in diamagnetic solid compounds [6-8]. For example, ^{7}Li NMR spectra of sec-BuLi in toluene showed a resonance at -2.47 ppm [6]. In the ^{7}Li NMR spectra of Li$_2$O and Li$_2$O$_2$, resonances at 2.7 ppm and -0.4 ppm respectively were reported [7]. A resonance at 0 ppm was also observed the ^{7}Li NMR spectra of LiCoO$_2$ [8]. In contrast, very large chemical shifts (100 ~ 2000 ppm) normally observed for lattice Li$^+$ ions in paramagnetic solids [3]. Thus the isotropic resonance at around 0 ppm, which show a long spin-lattice relaxation times (T$_1$ ~ 100 ms), are assigned to Li$^+$ ions in diamagnetic local environments in the form of impurities such as Li$_2$O and Li$_2$CO$_3$ possibly coating the surface of the MnO$_2$ particles.

The ^{7}Li MAS NMR spectrum of lithiated pyrolusite shows a second isotropic resonance at 110 ppm (see figure 1). A short spin-lattice relaxation time of this resonance (T$_1$ ~ 3.0 ms), which is similar to the value (T$_1$ ~ 2.5 ms) reported in the studies of paramagnetic LiMnO$_2$ compounds [3], suggests a Li$^+$ site with paramagnetic local environment. Thus, we assign the isotropic resonance at 110 ppm in figure 1 to the lithium ions in 1×1 tunnels of lithiated pyrolusite.

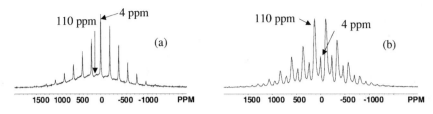

Figure 1. ^7Li MAS NMR spectra of pyrolusite (β-MnO$_2$) chemically lithiated with: (a) n-BuLi, & (b) LiI. Both spectra were acquired at a spinning speed of 14 kHz. Isotropic peaks are marked with the corresponding NMR shifts.

The isotropic resonance at 4 ppm is more intense in the NMR spectrum of the compound lithiated using n-BuLi (see figure 1a), which produces a more vigorous reaction condition than LiI. A side reaction, generating Li_2O and Mn_2O_3 as products, was suggested for lithiation of pyrolusite (β-MnO$_2$) with n-BuLi [5], which supports our assignment of the isotropic resonance at 4 ppm to lithium ions in diamagnetic impurities. Hence, figure 1 shows that LiI is a more effective lithiation reagent than n-BuLi.

Two more ^7Li isotropic resonances, other than the one arising at 0 ppm, were observed in lithiated ramsdellite at very different frequencies (600, 110 ppm) (see figure 2). Their intensities vary as a function of Li loading level. The spin-lattice relaxation times of the isotropic resonances at 600 and 110 ppm were 2.2 and 5.1 ms, respectively. The short T_1s of both resonances again suggest two different paramagnetic local environments for the lithium ions. We assign the isotropic resonance at 600 ppm, which is observed at low lithium intercalation levels, to the Li^+ ions in the 1×2 tunnels of ramsdellite (R-MnO$_2$).

Above a Li loading level of 0.5 per MnO$_2$ unit, an isotropic resonance at 110 ppm appears and becomes dominant as the Li loading level is increased. The intensity of the isotropic peak at 600 ppm decreases while the intensity of the resonance at 110 ppm increases dramatically above a Li intercalation level of 0.5(see figure 2e), where the manganese oxidation state Mn^{3+} predominates. What is noteworthy is, that instead of a gradual shift of the resonance at 600 ppm to lower frequencies as more Li^+ ions are intercalated into the structure, both the resonances at 600 & 100 ppm coexist from x = 0.5 to ~ 0.8, indicating the presence of two discrete local environments.

A cyclic voltammogram recorded during the redox cycling of a natural ramsdellite (R-MnO$_2$) in an alkaline electrolyte shows only one cathodic peak, suggesting one homogeneous reduction mechanism [9]. Another electrochemical study on a synthetic ramsdellite was also carried out in a lithium nonaqueous cell, where three distinct stages were observed in an open-circuit-voltage profile at lithiation levels of $0 \leq x \leq 0.3$, $0.3 \leq x \leq 0.9$, and $0.9 \leq x \leq 1.0$ [4]. In this work, the first stage corresponds to lithiation in the 1×2 tunnel structure with almost no structural change of the host compound [4], while the second and the third stages are due to lithiation accompanying the distortion of the 1×2 tunnel structure causing shears of hexagonally-closed-packed oxygen layers. An electron-filling mechanism was suggested for the electronic distribution of manganese ions to explain the distortion of the 1×2 tunnel structure observed during the elctrochemical reduction by H of an electrolytic γ-MnO$_2$ in a modified Leclanche cell [10]. According to the electron-filling mechanism, an electron entering the 1×2-tunnel structure

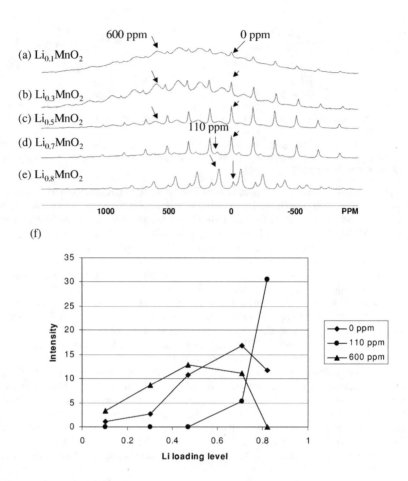

Figure 2. (a) - (e) ^{7}Li MAS NMR spectra of lithiated ramsdellite with different Li loading levels. Isotropic NMR shifts are marked with the corresponding NMR shifts. A spinning speed of ~14 kHz was used. (f) Relative intensities of the isotropic NMR resonances as a function of Li loading level.

during the first half of reduction is shared by two Mn^{4+} manganese ions nearby the inserted proton. Above x = 0.5, insertion of further protons results in electron localization and the formation of Mn^{3+} ions. In this work, the extent of Jahn-Teller distortion, expressed as the ratio of two edges of a MnO_6 octahedron, was also calculated as a function of the reduction level of manganese ions. The degree of Jahn-Teller distortion was shown to increase suddenly, after 0.7 electrons per MnO_2 were inserted to the tunnel structures. Based on the above discussion, we assign the ^{7}Li NMR isotropic resonance at 110 ppm observed at a Li loading level of above 0.5

per MnO_2 to Li^+ ions surrounded by Mn^{3+} ions in the distorted 1×2 tunnel structure of lithiated ramsdellite.

Although resonances at 110 ppm are observed for Li in both lithiated ramsdellite and pyrolusite, these resonances are associated with different spin-relaxation times (T1) and spinning sideband intensities, suggesting different local environments for Li. In other words, the two local environments are different crystallographically but similar magnetically.

Extensions of this work to study EMD following discharging of lithium primary cells are now underway.

CONCLUSIONS

A local environment for Li^+ ions is observed in lithiated pyrolusite, and assigned to the Li^+ ions residing in the 1×1 tunnels. Two ^7Li resonances with different local environments were observed in lithiated ramsdellite at very different frequencies (600 and 110 ppm). The resonance at 600 ppm, which is observed at low lithium intercalation levels, is assigned to Li^+ ions coordinated to both Mn^{3+} and Mn^{4+} ions in the 1×2 tunnels, while the resonance at 110 ppm is due to Li^+ ions coordinated to Mn^{3+} ions and appears at higher Li levels ($x \geq 0.5$). An isotropic resonance close to 0 ppm, observed in both ^7Li NMR spectra of lithiated pyrolusite and ramsdellite, is assigned to Li^+ ions in diamagnetic impurities such as Li_2O and Li_2CO_3 on the surface of the MnO_2 particles. LiI is a more effective reagent for the topochemical lithiation of pyrolusite than n-BuLi, since n-BuLi results in mostly side reactions and the formation of diamagnetic Li-compounds.

ACKNOWLEDGEMENTS

This work was performed with support from Duracell and the National Science Foundation (DMR 9901308).

REFERENCES

1. D. Linden, *Handbook of batteries,* 2nd ed. (McGraw-Hill, Inc., 1995)
2. P. M. De Wolff, *Acta Crystallogr.,* 12, 341 (1959)
3. Y. J. Lee, F. Wang, and C.P. Grey, *J. Am. Chem. Soc.,* 120, 12601 (1998).
4. M. M. Thackeray, M. H. Rossouw, R. J. Gummow, D. C. Liles, K. Pearce, A. De Kock, W. I. F. David, and S. Hull, *Electrochim. Acta,* 38, 1259 (1993).
5. A. R. West and P. G. Bruce, *Acta Cryst,* B38, 1891 (1982).
6. T. Zundel, C. Zune, P. Teyssie, and R. Jerome, *Macromolecules,* 31, 4089 (1998).
7. T. R. Krawietz, D. K. Murray, and J. F. Haw, *J. Phys. Chem. A.,* 102, 8779 (1998).
8. C. Marichal, J. Hirschinger, and P. Granger, *Inorg. Chem.* 34, 1773 (1995).
9. C. Poinsignon, M. Amarilla, and F. Tedjar, *Progress in Batteries & Battery Materials,* 13, 138 (1994)
10. W. C. Maskell, J. E. A. Shaw, and F. L. Tye, *Electrochim. Acta,* 26, 1403 (1981)

Mat. Res. Soc. Symp. Proc. Vol. 658 © 2001 Materials Research Society

A Composite Ionic-Electronic Conductor in the (Ca,Sr,Ba)-Bi Oxide System

James K. Meen, Oya A. Gökçen, I-C. Lin, Karoline Müller, and Binh Nguyen
Department of Chemistry and Texas Center for Superconductivity, University of Houston,
Houston, Texas 77204, U.S.A.

ABSTRACT

The rhombohedral alkaline earth-bismuth oxide phase, an oxygen ion conductor, does not coexist stably with electronic conductors in any of the three binary systems, Ca-Bi-O, Sr-Bi-O, Ba-Bi-O. A thermodynamically stable composite of a rhombohedral phase that contains Ba and Sr or Ca or both with the electronic conductor $BaBiO_3$ may be synthesized. The rhombohedral phase appears to have complete mutual miscibility of the alkaline earth elements. The compositions of rhombohedral phase that coexist with $BaBiO_3$ in the Sr-Ba-Bi ternary system and the Ca-Sr-Ba-Bi quaternary systems are described. The value of ionic conductivity of the rhombohedral phase (at a constant Bi: [Ca+Sr+Ba]) is not dependent on the relative amounts of Ca, Sr, and Ba. The temperature at which the rhombohedral phase undergoes a polymorphic transformation from a low-temperature (β_2) form that is a weak ion conductor to a high-temperature (β_1) form that is a much better oxygen ion conductor. The temperatures of the polymorphic transformation and of the upper stability limit of the rhombohedral phase both depend strongly on Ca: Sr: Ba. The β_1 form develops in the Ba-Bi system at the lowest temperatures and at the highest ones in the Ca-Bi system. On the other hand, the Ca-Bi phase has greater thermal stability than its Ba analogues. The temperature range over which a useful composite conductor can operate is, therefore, strongly dependent upon the bulk composition of the system.

INTRODUCTION

The rhombohedral AE-Bi oxide (where AE is an alkaline earth element, Ca, Sr, or Ba) phase is a well-known oxygen ion conductor. The Ba-Bi-O system also contains $BaBiO_3$, an electronic conductor. Rhombohedral Ba-Bi oxide and $BaBiO_3$ do not coexist stably [1]. Addition of Ca or Sr or both to the system results in production of a two-phase field in which rhombohedral AE-Bi oxide and $BaBiO_3$ coexist over a temperature range. A composite ionic-electronic conductor may be thus created. In this composite, the composition of the perovskite remains essentially constant but the rhombohedral phase has a wide range of alkaline earth contents. The two-phase field is terminated by three-phase fields that limit the range of compositions of rhombohedral phase that may be incorporated into a composite ionic-electronic conductor with the perovskite.

The properties of the composite ionic-electronic conductor are strongly influenced by the composition of the rhombohedral phase. The composition of the rhombohedral phase determines both the temperature of the polymorphic transformation and the temperature of decomposition of the biphasic assemblage. The low-temperature β_2 polymorph transforms to a high-temperature β_1 form through displacive transformation that involves no change in space group (R-3m) [2,3]. A sudden and marked increase in oxygen ion conductivity occurs at that transformation [4,5,6,7, 8]. The Ba-Bi rhombohedral phase melts at 753-790°C (depending on Ba: Bi) [1] and rhombohedral Sr-Bi oxide melts at 830->925°C [9]. Rhombohedral Ca-Bi oxides break down in the subsolidus with maximum temperature of stability of ~840°C [10]. A perovskite-rhombohedral conductor is most efficient if the rhombohedral phase is in β_1 form so the temperature range of its operation depends on the composition of the rhombohedral phase (and, thus, of the bulk composition). This paper investigates phase relations in part of the Ca-Sr-Ba-Bi quaternary to estimate both the compositional bounds on the perovskite-rhombohedral two-phase field and the thermal stability of that assemblage.

EXPERIMENTAL DETAILS

Starting materials were prepared by conventional solid-state synthesis techniques using dry high-purity BaO_2, $SrCO_3$, $CaCO_3$, and Bi_2O_3. They were dried, mixed in appropriate stoichiometric proportions, pressed into a 13 mm pellet, and calcined for 24 h at 700°C. Each pellet was finely ground, repelletized, and sintered; this step was repeated thrice and the final pellet ground to a fine powder. Some 20-30 mg of the powder was mounted on a loop of gold wire and experiments were performed in a Deltech VT-31 furnace in a flowing oxygen atmosphere with temperature controlled to within 2°C by S-type thermocouple. Experimental charges were drop quenched into an oil to preserve high-temperature phase assemblages to the maximum extent possible. After mounting in epoxy and polishing, samples were studied using a JEOL JXA-86(electron microprobe. X-ray analysis was performed using wavelength-dispersive spectrometer Standards employed were $CaMgSi_2O_6$ for Ca, $SrTiO_3$ for Sr, $BaTiSi_3O_9$ for Ba, and $Bi_{12}GeO_{20}$ for Bi and O. Counting times were 100 s on peaks and 50 s on each of two backgrounds. Inten sity data were processed by on-line Geller $\phi\rho Z$ software to obtain elemental concentrations.

Oxygen ion conductivities were determined on bars of pellets of powdered samples prepare using techniques identical to those discussed elsewhere [8]. Conductivities were measured in a as functions of temperature by a two-probe AC impedance method using gold electrodes. The data were collected between 300°C and an upper temperature dictated by the thermal stability of the rhombohedral phase through several heating and cooling cycles with equilibration time of at each experiment temperature. A HP 4192A Impedance Analyzer in the frequency range 5 H to 13 MHZ carried out impedance measurements.

EXPERIMENTAL RESULTS
Sr-Ba-Bi Oxide System at 725°C

Figure 1 is an isothermal section of the Sr-Ba-Bi oxide system at 725°C. There were considerable experimental difficulties in studying assemblages rich in alkaline earth metals, primar ily due to reaction of the Sr- or Ba-rich phases with the environment. The finely polished surfa required for electron microprobe analysis was highly reactive and phases hydrated or carbonate very rapidly. The phase relations of the more Sr- and Ba-rich parts of the phase diagram have not been determined.

The terminology for bismuth-barium oxides is adopted from a recent study of that system [1]. Bismuth oxide adopts a face-centered cubic (FCC) structure at this temperature and this structure can incorporate some Sr and Ba. There is complete solid solution from the Ba-Bi to t Ba-Sr joins near the Bi apex of the triangle. Alkaline-earth-saturated FCC coexists with Bi-

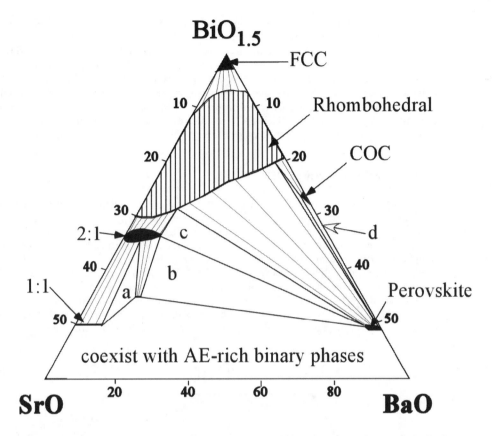

Figure 1. *Isothermal section at 725 °C demonstrating phase relations in the system Sr-Ba-Bi oxide under pure oxygen atmosphere. There are four three-phase fields lettered a, b, c, and d. The solid regions or lines denote the extent of each of the various solid solutions other than the rhombohedral phase. Thin lines are tie-lines across the two-phase fields. See text for explanation of terms.*

saturated rhombohedral phase and the rhombohedral phase also has complete Ba-Sr miscibility although the structure can accept more Sr than Ba at 725°C. The vertically-ruled field shows the extent of this solid solution. The other alkaline earth-bismuth oxides have more limited Sr-Ba miscibility. Ba-saturated rhombohedral phase coexists in the Ba-Bi oxide system with a phase with ~25% BaO that is denoted by COC as it occurs as low-temperature cubic, orthorhombic, and high-temperature cubic polymorphs. At 725°C, the phase dissolves very little Sr and only a narrow two-phase region between rhombohedral and COC phases occurs. An even narrower two-phase field exists between the COC phase and $BaBiO_3$, labeled "Perovskite". Both the Bi_2SrO_4 (2:1) and $Bi_2Sr_2O_5$ (1:1) phases can substitute appreciable Ba for Sr. The other phase with Sr>Ba and ~45% AE (all compositions in cat.%) is believed to be Ba-substituted body-centered tetragonal (BCT) phase. In the Sr-Bi oxide system, BCT has a lower thermal stability limit of 765±5°C [9] but that limiting temperature may be lowered in BCT with appreciable Ba. The three-phase field labeled "a" describes compositions of Ba-saturated 1:1, Sr-saturated BCT and a 2:1 that coexist at 725°C. Field "b" shows compositions of Sr-saturated perovskite, Ba-saturated 2:1, and Ba-saturated BCT. Those same perovskite and 2:1 coexist with the most Sr-rich rhombohedral phase in equilibrium with any perovskite are defined by field "c". The apice of field "d" define the most Ba-rich rhombohedral phase that occurs stably at 725°C with perov skite, that perovskite, and COC. Fields "c" and "d" are separated by a two-phase field in which rhombohedral phase with wide Sr: Ba range coexists with perovskite with little dissolved Sr.

Ca-Sr-Ba-Bi Oxide System at 725°C

Elucidation of phase relations in the quaternary system requires a much greater number of experiments than does determining phase relations in each ternary system and we have not at-tempted a full determination of those phase relations. Figure 2 summarizes in a schematic way the present state of our knowledge of the phases that coexist with the alkaline-earth-saturated rhombohedral phase. Only those solutions with very high Ba: (Ca+Sr+Ba) coexist with COC phases. Those with high Sr: Ba coexist with Bi_2SrO_4 (2:1). Rhombohedral phases in which Ca exceeds 40-45% of the alkaline earth content coexist with a quaternary oxide with approximate composition $Bi_4Ba(Ca,Sr)O_8$. This phase may be isostructural with a tetragonal Ba-Bi phase of the same AE: Bi [1]. It is currently under study. This figure also shows the approximate limits of compositions of the rhombohedral phase that coexist with the $BaBiO_3$ perovskite.

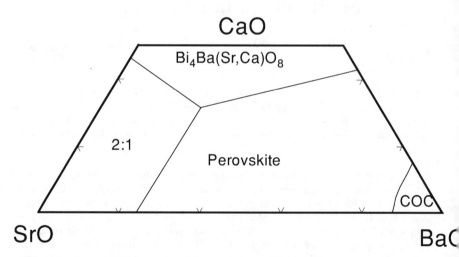

Figure 2. *Phases that coexist with the rhombohedral phase at 725°C as a function of the com-position of the rhombohedral phase in terms of Ca:Sr:Ba only. The perovskite coexists with a wide-range of quaternary (Ca,Sr,Ba)-Bi rhombohedral oxides but with no binary ones.*

Polymorphic Transformation Temperature of Rhombohedral Phase

The composition of the rhombohedral phase has a marked effect on the temperature at which it undergoes the β_2-β_1 phase transformation. As mentioned, the oxygen ion conduction of this phase increases greatly because of that phase transformation. The polymorphic transformation does not occur at a single temperature for most compositions of the rhombohedral phase. In the general case, there must exist a two-phase field in which β_2 and β_1 of different compositions co-exist. Only at the saturation limits of the rhombohedral solid solution does the β_2-β_1 phase transformation occur at a single temperature. Notably, the composite with perovskite must contain an AE-saturated rhombohedral phase so that the β_2-β_1 transformation is at a single temperature for each composition.

The conductivity-reciprocal temperature plots were evaluated by fitting straight lines through the three segments of the curves and calculating their intersections. By this method, the estimated range over which the polymorphic transformation occurs for $Bi_{80}Ba_{20}$ (that should transform at a single temperature) is 3°C. The estimate of the range for $Bi_{80}Sr_{10}Ba_{10}$ is nearly 50°C. In calculations below the polymorphic transformation temperature used is that defined for the midpoint of the range determined from each conductivity-reciprocal temperature plot. The polymorphic transformation temperature changes linearly with Ba: Sr in the rhombohedral phases of $Bi_{80}(Ba+Sr)_{20}$ [7,8]. Variations of this transition temperature with other composition parameters are, within our knowledge, close to linear and variations can be modeled as a simple linear combination. Minimizing residuals between calculated and measured temperatures yield a best-fit equation: $T = -1925 + 29.6*X_{Bi} + 10.7*X_{Ca} + 9.2*X_{Sr}$ (°C), where X_M is the cation percentage of M in the rhombohedral phase so $X_{Bi}+X_{Ca}+X_{Sr}+X_{Ba}=100$. Although not a perfect fit to the data, this equation predicts 10 of the 13 transition temperatures measured to within 30°.

This simple equation illustrates the influence of each element on the β_2-β_1 transformation. Increasing the bismuth content has the most marked influence in raising that temperature. Increasing the Ca content of the rhombohedral phase at the expense of the other alkaline earth elements raises the temperature of the β_2-β_1 transition. On the other hand, substitution of Ba for any other element lowers the temperature of the phase change.

The two-phase field in which β_2 and β_1 coexist is assumed to have neither maxima nor minima except at the saturation limits. (This is implicit in the assumption of linear variation of the transition temperature with composition.) In this case, the low-temperature β_2 phase always has higher Bi and lower Ba content than the coexisting β_1 form.

CONCLUSIONS

Perovskite with nominal composition $BaBiO_3$ may coexist with the alkaline earth-bismuth rhombohedral phase with a relatively wide field of compositions. Together they constitute a composite ionic-electronic conductor [11]. The variation in alkaline earth element contents of the high-temperature rhombohedral phase results in very little change in the ionic conductivity of that phase. The temperature at which the high-temperature forms upon heating is, on the other hand, a strong function of composition. The Bi content of the rhombohedral phase in a composite with $BaBiO_3$ cannot be independently fixed, as the rhombohedral phase must be saturated in alkaline earth elements. The pure Bi-Ba rhombohedral phase has the lowest transition temperature at a given Bi: AE but does not coexist with $BaBiO_3$. Rhombohedral phases with high Ba: (Ca+Sr) and transition temperatures below 550°C do coexist with that perovskite. That assemblage also melts at a relatively low temperature ($\approx$740-760°C) [8] and that places limits on the usefulness of the composite. Increasing the Ca or Sr contents of the rhombohedral phase increases the thermal stability of the assemblage but also results in increase in the transition temperature. The composite is thus useful in a higher temperature interval.

ACKNOWLEDGMENTS

We would like to thank Allan Jacobson for introducing us to some of the problems of this system and for making available the facilities for measuring conductivity. Sangtae Kim and Yuemei Yang provided invaluable assistance in those measurements. The microprobe work would not have been possible without the help of Kent Ross. Support was provided by Texas Advanced Research Program under Award Number 003652-886, by the MRSEC Program of the National Science Foundation under Award Number DMR-9632667, and by the Texas Center for Superconductivity at the University of Houston.

REFERENCES

1. K. Müller, J. K. Meen and D. Elthon, *Solid-State Chemistry of Inorganic Materials II*, ed. S. M. Kauzlarich, E. M. McCarron III, A. W. Sleight and H-C. zur Loye, (Materials Research Society Proceedings **547**, 1999) p. 261.
2. P. Conflant, J. C. Boivin, G. Nowogrocki and D. J. Thomas, *Solid State Ionics*, **9&10**, 925 (1983).
3. D. Mercurio, J. C. Champarnaud-Mesjard, B. Frit, P. Conflant, J. C. Boivin and T. Vogt, *Solid State Chem.*, **112**, 1 (1994).
4. T. Takahashi, H. Iwahara and Y. Nagai, *J. Appl. Electrochem.*, **2**, 97 (1972).
5. T. Takahashi, T. Esaka and H. Iwahara, *J. Solid State Chem.*, **16**, 317 (1976).
6. J. C. Boivin and D. J. Thomas, *Solid State Ionics*, **5**, 523 (1981).
7. O. A. Gökçen, S. Kim, J. K. Meen, D. Elthon and A. J. Jacobson, *Solid State Ionics V*, ed. G-A. Nazri, C. Julien and A. Rougier, (Materials Research Society Proceedings **548**, 1999) p. 499.
8. O. A. Gökçen, J. K. Meen, and A. J. Jacobson, *Mass and Charge Transport in Inorganic Materials: Fundamentals in Devices*, ed. P. Vincenzini and V. Buscaglia, Proceedings of the International Conference on "Mass and Charge Transport in Inorganic Materials: Fundamentals to Devices", Lido di Jesolo, Italy (2000) p. 91.
9. N. M. Hwang, R. S. Roth and C. J. Rawn, *J. Am. Ceram. Soc.* **73**, 2531 (1990).
10. O. A. Gökçen, V. J. Styve, J. K. Meen and D. Elthon, *J. Am. Ceram. Soc.*, **82**, 1908 (1999).
11. O. A. Gökçen, J. K. Meen, A. J. Jacobson, and D. Elthon [this volume].

Mat. Res. Soc. Symp. Proc. Vol. 658 © 2001 Materials Research Society

Manganese Vanadium Oxide Compounds As Cathodes For Lithium Batteries

J. Katana Ngala, Peter Y. Zavalij and M. Stanley Whittingham*
Institute for Materials Research and Department of Chemistry
State University of New York at Binghamton,
Binghamton, NY13902-6000, USA.

ABSTRACT

A pure form of the compound $[(CH_3)_4N]_{0.2}Mn_{0.06}V_2O_{5+\delta} \cdot 0.2H_2O$ has been synthesized by the hydrothermal technique. A manganese richer compound, $[(CH_3)_4N]_{0.18}Mn_{0.1}V_2O_{5+\delta} \cdot 0.35H_2O$, has also been synthesized. By comparison to $[(CH_3)_4N]_{0.17}Fe_{0.1}V_2O_{5+\delta} \cdot 0.17H_2O$, the structure of these compounds consists of double sheets of vanadium oxide with tetramethyl ammonium ions in the interlayer spaces. Both compounds display reversible lithium intercalation with sharp end points. The lithiation capacity is lowered by the increase in manganese content.

INTRODUCTION

Transition metal oxides may act as cathodes in lithium secondary batteries partly due to the ability of the metal ions to exhibit variable oxidation states, and their highly negative free energies of reaction with lithium ions, in contrast to other chalcogenides.

The commercially available layered $LiCoO_2$ has good cyclability. However, it finds limited use such as in lap tops and cellular phones, due to its high cost. The spinel compound, $LiMn_2O_4$, which was found to act as an intercalation host in the early 1980s, has been developed for commercial use as a replacement for $LiCoO_2$ in second-generation Li-ion batteries; the manganese is lower cost than cobalt. Unfortunately, only 0.5 Li per Mn may be inserted and de-inserted reversibly [1]. This corresponds to a capacity of about 110 Ah/kg in contrast to 130 Ah/kg for $LiCoO_2$.

Vanadium oxides have been extensively investigated as possible cathodes for lithium batteries. A class of compounds of particular interest for potential use in batteries is that consisting of the δ-V_2O_5 structure. This structure type contains double vanadium oxide sheets with the vanadyl groups on the outside of the sheets; the vanadium is in a distorted octahedral coordination rather than the square pyramid typical of V_2O_5 itself and the well known bronzes, such as LiV_2O_5. One of the most studied vanadium oxide cathodes, V_6O_{13}, contains these sheets and cycles lithium well [2] in secondary cells. Typical simple members of this class are $Ni_{0.25}V_2O_5$ [3], $[N(CH_3)_4]_{0.3}Fe_{0.1}V_2O_5$ and $Zn_{0.4}V_2O_5$ [4].

We report here on the preparation of a pure phase of $[N(CH_3)_4]_yMn_zV_2O_{5+\delta} \cdot nH_2O$ (compound A) by hydrothermal synthesis. Earlier we reported [5] this phase as one component of a two phase mixture. We also report on an isostructural material (compound B) with a higher manganese content. We compare their electrochemical behavior. Hydrothermal synthesis and other soft chemistry techniques such as sol-gel synthesis, have been widely applied for the low temperature syntheses of metal oxides. These techniques are useful in accessing metastable metal oxides.

EXPERIMENTAL

Compound A was hydrothermally synthesized by the treatment of V_2O_5 (Alfa-Aesar) with an aqueous solution of $HMnO_4$, in the presence of the templating cation $[(CH_3N)_4]OH$ (Alfa-Aesar). The reacting mole ratio was $V_2O_5:HMnO_4:[(CH_3)_4N]OH = 1:0.15:2$. Nitric acid was added to lower the pH. $HMnO_4$ was obtained by ion exchanging the potassium in $KMnO_4$ (Fisher) using an acidic cation exchange resin. The mixture was transferred to an autoclave after thorough stirring, placed in an oven at 165^0C and left to react for 3 days. The pH of the reaction mixture was initially 2.4 and 3.5 at the end of the reaction. This pH range along with the given reacting mole ratios ensured pure product. The brown clay-like solid was filtered and rinsed with distilled water.

Compound B was prepared by taking the above product (compound A) and adding to it LiOH and $HMnO_4$, in the mole ratio 3:1:1, respectively. The same heat treatment as above was followed. The reaction proceeded under similar pH conditions (2.2-3.5). A dark brown clay was isolated.

X-ray powder diffraction was performed using Cu $K\alpha$ radiation on a Scintag θ-θ diffractometer equipped with a Ge (Li) solid state detector. The data was collected from $2\theta = 2^0$ to $2\theta = 90^0$ with 0.02^0 steps and 10 sec. per step count time. The TGA data was obtained on a Perkin Elmer model TGA 7, the FTIR on a Bruker Equinox and electron microscopy on a JEOL 8900. A mixture of the sample, carbon black and polyvinylidene fluoride as the binder, was prepared in the weight ratio 8:1:1, respectively. Cyclopentanone was added to the mixture and thoroughly mixed to obtain a slurry that was coated on a 1 cm^2 aluminium disk using a dropper. After air–drying the disk was placed in an oven at about 150^0C for effective dehydration for at least 5 hours. Swagelok test cells were used with lithium metal, adhered to a nickel disk, as the anode. The electrolyte was 1M $LiPF_6$ dissolved in a 2:1 mixture of ethylene carbonate (EC) and dimethyl carbonate (DMC).

RESULTS AND DISCUSSION

Compound A:

EDS analysis on compound A indicated the presence of V and Mn in the ratio Mn:V = 0.06:2. This corresponds to $z = 0.06$ in the formula, $[N(CH_3)_4]_yMn_zV_2O_{5+8}\cdot nH_2O$ (standard deviation = ± 0.01). The amount of Mn obtained here is less than previously reported, probably due to extra manganese present in the other phases of the earlier product.

Powder x-ray analysis, shown in Figure 1, indicates a single phase product in contrast to that reported earlier [5] which had mixed phases. The pattern, which suggests low crystallinity, was indexed using a monoclinic unit cell in the space group C2/m, with the following parameters: $a = 11.66$ Å, $b = 3.61$ Å, $c = 13.91$ Å and $\beta = 108.8^0$. The low crystallinity is confirmed by the SEM image, which shows a fuzzy layered morphology.

The powder x-ray pattern gives a repeat distance of 13.1 Å, consistent with δ-V_2O_5 sheets having TMA ions sandwiched in the interlayer spacing. The in-plane spacing of 3.6 Å points to disordered organic cations as the diameter of these cations is >5Å. Unfortunately, Rietveld refinement could not be performed as the quality of the XRD pattern was insufficient. The structure was deduced from the similarity of its XRD pattern with that of $[(CH_3)_4N]_{0.17}Fe_{0.1}V_2O_{5+\delta}\cdot0.17H_2O$ whose structure we reported earlier [4]. From this analogy, we postulate that the structure of compound A consists of VO_6 distorted octahedra forming double sheets of V_2O_5 as shown in Figure 2.

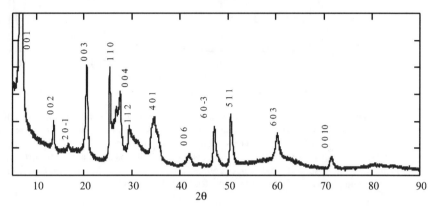

Figure 1: *Powder x-ray pattern for $[N(CH_3)_4]_yMn_zV_2O_{5+\delta}\cdot nH_2O;z = 0.06$.*

Figure 2: *The structure for $[N(CH_3)_4]_yMn_zV_2O_{5+\delta}\cdot nH_2O$.*

The FTIR spectrum (Figure 3) indicates the presence of a V=O group by the peak at 1012 cm^{-1}. The vanadyl groups are projected away from the double layers. The band at 3421 cm^{-1} is due to the presence of water, whereas those at 517 cm^{-1} and 754 cm^{-1} are attributed to V-O vibrations in the framework. The presence of the tetramethylammonium ion is indicated by the three bands situated at 2900 cm^{-1}, 1481 cm^{-1} and 950 cm^{-1}.

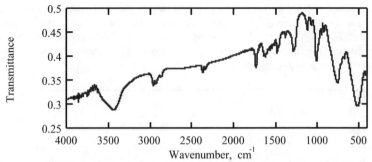

Figure 3: FTIR spectrum for $[N(CH_3)_4]_yMn_zV_2O_{5+\delta} \cdot nH_2O$; $z = 0.06$.

Thermal analysis, performed under oxygen atmosphere, results in two weight losses as displayed in Figure 4. The 1.8% weight loss in the range, 100^0C-250^0C is associated with water of solvation between the sheets. This corresponds to $n = 0.20$ in the formula: $[(CH_3)_4N]_yMn_zV_2O_{5+\delta} \cdot nH_2O$. The weight loss before 100^0C is associated with surface water. The 7.6% weight loss over the range 250^0C-300^0C is due to decomposition of the organic, corresponding to $y = 0.21$ in the formula: $[(CH_3)_4N]_yMn_zV_2O_{5+\delta} \cdot nH_2O$.

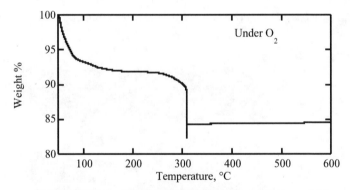

Figure 4: TGA profile for $[N(CH_3)_4]_yMn_zV_2O_{5+\delta} \cdot nH_2O$; $z = 0.06$.

The product obtained by subjecting the sample to a constant temperature of 200^0C for 4 hours shows an XRD pattern that is identical to that of the original sample. This implies that water is not crucial for the integrity of the structure. However, upon heating to 600^0C, the structure collapses as evidenced by the XRD pattern of the resulting product. The pattern is consistent with a mixture of MnV_2O_6 and V_2O_5.

The electrochemical data shows an open circuit potential of 3.62 V, consistent with vanadium in the +5 oxidation state. Figure 5 displays the charge–discharge profile of 7 cycles. It indicates initial capacity of 220 Ah/kg, corresponding to about 2 Li per formula unit. The same capacity was reported for the earlier product [5]. We still observe the plateau at about 2.8 V, suggesting some phase change during cycling.

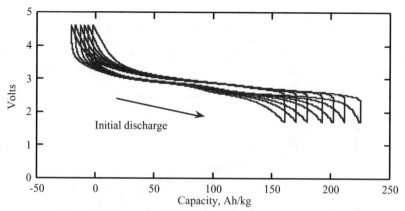

Figure 5: *Cycling of $[N(CH_3)_4]_yMn_zV_2O_{5+\delta}$; $z = 0.06$ at $0.1\ ma/cm^2$.*

Compound B

EDS analysis yields the approximate ratio Mn:V = 0.1:2, corresponding to $z = 0.1(1)$ in the formula $[(CH_3)_4N]_yMn_zV_2O_{5+\delta}.nH_2O$. On the other hand, the powder XRD pattern of this product is identical to that of compound A, so that that the additional Mn content does not alter the δ-phase structure.

Figure 6 gives the thermal analysis profile for this compound. The 3.0% weight loss over the temperature 82^0C-200^0C corresponds to $n = 0.35$ in the formula $[(CH_3)_4N]_yMn_zV_2O_{5+\delta}.nH_2O$, whereas the 6.3% loss in the interval, 200^0C-350^0C corresponds to $y = 0.18$. Thus, concomitant with increase in Mn content in the structure is a decrease in the amount of the organic cation and an increase in water of solvation. This suggests that probably the manganese ions, as the organic cations, are situated in the interlayer spaces in hydrated form. Thus the introduction of more manganese ions causes the displacement of the organic cations.

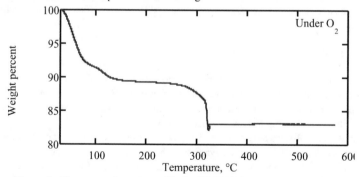

Figure 6: *Thermogravimetric profile for $[N(CH_3)_4]_yMn_zV_2O_{5+\delta}.\bullet nH_2O$; $z = 0.1$.*

The electrochemical behavior of this compound I shown in Figure 7. It exhibits an open circuit potential of 3.59 V, which is slightly less than that for compound A, due to the higher manganese content which will probably reduce the overall oxidation state of the oxide matrix. The cell was charged initially; essentially no capacity was observed during this charging process indicating no lithium in the sample, even though the reaction medium contained much lithium.

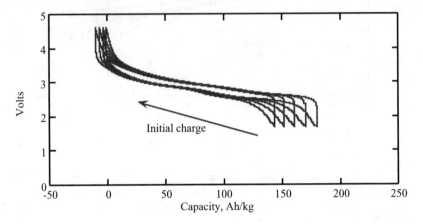

Figure 7: *Cycling of $[N(CH_3)_4]_yMn_zV_2O_{5+\delta}$; z = 0.1 at 0.1ma/cm².*

Figure 7 shows that the capacity of the cell, 180 Ah/kg, is significantly reduced from the 220 Ah/kg of sample A. The reason for this is not fully understood, as the 0.04 additional Mn ions cannot account for the drop of around 0.4 Li per formula unit simply based on oxidation state changes or site occupancy. It is possible that this loss is in part a kinetic effect, with the additional manganese ions pinning the layers together much as just a few per cent additional titanium ions in $Li_xTi_{1+y}S_2$ can cut the lithium diffusion coefficient by several orders of magnitude [6].

CONCLUSIONS

The pure form of the δ-phase compound $[(CH_3)_4N]_{0.2}Mn_{0.06}V_2O_{5+\delta}•0.2H_2O$ was hydrothermally synthesized. A more manganese rich phase, $[(CH_3)_4N]_{0.18}Mn_{0.1}V_2O_{5+\delta}•0.35H_2O$ was synthesized from the former compound. Both compounds display reversible cycling as cathodes, although the latter has lower capacity than the former.

We are currently investigating other δ-phase compounds of the series, to which the two compounds belong, with the aim of evaluating their performance as cathodes for lithium ion batteries. Work is also in progress to explore other types of pillars for the layers and to study the manganese oxide analogs.

ACKNOWLEDGEMENTS

We thank the Department of Energy, Office of Transportation Technologies, through Lawrence Berkeley Laboratory, and the National Science Foundation through Grant DMR–9810198 for partial support of this work. We also thank Bill Blackburn for the SEM studies.

REFERENCES

1. A. R. Armstrong, A. D. Robertson, R. Gitzendanner, and P. G. Bruce, *J. Solid State Chem.*, **145**, 549-550(1999).
2. Ö. Bergström, H. Björk, T. Gustafsson, J. O. Thomas, *J. Power Sources*, **81-82**(1999) 685.
3. Y. Oka, T. Yao, N. Yamamoto, *J. Solid State Chem.*, **132** (1997) 323.
4. F. Zhang, P. Y. Zavalij and M. S. Whittingham, *Mater. Res. Bull.*, **32** (1997) 701.
5. F. Zhang and M. S. Whittingham, *Electrochem. Comm.*, **2** (2000) 69.
6. M. S. Whittingham, Progress in Solid State Chemistry, **12** (1978) 41.

Mat. Res. Soc. Symp. Proc. Vol. 658 © 2001 Materials Research Society

Microporous Ruthenium Oxide Films for Energy Storage Applications

J.P. Zheng and Q.L. Fang
Department of Electrical and Computer Engineering
Florida A&M University and Florida State University
Tallahassee, FL 32310, USA

ABSTRACT

Highly porous ruthenium oxide films were prepared using ruthenium ethoxide solution at low temperatures. The specific capacitance of the ruthenium oxide film electrode made with ruthenium ethoxide solution is much higher than that made using the traditional ruthenium chloride solution. It was found that amorphous ruthenium oxide films with high porosity could be formed at temperatures of 100-200 °C. At temperatures above 250 °C, crystalline ruthenium oxide films were formed. Electrochemical capacitors were made with ruthenium oxide film electrodes and were tested under constant current charging and discharging. The maximum specific capacitance of 593 F/g was obtained from the electrode prepared at 200 °C. The interfacial capacitance increased linearly with increasing film thickness. A value of interfacial capacitance as high as 4 F/cm^2 was obtained from the electrode prepared at 200 °C.

INTRODUCTION

Powder forms of amorphous and hydrous ruthenium oxides ($RuO_2 \bullet xH_2O$) have been formed by the sol-gel method and found to be promising materials for electrochemical (EC) capacitors with both high power density and high energy density [1,2]. The advantages of amorphous ruthenium oxides include high specific capacitance, high conductivity, and good electrochemical reversibility [3]. A maximum specific capacitance of 768 F/g has been obtained from an amorphous ruthenium oxide [3,4]. The high specific capacitance leads to superior energy density for capacitors made with amorphous ruthenium oxides. The maximum power density of the EC capacitor is determined mainly by the internal resistance of the capacitor, including its electrical and ionic resistances. From an analysis of the distribution of resistance inside a capacitor made of powder electrode materials and aqueous electrolytes, it was previously determined that over 50% of the total internal resistance of the capacitor is due to the electrical contact resistance [5]. This includes the powder electrode material contact, the electrode and the current collector contact, the current collector and the current collector contact from the neighboring cell, and the current collector contact and the end electrode contact. In contrast with the powder electrode, the film electrode has extremely low contact resistance [6]. In this paper, we report a new method to prepare amorphous ruthenium film electrodes on a metallic tantalum (Ta) substrate. These film electrodes are shown to have a higher specific capacitance than that of crystalline films.

EXPERIMENTAL DETAILS

Tantalum (Ta) foil was used as the substrate. The surface of the Ta foil was abraded with a brush made of palladium wire to remove the oxide layer and at the same time to make the surface rough. This was followed by vacuum heat treatment at 900 °C for ½ hour. The procedure led to improved adhesion and greatly diminished the contact resistance of the Ta foil surface. The next step was vacuum deposition of a very thin layer of ruthenium metal on the Ta foil. The precursor for growth of amorphous ruthenium oxide film was made as follows:

$$RuCl_3 + 3Na(OC_2H_5) \rightarrow Ru(OC_2H_5)_3 + 3NaCl \qquad (1)$$

This reaction was carried out in ethanol at 78 °C, which is the boiling temperature of ethanol. When the reaction was completed, the ruthenium ethoxide, $Ru(OC_2H_5)_3$, dissolved, and NaCl precipitated in the ethanol solvent. Next, the substrate was heated to the desired temperature, and the ruthenium ethoxide solution was sprayed onto the substrate surface using an airbrush. After each coating, the film electrode was annealed for about 15 minutes and then dipped into boiling water for about 1 minute in order to remove any incompletely oxidized material and any residual NaCl. Five film electrodes were prepared at temperatures of 100, 150, 200, 250, and 300 °C, respectively. Different film thicknesses were obtained by different numbers of coatings. It was found that if the preparation temperature was below 100 °C, the film could not adhere to the substrate and was washed away when it was dipped into boiling water.

RESULTS AND DISCUSSION

The surface morphologies of ruthenium oxide film electrodes were studied using scanning electron microscope (SEM). Figure 1 shows SEM pictures of electrode surfaces grown using $Ru(OC_2H_5)_3$ precursor at temperatures of 100 °C, 200 °C, 250 °C, and 300 °C. It can be seen that at 100 °C, the film electrode was formed with irregularly shaped particles with an average size of about 1 µm. Each particle contains many crevices giving large surface area. At 200 °C, the electrode was still formed with porous particles but the particle size increased to about 10 µm. At 300 °C, however, the film formed with less porous flakes.

The crystalline structures of ruthenium oxide film electrodes were characterized using an x-ray diffractometer. The results showed that at lower temperatures ($\leq$ 200 °C), the material forms an amorphous phase. On increasing the annealing temperature, the film became crystalline.

EC capacitors were made with two ruthenium oxide film electrodes. The electrolyte was 0.5 mol/L H_2SO_4 solution. The typical size of each electrode was about 1-1.5 cm^2. The mass of the ruthenium oxide film was calculated from the mass difference before and after substrate coating procedure. It was found that the film thickness was difficult to measure due to the high porous nature of the ruthenium oxide films and the roughness of the substrate. Therefore, in this study, the material loading per unit area (mg/cm^2) was used instead of the film thickness. Figure 2 shows a typical dc charge/discharge cycle of the capacitor made with a film electrode prepared at 200 °C at a current of 2 mA and a voltage range 0-1 V. The average capacitance of the capacitor was calculated according to the equation of

$$C = \frac{I \times \Delta t}{\Delta V} \qquad (2)$$

where I is the constant current, Δt is the time duration of the discharge process, and ΔV is the voltage range. From Figure 2, a capacitance of 1.56 F was obtained. The energy efficiency of the

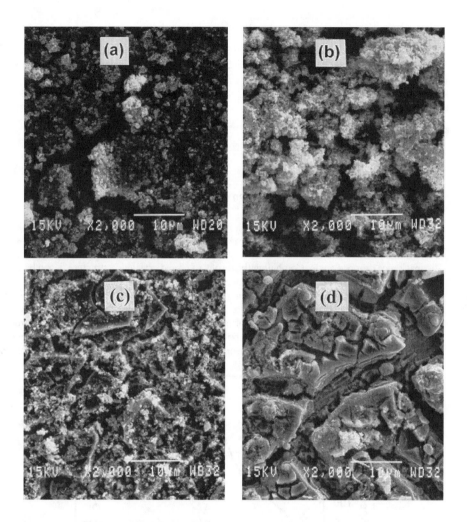

Figure 1 *SEM micrographs of the surface of a ruthenium oxide film prepared using $Ru(OC_2H_5)_3$ precursor at temperatures of (a) 100 °C, (b) 200 °C, (c) 250 °C,and (d) 300 °C.*

capacitor was calculated based on the difference in capacitances between the charge and discharge processes as 93%.

The specific capacitance (c_p) and interfacial capacitance (c_f) are defined as the capacitance per unit weight and per unit area, respectively. Figure 3 shows c_p as a function of the temperature at which the ruthenium film was grown. It can been seen that the specific capacitance increased initially with increasing annealing temperature and reached a maximum value of 593 F/g around 200 °C, but when the annealing temperature was higher than 250 °C, the value of specific

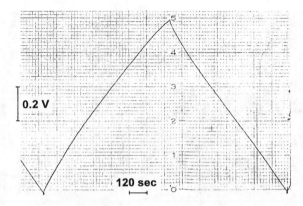

Figure 2 *DC charge and discharge curve of an EC capacitor made with two ruthenium oxide film electrodes H₂SO₄ electrolyte. The current was 2 mA and the voltage range was 0-1 V.*

capacitance dropped again. Combining the results of c_p and crystalline structure, it can be conclude that the maximum value of c_p can be obtained at the temperature just below the critical temperatu at which the crystalline phase of ruthenium oxide started to be formed. The result is consistent wit that observed from hydrous ruthenium oxide powder electrodes; however, values of c_p obtaine from film electrodes are still less than that obtained from hydrous ruthenium oxide electrodes (72 F/g) [3]. The high specific capacitance of powder ruthenium oxide is believed to be attributed to th water molecules in the structure, which favors proton intercalation into the bulk of ruthenium oxide

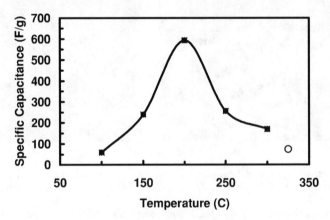

Figure 3 *The specific capacitance as a function of preparation temperature for ruthenium oxide film electrodes. T specific capacitance of a film electrode made using RuCl₃ precursor at 325 °C is also presented with open-circle.*

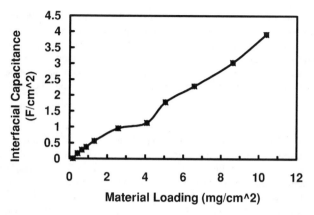

Figure 4 *The interfacial capacitance as a function of material loading of ruthenium oxide for film electrodes prepared at 200 °C.*

A previous study has demonstrated that proton diffusion in hydrous ruthenium oxide is much easier than that in anhydrous ruthenium oxide [7]. In addition, the specific surface area could also cause the difference in specific capacitance, but the values of specific surface area for film electrodes are not available; therefore, no comparison can be made.

Figure 4 shows c_f as a function of material loading for films prepared at 200 °C. It can be seen that c_f always increases with increasing material loading without showing a tendency to saturate up to a value as high as 4 F/cm^2. To our knowledge, this value is much higher than that observed from any other ruthenium oxide film electrodes. Observing high values of c_p and c_f from film electrodes prepared using $Ru(OC_2H_5)_3$ precursor is believed to be due to two reasons. The first is that the crystalline structure of ruthenium oxide films is different between this work and others. In this work, amorphous ruthenium oxide films were formed because a low preparation temperature was applied; in other reported studies, crystalline ruthenium oxides were formed due to high temperature processing. From our previous studies [4], it was found that for amorphous ruthenium oxides, the protonation reaction can occur not only at the surface of the ruthenium oxide material but also inside the bulk of the particle, because the protons can intercalate into the bulk of amorphous phase ruthenium oxide; however, for crystalline ruthenium oxides which are formed at high temperatures, the protonation reaction is restricted to the surface of the ruthenium oxide material. The second reason for high values of c_p and c_f obtained in this study is due to the high porosity of the film made with $Ru(OC_2H_5)_3$ precursor, because the c_p increased with increasing the specific surface area of the film [9].

It should be pointed out that the value of c_p from different film electrodes varied. For example, from the slope of Figure 4, an average c_p of about 400 F/g was obtained at 200 °C; but the highest c_p of 593 F/g was also obtained from a film electrode prepared at 200 °C. The large variation in c_p may be explained by the variation of preparation conditions during the ruthenium oxide film growth, such as the roughness of the substrate surface and material loading of each coating. It was found that the substrate surface roughness varies from different pieces of substrate and even on the same piece of substrate, and in general higher c_p values were obtained from the

substrates with rougher surfaces. It was also found that the variation of the material loading fro
each coating was quite significant, since the ruthenium ethoxide solution was sprayed on t
substrate by a manually-controlled airbrush. The uniformity and the material loading of ea
coating were therefore difficult to control. In addition, the concentration of $Ru(OC_2H_5)_3$ in soluti
may also have affected the value of c_p, because it was found that the concentration of $Ru(OC_2H$
in solution decreased with time due to the hydrolysis and alcoxolation processes [8].

CONCLUSION

Ruthenium oxide film electrodes were made using a novel organic precursor metho
Amorphous-form ruthenium oxide films were formed at temperatures from 100 °C to 200 °C.
temperatures above 250 °C, crystalline films were formed. The surface morphologies of the fil
were also dependent on the temperature used during the film growth. At low temperatures, t
ruthenium oxide film was formed with highly porous particles, while at high temperatures, t
film was formed with low porous flakes. It was found that the highest specific capacitance of 5
F/g and interfacial capacitance of 4.0 F/cm^2 were obtained from the film electrode grown at 2
°C, which was just below the critical temperature at which the crystalline phase of ruthenium oxi
starts to be formed.

ACKNOWLEDGMENTS

The authors would like to thank Mr. David A. Evans for providing Ta substrates for tl
work. The work was supported by the Air Force Research Laboratory under the STTR program

REFERENCES

1. J. P. Zheng and T. R. Jow, *J. Power Sources* **62**, 155 (1996).
2. J. P. Zheng, *Electrochem. and Solid-State Lett.* **2**, 359 (1999).
3. J. P. Zheng and T. R. Jow, *J. Electrochem. Soc.* **142**, L6, (1995).
4. J. P. Zheng, P. J. Cygan, and T. R. Jow, *J. Electrochem. Soc.* **142**, 2699 (1995).
5. J. P. Zheng, D. A. Evans, and S. L. Roberson, to be published.
6. I. D. Raistrick, in *The Electrochemistry of Semiconductors and Electrodes – Processes a
   Devices*, J. McHardy and F. Ludwig, Editors, p. 297, Noyes, Park Ridge, NJ (1992).
7. J. P. Zheng, T. R. Jow, Q. X. Jia, and X. D. Wu, *J. Electrochem. Soc.* **143**, 1068 (1996).
8. C. J. Brinker and G. W. Scherer, "Sol-Gel Science, The Physics and Chemistry of Sol-C
   Processing", Academic Press, Boston, (1990).
9. S. T. Mayer, R. W. Pekala, and J. L. Kaschmitter, *J. Electrochem. Soc.* **140**, 446 (1993).

Mat. Res. Soc. Symp. Proc. Vol. 658 © 2001 Materials Research Society

Two Different Interactions between Oxygen Vacancies and Dopant Cations for Ionic Conductivity in CaO-Doped CeO_2 Electrolyte Materials

Yuanzhong Zhou*, Xi Chen
Department of Physics, Huazhong University of Science and Technology, Wuhan, China
*Currently with Fairchild Semiconductor, South Portland, ME

ABSTRACT

The electrical measurement has demonstrated that conductivity of CaO-doped CeO_2 has higher activation energy for low temperature and lower activation energy for high temperature. A model with two different kinds of defect interactions between oxygen vacancy and doped cations has been used to interpret the phenomenon. Diffusion based on hopping of oxygen ions was assumed as the mechanism of electrical conduction. The analysis indicated that at high temperature free oxygen vacancies are dominant and the activation energy is only for oxygen ion hopping. At low temperature, however, oxygen vacancies associated with dopant calcium ions are dominant for high CaO content and the activation energy is the energy for hopping of an oxygen ion plus half of the association energy between one oxygen vacancy and one calcium ion. For low level doping, both free and associated oxygen vacancies are important.

INTRODUCTION

Ceria-based oxides, doped with ions of alkaline or rare earth elements, have attracted much interest for their high ionic conductivity and low activation energy [1,2]. Numerous experimental works on these materials have shown that maximum ionic conductivity occurs at around dopant level of 10~12mol% for ceria doped with oxides of two-valent metals and below 10mol% for rare earth doped ceria. Two slopes in Arrehenius plots of $log(\sigma T)$ versus reciprocal temperature have been observed in several works on CeO_2 doped with calcia and rare-earth oxides [3~5]. Similar results were also obtained in our experimental investigation on CaO-doped ceria, $(CeO_2)_{1-x}(CaO)_x$.

Several models have been proposed to describe ionic conductivity of CeO_2 based oxide materials (e.g. [6,7]). These models might involve multiple nearest neighbor dopant ions or long-range force, which could be too complicated to emphasize dominant defect status under different operation conditions, or did not clearly demonstrate the temperature dependence of the conductivity. In this paper we will demonstrate that a very simple model can be used to interpret the temperature dependence of the ionic conductivity of CaO-doped ceria materials. This model includes two different interactions between oxygen vacancies and dopant cations and could fit measure data well.

STRUCTURE AND CONDUCTIVITY

Ceria oxide has fluorite-type of structure. In this structure, anions form a simple cubic sublattice and cations form a face-centered cubic sublattice. Each oxygen ion neighbors with four cations and each cation is surrounded by eight oxygen ions. Only half of the body-centered sites in the oxygen sublattice are occupied by cations. This makes the structure be open and be quite favorable for oxygen ions to move in the materials. The diffusivity of oxygen was reported six order of magnitude higher than that of cations.

Pure CeO_2 is a mix conductor. There is small amount of oxygen vacancies in pure CeO_2 that come from Frenkel defects

$$O_O = O_i'' + V_O^{..}. \tag{1}$$

When doped with CaO, $(CeO)_{1-x}(CaO)_x$ forms a solid solution, large number of oxygen vacancies are generated

$$CaO = Ca_{Ce}'' + V_O^{..} + O_O. \tag{2}$$

Electronic conductivity is effectively suppressed in CaO doped CeO_2. Diffusion of oxyge ions is the major mechanism of electrical conduction. The electrical conductivity has been empirically written as

$$\sigma = \frac{C}{T} e^{\frac{-E_a}{kT}} \tag{3}$$

where C is a pre-exponential factor and E_a is the activation energy of electrical conduction. Fro diffusion model based on hopping of vacancies (or oxygen ions in opposite direction), equation (3) may be rewritten as

$$\sigma = \frac{C}{T} e^{\frac{-\Delta H_m}{kT}} = \frac{4}{k} q^2 n d^2 v_o e^{\left(\frac{\Delta S_m}{k}\right)} \cdot \frac{1}{T} e^{\frac{-\Delta H_m}{kT}} \tag{4}$$

where ΔS_m is the activation entropy of diffusion, ΔH_m is the activation enthalpy of diffusion, n the number of vacancies per unit volume, q the charge of a conducting ion, d is the jump distan of one hopping, and v_o lattice vibration frequency [2].

Since there are defect association, not all vacancies can conduct electricity. The vacancy density n in equation (4) should be replaced by the effective vacancy density n_{eff}.

Kilner and Steele pointed out there are two possible cases of defect associate for the divalent doped fluorite-type oxides [8]. The first case is that a dopant cation forms an associatic with an oxygen vacancy. For Ca doped CeO_2, the association reaction is

$$Ca_{Ce}'' + V_O^{..} = (Ca_{Ce}''V_O^{..})^\times. \tag{5}$$

Applying the law of mass to this reaction gives

$$K_A = \frac{\left[(Ca_{Ce}''V_O^{..})^\times\right]}{[Ca_{Ce}''][V_O^{..}]} = Ze^{\left(\frac{-\Delta S_A}{k}\right)}e^{\left(\frac{\Delta H_A}{kT}\right)} \tag{6}$$

where Z is the orientations of the associate, ΔS_A and ΔH_A are the entropy and the enthalpy of association, respectively. From the electrical neutral condition, we have

$$\left[V_O^{..}\right] = [Ca_{Ce}'']. \tag{7}$$

Then, the concentration of the oxygen vacancies becomes

$$\left[V_O^{..}\right] = \left[(Ca_{Ce}''V_O^{..})^\times\right]^{\frac{1}{2}} \left(\frac{1}{Z}\right)^{\frac{1}{2}} e^{\left(\frac{\Delta S_A}{2k}\right)} e^{\left(\frac{-\Delta H_A}{2kT}\right)}. \tag{8}$$

In the second case, all the oxygen vacancies are to be free. The concentration of charge carriers is equal to the concentration of the dopant Ca.

It is possible for an oxygen vacancy to form defect complex with more than one dopant ion. A simulation has be carried out by Chebotin and Mezrin [7] on defect clusters (VmM) where V represents the oxygen vacancy and mM represents m dopant ions. The calculation indicated that at low dopant level free vacancies and defect pairs (m=0,1) are dominant. Larger defect clusters are notable only when dopant content is as high as 20mol%. The solution limits of alkaline earth oxides in CeO_2 are usually pretty low. So we will only consider free vacancies and the defect associate of one vacancy with one CaO ion.

In the next two sections, we will discuss these two defect associate cases. Only the interaction between nearest defects or ions will be considered.

HIGH TEMPERATURE REGION

Before we start our discussion, we give two useful relations. When dopant concentration is x, concentration of oxygen vacancies, including associated and free vacancies, is $x/2$ and the number of oxygen vacancies per volume is $n = 0.5x/d^3$.

In high temperature region, we basically followed [9] assuming the defects are free. But we also assume that a vacancy will not move freely when it neighbors with a cation. An oxygen vacancy can move only when at least one of the nearest neighbor anion sites that is occupied by an oxygen ion. Including this, the effective number density of oxygen vacancies n_{eff} should replace n

$$n_{eff} = n\left(1-\frac{x}{2}\right) = \frac{1}{2}d^{-3}x\left(1-\frac{x}{2}\right). \qquad (9)$$

If any of the Z_1 nearest neighbor cation sites is occupied by a Ca^{2+}, the vacancy will be captured by Ca^{2+} to form a association ($Ca''_{Ce}V''_o$), so the effective vacancy density becomes

$$n_{eff} = \frac{1}{2}d^{-3}x\left(1-\frac{x}{2}\right)(1-x)^{Z_1}. \qquad (10)$$

Due to strong vacancy-vacancy repulse, the nearest vacancy-vacancy configuration is prohibited. Z_2 is number of nearest neighbor anion sites, then n_{eff} is further modified as

$$n_{eff} = \frac{1}{2}d^{-3}x(1-x)^{Z_1}\left(1-\frac{x}{2}\right)^{Z_2+1}. \qquad (11)$$

In fluorite structure, each anion site has four nearest neighbor cation sites ($Z_1=4$) and six nearest neighbor anion sites ($Z_2=6$). Then substituting the effective vacancy density into equation (4), the electrical conductivity is obtained as

$$\sigma = \frac{2e^2 v_o}{dkT}x(1-x)^4\left(1-\frac{x}{2}\right)^7 e^{\left(\frac{\Delta S_m}{k}\right)}e^{\left(\frac{-\Delta H_m}{kT}\right)}. \qquad (12)$$

Fig.1 shows the normalized pre-exponential factor of equation (12) vs. dopant content. The maximum conductivity occurs at $x=0.121$.

Although some oxygen vacancies are in the form of associated pairs, the conductivity only determined by free vacancies. The activation energy is the mobility energy of vacancie This is because that due to the low solution limit of CaO in CeO_2, the probability of a vacan in an associated pairs is smaller than that as a free defect and meanwhile the extra energy is needed to generate a charge carrier from associated pairs.

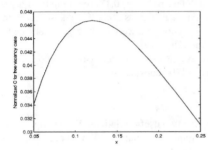

Fig. 1. Normalized pre-exponential factor for high T region vs. dopant content.

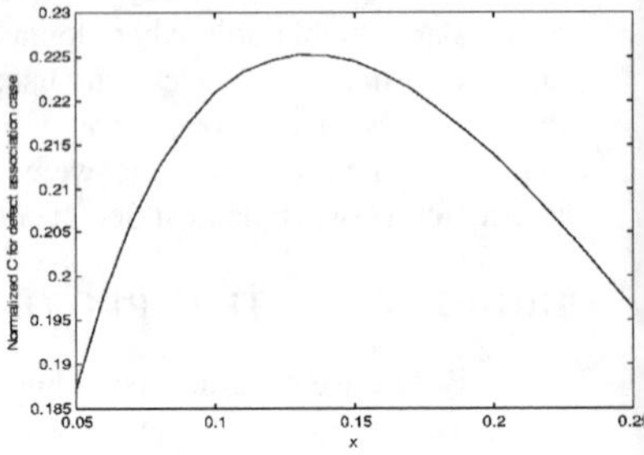

Fig. 2. Normalized pre-exponential factc for low T region vs. dopant conte

LOW TEMPERATURE REGION

At low temperature, the population of charge carrying defects is determined by the thermodynamic equilibrium between the free defects and the associated pairs [9]. Assumin, full association of defects, we have

$$\left[(Ca_{Ce}'' V_O^{\cdot\cdot})^x\right] = x. \tag{13}$$

Then from equation (9) gives the effective vacancy density:

$$n_{eff}' = x^{\frac{1}{2}}\left(\frac{1}{Z}\right)^{\frac{1}{2}} e^{\left(\frac{\Delta S_A}{2k}\right)} e^{\left(\frac{-\Delta H_A}{2kT}\right)}. \tag{14}$$

Considering the vacancy-vacancy repulse and the requirement on the nearest neighbo anion sites, similar to the free vacancy case, n_{eff}' is further modified as

$$n_{eff}' = x^{\frac{1}{2}}\left(1-\frac{x}{2}\right)^{Z_2+1}\left(\frac{1}{Z}\right)^{\frac{1}{2}} e^{\left(\frac{\Delta S_A}{2k}\right)} e^{\left(\frac{-\Delta H_A}{2kT}\right)}. \tag{15}$$

Substituting equation (15) into equation (4) and considering $Z_2=6$, we have the electri conductivity for fully associated defects as

$$\sigma = \frac{2e^2 v_o}{dkT} x^{\frac{1}{2}}\left(1-\frac{x}{2}\right)^7 e^{\left(\frac{2\Delta S_m+\Delta S_A}{2k}\right)} e^{\left(\frac{-\left(\Delta H_m + \Delta H_A/2\right)}{kT}\right)}. \tag{16}$$

The activation enthalpy is equal to the migration enthalpy plus half of the association enthalpy. Fig.2 shows the normalized pre-exponential factor of equation (16) vs. dopant co The maximum conductivity occurs at $x=0.133$.

As suggested by Wang et al. [10] and the calculation by [7], free vacancies exist at low dopant level, at which a vacancy may sit at a site that has no dopant ion on the nearest cation sites. The lower dopant content, the higher probability that a vacancy jumps to such a site. So at low temperature range, the conductivity can be written as

$$\sigma = \frac{C}{T} e^{\left(\frac{-(\Delta H_m)}{kT}\right)} + \frac{C'}{T} e^{\left(\frac{-\left(\Delta H_m + \Delta H_A/2\right)}{kT}\right)}. \tag{17}$$

The dopant content corresponding to maximum conductivity will shift to lower dopant level than that given by equation (16).

Using the previous analysis, we could get the relation of the activation enthalpy for mobility, ΔH_m, and the association enthalpy, ΔH_A, with the activation energies of conductivity in low and high temperature ranges, E_L and E_H.

$$E_H = \Delta H_m, \tag{18}$$

$$E_L = \Delta H_m + \frac{\Delta H_A}{2}. \tag{19}$$

J.D.Solier et al.[11] have used a statistical thermodynamics approach to deduce a function of temperature for the activation enthalpy in his work on ZrO-YO materials. Their equation is in the form of

$$E(T) = E_f + E_b \tanh\left[\beta(\frac{1}{T} - \frac{1}{T_f})\right]. \tag{20}$$

Though they only took account anionic sites in their model, equation (20) provides a clue for us to understand equations (18) and (19). E_L and E_H are simply the asymptotic values of $E(T)$ at low and high temperatures respectively. Since β is in the order of a few thousands, the transition range between the low and the high temperature ranges is small.

COMPARISON WITH EXPERIMENT

The $(CeO_2)_{1-x}(CaO)_x$ (0.05≤x 0.4) samples prepared from CeO_2 (99.9%) and $CaCO_3$ (99.0%) by solid state reaction were used for electrical conductivity and X-ray diffraction measurement. The ionic conductivity was measured by AC impedance spectroscope method at various temperatures between 300C° and 950C°.

The solubility limit of CaO in CeO_2 was found about 15mol%. The conductivity as a function of temperature for a given $(CeO_2)_{1-x}(CaO)_x$ consisted two linear lines. The transition occurred at about 600C°. As shown in Table 1, at high temperature the activation energies of all samples, except the one with x=0.3, were close to a constant of ~0.586 eV. At low temperature, the activation energies of the samples with dopant content of 0.12≤x ≤0.3 were close to a constant of ~0.92 eV. For low level doped samples (x< 0.12), the low temperature activation energy E_L increases as dopant CaO content increases. Extrapolating E_L to x=0, we could have an E_L that almost equals to E_H.

The temperature dependence of ionic conductivity observed in our experiment conformed well to our model for the samples with 0.12≤x ≤0.2. The higher E_L for low level doped is expected by equation (17). The model is not valid for x=0.3 since it is far higher than the solubility limit of CaO in CeO_2.

Table 1. Activation energy and pre-exponential term of conductivity in $(CeO_2)_{1-x}(CaO)_x$

x	E_L (eV)	$\sigma_{0L}\times10^{-6}$ (K/Ω·cm)	E_{II} (eV)	$\sigma_{0H}\times10^{-4}$ (K/Ω·cm)
0.05	0.730	0.1495	0.566	1.951
0.08	0.849	1.275	0.593	3.228
0.12	0.920	3.295	0.596	4.336
0.14	0.925	3.269	0.581	3.630
0.17	0.913	2.424	0.584	3.171
0.20	0.909	1.681	0.595	2.877
0.30	0.936	1.538	0.655	4.372

According to equations (18) and (19), the mobility enthalpy ΔH_m and the association enthalpy ΔH_A were obtained as $\Delta H_A=2(E_L-E_H)=0.668$eV and $\Delta H_m= E_H =0.586$eV, which were aligned well with the values reported by other authors [1].

Maximum conductivity occurred at x~0.14. It was close to the value predicted by the model. A slight shift of maximum conductivity to lower dopant level at low temperature, as suggested by equation (17), was also observed.

CONCLUSION

A diffusion model based on hopping of oxygen ions was used to interpret the temperature dependence of conductivity in divalent doped ceria. Two different kinds of defect interactions between oxygen vacancies and doped cations were used in the analysis. At high temperature free oxygen vacancies generated by dopant ions are dominant in electrical conductivity and the activation energy is only for oxygen ion hopping. At low temperature, however, most oxygen vacancies are associated with dopant calcium ions to form defect pairs for high CaO content an the activation energy is the energy for hopping of an oxygen ion plus half of the association energy between one oxygen vacancy and one calcium ion. For low level doping, both free and associated oxygen vacancies are important. This simple model can be used to extract mobility activation energy and association energy from the experimental conductivity data and are reasonably consistent with the experimental results of the electrical conductivity as a function c dopant content.

REFERENCES

[1] M. Mogensen, N. Sammes, and G. Tompsett, *Solid State Ionics*, vol. 129, 63 (2000)
[2] H. Inaba and H. Tagawa, *Solid State Ionics*, vol. 83, 1 (1996)
[3] T. Kudo and H. Obayashi, *J. Electrochem. Soc.*, vol. 122, 142(1975)
[4] D. K. Hohnke, *Solid State Ionics*, vol. 5, 531(1981)
[5] R. T. Dirstine, R. N. Blumenthal, and T. F. Kuech, *J. Electrochem. Soc.*, vol. 126, 264(197
[6] V. Bulter, C. R .A. Catlow, B. E. F. Fender, and J. H. Harding, *Solid State Ionics*, vol. 8, 1 (1983)
[7] V. N. Chebotin and V. A. Mezrin, *Phys. Status Solidi (a)*, vol. 89, 199(1985)
[8] J. A. Kilner and B. C. H. Steele, in : *Nonstoichiometric Oxides*, ed. O. T. Sorensen, Academic Press, New York, 1981
[9] J. A. Kilner and C. D. Waters, *Solid State Ionics*, vol. 6, 253 (1982)
[10] D. Y. Wang, D. S. Park, J. Griffith, and A. S. Nowick, *Solid State Ionics*, vol. 2, 95(1981)
[11] J.D.Solier, I.Cachadina, and A.Dominguez-Rodriguez, Phys. Rev. B48, 3704(1993)

Mat. Res. Soc. Symp. Proc. Vol. 658 © 2001 Materials Research Society

Nickel Determination by Complexation Utilizing a Functionalized Optical Waveguide Sensor

Erin S. Carter and Klaus-H. Dahmen
Florida State University, Dept of Chemistry, Tallahassee, FL 32312-4390, U.S.A.

ABSTRACT

This research focuses on the design of chemically functionalized optical waveguide sensors. The waveguide is an optically transparent sol-gel coated onto a glass substrate chip. By having a higher refractive index than the substrate, the waveguide internally reflects a laser beam to photodetectors at both ends of the chip. The adsorption of any species onto the waveguide surface changes the light propagation, and therefore its effective refractive index, N. The change in N is dependent upon the amount of analyte present. By covalently bonding specific chemical receptors onto its surface, it can be designed to target a particular analyte. This research involves functionalizing the surface of the waveguide with ED3A in order to complex out of solution Ni^{2+}. The change in N and the thickness of the adlayer will allow the concentration to be determined.

INTRODUCTION

The processes that occur at solid surfaces are used in many analytical techniques that involve adsorption and desorption, such as electrochemistry, gas chromatography, liquid chromatography, and capillary electrophoresis. However, the study of surface phenomena is very challenging due to the extremely small quantities of surface-analyte species. Fortunately, a relatively new class of detectors known as *optical waveguide sensors* (first developed during the early 1980's by Tiefenthaler and Lukosz at the Swiss Federal Institute of Technology, Zürich) has been used by several research groups to study reaction kinetics at solid/liquid and solid/gas interfaces.[1-4]

Optical waveguides are two-dimensional analogs of optical fibers [5]. They are thin (30-300 nm), non-crystalline films deposited on an optically transparent glass slide. The simplest waveguide chips are comprised of these two layers- the waveguide and the glass substrate. Because the waveguide has a higher refractive index than the glass substrate, it is possible to pass a plane-polarized laser beam through the substrate onto a diffraction grating inscribed within the chip and 'guide' it to a photodiode detector located at either end of the waveguide chip. The refractive index of the waveguide can be determined using the following theoretical relationship (Equation 1): [6]

$$n_{film} = \sin(\alpha_{in}) + (l\lambda/\Lambda) \qquad (1)$$

where n_{film} = refractive index of the film, α_{in} = angle of incoupling (angle at which there is total internal reflectance of the laser beam to the detector), l = diffraction order, λ = wavelength of the laser, and Λ = diffraction grating period (2400 lines/ mm).

The refractive index has been used to determine qualitatively and quantitatively the presence of the compound that is being analyzed. The analysis is carried out by measuring n_{fi} before and after the waveguide chip has been exposed to a gaseous or liquid medium that contains the analyte. The change in n_{film} is dependent upon the amount of analyte present. By covalently bonding specific chemical receptors onto the sensor surface (functionalizing), the waveguide chip can be designed to target a particular analyte.

EXPERIMENTAL

To investigate the ability of the waveguide sensor to quantitatively respond to Ni^{2+}, we measured the response of an IOS-1 integrated optics scanner (Artificial Sensing Instruments, Zürich) to dilutions of a 1000 ppm stock solution, which will be detailed further in the experimental section. A channeled cuvette was sealed to a waveguide chip's surface, and the entire assemblage was mounted into the instrumental goniometer, which allowed the incoupling angle to be determined (*Figure 1.*).

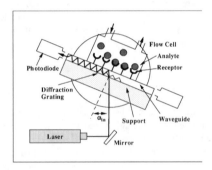

Figure 1. The configuration of the waveguide chip mounted on the goniometer. The surface the chip has been functionalized with a specific receptor. The analyte is introduced by flowing the solution at a rate of 0.01 mL/min by means of a peristaltic pump.

The NiCl$_2$.6H$_2$O, 99.8% pure, was purchased from Fisher Chemicals. A stock solution of Ni^{2+} was made 1000 ppm in the analyte metal ion. All metal ion solutions were kept in Nalgene bottles to minimize adsorption into the bottle's surface. The solvent used was 10mM HEPES (4-(2-hydroxyethyl)-1-piperazinethanesulfonic acid), obtained from Aldrich Chemical Company, 99% pure. All solutions were made and diluted with nanopure (18MΩ-resistance grade) H$_2$O. The waveguide chips, ASI 2400, were obtained from ASI Incorporated of Zurich, Switzerland. Previous to analysis or functionalization, the chips were dipped for 1 minute in concentrated H$_2$SO$_4$, triple-rinsed in nanopure (18MΩ-resistance grade) H$_2$O to hydrolyze the chip surface, and then soaked at least 24 hours in nanopure water. All chips were equilibrated in the reaction buffer for a further 24 hours before being utilized.The N-(trimethoxysilylpropyl)ethylenediamine triacetic acid, trisodium salt, 50% in water, hereafter abbreviated ED3A, was obtained from Gelest , Inc.

Functionalization with ED3A was accomplished by placing a chip in a Nalgene container and soaking it in 5 mL of ED3A along with 10 mL of 1 M HCl. The mixture was allowed to soak at 28° C for 24 hours. The chip was then triple-rinsed in nanopure H$_2$O, then allowed to equilibrate with HEPES buffer. *Figure 2* illustrates the analysis of Ni^{2+} with a ED3A-functionalized waveguide surface. The response of the optical waveguide chip to an aqueous solution containing Ni^{2+} is increased dramatically by functionalizing with a chelate such as ED3A.

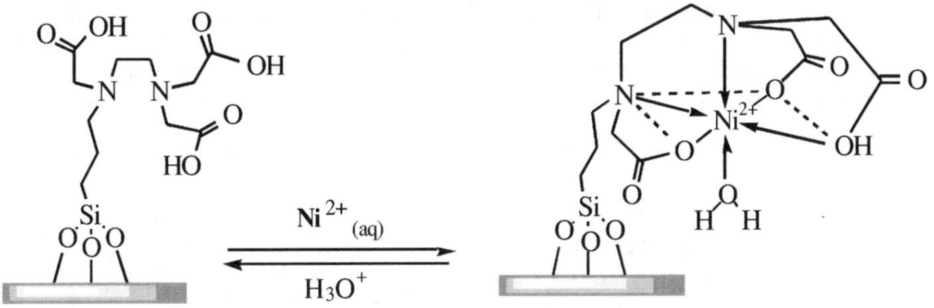

Figure 2. The mechanism of analyte recognition on a functionalized waveguide.

The HeNe laser was allowed to warm-up for 20 minutes before use. A ring stand was set up above the level of the goniometer/measuring head (to prevent pressure differences facilitating catastrophic air bubble formation in tubing); four rings were attached to the stand that held in place the solution bottles. The solution bottles contained .003 M HCl; 10 mM HEPES buffer; metal ion solution diluted from the 1000 ppm stock solution; and 10 mM HEPES buffer.

A preparatory chip was placed in the measuring head, and a small silicone rubber cuvette was mounted over the grating region. In the rubber is a three-sided groove that is placed against the chip forming a four-sided flow channel. The channel is 2mm x 8mm x 1mm, and the surface area of the chip exposed to solution is 16 mm^2. The head was inserted into its housing in the BIOS machine. Flow from the measuring head was split in two before the tubing was placed in the peristaltic pump, and the two cartridges pressing the tubing to the rollers were offset, in ord to dampen the peristaltic pulses. This facilitated a more even, smooth flow.

The peristaltic pump, a Cole-Parmer Masterflex L/S, was turned on and set to a flow rate o mL/min. Because the tubing through the measuring head was smaller than the standard tubing through the pump, this resulted in an actual flow of 0.282 mL/min. The flow was switched with the valve to each solution in turn, and an air bubble was allowed to flow through each of the tubes (by simply lifting the respective tube out of solution). The fluidity of the air bubbles' travel through the tubing served as a verification that all tubing connections, as well as the clamps on the peristaltic pump were correctly assembled. Once fluidity had been verified in each of the four tubes, the pump was stopped and the preparatory chip removed.

Using Nylon tweezers, the chip to be tested was removed from its storage bottle and dri with KimWipes. Once adequately dried, the chip's identification number was verified, and this corner was placed in the upper right hand corner of the measuring head. After inserting the hea into its housing, the pump and then the ASI computer program were started. After proceeding through left and right photodiode calibration, buffer was allowed to flow over the waveguide chip until a reliable baseline was established on the graph updated in real-time and displayed by the computer program. After a reliable baseline was established, the flow valve was rotated counterclockwise to the Ni^{2+} solution. After the graph responded and leveled out, the flow val was rotated in a counterclockwise direction two positions to the HEPES buffer, pausing briefly (approx. 3 seconds) in the HCl position. This step was necessary to rinse the metal ion off the ligand with excess protons. The chip was washed in buffer long enough for the graph to flatter out (signifying desired absence of non-specific binding) and the next metal solution was introduced. (*Figure 3*).

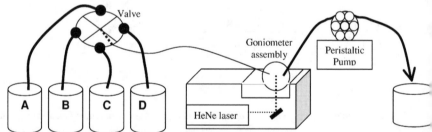

Figure 3. *Experimental set-up. A=0.003M HCl, B=HEPES buffer, C= Metal ion solution, D: HEPES buffer*

The data points, which were calculated every 13.4 seconds by the ASI program, were stored in a file on the computer's hard drive. The data was exported Microsoft Excel, where th graphing and data analysis took place. *Figures 4* and *5* are measurements of n_{film} taken over time as the chip was cycled through an aqueous Ni^{2+} solution followed by a non-Ni^{2+} containing aqueous buffer. They are drawn to the same scale.

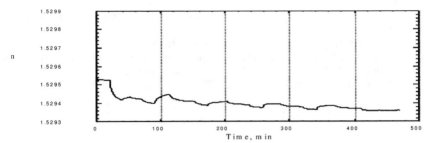

Figure 4. *Response of the OWGC sensor to aqueous NiCl$_2$ (3.0 x 10^{-5} M) prior to functionalization with either NTA or ED3A chelating ligand, pH 6.30, and 20 °C.*

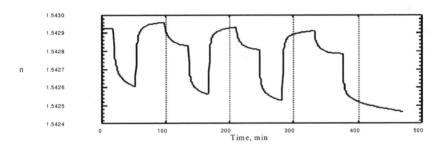

Figure 5. *Response of the OWGC sensor to aqueous NiCl$_2$ (3.0 x 10^{-5} M) after ED3A-functionalization, pH 6.30, and 20 °C.*

CONCLUSIONS

Our research requires adaptation of known analytical methods to waveguide technology. This instrument has not previously been used for inorganic analytical techniques to our knowledge, therefore our research is a novel application for the IOS-1. We have functionalized a sensor chip with a common chelate, ED3A, to detect and quantify the divalent metal ions Ni^{2+}. The response is significant and indicates that a functionalized chip is more sensitive to Ni^{2+} after functionalization. Further research will determine functional groups that are utilitarian for this method and their specificity for Ni^{2+} and other metal ions.

REFERENCES
1   L. Han, T.M. Niemczyk, Y. Lu, G.P. Lopez, *Applied Spectroscopy*, **52**, 119-122 (1998).

2 K. Kim, H. Minamitani, H. Hisamoto, K. Suzuki, S. Kang, *Analytica Chimica Acta*, **343**, 199-208 (1997).
3 K. Tiefenthaler, W. Lukosz, *Thin Solid Films*, **126**, 205-211 (1985).
4 L. Yang, S.S. Saavedra, N.R. Armstrong, *Analytical Chemistry*, **68**, 1834-1841 (1996).
5 R. Ulrich, H.P. Weber, *Applied Optics*, **11**, 428-434 (1972).
6 P.K. Tien, R. Ulrich, R.J. Martin, *Appl. Phys. Lett.*, **14**, 291 (1969).

Mat. Res. Soc. Symp. Proc. Vol. 658 © 2001 Materials Research Society

Competitive Adsorption of O_2 and H_2O at the Neutral and Defective SnO_2 (110) Surface

Ben Slater, C. Richard A. Catlow, David E. Williams[1] and A. Marshall Stoneham[2]
Davy Faraday Research Laboratory, The Royal Institution of Great Britain,
21 Albemarle St., London, W1S 4BS, U.K.
[1]Department of Chemistry, University College London, 20 Gordon St, London, WC1H OAJ,
U.K.
[2]Department of Physics and Astronomy, University College London, Gower St, London, WC1E
6BT, U.K.

ABSTRACT

Using first principles techniques we have examined the relative physisorption/chemisorption energetics of neutral molecular water and oxygen at the most thermodynamically stable surface (110) of tin dioxide. We find that water binds more strongly to the perfect surface at 5 co-ordinate tin sites than oxygen. However, binding of both water and oxygen at bridging oxygen vacancies in the defective surface is comparable. In the context of gas-sensing behaviour at moderate temperatures (~300K), we propose that the Mars and van Krevelen [1] re-oxidation reaction will slow when the partial pressure of water is high, since the number of favourable adsorption sites will effectively decrease. In addition, one would expect that the surface conductivity will increase, since the re-oxidation reaction will be hindered.

INTRODUCTION

Understanding the chemistry which governs the electrical conductivity of surfaces remains a challenge in the context of gas sensors. Although materials such as tin dioxide and tungsten oxide are widely used commercially, their effective exploitation is based on largely empirical findings. The aim of our calculations is to assist in the clarification of elementary reaction mechanisms which control the sensor response, specifically for SnO_2. Recently, Williams and Pratt [2] identified distinct sites at the (110) surface which were proposed as possible sites for adsorption of neutral or charged molecular or monoatomic oxygen species. Furthermore, Williams [3] has suggested a possible reaction mechanism between oxygen and SnO_2 which accounts for the observed decrease in conductivity upon re-oxidation.

Tin dioxide has been widely studied at the first principles level, in particular by Goniakowski et al.[4,5,6]. Recently, Oveido and Gillan [7] reported a systematic study of defect chemistry on the SnO_2 surface, which explored the energetics of line and point oxygen defect formation. Lindan [8] has also reported a comparative study of water adsorption characteristics at both SnO_2 and TiO_2 (110) surfaces. Additionally, we have recently reported [9] viable mechanisms for oxygen dissociation at the SnO_2 (110) surface. Clearly, it is desirable for these findings to be rationalised in the context of competitive adsorption characteristics of these simple molecules to elucidate the gas sensing mechanism. We have therefore carried out a combined study of oxygen and water adsorption using consistent, identical simulation conditions.

THEORY

Density functional, plane-wave based methods (CASTEP [10]) have been used throughout the following study. Ultra-soft pseudo-potentials were used for the tin, oxygen and hydrogen atoms with a kinetic energy cut-off of 300 eV, employing the gradient corrected PW91 local spin density functional. These conditions were found to give an excellent representation of the bulk geometry, where the relaxed a and c lattice parameters were calculated as 4.768 and 3.185 Å respectively, with an internal co-ordinate of 0.308. The corresponding experimental values are 4.737 and 3.186 Å, with an internal co-ordinate of 0.307. In the following calculations we found that a 2x2 surface repeat unit consisting of three layers, with a vacuum gap of 7Å gave a converged surface energy of 1.3 Jm^{-2}, in close agreement to those reported previously [4,5,6,9].

As noted in a previous study [9], it is imperative that we verify that the oxygen molecule is described correctly within this formalism, as the geometry and spin state are acutely sensitive to the computational conditions. We obtained an oxygen bond length of 1.230 Å and a binding energy per neutral oxygen molecule of 6.83 eV, which are in acceptable agreement with experiment and previous computational work.

DISCUSSION

Using an identical approach to that described in our recent work [9], we have investigated the end points of hypothetical reactions, where an oxygen molecule adsorbs to a neutral surface and where an oxygen molecule dissociates at a half reduced and fully reduced surface. Similarly, we calculated the total energies for adsorption of molecular water to an ideal surface and to a half and fully reduced atoms. Six scenarios are displayed in figure 1 (a-f). Dissociated oxygen atoms are shown in black, tin is shown in mid-gray, oxygen in dark gray and hydrogen in white. The orientation of the figures is askew to the (110) surface. Site A is an example of a bridging oxygen atom whilst site B indicates an in-plane tin site.

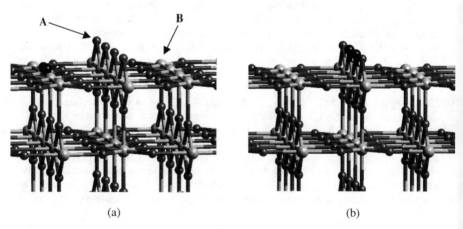

(a)   (b)

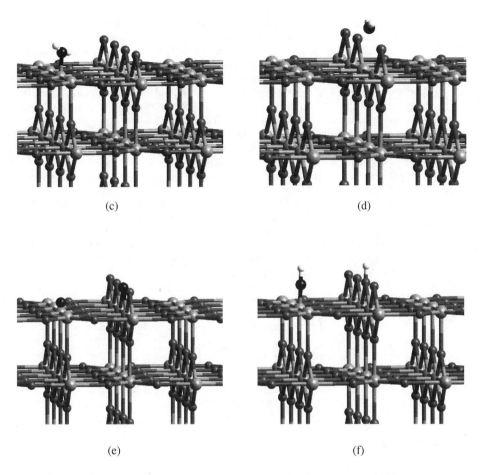

Figure 1: (a) Molecular oxygen on an ideal surface; (b) dissociated oxygen on a reduced surface, (c) a half reduced surface and (d) an ideal surface and (e) associated water on an ideal surface and (f) on a half reduced surface.

Using the numerical values for the total energy for the reactions considered, we calculated the binding or reaction enthalpies, shown in Table I. Firstly, we note that the energies for oxygen adsorption are similar to those obtained previously [9], where a different plane-wave basis set and independent code was used. In agreement with our previous findings, the enthalpies suggest that molecular oxygen physisorbs to the ideal surface at the in-plane 5 fold tin site. At point defects (neutral oxygen vacancies), oxygen dissociates with a net gain in energy of 0.7 eV. At line defects, oxygen would be expected to dissociate releasing approximately 5.1 eV of energy.

As noted previously, the dissociation mechanism at point defects is a predictive result which has been corroborated by experiment on the iso-structural TiO_2 (110) surface [11].

In the case of water adsorption, we find that H_2O can dissociate spontaneously during the geometry optimisation process, although in the initial configuration, molecular water (geometry optimised *in vacuo*) is introduced to the surface. As noted by Gianowski [4] *et al.* and Lindan [8], it is also possible to stabilise molecular water at an in-plane tin site in conjunction with a dissociated water molecule (the 'mixed' state). This phenomenon is indicative of the small energy barrier separating associated and dissociated states and is in agreement with previous studies reported in the literature. The optimum dissociation mechanism at the ideal surface is hydroxylation of an in-plane 5 fold tin site accompanied by protonation of the neighbouring bridging oxygen, releasing 1.80 eV. At point defects, dissociation can also occur, where hydroxylation of the bridging oxygen vacancy is followed by protonation by the neighbouring bridging oxygen giving a net energy of 0.50 eV. However, if we consider molecular adsorption of water at the vacancy, and at the in-plane tin site (in the presence of the vacancy), we find binding energies of 0.53 eV and 1.35 eV respectively. Clearly, the energy separating the dissociated and associated water states is very small and within the inherent errors associated with this methodology, it is not possible to unequivocally determine the most stable state. However, the binding energies suggest that adsorption/dissociation is far more likely to occur at the in-plane tin sites than the vacancy.

Table I. Binding energies of geometries indicated by the schematics in Figure 1 (a-f).

Geometry	Binding Energy / eV
Molecular Oxygen (ideal surface)	0.23
Dissociated Oxygen (fully reduced surface)	5.10
Dissociated Oxygen (half reduced surface)	0.72
Dissociated Water (ideal surface)	1.80
Associated Water (ideal surface)	1.35
Associated Water (half reduced surface)	1.48

CONCLUSIONS

The implications of our results are most readily understood by considering reaction scenarios, as the surface structure is strongly dependent on temperature and the relative partial pressures of pO_2 and pH_2O, in the context of the four reactive sites identified by Williams and Pratt.

If we firstly consider ambient conditions (low partial pressures of oxygen and water and temperature of 300 K), using our computed defect energies and TPD measurements [12] we can confidently assert that the principal defects at the surface will be of bridging oxygen type and that the concentration of this defect will be low. In terms of gas sensing and re-oxidation, we can regard the surface as close to 'ideal'; at equal pressures of water and oxygen, calculations suggest that water will exist in both an associated and dissociated state, but dominated by dissociated water, where in-plane tin sites become hydroxylated. Oxygen will bind far less strongly to any unoccupied in-plane tin sites than water, therefore the major effect on surface conductivity will be attributable to the hydroxylation and hydration process, which results in band bending at the surface, with a consequent *increase* in the surface conductivity.

Next we consider the same surface but with a higher concentration of bridging oxygen vacancies, which would be expected at higher temperatures. Considering first the competition between molecular water and oxygen for two sites (the vacancy and the in-plane tin site), our calculations suggest that the non-dissociative adsorption rate of either adsorbate is essentially equal at the vacancy, but since the in-plane tin sites will be unoccupied, molecular water would be bound far more strongly to these sites. Therefore, we would expect an increase in the surface conductivity due to the water donating density to the in-plane tin sites, increasing the carrier concentration which will be countered by the weakly chemisorbed oxygen molecule withdrawing density from the electron trap in the vacancy, formed by the loss of neutral oxygen. However, at higher temperatures water is unstable with respect to re-association and desorption and the conductivity would be expected to increase, due to the reduction of the surface by oxygen loss. Peroxide and superoxide ions have been observed at the surface and it is probable that these ions donate electrons to the in-plane tin sites but this process becomes less significant at high temperature. Another possibility is that oxygen can dissociate at the bridging oxygen vacancy [7], occupying the vacancy and releasing a monoatom which chemisorbs strongly to an in-plane tin site. This would also result in a net decrease in the number of available carriers present at the surface.

At even higher temperatures the surface can be regarded as fully reduced, with a very low concentration of bridging oxygen atoms. The surface exposes in-plane tin sites, but also sub-bridging oxygen tin sites. Upon formation of the bridging oxygen 'line' defects, the residual electron density remains localised on the tin sites, therefore these Sn^{2+} like cations are far more reduced than the Sn^{4+} in-plane sites. Molecular water and oxygen would therefore be expected to chemisorb to the 4+ and 2+ tin sites respectively. Changes in conductivity in this regime will be controlled by the rate of oxygen loss.

Beyond 700 K, the formation of in-plane oxygen vacancies becomes more favourable, which again results in the appearance of more surface states in the band gap and a substantial increase in the surface electron density. This increase in carrier concentration would clearly lead to increased surface conductivity as observed experimentally. There exists the possibility of dissociation of oxygen and water at such vacancies, but again at this elevated temperature it would be expected for both these reactions to be unstable with respect to re-association.

ACKNOWELDGEMENTS

We are grateful to EPSRC for funding and to the Materials Chemistry Consortium for time on the CRAY-T3E facility at Manchester.

REFERENCES

1. P. Mars and D. W. van Krevelen, *Chem. Eng. Sci*, **41,** 3 (1954).
2. D. E. Williams and K. F. E. Pratt, *J. Chem. Soc. Faraday Trans.*, **94** (23) 3493-3500 (1998).
3. D. E. Williams, in *Solid State Gas Sensing*, ed. Moseley P.T.; Tofield B.C., Adam Hilger, Bristol, (1987).
4. I. Manassidis, J. Goniakowski, L. N. Kantorovich and M. J. Gillan, *Surf. Sci.*, **339**, 258-271 (1995).
5. J. Goniakowski and M. J. Gillan, *Surf. Sci.*, 1996, **350**, 145-158 (1996).

6. J. Goniakowski, J. M. Holender, L.N. Kantorovich, M. J. Gillan and J. A. White, *Phys. Rev. B*, **53**(3), 957 (1996).
7. J. Oviedo and M. J. Gillan, *Surf. Sci.*, **467**: (1-3) 35-48 (2000).
8. P. J. Lindan, *Chem. Phys. Lett.*, **328**, 325-329 (2000).
9. B. Slater, C. R. A. Catlow, D. E. Williams and A. M. Stoneham, *Chem. Commun.*, **14**, 1235-1236 (2000).
10. CASTEP 4.2 Academic Version, Licensed Under the UKCP-MSI Agreement, 1999. *Rev Mod. Phys.*, **64**, 1045 (1992)
11. W. S. Epling, C. H. F. Peden, M. A. Henderson and U. Diebold, *Surf. Sci.*, **412/413**, 333 (1998).
12. V. A. Gercher and D. F. Cox, *Surf. Sci.*, **322**, 177-1 (1995).

Mat. Res. Soc. Symp. Proc. Vol. 658 © 2001 Materials Research Society

NO Decomposition Property of Lanthanum Manganite Porous Electrodes

Kazuyuki Matsuda, Takao Kanai, Masanobu Awano[1] and Kunihiro Maeda
Synergy Ceramics Laboratory, Fine Ceramics Research Association,
Nagoya, 463-8687, JAPAN
[1]National Industrial Research Institute of Nagoya, Nagoya, 463-8687, JAPAN

ABSTRACT

The NOx decomposition activity of electrochemical cells composed of YSZ solid electrolyte and $La_{1-x}Ca_xMnO_3$ porous electrodes was examined. $La_{1-x}Ca_xMnO_3$ powders were prepared by solid-state reaction using corresponding metal oxides and carbonates. The powders were dispersed in polyethylene glycol and screen-printed on both sides of the electrolyte pellet. The electrochemical decomposition of NO was carried out in the temperature range from 973 to 1173 K by passing a mixed gas of 1000 ppm NO and 2% O_2 in He through the cell. When the DC voltage was applied to the cell, NO was directly decomposed to N_2 and O_2 in the oxidizing atmosphere. The cells composed of the $La_{0.8}Ca_{0.2}MnO_3$ electrodes showed a higher NO decomposition ratio and current efficiency than that of other $La_{1-x}Ca_xMnO_3$ electrodes.

INTRODUCTION

The exhaust gases from lean-burn and diesel engines contain NOx and O_2. The emission of NOx causes serious environmental damage. Unfortunately, three-way catalysts cannot reduce NO to N_2 under oxidizing atmospheres. The selective decomposition of NOx from exhaust gases containing O_2 is an important subject. Recently, the electrochemical decomposition of NO using an electrochemical cell composed of a solid electrolyte and metal electrodes has been proposed [1,2]. In these cases, high current density is required to attain substantial NO decomposition. The manganite-based perovskite oxides are catalysts for NO decomposition [3], and these oxides have been widely used as the cathode for solid oxide fuel cells (SOFC) due to their high electrical conductivity, stability at high temperature and thermal expansion compatibility with YSZ. The purpose of this study is to evaluate the NO decomposition activity of electrochemical cells composed of 8mol% Y_2O_3 stabilized ZrO_2 (YSZ) solid electrolyte and $La_{1-x}Ca_xMnO_3$ ($x = 0.1$-0.7) electrodes.

EXPERIMENTAL DETAILS

Calcium substituted lanthanum manganite powders were prepared by solid-state reaction. La_2O_3 was fired at 1173 K for 3 hours before use, to remove adsorbed water. $CaCO_3$, La_2O_3 and Mn_3O_4 were mixed in a desired ratio and calcined at 1673 K for 5 hours. The products were ground in a ball mill for 24 hours. $La_{1-x}Ca_xMnO_3$ pastes were prepared by mixing the $La_{1-x}Ca_xMnO_3$ powders and polyethylene glycol. Electrochemical cells were constructed by screen-printing the pastes on both sides of the YSZ pellets (1 mm thick and 20 mm in diameter). The cells were sintered at 1373 K for 1 hour. The electrode surface area was 1.1cm². Platinum mesh

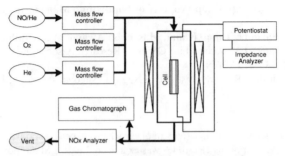

Figure 1. *Schematic arrangement of the experimental apparatus.*

and wire was attached to both sides of the electrodes, respectively, to connect with the power supply unit.

The experimental apparatus for NO decomposition is shown in Figure 1. The electrochemical decomposition of NO was carried out in the temperature range from 973 to 1173 K by passing a mixed gas of 1000 ppm NO and 2% O_2 in He through the cell at a flow rate of 50 ml/min. The applied voltage and current dependence of NO decomposition was investigated. The concentration of NO and N_2 in the outlet gas from the reactor was monitored by a NOx gas analyzer and by gas chromatography, respectively. The NO decomposition ratio was calculated as follows:

$$NO\ decomposition\ (\%) = ([NO]_{inlet} - [NO]_{outlet})/[NO]_{inlet} \times 100 \tag{1}$$

The specimens for electrical conductivity measurements were obtained by sintering at 1773 K for 5 hours. The conductivity of sintered samples (1x1x20 mm) was measured in the temperature range from room temperature to 1173 K by the four-probe method.

RESULTS AND DISCUSSION

X-ray diffraction patterns of prepared powders showed single phase perovskite structure without any impurities. Figure 2 shows the temperature dependence of the electrical conductivity for the $La_{1-x}Ca_xMnO_3$ sintered samples. All samples showed high conductivity, over 100 S/cm in

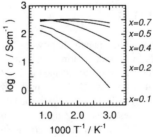

Figure 2. *Temperature dependence of the electrical conductivity for $La_{1-x}Ca_xMnO_3$.*

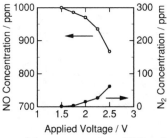

Figure 3. *Voltage dependence of the concentration of NO and N_2 in 2% oxygen at 1173 K.*

the temperature range from 973 to 1173 K. The conductivity is in the acceptable range for the cathode of an electrochemical cell. The activation energy of $La_{1-x}Ca_xMnO_3$ decreased with increasing x.

The NO decomposition reaction of the electrochemical cell composed of YSZ solid electrolyte and $La_{0.5}Ca_{0.5}MnO_3$ electrodes was investigated. Figure 3 shows the applied voltage dependence of NO and N_2 concentration in the outlet gas from the reactor at 1173 K. When a DC voltage over 1.5 V was applied to the cell, the concentration of NO decreased and that of N_2 increased with increasing voltage. The produced N_2 concentration was about one half of the decreased NO concentration. This shows that NO was electrochemically reduced to N_2, that is, the $La_{0.5}Ca_{0.5}MnO_3$ electrode directly decomposed NO to N_2 in the oxidizing atmosphere, when the DC voltage was applied to the cell. There is a possibility of platinum mesh and wire decomposing NO. In this research, platinum mesh and wire, used for a current collector and a current lead, do not have high specific surface area, so any effects of the presence of platinum on the NO decomposition are assumed negligible.

The influence of oxygen concentration on the NO decomposition behavior was evaluated for the cell composed of $La_{0.5}Ca_{0.5}MnO_3$ electrodes. Figure 4 (a) shows the voltage dependence of the current. Figure 4 (b) shows the current dependence of the NO decomposition in various oxygen concentrations at 973 K. The current - voltage curves for the cell were almost the same in different oxygen concentrations. That is, the resistance of the cell was independent of oxygen concentration. On the other hand, the NO decomposition ratio depended on the oxygen concentration. The cell showed some NO decomposition activity regardless of the oxygen concentration, however, NO decomposition ratios over 10% were attained at applied currents

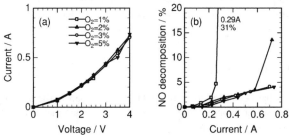

Figure 4. *(a) Voltage dependence of current, (b) Current dependence of NO decomposition for the $La_{0.5}Ca_{0.5}MnO_3|YSZ|La_{0.5}Ca_{0.5}MnO_3$ cell at 973 K.*

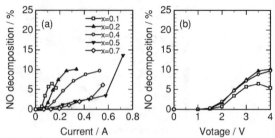

Figure 5. *(a) Current and (b) Voltage dependence of the NO decomposition for the $La_{1-x}Ca_xMnO_3|YSZ|La_{1-x}Ca_xMnO_3$ cells in 2% oxygen at 973 K.*

over 0.25 A in 1% oxygen and at over 0.7 A in 2% oxygen. It seems that the decrease of oxygen concentration around the electrode is necessary to attain high NO decomposition ratios.

The NO decomposition behavior of the cells composed of $La_{1-x}Ca_xMnO_3$ (x = 0.1-0.7) electrodes was measured to investigate the effect of electrode composition. Figure 5 (a) shows the current dependence and Figure 5 (b) shows the voltage dependence of the NO decomposition for different electrode compositions. As the content of Ca decreases, the NO decomposition ratio at the same current increased, that is, the current efficiency increased. The maximum NO decomposition ratio seemed to decrease with decreasing x. The $La_{0.8}Ca_{0.2}MnO_3$ electrode showed relatively high NO decomposition ratio and current efficiency. On the other hand, the onset voltage , at which the NO decomposition began, was independent of the electrode composition. Applying voltages higher than 1.0 V was necessary to decompose NO for all electrode compositions. From a point of view of electrical power, $La_{0.8}Ca_{0.2}MnO_3$ electrode is suitable for NO decomposition due to its high NO decomposition ratio and current efficiency in the low current region.

CONCLUSIONS

The NO decomposition activity of the cells composed of YSZ solid electrolyte and $La_{1-x}Ca_xMnO_3$ (x = 0.1-0.7) electrodes was evaluated. When applying the DC voltage to the cell, NO was directly decomposed to N_2 and O_2 in the oxidizing atmosphere. The cells composed of $La_{0.8}Ca_{0.2}MnO_3$ electrodes showed higher NO decomposition ratios and current efficiency than that of other $La_{1-x}Ca_xMnO_3$ electrodes. It seems that the decrease of oxygen concentration around the electrode is necessary to attain high NO decomposition ratios.

ACKNOWLEDGEMENTS

This work has been supported by METI, Japan, as part of the Synergy Ceramics Project. Part of the work has been supported by NEDO. The authors are members of the Joint Research Consortium of Synergy Ceramics.

REFERENCES

1. T.M.Gur and R.A.Huggins, *J.Electrochem.Soc.*, **126**, 1067 (1979).
2. T.Hibino, *Chem.Lett.*, 927 (1994).
3. Y.Teraoka, H.Fukuda and S.Kagawa, *Chem.Lett.*, 1 (1990).

Solid-State Ionics, Battery Materials, Energy Storage

Mat. Res. Soc. Symp. Proc. Vol. 658 © 2001 Materials Research Society

Bismuth Contribution to the Improvement of the Positive Electrode Performances in Ni/Cd and Ni/MH Batteries

V. Pralong, A. Delahaye-Vidal, B. Beaudoin and J-M. Tarascon
Laboratoire de Réactivité et de Chimie des Solides, Université de Picardie Jules Verne & CNRS, 33, rue St Leu 80039 Amiens, France

ABSTRACT

In this study we investigated the evolution of nickel hydroxide, which acts at the positive electrode of the Ni/Cd, Ni/MH and Ni/H₂ alkaline batteries. We found that the addition of bismuth oxide in the course of the active material preparation prevents the dissolution-re-crystallization processes of the nickel hydroxide that are harmful to the electrode efficiency. From XRD and SEM studies, it is shown that treatment of the bismuth doped-nickel hydroxide by hydrogen in 5 N KOH electrolyte prevents metallic nickel formation. Moreover, it appears to stabilize the α-type nickel hydroxide structure, preventing its transformation into the β-Ni(OH)$_2$ phase. Finally, an implementation of these findings towards the most efficient use of nickel positive electrodes is shown.

INTRODUCTION

Ni-based rechargeable alkaline batteries, Nickel-Cadmium and Nickel-Metal Hydride cells are widely used in portable and power tool applications despite a strong competition with Li-ion technology. Actually, this alkaline battery technology offers numerous advantages including low cost, high power capabilities and a long cycle life. Such a competition is an incentive for the improvement of both technologies in terms of autonomy, cycle life and reliability. For the Ni-MH technology such improvements could either be obtained on the Metal Hydride (MH) electrode or the Ni electrode.

The redox processes occurring at the Nickel Oxyhydroxide Electrode (NOE) mainly involve the Ni^{3+}-Ni^{2+} redox couple. The crystal chemistry complexity of this system was initially reported by Bode et al. [1](Figure 1). These authors established the occurrence of four phases in the oxydo-reduction processes of the NOE. In the discharged state of the battery, the nickel hydroxide Ni(OH)$_2$ usually adopts the β structural form. Occasionally, another structural form denoted as α can be obtained. Both forms crystallize in the hexagonal system with the brucite-type structure. Each Ni(OH)$_2$ layer consists of a hexagonal planar arrangement of octahedrally coordinated Ni^{2+} ions. The main difference between the two α and β-type structures resides in the stacking of the layers along the c axis. In the charged state of the batteries, namely the NiOOH nickel oxy-hydroxide, two structural forms (β and γ) are known to coexist in relative amounts depending upon the degree of charge and the amount of additive used to improve the performance of the electrode (usually Co based)[2-7]. Thus two redox processes may occur, a β/β transformation (reversible potential at about 1.25 V vs. Cd/Cd(OH)$_2$) and occasionally a γ/α transformation (reversible potential at about 1.17 V vs. Cd/Cd(OH)$_2$). The fact remains that most of the phase transformations occurring in the NOE during a charge-discharge process are solid-state processes. Even if very few reactions occur via the solution, however they could jeopardize the NOE performance. Indeed, when the electrode suffers from hydrogen generation and/or high temperature treatment, nickel monoxide NiO and nickel metal Ni° phases can be formed through a dissolution-re-crystallization process. Another problem is the α-type Ni(OH)$_2$ instability, known to transform into the more thermodynamically stable phase, β-Ni(OH)$_2$ (Ostwald ripening), when in concentrated alkaline media. Le Bihan and Figlarz [8] established that this transformation occurs via a dissolution-re-crystallization process. Stabilizing such a phase has been research focus for many years due to the higher capacity performance of the α/γ couple compared to the β/β–type phase, unfortunately without any success.

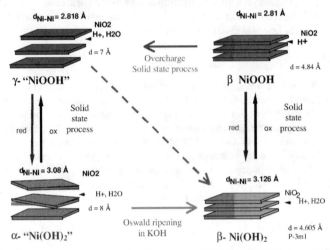

Figure 1: *Scheme showing the structures of the different phases Ni(OH)$_2$/NiOOH involving in the Nickel Oxyhydroxide Electrode during the redox process.*

As for the improvements due to the Co additives on the β/β couple performances, they depend on whether Co is co-precipitated in the active material preparation, leading to a $Ni_{1-x}Co_x(OH)_2$ compound, or whether $Co(OH)_2$ is simply incorporated within the positive composite electrode under specific conditions (dispersed or coated). The co-precipitated cobalt is recognized to limit, on charge, the γ–type phase formation that could cause some damage to the positive electrode due to an important swelling. Gautier et al.[5] clearly showed that cobalt ions could migrate when co-precipitated in the α-nickel hydroxide, finally going into solution at high temperature and then crystallizing as CoOOH phase. This phenomenon, currently named demixion, jeopardizes the $Ni_{1-x}Co_x(OH)_2$ performance. Concerning the coated cobalt, detailed electrochemical studies [6, 7] have recently thrown some light on the $Co(OH)_2$ redox chemistry (Figure 2). In short, we have shown that it is possible to control the physico-chemical characteristics of the electrochemically formed phases by adjusting the charge-discharge rates and potential limits both of which govern the dissolution processes. On charge, a highly conductive cobalt oxyhydroxide is formed when using a high rate, due to a pure solid state process. On discharge, the $Co^{3+} \rightarrow Co^{2+}$ reduction that occurs at 0.7 V is associated with the dissolution of Co^{2+} species, and followed by a migration of cobalt towards the current collector, resulting overall in an electrode degradation. Moreover, we recently demonstrated that the electrochemically inactive Co_3O_4 phase could appear through a dissolution-re-crystallization process from both $Co(OH)_2$ and CoOOH when the electrode was submitted to a floating (keeping the cell at a constant potential to compensate for the internal losses) during which the cell temperature can rise up to 60°C [9].

So any chemical/physical means that could affect, decrease or block the dissolution processes in the NOE, like the cobalt demixion or merely the formation of NiO, Ni^0, Co_3O_4, Co^0 and $Co(OH)_2$ phases, should help improve the performances of the Co-containing nickel positive electrodes. This way, we recently showed that adding bismuth-based salts or bismuth oxide to the NOE prevents the $Co(OH)_2$ and Co_3O_4 phase formation[10].

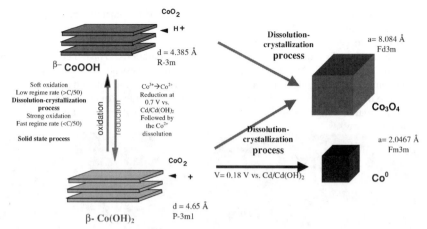

Figure 2: *Schematic representation showing the structures of the different phases Co(OH)₂/CoOOH/Co₃O₄ involved in the Nickel Oxyhydroxide Electrode during the redox process when the cobalt additive is post-added in the electrode.*

From chemical and electrochemical studies, coupled with potentiometric titration, UV-visible spectroscopy and cyclic voltammetry, we concluded that the positive effect of these bismuth additives is twofold. First, bismuth adsorption onto the Co-surface alters the CoOOH chemical reactivity and thereby its dissolution. Second, as bismuth forms Bi-Co-oxohydroxo complexes in solution, it delays the precipitation of the Co_3O_4 and $Co(OH)_2$ phases during cycling (as experimentally proven). Thus, based on the crystal chemistry similarity between the Co- and Ni-based oxide systems, we undertook an investigation of the effect of Bi additives on the chemical stability of $Ni(OH)_2$ with the hope of preventing its transformations via solution processes.

EXPERIMENTAL

Commercial $Co(OH)_2$, coated $Ni(OH)_2$ and home-made $Ni(OH)_2$ denoted as $\beta(II)$, both crystallizing in the P-3m1 space group with a = 4.65 Å and c = 3.12 Å, were used for all the experiments. High purity bismuth oxide Bi_2O_3 (Prolabo, 99% min.) was added to the active material (1 to 5 wt%)[9]. The electrochemical cells used for this study were described in a previous paper [7]. Phase identifications were made by X-ray diffraction (XRD) using a Philips PW 1710/1729 diffractometer with CuKα radiation (λ=1.5418 Å) and a diffracted beam monochromator. Scanning electron microscopy (SEM) observations were made by means of a Philips XL-30 Field Emission Gun microscope.

RESULTS AND DISCUSSION

As previously shown [9], a bismuth treatment improves the chargeability of the Co-coated NOE by increasing the formation of a highly conductive cobalt oxyhydroxide during the first charge. Then, the question is does bismuth only preserve the cobalt coating or does it act directly on the nickel hydroxide?

First, SEM was used to observe the surface of a spherical cobalt-coated nickel hydroxide after aging for 48 h in a concentrated KOH solution at 90°C. The SEM images (Figure 3) revealed that drastic dissolution crystallization processes might occur in these conditions and are limited by the presence of bismuth. Unfortunately, we are unable to characterize the phase, which has precipitated on the surface of the sample Bi-free. But the above results demonstrate that the $Co(OH)_2$ coated $Ni(OH)_2$ sample that we previously coated with bismuth is far more stable under these extreme conditions. Thus, these results stress the fact that the bismuth treatment restricts the dissolution process occurring during storage at high temperature in KOH solution as previously shown for pure cobalt hydroxide [9]. In order to directly evidence the bismuth effect on pure nickel hydroxide, we started a fundamental investigation of the nickel hydroxide transformations possibly occurring via solution processes.

Figure 3: *Scanning electron microscopy image of the cobalt coated nickel hydroxide. From left to right: starting material, then after 48h in 5 N KOH at 90°C without bismuth additive and with 3.4 wt.% Bi_2O_3 addition.*

Chemical reduction of NiOOH by NaBH₄

When working with a metal hydride or hydrogen as negative electrode, diffusion of hydrogen towards the positive electrode occurs. In order to check the bismuth effect on the NOE under such hydrogen evolution, we studied the effect of a chemical reduction involving such hydrogen evolution. Treatment of the NiOOH phases with an aqueous solution of $NaBH_4$ (Aldrich 99%, 50 g/L) allowed the simulation of a drastic reduction reaction, firstly because of the very low potential of the redox couple $BH_4^-/H_2BO_3^-$, and secondly because of the hydrogen evolution in the solution process: The reduction of H_2O by sodium borohydride (see reactions below) releases hydrogen [10, 11].

$$BH_4^- + 8\ OH^- \Leftrightarrow H_2BO_3^- + 5\ H_2O + 8\ e^- \qquad - 1.240\ V\ vs.\ NHE$$
$$2\ H_2O + 2e^- \Leftrightarrow H_2 + 2OH^- \qquad - 0.8277\ V\ vs.\ NHE$$
$$NiO_2 + 2H_2O + 2e^- \Leftrightarrow Ni(OH)_2 + 2OH^- \qquad 0.490\ V\ vs.\ NHE$$

The XRD patterns of the fully reduced materials are reported in Figure 4. The bismuth added material exhibits the typical β-$Ni(OH)_2$ peaks only, contrary to the non treated material for which NiO and Ni^0 peaks are also observed.

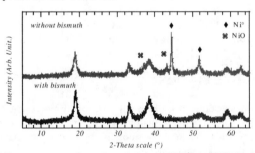

Figure 4: *X-ray powder diffraction pattern of samples reduced with NaBH₄ in 5N KOH at 60°C with bismuth and without bismuth.*

This result, combined with the fact that a solution process forms a metallic nickel powder, confirms the effect of bismuth regarding the surface protection. Although we do not have any proof yet, it is highly probable that the protection mechanism from the dissolution of the nickel is the same as the one we showed to be involved in the cobalt phases [9]: a bismuth hydroxo-complexes adsorption at the nickel hydroxide surface.

Within the context of this study, it is worth recalling several Japanese patents [13-16] that quoted the use of Bi additives in the NOE as a way of pushing the water oxidation front to higher potentials, and thereby enhancing the chargeability of the NOE at high temperature. In a previous paper [9], we already dwelt on the improvement of the conductivity of a non-stoichiometric cobalt oxyhydroxide formed during the first charge, leading to a better chargeability of the cobalt-coated positive electrodes. As an attempt to check this effect on a pure nickel hydroxide compound, we studied the electrochemical behavior of a pure β-Ni(OH)$_2$ electrode with and without bismuth oxide. However, no significant improvement was observed. Finally, we should mention that the oxidation of β-type nickel hydroxide occurs in the solid state and, as indicated above, the bismuth additive seems to act only on the dissolution processes. Thus, it is easily understood that it cannot have any influence on the nickel hydroxide oxidation. And the chargeability improvement of the NOE is essentially due to the production of the highly conductive cobalt oxy-hydroxide coating phase during the first charge.

$\alpha \rightarrow \beta$-Ni(OH)$_2$ transformation

One of the most well-known transformations in the nickel system remains the Oswald ripening corresponding to the $\alpha \rightarrow \beta$ Ni(OH)$_2$ transformation in highly alkaline media. Through XRD, we looked at the effect of bismuth addition to this transformation. In Figure 5 are reported the patterns of a pure α-Ni(OH)$_2$ phase, and those of samples after aging in 5 N KOH at 80°C for one week without (b) and with (c) bismuth.

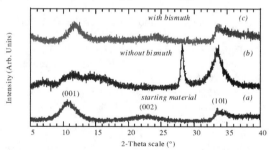

Figure 5: XRD patterns of (a) as prepared α-Ni(OH)$_2$ and of α-Ni(OH)$_2$ aged in 5N KOH at 80°C for one week (b) without bismuth in solution (c) with bismuth in solution.

For both samples the XRD patterns of the materials aged without and with bismuth correspond to a poorly crystallized phase. However, these two samples have two different structures. In the bismuth treated sample, the α-type stacking of the layers along the c-axis is characterized by the dissymmetrical (001) and (10ℓ) reflections whereas the non treated sample exhibits an interstratified nickel hydroxide stacking. That is to say that both α and β layers co-exist and are stacked along the c-axis. It is clear that the bismuth addition prevents both the α–Ni(OH)$_2$ from dissolution and its evolution towards the β-type; since bismuth doping also stabilizes the γ-NiOOH phase as shown by Tuomi et al.[12]. Then, we now know how to stabilize the α/γ redox couple during the cycling and prevent its transformation into the β/β redox couple.

CONCLUSION

We have shown that the incorporation of Bi-additives to Co-coated nickel oxy-hydroxide electrodes slows down the phase transformations occurring via dissolution re-crystallization processes. Therefore this must be considered as a means to enhance the efficiency of the nickel-based positive electrode in alkaline batteries. And henceforth, due to its stabilization effect on the α and γ phases, we can shed light on the possible utilization of the γ/α couple as the positive electrode redox process that will give a larger specific capacity than that of the β/β couple.

ACKNOWLEDGEMENTS

The authors thank J. Scoyer, A. Audemer, M. Nelson and T. Kerr for enlightening discussions. They are grateful to Union Minière (UM) and the Agence Nationale pour la Recherche et la Technologie (ANRT) for their financial support.

REFERENCES

1. H. Bode, K. Dehmelt, J. Witte, *Electrochem. Acta*, **11**, 1079 (1966).
2. M. Oshitani, H. Yufu, K. Takashima, S. Tsuji and Y. Massumaru, *J. Electrochem. Soc.*, **136**, 1590 (1989).
3. P. Benson, G. W. D. Briggs and W. F. K. Wynne-Jones, *Electrochim. Acta*, **9**, 275 (1964).
4. Al. H. Zimmerman and R. Seaver, *J. Electrochem. Soc.*, **137**, 2662 (1990).
5. M. Butel, L. Gautier and C. Delmas, *Solid State Ionics*, **122**, 271 (1999) ; L. Gautier, Thèse de doctorat, Bordeaux I (1995).
6. V. Pralong, A. Delahaye-Vidal, B. Beaudoin, B. Gérand, J-M. Tarascon, *J. Mat. Research*, **9**, 955 (1999).
7. V. Pralong, A. Delahaye-Vidal, B. Beaudoin, J-B. Leriche, J-M. Tarascon, *J. Electrochem. Soc.*, **147**, 1306 (2000).
8. S. Le Bihan and M. Figlarz, *Electrochim. Acta*, 18, 123 (1973).
9. V. Pralong, A. Delahaye-Vidal, B. Beaudoin, J-B. Leriche, J. Scoyer, J-M. Tarascon, *J. Electrochem. Soc.*, **147** , 2096 (2000).
10. D. W. Murphy, S. M. Zahurak, B. Vias et al. *Chem. Mat.*, **555**, 767 (1993).
11. N. SacEpée, B. Beaudoin, V. Pralong, T. Jamin, J.-M. Tarascon and A. Delahaye-Vidal, *J. Electrochem. Soc.*, **146**, 2376 (1999).
12. P. Tuomi, *J. Electrochem. Soc.*, **1**, 112, (1965).
13. Hanabusa, Patent Application N° Jp 8-195198 (1996) Furukawa.
14. K. Miyamoto, T. Fukuju, K. Sugimoto, U.S. Patent, Serial N° 557.394 (1995).
15. M. Inoue, Japanese patent, Application N° Hei 8-331055 (1996) Sanyo.
16. K. Niiyama, and R. Maeda, Y. Matsusita, M. Nogami, I. Yonetsu T. Nishio, Japanese patent, Application, N° Hei 9-278117 (1997).

Mat. Res. Soc. Symp. Proc. Vol. 658 © 2001 Materials Research Society

Li-insertion Behavior in Nanocrystalline TiO$_2$-(MoO$_3$)$_z$ Core-Shell Materials

Gregory J. Moore[1], Dominique Guyomard and Scott H. Elder[2]
Institut des Matériaux Jean Rouxel, CNRS - University of Nantes
BP 32229 - 44322 Nantes Cedex 3 – France
[1] Argonne National Laboratory
Argonne, IL 60439, U.S.A.
[2] Pacific Northwest National Laboratory
Richland, WA 99352, U.S.A.

ABSTRACT

A fundamental study of the Li insertion behavior of a series of materials consisting of a TiO$_2$ core having MoO$_3$ on the surface has been carried out in order to determine the influence of the shell. These TiO$_2$-(MoO$_3$)$_z$ materials, where (z) denotes the fraction of coverage from a partial to a double layer, range in diameter from 40-100 Å. Calculations have been done on their theoretical lithium capacity using a maximum of Li$_{0.5}$TiO$_2$ for the core, and Li$_{1.5}$MoO$_3$ at the TiO$_2$/MoO$_3$ interface, and they have been compared to that found experimentally. The reversible Li-insertion capacity was shown to increase from 0.34 per Ti for the pure TiO$_2$ sample, to 0.91 Li per transition metal when the MoO$_3$ coverage increased to one monolayer. There was only one plateau observed in the electrochemical scans for the samples showing that they function as a single-phase material making them interesting for electrodes. The redox voltage of the TiO$_2$/Li$_{0.5}$TiO$_2$ biphasic transformation increased 60 mV from the pure TiO$_2$ to the sample containing one monolayer of MoO$_3$. This effect was interpreted as due to a change in TiO$_2$ surface charge coming from an inductive effect of   Ti-O-Mo bonds.

INTRODUCTION

Nanocrystalline TiO$_2$ has shown potential as a positive electrode in a lithium cell as well as holds promise as a negative electrode material for rocking chair battery assemblies [1]. When cycled either against Li as a positive cathode, or against LiCoO$_2$ as a negative electrode, it has demonstrated its ability to insert up to over 0.5 Li/Ti [2-7]. However, during cycling, the cyclability tends to fall off due to a phase transformation. TiO$_2$, as a mesoscopic insertion compound, has been further investigated using zirconium as a stabilization component [8]. TiO$_2$ has a tendency to undergo thermal induced crystalline growth, accompanied by partial phase transformation to rutile, which is an inactive phase at room temperature. The nanoarchitectured TiO$_2$/zirconium materials proved to be resilient against these transformations, and could potentially make them amenable for battery applications. It is therefore of interest to exploit the potential of TiO$_2$ as a Li-insertion compound, as well as exploring nanocomposite architectures as a means to stabilize and enhance its electrochemical properties.

A novel type of nanoarchitectured material has recently been reported having the composition of TiO$_2$-(MoO$_3$)$_z$, with a core-shell structure [9]. These materials are composed of a nanocrystalline TiO$_2$ (anatase) core with MoO$_3$ as a shell. These materials were synthesized by a novel surfactant induced nucleation reaction. The subscript z in TiO$_2$-(MoO$_3$)$_z$ indicates the fraction of shell coverage on the TiO$_2$ core. At low concentrations (z = 0.18, 0.36 and 0.54) of MoO$_3$, the shell only partially covers the TiO$_2$ core and can be viewed as existing as discrete Mo-O octahedra. At 38 wt% of MoO$_3$ (z = 1.1), the shell is a monolayer of corner sharing octahedra with partial development of a second layer, and at 44 wt% (z = 1.8) it is composed of nearly a full bilayer. It was our goal to understand how the MoO$_3$ shell influences the electrochemical performance and properties of these materials; to determine the influence of capacity, redox voltage, and the contribution of each phase to the Li-insertion.

EXPERIMENTAL

Composite electrodes were prepared by mixing the active material (80% by mass) with carbon black (Super P. from Chemetals) (10%) and an organic binder (PVDF) (10%) dissolved in cyclopentanone. The films were cast on Al disks and dried at 160°C in an oxygen atmosphere prior to their use. The Li insertion behavior was studied in two-electrode Swagelok-type cells versus lithium metal, using 1 M LiPF$_6$/EC/DMC as an electrolyte. A Mac-Pile galvanostat/potentiostat was used for all of the electrochemical experiments.

RESULTS AND DISCUSSION

Material characteristics: These materials were thoroughly characterized after synthesis [9] and were shown to truly have a TiO$_2$ anatase core by powder x-ray diffraction and HRTEM. The particle sizes were determined both by the Scherrer formula using the x-ray patterns as well as observation with the electron microscope. The absence of any long range MoO$_3$ ordered structure was also verified by these techniques. The molybdenum being in an octahedral environment was shown by XANES, and Raman spectroscopy showed that it had coordination similar to that found in α-MoO$_3$. α-MoO3 is made up of bilayers of corner sharing MoO$_6$ octahedra with Li-ion sites between successive layers. Some of the properties of the as synthesized materials are shown in table I.

Table I. Properties of synthesized TiO$_2$-(MoO$_3$)$_z$.

z in TiO$_2$-(MoO$_3$)$_z$	Particle Size (Å)	Wt. % MoO$_3$	Theoretical Li-Capacity (Li/T.M.)
0	100	0	0.5
0.18	80	4	0.58
0.36	70	10	0.69
0.54	60	17	0.83
1.10	50	38	1.13
1.80	40	44	1.27

The size of these materials varied as a direct function of the Mo/Ti ratio in the reactants, which also controlled the extent of coverage. Figure 1 is included to help visualize these structures. Only the materials containing the 0.18 and 1.8 shell fraction coverage are included in this schematic. All of the structures are spherical, with the lower concentrations providing isolated islands of MoO$_6$ octahedra, up to the 44% mass fraction that provides nearly a full bilayer.

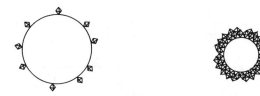

Figure 1. Schematic of the core –shell structure for TiO_2-$(MoO_3)_z$ for the $z = 0.18$ and 1.8 materials.

Cycling data of Li-cells: Galvanostatic experiments have been performed for all samples. Figure 2 shows the cycling scans run at a rate of C/100 for both the material containing 0.18 and 1.8 shell fraction coverage. The capacity is normalized per total amount of transition metal (T.M.). These scans show the increase of the Li-capacity of the TiO_2 core by the addition of this MoO_3 shell interface. For the α-MoO_3 structure, there are available sites for lithium between every bilayer of MoO_6 octahedra [10-12]. Since none of these materials exceed the bilayer, at the greatest coverage there is at most a 1.8 layer thickness, there are no actual sites provided between corresponding MoO_3 layers. The only sites provided by the MoO_3 lie at the interface of the core and shell. The Li-insertion in figure 2 therefore shows the great increase of the capacity of these materials due to the addition of small amounts of Mo to the surface. Further seen in these curves is a single plateau of insertion. These materials behave as a single phase, absent of individual plateaus for both the TiO_2 and MoO_3. Therefore it was found that these materials had greatly enhanced capacities along with maintaining one insertion voltage range.

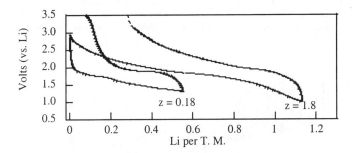

Figure 2. Cycling data of TiO_2-$(MoO_3)_z$ showing the $z = 0.18$ and 1.8 samples.

Redox potential: Of further interest was the effect of the shell on the working voltage of the TiO_2 core. This was observed by running the materials using a potential sweep. Figure 3 again shows the materials containing 0.18 and 1.8 shell fraction of coverage in an I-V curve. It is shown in this diagram that the TiO_2 core with only a small amount of Mo has a reduction peak at 1.72 and the material with over full coverage has a peak at 1.79 V. This reduction peak at 1.79 V was also seen for the material with the 1.1 fraction of coverage. This showed that the materials had reached their full effect of the Mo once the shell was complete. The increase of this reduction potential is assumed to be due to a surface inductive effect occurring between the Ti-O-Mo link. This inductive effect has been previously seen in Fe-O-X where X = Mo, W, S or P. The linked component causes a polarization of the oxygen atoms and a lowering of the covalency of the Fe-O bond, causing a stabilization of the redox couple and a positive increase of the redox potential [13].

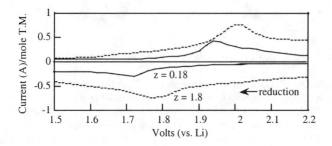

Figure 3. *I-V curves for the TiO_2-$(MoO_3)_z$ materials for $z = 0.18$ and 1.8.*

Improved cyclability: To investigate cycling behavior of these materials, to see if the shell can prevent irreversible losses of lithium capacity, several cycles were performed. This was done by cycling the material with a 1.8 shell fraction coverage at a rate of C/15 for 9 cycles. The reversible capacity shown at this rate was 110 mAh/g. The results of this cycling are shown in figure 4. Other cycling has been carried out on this material at C/100 and C/25. The reversible capacity of the material at C/100 was 240 mAh/g for 4 cycles and 140 mAh/g for 8 cycles at C/25. The slow rates were used because it was found that these cells had a kinetics limitation. No further investigations were carried out as to the cause of this but it is assumed that there may have been a mismatch between the carbon and small particle size of the active materials. Further investigations will be carried out in the consideration of different carbon or ball-milling the mixture to have a more intimate mixture.

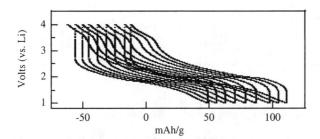

Figure 4. *Cycling of the TiO_2-$(MoO_3)_{1.8}$ material at a rate of C/15.*

In figure 4, a slight shift towards negative capacity is shown for the successive charging. These materials occasionally showed side reactions upon charging, therefore later scans for these materials were terminated at 3.5 V. However, the curves in the 9 cycles are superimposable and demonstrate the reversibility.

CONCLUSION

This fundamental study of these novel nanoarchitectured materials has shown interesting results. It was seen that the small addition of Mo oxide to the surface of TiO_2 has a tremendous effect on the prospective of TiO_2 in lithium cells. There was an important increase in the capacity of the bulk along with a constant increase of redox potential due to the interface, up to the full layer coverage. It was also seen that for repetitive cycling the MoO_3 coverage has an effect on stabilizing the reversible capacity. Further studies are of interest in searching for a material to use as a shell that may cause a negative inductive effect, i.e. reduce the working potential versus lithium, which could lead to the increased importance of TiO_2 as a negative electrode in Li-ion batteries.

ACKNOWLEDGMENT

Gregory J. Moore is grateful to the Conseil General des Pays de la Loire for his post-doctoral fellowship and the opportunity to study at the Institute of Materials of Nantes, France.

REFERENCES

1. S. Y. Huang et al., *J. Electrochem. Soc.*, **142**, L142 (1995).
2. R.J.Cav, D.W.Murphy, S.Zahurak, A.Santoro and R.S.Roth, *J. Solid State Chem.*, **53**, 64 (1984).
3. B. Zachau-Christiansen, K. West, T. Jacobsen, S. Skaarup, *Solid State Ionics*, **53-56**, 364 (1992).
4. T. Ohzuku, Z. Takehara and S. Yashizawa, *Electrochimica Acta*, **24**, 219 (1979).
5. F. Bonino, L. Busani, M. Lazzari, M. Manstretta, B. Rivolta and B. Scrosati, *J. Power Sources*, **6**, 261 (1981).
6. T. Ohzuku, T. Kodama and T. Hirai, *J. Power Sources*, **14**, 153 (1985).
7. W.J.Macklin and R.J.Neat, *Solid State Ionics*, **53-56**, 694 (1992).
8. L. Kavan, A. Attia, F. Lenzmann, S.H. Elder and M. Gratzel, *J.Electrochem. Soc.*, **147**, 2897 (2000).
9. S.H.Elder, F.M.Cot, Y.Su, S.M.Heald, A.M.Tyryshkin, M.K.Bowman, Y.Gao, A.G.Joly, M.L.Balmer, A.C.Kolwaite, K.A.Magrini and D.M.Blake, *J.Am.Chem. Soc.*, **122**, 5138 (2000).
10. C. Julien, G.A. Nazri, J.P.Guesdon, A.Gorenstein, A.Khefla and O.M.Hussain, *Solid State Ionics*, **73**, 319 (1994).
11. I.Boschen and B.Krebs, *Acta. Cryst.*, **B30**, 1795 (1974).
12. H.R.Oswald, J.R.Gunter and E.Dubler, *J. Solid State Chem.*, **5**, 354 (1975).
13. A.K. Padhi, K.S. Nanjundaswamy, C. Masquelier, S.Okada, and J.B. Goodenough, *J.Electrochem Soc.*, **144**, 1609 (1997).

Mat. Res. Soc. Symp. Proc. Vol. 658 © 2001 Materials Research Society

Vanadium Oxide Frameworks Modified with Transition Metals

Peter Y. Zavalij and M. Stanley Whittingham
Institute for Materials Research and Chemistry Department,
State University of New York at Binghamton,
Binghamton, NY, 13902-6016, USA

ABSTRACT

It is well known that vanadium oxides readily form open structures that defines the potential use of vanadium compounds as cathode materials for lithium batteries. In order to stabilize open frameworks during electrochemical cycling and improve their performance, vanadium oxide can be modified with other transition metals. This work presents the structural analysis of vanadium oxide frameworks modified with Mn and Zn as well as reporting on several interesting new structures, such as $TMA_4[Zn_4V_{21}O_{58}]$ and $NaMn(VO_3)_4 \cdot 2H_2O$.

INTRODUCTION

The great interest in vanadium compounds is due to their potential use both as cathode materials for advanced lithium batteries and as catalysts. The special characteristics of vanadium that distinguishes it from other transition metals consists in wide range of oxidation states (from 5+ to 3+ and even 2+), the often present vanadyl group, its red-ox properties and whole spectra of coordination polyhedra (tetrahedra (T), trigonal bipyramid (TB), distorted (O^D) and regular (O) octahedra). Naturally, such great variety of oxidation states and coordination polyhedra results in even greater variety of vanadium oxide frameworks: from chains and tunnel structures through single and double sheets layers to 3D nets. Some of our earlier work on the pure vanadium oxide frameworks has been published as a review [1].

The vanadium oxide framework can be modified with such transition metals as Mn, Fe, Co, Ni, Cu, Zn and others. Interacting with the vanadium oxide framework transition metals modify the electronic configuration and therefore the compound stability and properties that provides an additional variable in the search for new materials.

This work presents the structural analysis of vanadium oxide frameworks modified with manganese and zinc as well as reporting on several interesting new structures, such as $TMA_4[Zn_4V_{21}O_{58}]$ and $NaMn(VO_3)_4 \cdot 2H_2O$.

EXPERIMENTAL

In the last decade hydrothermal synthesis at mild conditions was successfully used to obtain various often-metastable structures with open framework. A review on vanadium oxide synthesis using hydrothermal method has been published in [2]. pH also plays an important role in the framework formation [3] – lower pH increases coordination number of vanadium and therefore increases dimensions and connectivity of the framework.

Our results of vanadium oxides prepared hydrothermally in presence of Mn and Zn are summarized in Table I and II respectively, where initial pH is given along with the polyhedra present and their connectivity in framework.

Table I. Mn compounds prepared hydrothermally

Mn compound	Initial pH	Polyhedra		Connectivit
		Mn	V	
$Mn_7(OH)_3(VO_4)_4$	11	O	T	3D; Mn-3D
γ-MnV_2O_5	6	O	SP/O	3D; V-2D
$NaMn(VO_3)_4 \cdot 2H_2O$	5	O	SP/O	3D; V,Mn-
β-MnV_2O_6	4-5	O	O^D	3D; V-2D
δ-$TMA_xMn_yV_4O_{10} \cdot nH_2O$	3-4	O	O^D	3D; V-2D

Table II. Zn compounds prepared hydrothermally

Zn compound	$ZnCl_2$:V_2O_5:TMA	Polyhedra		Connectivit
		Zn	V	
$Zn_2(OH)_3VO_3$	1:1:4	O	T	3D; Zn-2D
$Zn_3(OH)_2V_2O_7 \cdot 2H_2O$	1:1:3	O	T	3D; Zn-2D
α-$Zn_2V_2O_7$	2:1:2	TB	T	3D; Zn-2D
$TMA_4[Zn_4V_{21}O_{58}]$	1:1:2*	T	O^D	2D; Zn/V-2
δ-$Zn_{0.8}V_4O_{10} \cdot 0.6H_2O$	1:1:1	O	O^D	3D; V-2D

* pH adjusted with acetic acid to 3.5-4

DISCUSSION

In the following sections we discuss the mixed vanadium oxide structures grouped by vanadium polyhedra and type of framework.

Mn and Zn octahedra framework with V tetrahedra

This very common type of structure forms at higher pH. It has Mn or Zn main framewor[k] whereas V tetrahedra work like pillars or simply fill the cavities. It has to be mentioned that tetrahedral V can only have the 5+ oxidation state and as soon as the vanadium is reduced the framework collapses. This fact limits possible electrochemical applications to Mn containing compounds since Zn cannot cycle accept through complete reduction to zinc metal.
The $Mn_7(OH)_3(VO_4)_4$ structure [4] shown in Figure 1 is constructed from quarto-chain of edge sharing Mn(II) octahedra. The chains share corners to form 3D framework with V tetrahedra a[nd] Mn(III) single octahedra. Figure 2 depicts the $Zn_2(OH)_3VO_3$ structure [5] contains hexagonal closest packed oxygen atoms where the Zn and V atoms partially occupy the octahedral and tetrahedral cavities.
Another example of Zn structure is $Zn_3(OH)_2(V_2O_7) \cdot 2H_2O$ [6] shown in Figure 3a. Here Z[n] octahedra form a layer (see Figure 3b) with ¾ occupancy whereas divanadate V_2O_7 groups wo[rk] as pillars between the layers. H-bonded water molecules fill the interlayer space and can be easily removed on heating. Thus, this structure would be interesting for Li redox reactions if t[he] inactive zinc can be replaced with other metals such as Mn or V that can cycle electrochemical[ly].

Figure 1. $Mn_7(OH)_3(VO_4)_4$: $Mn(II)$ octahedra share edges and corners to form 3D framework with trigonal tunnels that are filled with V tetrahedra and Mn(III) octahedra.

Figure 2. $Zn_2(OH)_3VO_3$: closest packing of oxygen atoms where Zn and V occupies octahedral and tetrahedral cavities respectively.

Figure 3. $Zn_3(OH)_2(V_2O_7)\cdot2H_2O$ (a): Zn octahedra layers (b) pillared by V_2O_7 groups.

a)   b)

V square-pyramid chains and layers separated with Mn octahedra

In the γ-MnV_2O_5 structure [7] (see Figure 4) chains of edge sharing V square pyramids form zigzag layer by sharing corners, whereas Mn occupies octahedral cavities between the layers. This structure is isotypical to γ-LiV_2O_5.

Our new compound $NaMn(VO_3)_4\cdot2H_2O$ synthesized hydrothermally is formed by unusual double chain of edge sharing V square pyramids where all SP's are equally directed. The chains are linked into layers by Mn octahedra as shown in Figure 5. The sodium atoms form links between the layers and are bonded to the water molecules.

Essentially the same type of layer was recently discovered also in $Li_2Mn(VO_3)_4\cdot2H_2O$ [8]. However, stacking of the layers is different as well as oxidation state of Mn that is 2+ in Li compound versus 3+ in our Na compound.

Mn and Zn vanadates with brannerite and δ-phase type of structure

Vanadates with composition MV_2O_6 often have the brannerite type of structure that is shown for Mn and Zn [9] in Figure 6. In this structure V distorted octahedra share edges to form a layer

and M atoms fill octahedral cavities between the layers. This is the high temperature modifyca-
tion and at low temperature often the pseudo-brannerite structure is formed, which is deformed
so the coordination polyhedron of vanadium converts to SP. An example is the α-$Mn_{0.5}MoVO_6$
[10] structure shown in Figure 7.

Figure 4. *γ-MnV_2O_5: layers of V square pyramids linked with Mn octahedra.*

Figure 5. *$NaMn(VO_3)_4 \cdot 2H_2O$: V square pyramids form double chains that are linked with Mn octahedr to form layers held together by Na octahedra.*

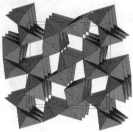
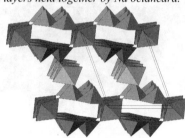

Figure 6. *β-MV_2O_6 (M=Mn, Zn, etc.): edge sharing V octahedra form the layer whereas M occupies octahedral cavities in between.*

Figure 7. *α-$Mn_{0.5}MoVO_6$: deformation of the β-form so that the V coordination becomes square pyramidal.*

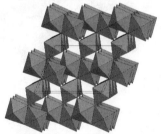
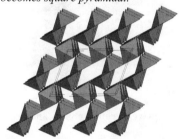

Another example of a structure type often found with many different metals is the so calle
layered δ-phase. Here distorted VO_6 octahedra share edges to form double layer and other
species such as transition metal, organic cation and water reside between the layers. Due to low
crystallinity of the δ-phases exact location of the intercalated species in $TMA_xMn_yV_4O_{10} \cdot nH_2O$
[11] and $Zn_{0.8}V_4O_{10} \cdot 0.6H_2O$ [12] were not determined.

Mixed V octahedra and tetrahedra layer

The major building block for this type of structure is the wavelike chain (W) of edge sharin
V distorted octahedra. In $Zn_{0.25}V_2O_5 \cdot H_2O$ [13] the W chains share corners between themselves
and with V tetrahedra and square pyramids to form double layer. Zn atoms coordinated with
water molecules link the vanadium oxide layers into 3D framework (Figure 8). In ZnV_3O_8 [14]
the same W chains share corners only with V tetrahedra yielding layers that are linked with pai
of V trigonal bipyramids into 3D framework (Figure 9). The interesting fact here is Zn atoms

replace half of the V atoms alternating along the W chains. This structure type was found with other metals also, for example NiV_2O_6 [15], $FeV_2O_6H_{0.5}$ [16], MgV_3O_8 [17] and α-CoV_3O_8 [18]. The last is isotypical with ZnV_3O_8 while Co atoms replace V in disordered manner.

Figure 8. $Zn_{0.25}V_2O_5\cdot H_2O$: edge sharing V octahedra form the W chains which share corners with V tetrahedra and SP's to form the layers held together with Zn octahedra.

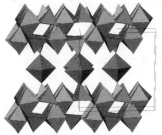

Figure 9. ZnV_3O_8: alternating V and Zn octahedra share edges and form the W chains which share corners with each other and V tetrahedra to form the layers linked with pairs of V trigonal bipyramids.

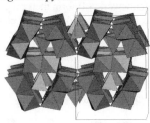

V octahedral framework with incorporated Zn tetrahedra

Sometimes additional transition metal yields incorporates into vanadium framework as we found in the new compound $TMA_4[Zn_4V_{21}O_{58}]$. V distorted octahedra sharing edges form double sheet layer as depicted in Figure 10. In this framework four V octahedra are missing (two from the lower and two from the upper sheets) creating space for Zn atom that resides in tetrahedral cavity. Tetramethyl ammonium ions separate the layers. This type of double sheet layer was found in $BaV_7O_{16}\cdot nH_2O$ [19] (see Figure 11), where V atoms occupy tetrahedral cavities yielding pure V framework.

Figure 10. $TMA_4[Zn_4V_{21}O_{58}]$: double sheet layer of edge sharing VO_6 octahedra and Zn tetrahedra.

Figure 11. $BaV_7O_{16}\cdot nH_2O$: double sheet layer of edge sharing VO_6 octahedra and an additional corner sharing V tetrahedra.

The electrochemical cycling of $TMA_4[Zn_4V_{21}O_{58}]$ is quite good with an average voltage around 2.5 volts and little capacity loss after multiple lithium insertions and removal. Thus, about 0.55 Li above a cut off of 2 V could be inserted into the structure on discharge.

CONCLUSIONS

In conclusion, 3d metals (M) interact with the vanadium-oxygen framework in the following ways: (1) occupy cavities or work as pillar in known or new V-O frameworks; (2) reside in the same sites as V in ordered or disordered manner; (3) yield new type of mixed V-M O framework; (4) form their own framework linked by V tetrahedra.

Particularly promising for electrochemical applications are those vanadium oxides with octahedral or square pyramidal coordination that form 3D framework or layers with pillars or other spacers. These compounds can be obtained by hydrothermal synthesis at low pH.

ACKNOWLEDGEMENT

This work was supported by the National Science Foundation through grant DMR-981019

REFERENCES

1. P. Y. Zavalij and M. S. Whittingham. *Acta Cryst.,* **B55**, 627-663 (1999).
2. T. Chirayil, P. Y. Zavalij and M. S. Whittingham. *Chem. Mater.,* **10**, 2629-2640 (1998).
3. T. G. Chirayil, E. A. Boylan, M. Mamak, P. Y. Zavalij, and M. S. Whittingham. *Chem. Commun.,* 33-34 (1997).
4. F. Zhang, P. Y. Zavalij, and M. S. Whittingham. *J. Mater. Chem.,* **9**, 3137-3140 (1999).
5. F. Zhang, P. Y. Zavalij, and M. S. Whittingham. *J. Phys. Chem. Solids.,* (2000) (in press).
6. P. Y. Zavalij, F. Zhang, and M. S. Whittingham. *Acta Cryst.,* **C53**, 1738-1739 (1997).
7. F. Zhang, P. Zavalij, and M. S. Whittingham. *Electrochem. Commun.,* 564-567 (1999).
8. V. Legagneur, J.-H. Liao, Y. An, A. Le Gal La Salle, A. Verbaere, Y. Piffard, and D. Guyomard. *Solid State Ionics,* **133**, 161-170 (2000)
9. G. D. Andreetti, G. Calestani, A. Montenero, and M. Bettinelli. *Z. Kristallogr.,* **168**, 53-58 (1984).
10. R. Kozlowski and K. Stadnicka. *J. Solid State Chem.,* **39**, 271-276 (1981).
11. K. Ngala, P. Zavalij, and M.S. Whittingham. *Mater. Res. Soc. Proc.,* (2001) (in press).
12. F. Zhang, P. Y. Zavalij, and M. S. Whittingham. *Mater. Res. Soc. Proc.,* **496**, 367-372 (1998).
13. Y. Oka, O. Tamada, T. Yao, and N. Yamamoto. *J. Solid State Chem.,* **126**, 65-73 (1996).
14. D. J. Lloyd and J. Galy. *Cryst. Struct. Commun.,* **2**, 209- 211 (1973).
15. H. Mueller-Buschbaum and M. Kobel. *Z. Anorg. Allg. Chem.,* **596**, 23-28 (1991).
16. J. Muller, J. C. Joubert, and M. Marezio. *J. Solid State Chem.,* **27**, 367-382 (1979).
17. M. Saux and J. Galy. *Comptes Rendus Hebdomadaires des Seances de l'Academie des Sciences, Serie C,* **276**, 81-84 (1973).
18. Y. Oka, T. Yao, N. Yamamoto, and Y. Ueda. *J. Solid State Chem.,* **141**, 133-139 (1998).
19. X. Wang, L. Liu, R. Bontchev, and A. J. Jacobson. *Chem. Commun.,* **1989**, 1009-1010 (1998).

Mat. Res. Soc. Symp. Proc. Vol. 658 © 2001 Materials Research Society

Study of Fluoride Ion Motions in PbSnF$_4$ and BaSnF$_4$ Compounds With Molecular Dynamics Simulation and Solid State NMR Techniques

Santanu Chaudhuri, Michael Castiglione[1], Francis Wang, Mark Wilson[1], Paul A. Madden[1], and Clare P. Grey,
SUNY Stony Brook, Dept. of Chemistry, New York, USA
[1]Physical and Theoretical Chemistry Laboratory, University of Oxford, UK

ABSTRACT

A combined approach, using solid state NMR and Molecular Dynamics (MD) simulations, has been employed in this work to investigate fluoride-ion motion in the PbSnF$_4$ family of anionic conductors, materials that contain double layers of Sn^{2+} and M^{2+} cations. ^{19}F MAS NMR spectra of PbSnF$_4$ and BaSnF$_4$ show that the fluoride ions are mobile on the NMR timescale (10^{-4} s), even at room temperature. In the case of BaSnF$_4$, two different groups of fluoride ions were observed, one group corresponding to fluorine atoms between the layers of Ba^{2+} cations, and the other set, corresponding to mobile fluoride ions undergoing exchange between sites in the Ba-Sn and Sn-Sn layers. The ^{119}Sn NMR suggests a highly distorted Sn environment in these compounds, consistent with the presence of stereoactive Sn lone pairs. MD simulations, using the Polarizable Ion Model, have been carried out to probe the conduction mechanism. These simulations are able to reproduce elements of the structure such as the reduction in the occupancy of the fluorine ions between the Sn-Sn layers. Anisotropic conductivity, involving primarily motion in the M-Sn layers, is predicted, consistent with the NMR results. In the case of BaSnF$_4$, no motion involving the fluoride ions in the Ba-Ba layers is observed on the simulation timescale (10^{-12} s) and a cyclic mechanism of fluoride-ion motion involving two types of fluoride ions in the Ba-Sn layers is proposed.

INTRODUCTION

PbSnF$_4$ and related compounds show the highest room temperature anionic conductivity measured to date [1]. The wide range of temperature and pressure in which the PbSnF$_4$ remains super-ionic has made it a viable solid electrolyte in amperometric solid-state oxygen sensors [2]. PbSnF$_4$ and members of this family, BaSnF$_4$ and SrSnF$_4$, all crystallize in a layered structure derived from the fluorite-type structure with alternating double layers of cations (Figure 1). There are several allotropic forms of PbSnF$_4$ that include a room temperature α-phase, a tetragonal β-phase and a high temperature cubic γ-phase [3]. Neutron diffraction refinement data show that β-PbSnF$_4$ adopts a tetragonal space group *P 4/ nmm* [4, 5]. BaSnF$_4$ crystallizes in the same tetragonal space group but shows noticeably lower anionic conductivity. This may be caused by the difference in cation size (Ba^{2+} ionic radii 1.49Å compared to 1.33Å for Pb^{2+}) and polarizability. Mössbauer results suggest the existence of stereo-active Sn lone pairs [6, 7] in all these compounds, which is associated with a reduction in the population of the (F1) fluoride ions between the two Sn-layers. A model of the Sn coordination environment obtained from the neutron refinement data of ref. [5] is shown in Figure 1b. The interstitial site (F4) and the fluoride ions in distorted fluorite-type sites (F2) are thought to play a major role in the conduction process.

In this paper we present the results from a joint NMR and molecular dynamics (MD) simulation of the MSnF$_4$ compounds. Solid state MAS ^{19}F NMR of fluoride-ion conductors is employed [8, 9] to characterize the local order in fluorine sublattices and to determine the rate of fluoride-ion motion. In addition, ^{119}Sn NMR has been used to probe into the local electronic environment of Sn.

(a)

(b)

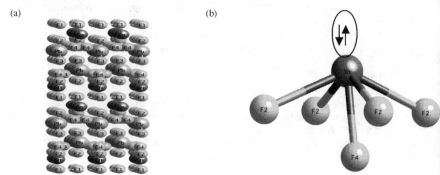

Figure 1. (a)PbSnF$_4$ layered structure in fluorite type lattice showing the cation double layers (b) Average Sn co-ordination environment in PbSnF$_4$ showing the position of the additional interstitial site (F4) between the Pb and Sn layers.

MD simulations based only on a simple ionic model will not be able to predict the formation of these MSnF$_4$ cation double layer structures and the presence of the so-called stereo active Sn lone pairs. The Polarizable Ion model (PIM) [10] which includes polarizability effect accounts for the ionic and dipolar interactions, can be used to determine the roles played by the highly polarizable cations [11] like Sn^{2+}, Ba^{2+}, and Pb^{2+} in the formation of the double layered structure and in the conduction process. The fluoride positions are poorly determined in these materials due to the considerable local disorder of the fluoride sublattice. Thus, both the simulations and NMR spectroscopy can be used [12, 13] to determine the fluorine environment and to investigate the factors that control the local structure and anionic mobility. Models of dynamic processes proposed by solid-state NMR can be verified by MD simulation and vice versa.

EXPERIMENTAL DETAILS

Preparation: The samples were prepared from a 1:1 mixture of the respective metal fluorides. They were mixed and ground thoroughly in an agate mortar and pelletized in a conventional hydraulic press. They were put into copper tubes, crimped under a nitrogen environment and then sealed under vacuum in quartz glass tubes. They were heated at 250°C for 8 hours followed by an 1 hour heating at 500°C and then quenched to room temperature. The samples were characterized using x-ray powder diffraction and no impurities such as unreacted BaF$_2$ were detected.

NMR experiments were performed on a 360 MHz Chemagnetics spectrometer with operating frequencies of 338.75 and 134.12 MHz for ^{19}F and [119]Sn, respectively. The NMR spectra were collected with double resonance Chemagnetics 4 mm and 3.2 mm pencil probes at different spinning speeds. [19]F spectra were referenced to liquid CFCl$_3$ at 0 ppm and [119]Sn spectra were referenced to solid SnO$_2$ at –604.3 ppm. Standard one-pulse and rotor synchronized Hahn-echo pulse sequences were employed for data acquisition. 2D magnetization exchange NMR spectroscopy was used to study the exchange between fluorine sites in BaSnF$_4$. Experiments were performed as a function of mixing time intervals (1 ms, 10 ms and 50 ms) where the mixing time is defined as τ_m in the following schematic 2D magnetization exchange pulse sequence:

$$(\pi/2)\xrightarrow{T_1}(\pi/2)\xrightarrow{\tau_m}(\pi/2)\text{--(Data Acquisition)}\quad T_2$$

MD simulations: Potentials of the type described in ref. [11] were constructed. Relativistic *ab initio* values are available for PbF$_2$ [14], which can be used to extract [12] reliable short-range repulsion (B_{ij}), quadrupolar terms (C^6 and C^8) and damping parameters (b) for Pb-Pb, F-F and Pb-F pairs. The Sn-F, Sn-Sn, Pb-Sn short-range interaction parameters were calculated by scaling the parameters based on the differences in ionic radii (σ_i, σ_j) by using relationships such as [12]:

$$B_{ij} \propto e^{(\sigma_i + \sigma_j)a_{ij}}\qquad(1)$$

The Ba-Ba, Ba-F and Ba-Sn parameters for BaSnF$_4$ were scaled with a similar procedure. Polarizabilities of 17.9, 15.8, 10.1, and 7.7 au were used for the Pb, Sn, Ba, and F ions, respectively. The simulation of PbSnF$_4$ and BaSnF$_4$ was started with a tetragonal cell with cations and anions in positions similar to those shown in Figure 1a. Each step in simulation corresponds to 10^{-15} s and simulations were carried out in the temperature range of 300 to 800 K.

RESULTS AND DISCUSSIONS

The room temperature ^{19}F MAS NMR spectra of PbSnF$_4$ show a single narrow resonance at −44 ppm with no spinning sidebands. As the temperature is lowered the resonance broadens and spinning sidebands appear as shown in Figure 2. The single resonance at room temperature has a Lorentzian peak-shape signifying fast exchange (i.e., processes faster than 10^{-6} s) between different fluoride-ion sites. This result is consistent with the high room-temperature conductivity of PbSnF$_4$. The broad overlapping distribution of fluoride-ion resonances at lower temperatures signifies a marked slow-down of the exchange process. But, due to the probable similarity in the chemical shift values of the different fluoride-ion sites, the resonances overlap giving an asymmetric line-shape. The presence of spinning sidebands indicates strong dipolar coupling between fluoride-ion sites below −120°C.

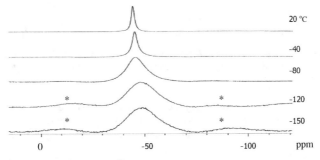

Figure 2. *Variable temperature ^{19}F MAS NMR of PbSnF$_4$ showing considerable slowing down of the mobile fluoride ion at lower temperatures. The spectrum was acquired at a MAS speed of 10 kHz. * denotes spinning sidebands.*

BaSnF$_4$ has a lower ionic conductivity compared to PbSnF$_4$ [1] and at room temperature the ^{19}F MAS NMR spectrum shows two distinctly different fluoride-ion environments at –3 and –48 ppm consistent with earlier wideline ^{19}F MAS NMR results[15]. Unlike the resonance at –. ppm, the resonance at –48 ppm has a Lorentzian lineshape, and no associated spinning sidebands. Given the chemical shifts for BaF$_2$ (-13.6 ppm) and SnF$_2$ type environments (-40 to 100 ppm) it can be inferred that a fluoride-ion in the Ba-Ba layer will resonate at a lower frequency compared to a fluoride-ion in Ba-Sn or Sn-Sn layers. Thus, the –3 ppm resonance is assigned to the rigid fluoride ions in the Ba-Ba layers. Only one average resonance is detected a around –48 ppm due to the rapid exchange between fluoride ions in Ba-Sn and Sn-Sn layers. Th assignment is also supported by the fact that, in a fluorite-type structure, there are three times more fluoride ions in the Ba-Sn plus Sn-Sn layers than in the Ba-Ba layers. The ratio of the pea areas calculated from the one pulse ^{19}F NMR spectrum is 1.0: 3.44, which is very close to that predicted from a perfect fluorite model value. A ^{19}F to ^{119}Sn cross-polarization experiment does not show any transfer of polarization from the mobile ^{19}F spins to the ^{119}Sn spins at room temperature.

The ^{19}F NMR spectra of BaSnF$_4$ at low temperatures consist of several overlapping resonances. The resonances at –3, -28, -48 and –72 ppm were identified as the isotropic resonances. Using the crystallographic model (Figure 1) as a guide to the most probable fluoride ion positions, we propose that the resonance at –3 ppm is due to the F3 sites in Ba-Ba layer. The -28 ppm resonance can be tentatively assigned to F4 sites in the Ba-Sn layers and the –48 ppm resonance to F2 sites in the Ba-Sn layer. The neutron refinement of PbSnF$_4$ showed [5] that fluoride occupancy between the Sn-Sn layers increases at lower temperatures and on this basis the resonance at –72 ppm is assigned to F1 sites in the Sn-Sn layers. Note that the absence of an resonances in the –70 ppm region of the spectra of PbSnF$_4$, indicates very low occupancy of fluoride-ion sites in the Sn-Sn layers.

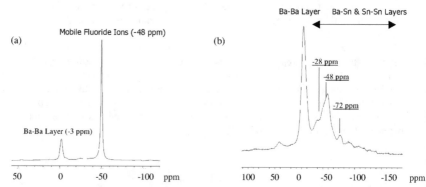

Figure 3. *Variable temperature ^{19}F MAS NMR of BaSnF$_4$ showing (a) The room temperature NMR with two distinct fluoride ion environments and (b) the disordered fluoride ion environments at –150 °C with very little sign of ionic motion.*

To probe the slower exchange process between the two fluoride sites in BaSnF$_4$, 2D-magnetization exchange NMR were acquired. The data show (Figure 4a) that there are cross-peaks between the resonances at –3 and –48 ppm indicating exchange between fluoride ions in

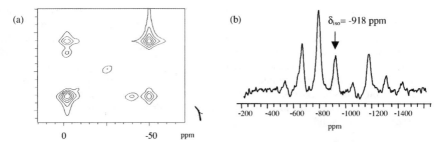

Figure 4. (a) Room temperature 2D magnetization exchange spectrum for BaSnF₄ showing cross peaks for a 24 ms mixing time interval. (b) ¹¹⁹Sn MAS NMR spectrum showing the large spinning sideband manifold.

Ba-Ba layers and Ba-Sn layers on the 24ms time scale. The ¹¹⁹Sn spectra show large spinning sidebands manifold indicating an anisotropic coordination environment for Sn (Figure 4b).

The PbSnF₄ simulation results show fluoride-ion motion at room temperatures. The motion increases steadily with temperature and there is a complete exchange between fluoride ions in the different layers. The values determined from the diffusion coefficient and mean-square displacements of fluoride ions in the crystal lattice obtained from simulations give ionic conductivity values of 2.8 x 10⁻³ Ω^{-1} cm⁻¹ at 300 K and 2.4 Ω^{-1} cm⁻¹ at 700 K. These values are consistent with experimental results for a β-PbSnF₄[1]. The fluoride ions show a marked tendency to vacate the Sn-Sn layers. There is no long-term occupation of the fluoride-ion sites (F1) for the Sn-Sn layers. The fluoride-ion population in the Sn-Sn layers significantly increases at 700 K (Figure 5a) compared to 600 K. The overall result is consistent with NMR findings. The existence of a few jumps in the room temperature simulation in a picosecond time scale should correspond to a total exchange of fluoride ions in the milliseconds time scale of NMR experiment. So, the single peak observed in ¹⁹F NMR spectrum is consistent with the calculated diffusion rate.

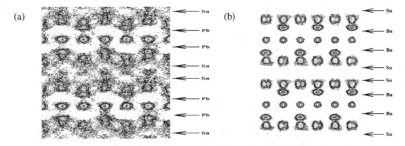

Figure 5. (a) Contour plot showing the distribution of fluoride ions in a simulation of PbSnF₄ at 700K. ; (b) BaSnF₄ results showing exchange between fluoride ions in Ba-Sn layers at 600 K simulation. Arrows are drawn as a guide to the eye, showing the cation layers.

The contour plot for the BaSnF₄ simulation (Figure 5b) shows considerable exchange between fluoride ions within the Ba-Sn layers. The fluoride ions in the Ba double layers do not exchange with fluorides in other layers or within the same layer. So, the simulations predict that the Ba-Ba layer acts as a barrier to full 3D conduction of fluoride ions. The exchange between the fluoride ions in Ba-Sn layer can be explained in terms of a cyclic mechanism involving one F4 site and two F2 sites. This cyclic exchange process is responsible for the 2D conductivity within the Ba-Sn layers. The two resonances seen in the ^{19}F room temperature NMR spectra support the MD simulations where two distinct sets of fluorines are observed. The NMR results show slow exchange between fluoride ions in the Ba-Ba and Ba-Sn layers on a 24ms time scale. Simulation shows no evidence of such an exchange process. This may be due to the much shorter time-scale of simulation or to some impurity or defect driven exchange that becomes possible in the case of a real solid.

CONCLUSIONS

The current work represents a detailed study of the dynamic processes and local ordering using two methods with very different time-scales of NMR and MD simulations. The two materials investigated, $PbSnF_4$ and $BaSnF_4$, show anisotropic conductivity, high conductivity being observed in the a-b plane. Distinct differences in the conduction processes are, however, observed. The fluoride anions in $BaSnF_4$ are considerably more ordered than those of $PbSnF_4$, which is associated with a reduced mobility for this compound. The fluoride ions between the Ba^{2+} layers undergo very slow exchange with the more mobile ions. In contrast, all the ions are mobile in $PbSnF_4$ in both the simulation and NMR timescale. This is ascribed to the considerable disorder of the fluoride ions within the Sn-Pb layers, the small population of the Sn Sn layers, and the presence of the more polarizable Pb^{2+} cation.

ACKNOWLEDGEMENTS

The authors acknowledge funding from NSF (DMR0074858 and DMR9901308) and NATO grant CRG972228.

REFERENCES

[1]   A. V. Chadwick, E. S. Hammam, D. Van der Putten, J. H. Strange, *Cryst. Latt. Def. and Amorph. Mat.,* **15,** 303 (1987).
[2]   T. Eguchi, S. Suda, H. Amasaki, J. Kuwano, Y. Saito, *Solid State Ionics,* **121,** 235 (1999).
[3]   G. Perez, S. Vilminot, W. Granier, L. Cot, C. Lucat, J. M. Reau, J. Portier, P. Hagenmuller, *Mat. Res. Bull,* **15,** 587 (1980).
[4]   S. V. Chernov, A. L. Moskvin, I. V. Murin, *Solid State Ionics,* **47,** 71 (1991).
[5]   R. Kanno, K. Ohno, H. Izumi, Y. Kawamoto, T. Kamiyama, H. Asano, F. Izumi, *Solid State Ionics,* **70/71,** 253 (1994).
[6]   T. Birchall, G. Denes, K. Ruebenbauer, J. pennetier, *Hyperf. Interact.,* **29,** 1331 (1986).
[7]   G. Denes, Y. H. Yu, T. Tyliszczac, A. P. Hitchcock, *J. Solid State Chem.,* **91,** 1 (1991).
[8]   F. Wang, C. P. Grey, *Chem. Mater.,* **9,** 1068 (1997).
[9]   F. Wang, C. P. Grey, *J. Am. Chem. Soc.,* **120,** 970 (1998).
[10]   P. A. Madden, M. Wilson, *Chem. Soc. Rev.,* **25,** 339 (1996).
[11]   M. Wilson, P. A. Madden, S. A. Peebles, P. W. Fowler, *Molec. Phys.,* **88,** 1143 (1996).
[12]   Michael Castiglione, *Computer Simulation of Superionic Fluorides,* D.Phil. Thesis, University of Oxford, Oxford, UK (2000).
[13]   F. Wang, C. P. Grey, *J. Am. Chem. Soc.,* **177,** 6637 (1995).
[14]   N. C. Pyper, *Phil. Trans. Roy. Soc. A,* **320,** 107 (1986).
[15]   M. D.-L. Floch, J. Pannetier, G. Denes, *Phys. Rev. B,* **33,** 632 (1986).

Thermo Power, Thermal
Expansion, Optical Materials

Mat. Res. Soc. Symp. Proc. Vol. 658 © 2001 Materials Research Society

"Open structure" semiconductors: Clathrate and channel compounds for low thermal conductivity thermoelectric materials

George S. Nolas
R&D Division, Marlow Industries, Inc., 10451 Vista Park Road, Dallas, Texas 75238

ABSTRACT

In a good semiconductor the electrons (or holes) propagate through the lattice structure of well ordered atoms without being scattered by the coherent vibrations of the crystal. Thus semiconductors are good conductors of electrons (or holes) and as such have given rise to modern microprocessors that are revolutionizing the way we live. In the same semiconductors the vibrations of the lattice atoms mainly carry the heat. Due to the covalent nature of the bonding in these materials the thermal conductivity is very large. These are therefore poor materials for thermoelectric applications. If the atomic vibrations, or phonons, can be localized so that the heat transfer is essentially an atom-to-atom propagation, then the thermal conduction can be drastically reduced. A semiconductor can, in principle, have the thermal conductivity of a glass. Amorphous semiconductors are of course very poor conductors of electricity therefore one does not want the electrons to propagate through a glass-like material. Instead one wants the electrons to travel as though they only "see" the well-ordered, periodic structure of a crystal while the phonons are scattered by localized disorder within the covalently bonded lattice. "Open structure" semiconductors do, in fact, exist and recent research has given rise to new thermoelectric materials.

INTRODUCTION

Among the varying approaches currently underway in an effort to identify promising materials with potential for thermoelectric applications are compounds that display semiconductor properties but also possess low thermal conductivity. Some of these novel materials even possess "glass-like" thermal conductivity [1]. The crystal structure of these materials can be considered as being "open" in the sense that they possess voids whereby interstitially placed atoms are bounded loosely thereby creating localized disorder in an otherwise well-ordered, covalently bonded lattice. The optimum situation occurs if the intrinsic mobility is relatively high due to the well-ordered, periodic structure of the crystal framework while the phonons are scattered by localized disorder. "Open-structure" semiconductors do, in fact, exist and recent research has demonstrated why they are of increasing interest for thermoelectric applications [1].

There are a fairly large number of molecules and compounds that entrap other atoms or molecules within their structures. A material with an "open structure" crystal lattice is one with voids in the host lattice that can accommodate guest atoms or molecules. These are also called clathrate compounds. Many examples can be found in the oxide zeolites based on Si-Al-O networks, however these compounds are electrical insulators. Two examples that have received great interest and represent two of the main new developments in the field of thermoelectric materials research are the semiconducting skutterudites (i.e. $CoSb_3$-based compounds) [2] and clathrates (e.g. $Sr_8Ga_{16}Ge_{30}$) [3]. The search for semiconductors with glass-like thermal properties ("phonon-glass, electron crystals" or PGEC's) was initiated by Slack [4] and motivated by thermoelectric applications which could be substantially improved by the higher figure of merit such hypothetical materials would possess. In the PGEC approach to thermoelectric materials research, the best material would be "engineered" such that the thermal properties were similar to that in an amorphous material, "a phonon-glass", and the electronic properties were similar to that of a perfectly ordered, highly covalent single crystal, "an electron single crystal". The importance of this approach emerges very clearly from the definition of the figure of merit, Z, with $Z=S^2\sigma/\lambda$. In this equation S is the Seebeck coefficient, σ the electrical conductivity and λ the thermal conductivity. From the expression for Z the PGEC approach is clear and simple. In practice, however, the development of such a material system is not simple. We do have examples that may provide guidelines in the search for such a material. In addition some materials have been reported to have PGEC properties [1, 2]. In this report I present examples of "open structure" semiconductors with low thermal conductivity and discuss the mechanisms that influence their phonon transport.

LOW THERMAL CONDUCTIVITY SEMICONDUCTORS

Borides

There has been substantial work on boride compounds, many of which display low thermal conductivity in a certain temperature range [1]. In lanthanide hexaborides, such as SmB_6 or EuB_6 for example, there is one formula unit per unit cell and the Einstein frequency of the local lanthanide vibrational mode is well below the optic mode frequencies of the boron vibrations. At temperatures below 100 K, SmB_6 is a semiconductor and its thermal conductivity is primarily due to phonons. SmB_6 exhibits a broad minimum at 55 K in the thermal conductivity, which is caused by the (lone) vibrational Einstein mode at 84.5 cm^{-1}. The effect on the lattice thermal conductivity, λ_L, extends from 10 K to 300 K. This is a very interesting phonon scattering mechanism. By contrast a simple boride crystal with no low-frequency resonance is As_2B_{12}. This compound displays a temperature dependent λ_L that is typical of simple solids.

One particular example of a boride compound with low, glass-like λ_L is LnB_{68}, where Ln represents a lanthanide atom. In this compound there are 1652 atoms in the cubic unit cell and 413 atoms in the primitive unit cell. The optic modes extend over the range from 80 cm^{-1} to 1100 cm^{-1}, and the yttrium vibration, YB_{68}, is at 140 cm^{-1}. The yttrium vibration therefore may couple to the optic modes but does very little to scatter the acoustic phonons that lie below 80 cm^{-1}. The low thermal conductivity of LnB_{68} compounds, shown in Figure 1, is like that of a glass, but is mainly a result of the very large number of atoms in the unit cell [1].

Skutterudites

The major effort in the investigation of the skutterudite material system for thermoelectric applications centers on the fact that atoms placed in the voids of this structure have a large effect on the phonon propagation[2]. The smaller and more massive the ion in the skutterudite voids results in the lowest λ_L. This is due to the dynamic

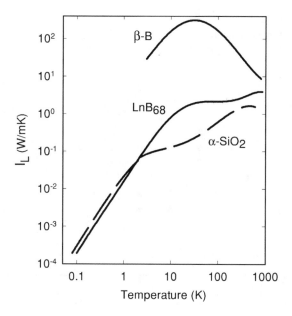

Figure 1. λ_L *versus temperature of LnB$_{68}$ (where Ln represent lanthanide ions), β-boron and amorphous silica (α-SiO$_2$).*

disorder induced by the ions and thereby able to interact with lower-frequency phonons than in the case of larger ions. This guest ion-phonon coupling is an effective phonon-scattering mechanism and one that illustrates the potential of these materials for thermoelectric applications [2]. This result has been corroborated by inelastic neutron scattering [5] and Raman scattering [6] studies.

A most prominent feature of the skutterudite material system is the fact that there exist two relatively large voids at the a positions of the unit cell that can be interstitially filled with atoms. The cubic unit cell can then be written in a general way as $[]_2M_8X_{24}$, where [] represents a void. Filled skutterudites have been formed with group-III, group-IV, lanthanide, actinide and alkaline-earth ions interstitially occupying these voids [2]. The interstitial ion in this structure is six-fold coordinated by the X-atom planar groups and is thereby enclosed in an irregular dodecahedral (12 coordinated) "cage" of X atoms. An illustration of the unit cell of a filled skutterudite centered at the position of one of the interstitial "guest" ions is shown in Figure 2. This configuration of filled skutterudites is one of the most conspicuous aspects of the structure and directly determines many of their physical properties, as will be discussed below.

In Figure 3 λ_L versus temperature is plotted for three lanthanide-filled polycrystalline skutterudites [7]. Also shown in the figure is the calculated minimum thermal conductivity, λ_{min}, of IrSb$_3$ [7]. As seen in this figure there is an order-of-magnitude decrease in λ_L at room temperature as compared to IrSb$_3$ and an even larger decrease at lower temperatures. In addition, the smallest most massive lanthanide ion results in the lowest λ_L. These smaller ions are freer to "rattle" inside the voids of the skutterudite structure as compared to La^{3+}, and are thereby better able to interact with lower-frequency phonons. This "guest ion"-phonon coupling is

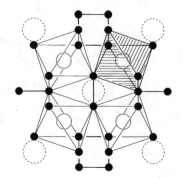

Figure 2 *The skutterudite unit cell centered at a "guest" atom position.*

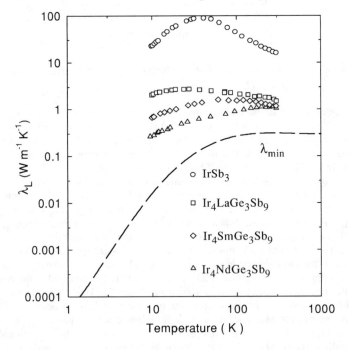

Figure 3. *λ_L versus temperature of three polycrystalline lanthanide-filled skutterudites and IrSb3 and λ_L calculated for IrSb3.*

an effective phonon scattering mechanism. In addition, the low-lying 4f electronic levels in the case of Nd^{3+} and Sm^{3+} also produce additional phonon scattering, further reducing λ_L [7]. This effect is most prominent in Nd-filled skutterudite of the three samples shown in Figure 3 due to the low lying ground-state energy levels of Nd^{3+}. Although the f-shell transitions are relatively weak phonon scatterers, as compared to the disorder introduced by the lanthanide ion in the void of this structure, the effect is quite clear, as shown in Figure 3, particularly at lower temperatures.

Even a small concentration of interstitial void filler results in a large reduction in λ_L, as has been reported previously [2]. This is illustrated in Figure 4 which shows the thermal resistivity, W, at room temperature as a function of La concentration, x, in La and Ce-filled skutterudites. Note the initial steep increase in W from only a small amount of La, or Ce, in the voids. Of particular note is that the maximum occurs at approximately $x=0.5$ or 50 % void filling. This is similar to the mass-fluctuation scattering as outlined by Abeles [8]. Apparently the random distribution of La and voids in the skutterudite lattice introduces additional phonon scattering, as opposed to an ordered system where all the voids are filled or all are empty.

Clathrates

In the 1960s Cros and co-workers [9] reported the existence of two clathrate phases, Na_8Si_{46} and $Na_{24}Si_{136}$, isomorphic with that of clathrate hydrates. Silicon, germanium and later tin [2] were found to form clathrate structures in which the guest atoms are alkaline atoms. In these materials, the host lattice is formed by atoms of one kind bonded by strong covalent forces with bond lengths similar to those in diamond-structured Si, Ge or Sn.

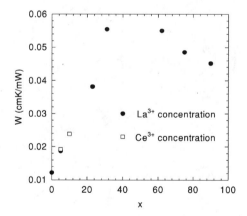

Figure 4. *Room temperature thermal resistivity, W, as a function of lanthanum and cerium concentration, x, in the voids of CoSb₃.*

With the exception of very recent results, [10] most of the work on clathrates for thermoelectric applications has focused on materials with the type I clathrate hydrate crystal structure. The type I structure can be represented by the general formula X_8E_{46}, where X are typically alkali-metal or alkaline-earth atoms and E represents a group-IV element Si, Ge, Sn although Zn, Cd, Al, Ga, In, As, Sb or Bi can also be substituted for these elements to some degree. The key characteristic is that the framework is formed by covalent, tetrahedrally bonded E atoms comprised of two different face-sharing polyhedra. This is one of the most conspicuous aspects of these compounds and directly determines many of their interesting and unique properties, including their thermoelectric properties as will be described in detail below. The type I structure (Figure 5) is comprised of two pentagonal dodecahedra, E_{20}, creating a center with $3m$ symmetry, and six tetrakaidecahedra formed by 12 pentagonal and 2 hexagonal faces, E_{24}, creating a center with $4m2$ symmetry. The corresponding unit cell is cubic with space group $Pm3n$.

One of the more interesting discoveries in terms of thermoelectrics was the magnitude and temperature dependence of λ_L of semiconducting Ge clathrates, as shown by Nolas et al. [3]. Figure 6 shows λ as a function of temperature from 0.6 to 300 K for a typical phase-pure polycrystalline $Sr_8Ga_{16}Ge_{30}$ specimen with large average grain size and high resistivity (i.e. assuming the Wiedemann-Franz relation, essentially $\lambda \cong \lambda_L$ for this sample). Also shown in Figure 6 is λ for single crystal Ge, amorphous Ge (a-Ge), amorphous SiO_2 (a-SiO_2) and λ_{min} calculated for Ge. We note that λ is lower than that of a-SiO_2 above 100 K, it is close to that of a-Ge at room temperature, and it exhibits a temperature dependence that is reminiscent of amorphous materials. The low temperature (<1 K) data indicate a T^2 temperature dependence, as shown by the straight line fit to the data in Figure 6. Higher temperature data show a minimum, or dip, in the 4 to 35 K

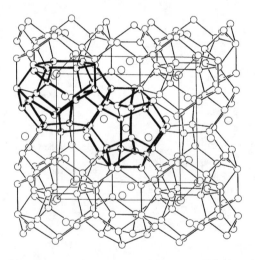

Figure 5. *The type I clathrate crystal structure with the two polyhedra (dodecahedra, center, and tetrakeidecahedra) outlined. The open circles represent the framework atom positions.*

range indicative of possible resonance scattering. The low temperature data also have a T^2 dependence on temperature similar to that found for amorphous material, and similar to the T^2 dependence calculated for λ_{min}. It is clear from these data that in the $Sr_8Ga_{16}Ge_{30}$ compound the traditional alloy phonon scattering, which predominantly scatters the highest frequency phonons, has been replaced by one or more much lower frequency scattering mechanisms. The highest frequency optic phonons in the clathrate structure have very low or zero group velocity and contribute little to λ while the low frequency acoustic phonons have the highest group velocity and contribute most to λ. The scattering of these low frequency acoustic phonons by the encaged ions results in low thermal conductivities. These low frequency optic "rattle" modes of Sr, as well as other type I clathrates, have been predicted theoretically [11] and recently observed employing Raman spectroscopy [12].

As seen in Figure 7, several Ge-clathrates exhibit λ_L that is typical of amorphous solids. The values are well below that of a-SiO$_2$ and close to that of amorphous germanium, a-Ge, near room temperature. In addition, the temperature dependence of the Ge-clathrates is much like that of a-SiO$_2$. In $Sr_4Eu_4Ga_{16}Ge_{30}$ there are two different types of atoms in the voids of the crystal structure that introduces an even greater disorder -- six "rattle" modes instead of three. This compound exhibits the lowest λ_L values in the temperature range shown and the most "glass-like" temperature dependence of λ_L.

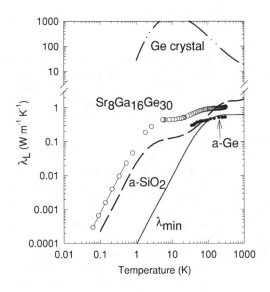

Figure 6. λ_L versus temperature of polycrystalline $Sr_8Ga_{16}Ge_{30}$, single crystal diamond structured Ge, amorphous Ge (a-Ge), a-SiO$_2$ and λ_{min} calculated for Ge.

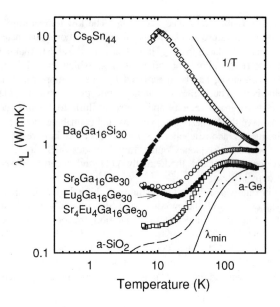

Figure 7. λ_L *versus temperature of several polycrystalline type I clathrates, a-Ge, α-SiO_2 and λ_min calculated for Ge*

The structural properties of these interesting compounds directly influences their transport properties [1]. In the case of Cs_8Sn_{44}, λ_L has a temperature dependence typical of crystalline materials (a T^{-1} temperature dependence as shown in Figure 7) although the magnitude of λ_L is also rather low. This temperature dependence is due to the fact that there are two Sn vacancies resulting in reducing the size of the cages encapsulating the Cs^+ ions [10, 13].

In order of decreasing radii, the "guest" ions in these clathrates are Cs^+, Ba^{2+}, Eu^{2+} and Sr^{2+}. The diameter of Ba^{2+} is approximately equal to the radius of the smaller Si_{20} cage. Although Ba is much more massive than the elements that make up the host matrix (i.e. Ga, Si or Ge), a prerequisite for glass-like λ_L [3], the temperature dependence of λ_L is not similar to that of the Sr and Eu-filled Ge-clathrates. This is due to the fact that Ba^{2+} is much too large an ion. The Ge_{20} and Ge_{24} cages are larger than the Si_{20} and Si_{24} cages. Also, since the cage sizes do not change substantially with filler ion, the Sr^{2+} and Eu^{2+} ions have much more room to "rattle" in the voids of the Ge-clathrate than Ba^{2+} in the Si or Ge-clathrate. The result is a much larger disorder in the Ge-clathrates, even though Ba^{2+} is the more massive ion. The interaction between the guest ions and the host lattice forming the relatively large cages creates disorder due to the fact that the guest ions are entrapped in oversized cages. These ions resonantly scatter phonons via localized low-frequency vibrations. This "rattling" of the trapped atoms in their oversized cages creates a marked reduction in λ_L. In addition to this dynamic disorder, a static disorder also results at low temperatures from a random displacement of these ions from the center of the 20 and 24 CN polyhedra [14]. In both cases the smaller and more massive the ion in a particular cage, the larger the reduction in λ_L. This effect is similar to that observed in the skutterudite system. In the case of the clathrates however the disorder results in a glass-like λ_L.

CONCLUSIONS

The above examples serve to illustrate how low thermal conductivities can be obtained in semiconducting compounds. In addition, they emphasize the importance of the PGEC approach to thermoelectric materials research. Ongoing research on skutterudite, clathrate, and other "open-structure" material systems currently underway will no doubt lead to greater improvements in material performance in the future.

ACKNOWLEDGEMENTS

I gratefully acknowledge the support of the U.S. Army Research Laboratory under contract # DAAD17-99-C-006 and Marlow Industries, Inc.

REFERENCES

1. For a recent review see G.S. Nolas, G.A. Slack and S.B. Schujman, *Semiconductors and Semimetals*, Vol. 69, edited by T.M. Tritt (Academic Press, San Diego, 2000), p. 255 and references therein.
2. Two review articles: G.S. Nolas, D.T. Morelli and T.M. Tritt, *Annu. Rev. Mater. Sci.*, Vol. 29 (1999) p. 89 and C. Uher, *Semiconductors and Semimetals*, Vol. 69, edited by T.M. Tritt (Academic Press, San Diego, 2000), p. 139.
3. G. S. Nolas, J. L. Cohn, G. A. Slack and S. B. Schujman, *Appl. Phys. Lett.*, Vol. 73 (1998) p. 178; J.L Cohn, G.S. Nolas, V. Fessatidis, T.H. Metcalf and G.A. Slack, *Phys. Rev. Lett.*, Vol. 82 (1999) p. 779; G.S. Nolas, T.J.R. Weakley, J.L. Cohn and R. Sharma, *Phys. Rev. B*, Vol. 61 (2000) p. 3845.
4. G.A. Slack, *Mat. Res. Soc. Symp. Proc.*, Vol. 478 (1997) p. 47.
5. V. Keppens, et al., *Nature*, Vol. 395 (1998) p. 876
6. G.S. Nolas and C.A. Kendziora, *Phys. Rev. B*, Vol. 59 (1999) p. 6189.
7. G.S. Nolas, G.A. Slack, D.T. Morelli, T.M. Tritt and A. Ehrlich, *J. Appl. Phys.*, Vol. 79 (1996) p. 4002.
8. B. Abeles, *Phys. Rev.*, Vol. 131 (1963) p. 1906.
9. See for example C. Cros, M. Pouchard and P. Hagenmuller, *J. Solid State Chem.*, Vol. 2 (1970) p. 570 and references therein.
10. G.S. Nolas, J.L. Cohn, M. Kaeser and T.M. Tritt, *Mater. Res. Soc. Symp. Proc.*, Vol. 626 (2001) p. Z13.1.1.
11. J. Dong, O.F. Sankey, G.K. Ramachandran and P.F. McMillan, *J. Appl. Phys.*, Vol. 87 (2000) p. 7726.
12. G.S. Nolas and C.A. Kendziora, *Phys. Rev. B*, Vol. 62 (2000) p. 7157.
13. G.S. Nolas, B.C. Chakoumakos, B. Mahieu, J. Long and T.J.R. Weakley, *Chem. Mater.*, Vol. 12 (2000) p. 1947.
14. B.C. Chakoumakos, B.C. Sales, D.G. Mandrus and G.S. Nolas, *J. Alloys and Comp.*, Vol. 296 (2000) p. 80.

Mat. Res. Soc. Symp. Proc. Vol. 658 © 2001 Materials Research Society

Spectral Properties of Various Cerium Doped Garnet Phosphors for Application in White GaN-based LEDs

Jennifer L. Wu[1], Steven P. Denbaars[2], Vojislav Srdanov[3], Henry Weinberg[4]
[1]Department of Chemical Engineering, [2]Department of Materials, [3]Center for Polymers and Organic Solids, University of California, Santa Barbara
Santa Barbara, CA, 93106
[4]Symyx Technologies, 3100 Central Expressway
Santa Clara, CA, 95051

ABSTRACT

Recently, a renewed research interest has emerged for the development of visible light, down-converting phosphors for application in white light emitting diodes (LEDs). In such devices, a blue GaN LED can act as the primary light source, exciting photoluminescence in a phosphor with subsequent broad-band emission occurring at lower energies. It was recently reported that use of a combinatorial approach to synthesize and screen potential inorganic phosphors for such an application could aid in identifying improved phosphors for blue to yellow down conversion. Using solution chemistry techniques developed by Symyx Technologies, solid state thin-film arrays (libraries) based on the garnet structure $(A_{1-x}B_x)_3(C_{1-y}D_y)_5O_{12}:Ce^{3+}$, where x and y = 0 to 1.0 and A, B = Y, Gd, or Lu; C, D = Al or Ga, were synthesized. X-ray diffraction was used to select library samples of the crystalline garnet phase. Libraries of these various garnets were then characterized, and their spectral properties compared to traditionally prepared bulk powder phosphors of similar composition. Emission and excitation trends show that as larger cations are substituted for Y (A = Y), emission and excitation are red-shifted and as larger cations are substituted for Al (C = Al), emission and excitation are blue-shifted. If smaller cations are substituted for Y and Al, an opposite trend is observed.

INTRODUCTION

Several methods have been used to develop white LEDs, which include use of conjugated polymers [1], organic dye molecules [2], and inorganic phosphors [3] as luminescent converters of blue LED light. In such a device, an InGaN blue LED can serve as the primary light source, acting as a pump to generate photoluminescence (PL) in organic or inorganic luminescent materials in which subsequent photon emission occurs at lower energies [1-5], cf., Fig. 1.

White LEDs using the inorganic phosphor $Y_3Al_5O_{12}:Ce^{3+}$ (YAG:Ce^{3+}) have been realized. YAG:Ce^{3+} emission is well suited for such applications, since when properly mixed its yellow emission under blue-light excitation yields white light [2,3]. However, improvements in quantum efficiency and color point are needed to penetrate into the incandescent lighting market. The excitation and emission of YAG:Ce^{3+} have been well studied [6,7]. It is known that substitution of Gd^{3+} and Ga^{3+} for Y^{3+} and Al^{3+}, respectively, in the garnet host shifts the emission of YAG:Ce^{3+} so that different shades of white light can be realized [2-5]. The addition of the larger ion Gd^{3+} for Y^{3+} red-shifts YAG:Ce^{3+} emission and substitution of Ga^{3+} for Al^{3+} tends to blue shift the characteristic yellow emission. A complex compositional space, such as the substituted YAG host or ternary systems, lends itself well to the combinatorial chemistry approach.

The purpose of this work is to use the combinatorial chemistry technique to understand and examine spectral trends in phosphors for application in blue to yellow conversion LEDs. Both

YAG:Ce^{3+} and various garnet hosts ($A_{3-x}B_xC_{5-y}D_yO_{12}$:$Ce^{3+}$) were synthesized and characterized. Library and bulk phosphor excitation and emission trends, as well as color point are compared demonstrate that the combinatorial approach can be used to rapidly screen potential phosphors for use in luminescence down conversion LEDs.

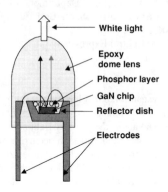

White light
Epoxy dome lens
Phosphor layer
GaN chip
Reflector dish
Electrodes

Figure 1. Structure of a GaN-based white light emitting diode.

EXPERIMENTAL

The thin film garnet libraries were prepared via solution precursors of Y, Gd, Lu, La, Al, Ga, and Ce and deposited on silicon wafers using methods developed by Symyx Technologies [8]. These metal precursor solutions were synthesized and then dispensed in the desired ratios 96-well microtiter plates. Volumes of 2.50 μL were then transferred in a predetermined library design onto silicon wafers, cf., Fig. 2. Multiple libraries were synthesized of ($(A_{1-x}B_x)_3(C_{1-y}D_y)_5O_{12}$:$Ce^{3+}$ with $x = y = 0$ to 1.0 and a constant Ce^{3+} dopant concentration of 1 mol%. The A/B site of the garnet contained Y, Gd, La, or Lu, while the C/D site was a combination of Al and Ga. Along the x-axis of a library, element B is substituted at 10 % increments for element such that the first column of the library is pure A, the second column is 90 % A and 10 % B, th third column is 80 % A and 20 % B, until finally the last column is pure B. This same trend occurs down the y-axis where D is substituted for C; the ratio of C:D is constant along each row of the library. These libraries were processed in air at temperatures from 900°C to 1100°C for hours. Bulk garnet powders of various substituted YAG-compositions were also synthesized using a mixed oxide solid-state approach [9]. The starting materials were Y_2O_3, Gd_2O_3, Lu_2O_3 Al_2O_3, Ga_2O_3, and CeO_2. These were milled in the appropriate stoichiometric amounts and fire at 1450°C for 6 hours in air. The product was then reground. X-ray diffraction was performed on the bulk samples and on various elements of the thin film library arrays to verify crystallization of the garnet structure.

The combinatorial libraries were characterized using a high throughput screen that illuminated the library with blue light to rapidly screen for potential hits. Light from a Hg/Xe lamp was passed through a series of bandpass filters: 430, 450, and 465-nm light (FWHM=25-nm) for phosphor excitation. Emitted light from the libraries was collected through a 505-nm long wave pass filter to block the reflected blue light from the excitation source and then image onto a charge-coupled device (CCD). After preliminary screening of the libraries, emission an excitation spectra were collected on a scanning fluorimetry setup, employing two SPEX model

1680 double spectrometers and a Xe lamp excitation source. The collected signal was detected with a photon counting Hamamatsu R928HA photomultiplier tube, along with a Pacific Instruments AD-126 amplifier/discriminator and Model 126 photometer.

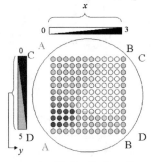

$$A_{3-x}B_xC_{5-y}D_yO_{12}:Ce$$

Figure 2. *Schematic map of the combinatorial garnet libraries of the form $A_{3-x}B_xC_{5-y}D_yO_{12}$:Ce libraries. (A, B = Y, Gd, Lu, La; C, D = Al, Ga).*

RESULTS AND DISCUSSION

In a previous study [10], we showed that YAG:Ce^{3+} could be synthesized in thin-film library format and that its properties matched that of a bulk-synthesized YAG phosphor powder. X-ray diffraction data confirm that as larger cations are substituted for A and C in the garnet lattice, the d-spacing increases—effectively showing that the lattice constants are increasing and that substitutions are occurring in this thin-film combinatorial approach, cf., Fig. 3. The converse is also evidenced for the substitution of smaller elements into the lattice. Using a Gd$_{3-x}$Lu$_x$Al$_{5-y}$Ga$_y$O$_{12}$:Ce library as an example, we see that substitution of the smaller cation Lu^{3+} for Gd^{3+} decreases the lattice constant, while substitution of the larger Ga^{3+} ion for Al^{3+} increases the lattice constant, cf., Fig. 4. We must note however that in using the solution thin-film synthesis approach for library elements near the composition of Gd$_3$Al$_5$O$_{12}$, it is difficult to achieve a pure garnet phase.

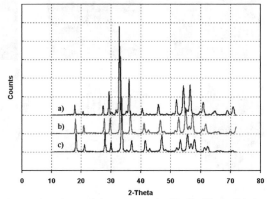

Figure 3. *X-ray diffraction patterns of several elements from the thin-film library Gd$_{3-x}$Lu$_x$Al$_{5-y}$Ga$_y$O$_{12}$:Ce: a) Lu$_3$Ga$_5$O$_{12}$:Ce, b) Lu$_3$Al$_{2.5}$Ga$_{2.5}$O$_{12}$:Ce, and c) Lu$_3$Al$_5$O$_{12}$:Ce.*

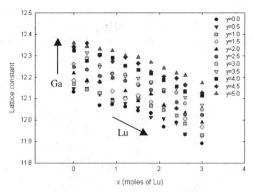

Figure 4. Lattice constant changes in $Gd_{3-x}Lu_xAl_{5-y}Ga_yO_{12}$:Ce library.

The preliminary blue to yellow phosphor screening using CCD imaging was utilized to examine the intensities of the various garnet phosphors. High throughput screens of the $A_{3-x}B_xC_{5-y}D_yO_{12}$:Ce libraries show variations in luminescence intensity due to compositional changes. The most intense elements of a library are dependent on the excitation wavelength. This is observed in the $Gd_{3-x}Lu_xAl_{5-y}Ga_yO_{12}$:Ce library shown in Fig. 5. At 465-nm excitation, $Gd_{2.4}Lu_{0.6}Al_5O_{12}$, element (1,3) using a (row, column) notation, and $Gd_{2.7}Lu_{0.3}Al_{4.5}Ga_{0.5}O_{12}$, element (2,2), are the brightest compositions. As the excitation shifts to lower wavelengths, Ga substituted elements increase in intensity and element (4,1), $Gd_3Al_{3.5}Ga_{1.5}O_{12}$:Ce, becomes the most intense at 430 nm. Similarly, other garnet libraries were characterized and the most intense garnet compositions from those libraries were identified. The most intense library elements were then synthesized in bulk to compare bulk and thin-film trends and to more closely examine the spectroscopic properties. Those results will be reported in a future publication [11].

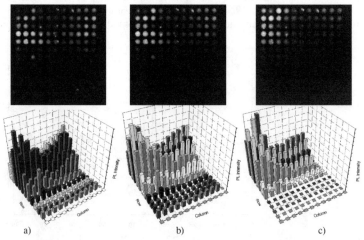

Figure 5. High throughput images of $Y_{3-x}Gd_xAl_{5-y}Ga_yO_{12}$:Ce library along with histograms of corresponding intensities at a) 430-nm, b) 450-nm and c) 465-nm excitation.

Excitation and emission spectra were also collected from the various garnet libraries and show that as larger cations are substituted in the A/B garnet site, the emission maximum red shifts. Conversely, as larger ions are substituted for the C/D garnet site, the emission blue shifts. These findings are in agreement with earlier reports in the literature for Gd^{3+} and Ga^{3+} substituted YAG [2,4,5]. Emission and excitation spectra of several library samples are shown in Fig. 6. In general, the trends found in library format hold for the bulk synthesized powder samples.

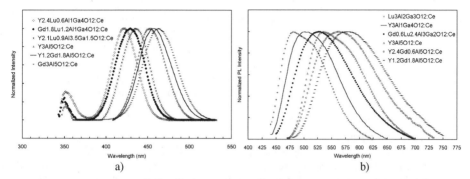

a)   b)

Figure 6. a) Excitation and b) emission spectra from various cerium doped garnets.

Data show that emission shifts are not purely a function of ionic radii; the emission maximum clearly depends on the crystalline site and degree of host lattice substitution. Along with shifting peak emission, $\lambda_{ex,max}$ decreases as larger cations are substituted in the Al site of YAG, while substitution in the Y site slightly increases $\lambda_{ex,max}$, cf., Fig. 7. The opposite is true when smaller cations are substituted in those respective garnet sites. Figure 8 shows a CIE plot of the various garnet libraries synthesized, illustrating the range of colors we can achieve through appropriate substitution in the garnet lattice.

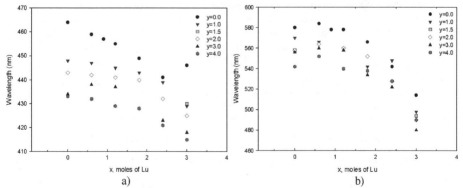

a)   b)

Figure 7. a) Excitation and b) emission maxima from various compositions of $Gd_{3-x}Lu_xAl_{5-y}Ga_yO_{12}$:Ce library samples.

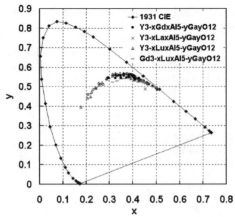

Figure 7. *CIE chromaticity coordinates corresponding to emission from $A_{3-x}B_xC_{5-y}D_yO_{12}$:Ce garnet libraries.*

CONCLUSIONS

The thin film combinatorial approach was demonstrated to be effective in qualitatively evaluating the luminescence properties of blue to yellow down-conversion phosphors. Emission and excitation trends show that as larger cations are substituted for the Y site in YAG, emission and excitation are red-shifted and as larger cations are substituted for the Al site, emission and excitation are blue-shifted. If smaller ions are substituted for those respective sites an opposite trend is observed. By varying the garnet composition, which controls the color of the white LED, and by taking advantage of the ability to tune the wavelength of blue excitation light through changes in the indium composition of InGaN, a more efficient white LED may be realized.

ACKNOWLEDGMENTS

The authors would like to thank Eric McFarland, Marty Devenney, Earl Danielson, and Damadora Poojary for their assistance, Symyx Technologies where the combinatorial work was conducted, Lawrence Livermore National Lab for use of their spectroscopy facilities, and Materials Data, Inc. for their assistance in analyzing the x-ray diffraction data. Graduate research funding was supplemented by the National Science Foundation and HOTC.

REFERENCES

1. F. Hide, P.Kozodoy, S.P. Denbaars, A.J. Heeger, *Appl. Phys. Letters*, **70** (20), 2664-2666 (1997).
2. P. Schlotter, R. Schmidt, J. Schneider, *Appl. Phys. A*, **64**, 417-418 (1997).
3. J. Baur, P. Schlotter, J. Schneider, *Festkoerprobleme*, **37**, 67-78 (1998).
4. S. Nakamura, in *Light-Emitting Diodes: Research, Manufacturing, and Applications*, (Proceedings of SPIE, F 13-14, 1997), p. 26.
5. S. Nakamura, G. Fasol, *The Blue Laser Diode*, (Springer, Berlin, 1997), pp. 216-219.
6. W. W. Holloway, Jr., *J. Opt. Soc. Am.*, **59** (1), 60-63 (1969).
7. T. Y. Tien, E.F. Gibbons, R.G. DeLosh, P.J. Zacmanidis, D.E.Smith, H. L. Stadler, *J. Electrochem. Soc.*, **120** (2), 278-281 (1973).
8. U.S. Patent 6,013,199.
9. R.C. Ropp, *The Chemistry of Artificial Lighting Devices*, (Elsevier Science, New York, 1993), pp.502-504.
10. J. L. Wu, M. Devenney, E. Danielson, D. Poojary, H. Weinberg, *Mater. Res. Soc. Symp. Proc.*, **560** (Luminescent Materials), 65-70 (1999).
11. J. L. Wu, S. P. Denbaars, P. Ford, S. Massick, V. Srdanov, H. Weinberg, in preparation.

Mat. Res. Soc. Symp. Proc. Vol. 658 © 2001 Materials Research Society

Origin of Ferroelectricity in Aurivillius Compounds

Donají Y. Suárez Ian M. Reaney, and William E. Lee
Department of Engineering Materials, University of Sheffield,
Mappin St, Sheffield, S1 3JD, United Kingdom

ABSTRACT

The structures and microstructures of a range of Aurivillius phases have been investigated using transmission electron microscopy. Superlattice reflections arising from rotations of octahedra around the c-axis have been identified and their intensity at room temperature has been shown to diminish as the tolerance factor of the perovskite blocks increases. The paraelectric to ferroelectric phase transition temperature (T_c) has been monitored using permittivity measurements as a function of temperature and T_c has also been shown to decrease as tolerance factor increases. It is proposed that the onset of octahedral tilting and T_c are related in Aurivillius phases.

INTRODUCTION

Aurivillius phases, general formula $Bi_2O_2^{2+}(M_{n-1}R_nO_{3n+1})^{2-}$, were first characterised in the 1950's and are composed of perovskite blocks, $(M_{n-1}R_nO_{3n+1})^{2-}$, separated and sheared along 1/2[111] by rock salt structured $Bi_2O_2^{2+}$ layers [1]. The archetype compound is $Bi_4Ti_3O_{12}$, which has two perovskite/three octahedral units in between the Bi-oxide layers. It has tetragonal prototype symmetry, space group I4/mmm (Figure 1a) and undergoes a paraelectric-ferroelectric (PE-FE) phase transition (T_c) at ~672°C on cooling [2]. Room temperature structural refinements for ferroelectric $Bi_4Ti_3O_{12}$ fit well to an orthorhombic cell with space group, B2cb, and lattice parameters, $a = 0.545$ nm, $b = 0.541$ nm and $c = 3.28$ nm, but optical microscopy has revealed a domain structure only possible if $Bi_4Ti_3O_{12}$ is monoclinic (Pc) [2]. The main polarisation vector is parallel with the a axis, but there is also a small component along c (long axis) [3]. The addition of $MTiO_3$, e.g. $SrTiO_3$ and $BaTiO_3$ to $Bi_4Ti_3O_{12}$ results in the formation of compounds with a larger number of perovskite units [1,4]. Compounds with fewer perovskite units can also be obtained by substitutions of cations onto the Ti-site with a higher valence state than Ti^{4+} such as Nb^{5+} and W^{6+}, e.g. $PbBi_2Nb_2O_9$ and Bi_3TiNbO_9 [2].

In the early 1970's, Newnham et al. [2], and Dorrian et al.[3] refined the structure of $Bi_4Ti_3O_{12}$ and related compounds and demonstrated that the unit cell not only contained, FE cation displacements but that the octahedra in between the Bi-oxide layers were tilted. In $Bi_4Ti_3O_{12}$, n=3 (odd), tilting occurs in antiphase around the c-axis and in phase around the a-axis [5]. Antiphase rotations (+/-7.5°) are not present in the central octahedra but in those adjacent to the Bi oxide layer. This structural effect results in a doubling of the ab plane such that a and b of the room temperature orthorhombic/monoclinic cell are approximately $\sqrt{2}$ times and at 45° to the

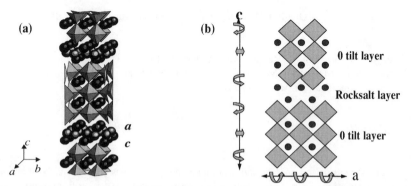

Figure 1. *(a) Prototype unit cell of $Bi_4Ti_3O_{12}$ (b) Schematic showing the tilting of octahedra.*
prototype tetragonal lattice parameters (~0.38 nm). In phase tilting around the a-axis does not
double the unit cell dimensions. The tilting is schematically illustrated in Figure 1.b. For Aurivilli
compounds where n is even, tilting around c is in-phase and the symmetry is $A2_1am$.

Aurivillius phases also occur as mixed-layer compounds, general formula $Bi_4A_{2n-1}B_{2n+1}O_{6n+9}$
[6]; and these phases are built up of regular intergrowths of two half-unit cells of conventional
Aurivillius phases [6-7]. For these compounds a high temperature prototype symmetry of
P4/mmm has been suggested [7].

Subbarao [9] investigated trends in crystal chemistry associated with the onset of the FE
phase transition in Aurivillius phases. It was noticed that increasing the size of the A cation in
$ABi_4Ti_4O_{15}$ compounds reduced the T_c [9]. In addition, if compounds such as Bi_3TiNbO_9,
$Bi_4Ti_3O_{12}$, $SrBi_4Ti_4O_{15}$ and $Sr_2Bi_4Ti_5O_{18}$ are taken as a series then the T_c seems to decrease as th
number of perovskite/octahedral units increases [2]. So far no comprehensive crystallochemical
description of the onset of ferroelectricity in Aurivillius phases has been made. Here, electron
diffraction and dark-field imaging in the TEM have been used to monitor the appearance and
intensity of superlattices and planar defects associated with the onset of octahedral tilting. These
results have been correlated with T_c and tolerance factor (t) of the perovskite blocks, given by e
(1), where $R_{M,R}$ and $_O$ are the average effective ionic radii relative to oxygen of the M and R
cation and oxygen in the general Aurivillius formula unit.

$$t = (R_M + R_O)/\sqrt{2}(R_R + R_O),$$ (1)

RESULTS AND DISCUSSION

Sample preparation, electrical and structural characterisation are described in detail in
reference [10]. Abbreviations of the compounds studied are given in the Table I and for simplici
will be used from this point on. Table I also summarises the FE phase transition temperature, T_c
crystal system and the tolerance factor, t of the perovskite blocks for each compound studied in
this work. The tolerance factor, t, has been calculated based on average effective ionic radii
relative to oxygen from Shannon [11].

Table I. *Compounds studied, abbreviations, crystal system, T_c, and t of the perovskite blocks.*

Compound	Abbreviation	Crystal system	T_c (°C)	t
$Pb_2Bi_4Ti_5O_{18}$	(Pb_{18})	Tetragonal	262	0.986
$Sr_2Bi_4Ti_5O_{18}$	(Sr_{18})	Tetragonal	261	0.979
$PbBi_4Ti_4O_{15}$	(Pb_{15})	Orthorhombic	555	0.976
$SrBi_4Ti_4O_{15}$	(Sr_{15})	Orthorhombic	512	0.971
$BaBi_4Ti_4O_{15}$	(Ba_{15})	Orthorhombic	396	0.986
$CaBi_4Ti_4O_{15}$	(Ca_{15})	Orthorhombic	789	0.963
$Bi_4Ti_3O_{12}$	(BTO_{12})	Monoclinic	675	0.958
$PbBi_8Ti_7O_{27}$	(Pb_{27})	Tetragonal	505	0.967
$SrBi_8Ti_7O_{27}$	(Sr_{27})	Tetragonal	600	0.963
$BaBi_8Ti_7O_{27}$	(Ba_{27})	Orthorhombic	490	0.971
$CaBi_8Ti_7O_{27}$	(Ca_{27})	Orthorhombic	668	0.960
$PbBi_2Nb_2O_9$	(Pb_9)	Orthorhombic	548	0.975
$SrBi_2Nb_2O_9$	(Sr_9)	Orthorhombic	432	0.967

Figure 2 is a plot of the number of octahedral units versus T_c and Figure 3 is a plot of T_c versus tolerance factor. A limited number of Aurivillius compounds such as BTO_{12}, Pb_{15}, Sr_{15}, Pb_{18}, Sr_{18} and Sr_{27} lie on a straight line whereby T_C diminishes as the number of octahedral units increases. However, Sr_9, Ca_9, Pb_9, Ba_{27}, Pb_{27}, Ca_{27} and Ca_{15} do not fit the trend. A better correlation for the broad range of Aurivillius compounds is observed between t and Tc in Figure 3. As with all plots of this nature, correlation of a property with a crystallochemical factor is not proof of cause and effect. However, the close relationship between T_c and t is worthy of further consideration. If complex or simple perovskites rather than layered structures are considered, then t is known to influence the crystal structure of the compound and in particular the degree of octahedral tilting present at room temperature [12]. When $0.99 < t < 1.04$, the lattice is untilted [12]. When $t \leq 0.985$ a series of phase transitions occurs initially involving tilting of the octahedra in antiphase and usually followed by in phase rotations around principle axes as t drops to around 0.96 [13]. The resulting structures exhibit superlattice reflections in electron diffraction patterns of the type, 1/2{hkl} for antiphase and 1/2{hk0} for in-phase rotations [12]. The phase transitions are complicated by an antiparallel, A-site cation displacement often associated with in-phase tilting that gives rise to 1/2{h00} reflections [13].

Crystal structure refinements of several Aurivillius phases have shown that at room temperature rotations of the octahedra are present that double the unit cell in the *ab* plane [2,3].

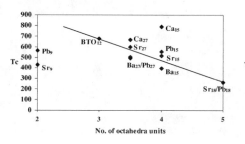

Figure 2. *Tc versus the number of octahedral units for Aurivillius compounds.*

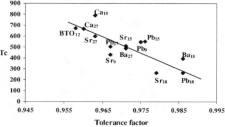

Figure 3. *Tolerance factor versus T_c for Aurivillius compounds.*

The question is whether the onset of tilting and the degree or amplitude of tilting at room temperature is influenced by t of the perovskite units. To answer this question, reflections in electron diffraction patterns unique to tilting need to be identified.

The [100] and [010] in Aurivillius compounds are equivalent zone axes to the simple perovskite <110>. Figures 4 a to c show [100] zone axis diffraction patterns, for (a) Pb_9 (b) Pb and (c) Pb_{18} samples. Alternate systematic strong and weak intensity rows are observed in all patterns. The superlattice reflections are at half integer positions with respect to the basal plane the high temperature prototype tetragonal cell. In conventional perovskites, the amplitude of tilting and thus the induced distortions are sensitive to t [12]. Therefore, as t increases from Pb_9 Pb_{18}, the superlattice rows become weaker in intensity, and the amplitude of tilting of the octahedra around the c axis diminishes. The intensities of the superlattice rows in Pb_9 (0.76) are qualitatively similar to those in Pb_{15} (0.975) and the tolerance factor for each compound is virtually the same.

Figures 4 d to f show [010] orthorhombic ([1̄10] pseudotetragonal) zone axis diffraction patterns from Pb_9, Pb_{15} and Pb_{18}, respectively, obtained from the same grains and taken under th same conditions as those in Figure 4 a to c, except the selected area aperture has been displaced a different region within the pseudotetragonal grain. Superlattice reflections are absent. These patterns indicate that there is domain variance within pseudotetragonal grains and superlattice reflections are present in [100] but not in [010] orthorhombic zone axes. These define the structural instability involving octahedral tilting as cell doubling with a change in crystal class. Dark-field images using the superlattice reflections in Figures 4a-c should also demonstrate the domain variance, illuminating areas from which superlattice arises. However, the phase transitic giving rise to these spots is cell doubling and the possibility remains that anti-phase boundaries may be present as a result of the impingement of regions of tilt that have nucleated out-of-phase Figure 5 is a dark-field TEM image of a grain in Pb_9 in which a domain variant is illuminated an ribbon like planar defects typical of APB's are present. Similar results have been demonstrated previously for Sr_{15} and Sr_{27} [14,15].

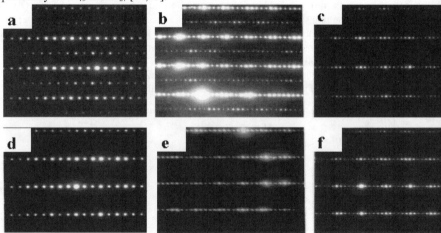

Figure 4. *[100] zone axis diffraction patterns from (a) Pb_9, (b) Pb_{15} and (c) Pb_{18} and [010] zone axes diffraction patterns from (d) Pb_9, (e) Pb_{15} and (f) Pb_{18}.*

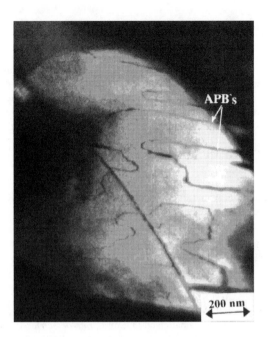

Figure 5. *Dark field image of Pb₉ obtained using the systematic row of superlattice reflections shown in Figure 4.*

The evidence presented here is clearly circumstantial and it may merely be coincidence that the degree of octahedral tilting decreases along with T_c. However, certain key pieces of evidence linking T_c, t and tilting can still be gleaned from the data. Consider, the T_c of Pb_{18} and Sr_{18}; they are virtually identical at approximately 260°C. Within the perovskite units of Sr_{18} and Pb_{18}, 3:2 mixtures of $Bi_4Ti_3O_{12}$ with $SrTiO_3$ and $PbTiO_3$ respectively are present. The T_c's of these compounds are 672°C for $Bi_4Ti_3O_{12}$, and 495°C for $PbTiO_3$ whereas $SrTiO_3$ never becomes FE. A phase transition temperature of 260°C is reasonable therefore for Sr_{18} using a simple mixture rule, but Pb_{18} should have a T_c between 495 and 672°C. The addition of $CaTiO_3$ to Aurivillius phases increases T_c which again is counter intuitive. $CaTiO_3$, however, undergoes its first tilt transition at approximately 1300 °C, higher even than Bi based compounds [16]. Ca does not significantly alter t when substituted into Aurivillius compounds but will increase the temperature of any tilt transition. The T_c's reported here are reproducible and consistent with other authors [8]. Given the data presented above, it is proposed that T_c in Aurivillius phases is driven not uniquely by the polarisablity of the cations but is linked to structural instabilities that involve rotations of the octahedra.

CONCLUSIONS

T_c in Aurivillius phases decreases as t increases. All Aurivillius phases examined in this stu show weak and strong systematic rows in [100]/[010] electron diffraction patterns arising fr rotations of the octahedra around the long c axis Compounds with higher t's in the perovsk units show weaker reflections at room temperature. It is tentatively suggested that T_c in Aurivill is related to structural instabilities associated with octahedral rotations.

ACKNOWLEDGEMENTS

The authors would like to acknowledge the support of CONACyT (Consejo Nacional Ciencia y Tecnología, México) for sponsoring the project.

REFERENCES

1. B. Aurivillius, *Arki Kemi*, 1949, **59**, 499.
2. R.E. Newnham, R.W. Wolfe, and J.F. Dorrian, *Mat. Res. Bull,* 1971, **6**, 1029.
3. J.F. Dorrian, R.E. Newnham, and D. Smith, *Ferroelectrics* 1971, **3**, 17.
4. I.S. Yi, and M. Miyayama, *J. Ceram. Soc. Japan*, 1988, **106**, 285.
5. R.L. Withers, J. G. Thompson, and A.D. Rae, *J. Sol. State. Chem.*, 1991, **94**, 404.
6. T. Kikuchi, *J. Less-Comm. Met*, 1976,**48**, 319.
7. I.M. Reaney, M. Roulin, H. S. Shulman, and D. Damjanovic, *Conf. Proc., Elctroceram* 1994, 131.
8. B. Frit, and J. P. Mercurio, *J. Alloys Comp.*, 1992, **188**, (1-2), 27.
9. E.C. Subbarao, *Integrated Ferroelectrics*, 1996, **12**, 33.
10. D.Y. Suárez, I. M. Reaney, and W. E. Lee, *(to be published in J. Mat. Res.)*
11. R.D. Shannon, and C.T. Prewett, *Acta Cryst*, 1976, **A32**, 751.
12. I. M. Reaney, E. Colla and N. Setter, *Jap. J. Appl. Phys*, 1994, **33** (7,1), 3984.
13. A.M. Glazer, *Acta Cryst.,* 1972, **B28**, 3384.
14. I. M. Reaney, M. Roulin, H. S. Shulman and N. Setter, *Ferroelectrics*, 1995, **165**, 295.
15. D. Damjanovic, and I. M. Reaney, *J. Appl. Phys.*, 1996, **80** (7), 4223.
16. F. Guyot, P. Richet, and Ph. Gillet, *Phys. Chem. Minerals*, 1993, **20**, 147.

Addendum

Mat. Res. Soc. Symp. Proc. Vol. 658 © 2001 Materials Research Society

FORMATION AND EROSION OF WC UNDER W⁺ IRRADIATION OF GRAPHITE

J. Roth[a], U. v. Toussaint[a], K. Schmid[a], J. Luthin[a], W. Eckstein[a], R. A. Zuhr[b], D.K. Hensley[b]
[a]Max-Planck-Institut für Plasmaphysik, EURATOM Association, Garching, GERMANY;
[b]Solid State Division, Oak Ridge Nat. Lab., Oak Ridge, TN., USA

ABSTRACT

The bombardment of C with 100 keV and 1 MeV W at normal incidence is studied as a function of the incident W fluence experimentally and by computer simulation with the program TRIDYN. Calculated oscillations in the amount of retained W and in the target weight change are confirmed experimentally for 100 keV at room temperature. XPS investigations show W_2C formation during ion implantation already at room temperature. RBS depth profiles for 1 MeV bombardment show W mobility and surface segregation even at liquid nitrogen (LN_2) temperatures. At elevated temperatures W clusters to form nano-particles at the surface and the oscillations in the retained amount of W disappear.

INTRODUCTION

In a future large fusion machine, ITER [1], three elements have been proposed to serve as first wall materials in different locations; Be for the main vessel wall, and C and W in the divertor. These materials will interact on plasma-facing surfaces due to erosion and re-deposition. As shown earlier [2], in the case of bombardment of W with C, erosion occurs at small ion fluences followed by net deposition as C covers the W surface. Recently it has been found in computer simulations, that oscillations in the sputtering yield as a function of fluence can occur [3]. Due to the strong implications of these kinds of effects in fusion plasma machines, it is the aim of this paper to show that these oscillation effects can also be observed experimentally. At temperatures of 100°C and above the oscillations disappear.

EXPERIMENTAL PROCEDURE

The substrates used in this work were 12 x 15 x 0.5 mm^3 rectangles of oriented high purity pyrolytic graphite supplied by Union Carbide Corp. that were cut with the graphite planes either parallel or perpendicular to the sample surface. The blanks were mechanically polished to produce a mirror finish on the active surface.

Bombardment with 100 keV W ions was carried out on an Extrion model 200-1000 ion implantation accelerator at room temperature and elevated temperatures up to 700°C. 1 MeV implantations were performed on a 1.7 MeV General Ionics Tandetron at room temperature and at liquid-nitrogen (LN_2) temperature. Current densities were on the order of 1 $\mu A/cm^2$ at 100 keV and 0.1 $\mu A/cm^2$ at 1MeV. Temperatures were monitored by chromel-alumel thermocouples mounted in the sample holders. For room temperature runs the sample holder was actively cooled to between 0° and –20° C to keep the sample from warming to above room temperature. The vacuum in the chamber was typically in the low 10^{-7} mbar range during operation. For all implants the beam was rastered so that the beam was reduced from its unrastered value by approximately 30% in each direction to give reasonable uniformity over the implanted region.

The samples were weighed before and after each implantation using a Mettler micro-balance with a sensitivity of 1 microgram. The balance was zeroed before and after each measurement, but a drift could not be totally eliminated and the error in the measured masses was estimated to be ±3 micrograms. The amount of W retained in the samples was measured quantitatively after each dose by using Rutherford backscattering spectrometry (RBS) with 2.3 MeV ^{4}He ions. Th detector angle was 160° and the sample was tilted at 60° to the incident beam in the direction of the detector to improve the depth resolution. Accurate depth profiles were extracted from the ra data using Bayesian statistics as described in reference [4].

XPS analysis was done in a Perkin-Elmer ESCA 6000 Chamber.

SIMULATION

The Monte Carlo program TRIDYN [5,6] is used for the calculations. It is based on the stati Monte Carlo program TRIM.SP [6,7], but takes dynamic target changes into account. A randomized target structure is assumed, and the atomic interactions are treated as a sequence of binary collisions. In all calculations the WHB (Kr-C) potential [8] is applied. The inelastic energy loss is described by an equipartition of the continuous Lindhard-Scharff [9] and the loca Oen-Robinson [10] models.

A pseudo-projectile corresponds to an incoming fluence. After the collision cascade of each pseudo-projectile is finished, the target composition and density are updated [5]. The surface binding energy is chosen according to the surface composition as given in ref . [11], where the surface binding energy is linearly interpolated between the corresponding values of the pure elements. TRIDYN allows the determination of sputtering yields, reflection coefficients, composition, depth profiles of the implanted species, and related values as a function of the incident fluence. The target weight changes according to the difference between the implanted amount and the amount removed from the target.

RESULTS AND DISCUSSION

100 keV W$^+$ implantation at different temperatures
Room temperature

The retention of W at 100 keV increases linearly at low fluences with a slope close to 100% At high fluences it reaches a steady state as the selfsputtering yield of W is larger than unity and no solid W layers can build up. The transition from the linear increase to steady state proceeds through a number of oscillations which have been predicted in computer simulations [3] and demonstrated experimentally [4] (Figure 1a). These oscillations can best be understood from th implantation profiles of W at different fluences. Experimental and calculated profiles show good agreement at the lowest fluences for the implantation profile of W into the pure carbon substrate The mean range is about 40 nm and the width of the distribution at 10 nm is much smaller than the mean range. This is typical for heavy ion implantation into low-Z substrates, as the heavy ions cannot be scattered by angles larger than $\arcsin(M_2/M_1)$. With increasing fluence carbon surface atoms are sputtered, bringing previously implanted W atoms closer to the surface. In the present case the carbon sputtering yield, Y_C, due to 100 keV W ions is of the order of 2, increasing to values around 4 due to the accumulation of W near the surface [3a]. With continuous sputtering the resulting W profile flattens, and the maximum concentration in the bulk gets close to 20 at%. At a fluence of 1×10^{17} atoms/cm^2, the erosion has brought the distribution close to the surface and the W surface concentration increases. In addition to carbon sputtering, W sputtering also has to be considered. As the self-sputtering yield of pure W, Y_W, is about 3, the same amount of W is eroded as implanted for a surface concentration of $1/Y_W$, i.e., at 33 at%. In order to reach steady state conditions, preferential sputtering of carbon atoms

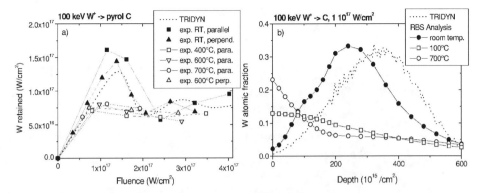

Figure 1, a) Retained W at different target temperatures. b) W depth profiles at $1x10^{17}$ W/cm^2 for different target temperatures. The dotted lines represent TRIDYN computer simulations.

will increase the W concentration within the first monolayers to 33%. The concentration within the ion range is determined by the total sputtering yield to be $1/Y_{tot} = 0.2$. These features are clearly seen in the calculated steady state W profiles and are reproduced experimentally [4].

These zero-order processes alone would lead to a continuous increase of the implanted amount of W with fluence until steady state is reached. However, the different locations of W deposition, at the end of range, and W erosion, at the surface, lead to a non-monotone approach to saturation. Initially, with no W near the surface, sputtering proceeds slowly and high W concentrations build up within the ion range while W sputtering is still negligible. This over-saturation of W leads transiently to an implanted amount much larger than in saturation. As soon as this high W concentration reaches the surface, self-sputtering will rapidly reduce the W amount until a surface concentration of 33% is reached. At this stage, the narrow W profile contains fewer W atoms than steady state. During further implantation, the profile broadens due to the lower electronic stopping and the steady state W concentration is reached after several successively smaller oscillations. The calculated and measured dependences of the implanted W amount with fluence at room temperature are compared in Figure 1a and show excellent agreement.

Another quantity, which gives direct information about the erosion of the carbon substrate, is the weight change of the probe. As W atoms are about 15 times heavier than carbon atoms, the probe will initially gain weight in oscillations similar to those of the implanted amount of W. Once steady state is reached, i.e. the implanted W amount stays constant, the additional weight loss becomes a measure of the carbon sputtering yield [4]. From the weight decrease with fluence at saturation a carbon sputtering yield of 3.5 is deduced.

Elevated temperatures

At elevated temperature diffusional effects may alter the implantation profiles and the erosion behavior without being predictable using the TRIDYN code. Figure 1a shows the fluence dependence of the W retention at 400, 600 and 700°C in comparison with the calculated retention curve. Clearly, at elevated temperatures the oscillations in the retained W amount have disappeared, and a monotonic increase and smooth transition toward the steady state value are observed. Data at 600 C show that this behavior is independent of the direction of the crystal lattice in the pyrolytic graphite, as damage due to the W implantation amorphises the carbon material [12].

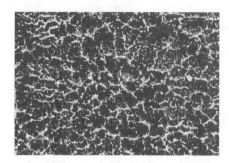 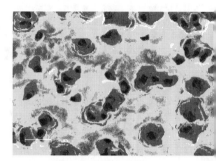

Figure 2, a) Surface topography after $3x10^{17}$ W/cm^2 at 100 keV and 600°C.
b) Surface topography after $1.3x10^{18}$ W/cm^2 at 1MeV and room temperature.

The change in the retention behavior must reflect in the depth profiles at room temperature and elevated temperatures. While at room temperature a W implantation profile is found with a maximum correlated to the ion range and negligible surface concentration, the profile at 700°C

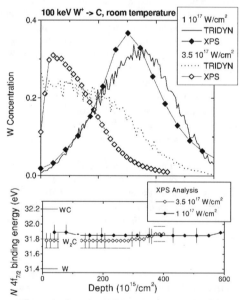

Figure 3, Depth profiles of W concentration and W $4f_{7/2}$ electron binding energy at 20°C.

has its maximum at the surface (Figure 1b). Already at 100°C the profile broadens towards the surface and the oscillations disappear. Evidently, diffusional effects and surface segregation tend to broaden the profile. As literature values for carbon self-diffusion and diffusion of carbon in W at room temperature are much too low to explain these effects, ion-enhanced diffusion of displaced carbon atoms must be responsible for the observed mobility [13].

The mobility of implanted W atom not only leads to segregation at the surface as revealed by the depth profiles, but also to clustering of W atoms into particles of sizes up to 100 nm. Figure 2a shows the surface topography after 1×10^{17} W/cm^2 implantation at 600°C.

A network of small clusters decorates the grain structure of the pyrolytic graphite. The Z-contrast of backscattered electrons reveals the high W content of most of these particles.

Depressions between clusters, due to erosion, indicate that W rich clusters, most probably W_2C, protect the surface from erosion, while unprotected graphite is preferentially sputtered. Chemical phase investigation using X-ray photoelectron spectroscopy (XPS) demonstrates that evaporated carbon layers on W transform quantitatively to W_2C or WC at temperatures of 700°C [14,15]. The present investigations show that during ion implantation tungsten carbide is formed already at room temperature. In Figure 3 the depth profiles of W at room temperature for fluences of 1×10^{17} W/cm^2 and in saturation is shown as obtained from 5 keV Ar sputtering of the sample after transfer to the XPS system. From the shift of the binding energy of the $W4f_{7/2}$ electrons, the carbide can be identified as W_2C (Figure 3). This result is in agreement with arguments that carbide formation in the W-C system occurs endothemically as soon as C and W atoms come into contact, and enhanced temperatures are only necessary in layered systems requiring atom mobility [13].

1 MeV W+ implantation at 20°C and LN$_2$ temperatures
Room temperature

At 1 MeV the W range is about 10 times longer, and the sputtering yield is lower than at 100 keV. Consequently, predicted oscillations occur at much higher fluences and are more pronounced. However, already at room temperature no such oscillations are observed (Figure 4a). At the same time, depth profiles of W show a broad indistinct distribution already at fluences of 5 $\times 10^{17}$ W/cm^2.

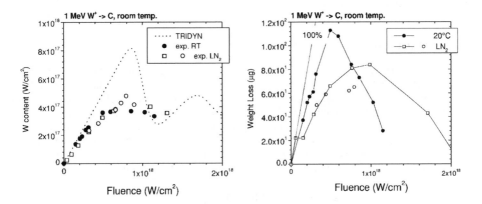

Figure 4, a) Fluence dependence of retained W. b) Fluence dependence of weight loss.

Scanning electron micrographs (SEM) of the implanted surface at a fluence of 1 $\times 10^{18}$ W/cm^2 show strong non-uniform erosion of the surface. Parts of the surface are covered with a W-rich layer, as revealed by backscattered electron imaging, while unprotected parts show deep holes of about 1 μm diameter (Figure 2b). Evidently, in contrast to the case of 100 keV implantation, for 1 MeV bombardment at room temperature, W mobility leads to surface accumulation and clustering.

Liquid nitrogen temperatures

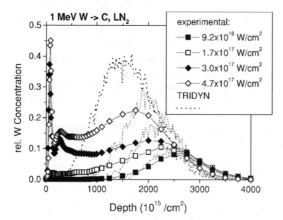

experimental:
- ■— 9.2×10^{16} W/cm^2
- □— 1.7×10^{17} W/cm^2
- ◆— 3.0×10^{17} W/cm^2
- ◇— 4.7×10^{17} W/cm^2

TRIDYN
......

Figure 5, Comparison of measured depth profiles at LN$_2$ with corresponding TRIDYN calculations.

In order to reduce the W mobilit the same implantation sequence was performed at LN$_2$ temperatures. The retention of W increases further than at room temperature but there is onl vague indication of an oscillation (F 4a). At LN$_2$ temperatures SEM show no surface roughening or W clustering. However, the depth profi of W as determined from RBS show a very distinct change (Figure 5). At fluences up to 1×10^{17} W/cm^2 the profile still agrees well with TRIDY calculations. At higher fluences, however, the W profile stretches towards the surface and at 2×10^{17} W/cm^2 a sharp, narrow surface peak develops, indicating a uniform surfa coverage of about 10 monolayers of segregated W atoms. As this fluence is reached, W sputteri sets in and prevents the development of a pronounced oscillation peak. This behavior is clearly reflected in the development of the weight change of the sample. At 2×10^{17} W/cm^2 the linear increase due to W implantation deviates from the predicted 100% line (Figure 4b).

CONCLUSION

For 100 keV W bombardment of graphite, the W implantation profiles in carbon, the retaine amount of W, and the carbon erosion at room temperature can be reproduced well by dynamic implantation codes such as TRIDYN. Since the self-sputtering of W above an energy of 1 keV i larger than unity, a saturation in the implanted amount is reached at high fluences. The oscillatory behavior in approaching this steady state is due to changes in the implantation profil during the implantation and erosion process. At temperatures slightly above room temperature, ion induced mobility and carbide phase formation move the implantation profile towards the surface, and SEM imaging reveals surface clustering of W$_2$C precipitates. The steady state is reached smoothly without oscillation.

At 1 MeV the larger separation of the range profile from the surface facilitates the detection of surface segregation effects. Indeed, it has be shown that, even at LN2 temperatures, several monolayers of W segregate uniformly at the surface without drastic changes of the surface topography. At room temperatures this segregated layer clusters into non-uniform W$_2$C particle leading to a strongly non-uniform erosion and leaving, at fluences above 1×10^{18} W/cm^2, a porous surface with deep holes.

ACKNOWLEDGEMENT

Research sponsored in part by the U.S. Department of Energy under contract DE-AC05-00OR22725 with the Oak Ridge National Laboratory, managed by UT-Batelle, LLC.

REFERENCES

1. ITER Physics Expert Group on Divertor, ITER Physics Expert Group on Divertor Modelling and Database, ITER Physics Basis Editors, Nucl.Fusion **39**, 2391 (1999).
2. W. Eckstein, J. Roth, Nucl. Instrum. Meth. **B 53**, 279 (1991).
3. W. Eckstein, Nucl. Instrum. Meth. B, accepted for publication.
4. R.A. Zuhr, J. Roth, W. Eckstein, U. v. Toussaint, J. Luthin, J. Nucl. Mater. **290-293**, 162 (2001).
5. W. Moeller, W. Eckstein, J. P. Biersack, Comp. Phys. Commun. **51**, 355 (1988).
6. W. Eckstein, Computer Simulation of Ion-Solid Interaction, Springer Series in Materials Science, Vol.10, (Springer-Verlag, Berlin, Heidelberg, 1991).
7. J. P. Biersack and W. Eckstein, Appl. Phys. **A 34**, 73 (1984).
8. W. D. Wilson, L. G. Haggmark, J. P. Biersack, Phys. Rev. **B 15**, 2458 (1977).
9. J. Lindhard, M. Scharff, Phys. Rev. **124**, 128 (1961).
10. O. S. Oen, M. T. Robinson, Nucl. Instrum. Meth. **132**, 647 (1976).
11. W. Eckstein, M. Hou, V. I. Shulga, Nucl. Instrum. Meth. **B 119**, 477 (1996).
12. J. Roth, R. A. Zuhr, S. P. Withrow, and W. P. Eatherly, J. Appl. Phys. **63**, 2603 (1988).
13. B. Soeder, J. Roth and W. Moeller, Phys. Rev. **B 37**, 815 (1988).
14. J. Luthin, Ch. Linsmeier, Surf. Science **454-456**, 78 (2000).
15. D. Hildebrandt, P. Wienhold, W. Schneider, J. Nucl. Mater. **290-293**, 89 (2001).

AUTHOR INDEX

Abramowski, Cobey, GG3.27
Aksay, Ilhan A., GG7.5
Alario-Franco, M.A., GG2.3
Alvarez, Rodolfo A., GG5.6
Amador, U., GG2.3
Amano, Muneyuki, GG9.1
Arunmozhi, G., GG3.5
Attfield, J. Paul, GG2.6
Attfield, Martin P., GG6.31
Augrain, François-Xavier, GG8.5
Awano, Masanobu, GG6.1, GG9.36

Barnes, Amy S., GG3.22
Battle, P.D., GG2.4
Beaudoin, B., GG10.1
Becker, Olav, GG4.2
Bedel, L., GG6.22
Bertsch, Paul M., GG3.36
Bieringer, Mario, GG4.1
Bonnell, Dawn A., GG5.6
Bontchev, Ranko P., GG1.7
Bottero, Jean Yves, GG3.36
Bouhedja, Laure, GG8.3
Bowden, William, GG9.11
Brodersen, S., GG4.9

Calles, A.G., GG6.15
Carlier, D., GG9.9
Carter, Erin S., GG9.29
Caruntu, Gabriel A., GG8.5
Castiglione, Michael, GG10.9
Catlow, C. Richard A., GG9.33
Chaudhuri, Santanu, GG10.9
Chebiam, R.V., GG8.12
Chen, Xi, GG9.19
Chotard, Jean-Noël, GG8.5
Constantinescu, A.M., GG6.32
Costa, M. Margarida R., GG3.5
Criado, A., GG3.5
Culp, Jeffrey T., GG5.2
Currie, David B., GG9.5
Cussen, E.J., GG2.4

Dabbs, Daniel M., GG7.5
Dahmen, Klaus-H., GG9.29
Dai, D., GG5.3
Darriet, J., GG1.4
Delahaye-Vidal, A., GG10.1

Delmas, C., GG9.9
de Matos Gomes, E., GG3.5
Denbaars, Steven P., GG11.8
Dixit, A., GG6.19
Do, Junghwan, GG1.7
Doan, Trong-Duc, GG3.27
Dobal, P.S., GG6.19
Dobley, Arthur, GG6.16
Doelsch, Emmanuel, GG3.36
Dong, Z.L., GG6.33
Dresselhaus, Mildred S., GG6.18
Dvorak, Matthew R., GG3.21

Ebbinghaus, Stefan, GG4.2
Eckstein, W., O5.31
El Abed, A., GG1.4
Elder, Scott H., GG10.4
Elthon, Don, GG4.7
Endo, Morinobu, GG6.18
Estournès, Claude, GG6.4, GG6.22

Fang, Q.L., GG9.17
Franke, M.E., GG7.4

García-Martín, S., GG2.3
Gibson, James M., GG3.22
Gilbert, Erin W., GG8.5
Gökçen, Oya A., GG4.7, GG9.14
Goodenough, John B., GG2.2
Gouma, Perina, GG6.29
Greedan, John E., GG4.1
Green, M.A., GG4.5
Greenblatt, Martha, GG5.6
Grey, Clare P., GG6.29, GG9.11,
    GG10.9
Guyomard, Dominique, GG10.4

Hamano, Kenya, GG3.17
Harbison, Gerard S., GG3.21
Harvey, Howard G., GG6.31
Hector, Andrew L., GG9.5
Henry, Paul F., GG3.31, GG7.2
Hensley, D.K., O5.31
Herman, Herbert, GG6.29
Hidalgo, A., GG6.19
Honda, Takahisa, GG6.28
Hong, Jeong-Oh, GG3.25
Honig, J.M., GG4.3

Hozumi, Atsushi, GG6.14
Huang, Xiaoying, GG6.12, GG6.17
Hughes, Robert W., GG7.2

Inagaki, Masahiko, GG6.14
Itagaki, Kimio, GG1.10

Jacobson, Allan J., GG1.7, GG4.7,
    GG8.1
Jockel, J., GG4.9

Kalinin, Sergei V., GG5.6
Kameyama, Tetsuya, GG6.14
Kanai, Takao, GG9.36
Kanoh, Hirofumi, GG6.25
Katiyar, R.S., GG6.19
Katz, Howard E., GG7.5
Kauzlarich, Susan M., GG4.1
Kawamoto, Hiroshi, GG3.17
Kawasaki, R., GG5.9
Keller, Nicolas, GG6.4
Kiennemann, A., GG6.22
Kim, Hyung H., GG3.14
Kinomura, N., GG8.7
Kitaoka, Satoshi, GG3.17
Ko, Kyung H., GG3.14
Kodenkandath, Thomas A., GG8.5
Koizumi, Akihisa, GG5.10
Koo, H.-J., GG5.3
Krasovskii, E.E., GG4.9
Kubota, Yoshihiro, GG6.28
Kuenhold, K.A., GG6.19
Kumada, N., GG8.7
Kuwahara, H., GG5.9

Lanceros-Mendez, S., GG3.5
Lavela, Pedro, GG9.7
Lawandy, Michael A., GG6.12
Ledoux, Marc J., GG6.4
Lee, William E., GG11.9
Lee, Young J., GG9.11
Le Gal La Salle, Annie, GG10.4
Li, Dong, GG3.4
Li, Jing, GG6.12, GG6.17
Lin, C.L., GG6.12
Lin, I.-C., GG9.14
Liu, Haiming, GG6.29
Livage, Jacques, GG8.3
Luthin, J., O5.31

Madden, Paul A., GG10.9
Maeda, Kunihiro, GG6.1, GG9.36
Malik, Abds-Sami, GG7.5
Mandrus, D., GG4.4
Manivannan, A., GG6.32
Mano, J.F., GG3.5
Manthiram, A., GG8.12
Maquet, Jocelyne, GG8.3
Margolis, Joshua, GG6.29
Masion, Armand, GG3.36
Matsuda, Kazuyuki, GG9.36
Mc Nelis, M.A., GG6.15
Meen, James K., GG4.7, GG9.14
Meisel, Mark W., GG5.2
Melgarejo, R.E., GG6.19
Ménétrier, M., GG9.9
Miyagawa, Hayato, GG5.10
Miyazaki, Takashi, GG6.18
Mizumaki, Masaichiro, GG5.10
Moore, Gregory J., GG10.4
Morán, E., GG2.3
Morata-Orrantia, A., GG2.3
Morgan, A. Nicole, GG5.2
Mueller, Karl T., GG3.20, GG3.22
Müller, Karoline, GG9.14
Müller-Buschbaum, Hanskarl, GG1.10

Nakagawa, Zenbe-e, GG3.17
Nakamura, Tetsuya, GG5.10
Nakazawa, Tatsuo, GG6.18
Nanao, Susumu, GG5.10
Ngala, J. Katana, GG9.16
Nguyen, Binh, GG9.14
Noda, K., GG5.9
Nogueira, E., GG3.5
Nolas, George S., GG11.1
Nonaka, Takamasa, GG9.6

Ohya, Yutaka, GG3.17
Okamoto, Tokuhiko, GG9.6
Okubo, Tatsuya, GG6.28
Okuda, Chikaaki, GG9.6
Ooi, Kenta, GG6.25
Oshida, Kyoichi, GG6.18
Owen, John R., GG9.5
Ozawa, Kiyoshi, GG9.1
Ozawa, Tadashi C., GG4.1

Paik, Younkee, GG9.11

Pan, Long, GG6.12, GG6.17
Parise, John B., GG6.29
Park, Yong H., GG3.14
Parukuttyamma, Sujatha D., GG6.29
Pérez-Vicente, Carlos, GG9.7
Petit, C., GG6.22
Pham-Huu, Cuong, GG6.4
Plévert, Jacques, GG6.28
Popov, Guerman, GG5.6
Prabakar, S., GG3.20
Prado, F., GG8.12
Pralong, V., GG10.1
Provendier, H., GG6.22

Radaelli, Paolo G., GG2.6
Raekelboom, Emmanuelle A., GG9.5
Reaney, Ian M., GG11.9
Reller, Armin, GG4.2
Renner, Bernd, GG4.2
Roger, A.C., GG6.22
Rose, Jérôme, GG3.36
Rossi, Paolo, GG3.21
Roth, J., O5.31

Sakai, Nobuhiko, GG5.10
Sakka, Yoshio, GG9.1
Sakurai, Yoshiharu, GG5.10
Salvador, Paul A., GG3.27
Sampath, Sanjay, GG6.29
Sauer, J., GG7.4
Schattke, W., GG4.9
Schmid, K., O5.31
Schulte, Jürgen, GG6.16
Seehra, M.S., GG6.32
Sevov, Slavi C., GG7.8
Shin, Moo Y., GG3.14
Shirata, Masashi, GG6.24
Sierka, M., GG7.4
Simon, U., GG4.9, GG7.4
Sims, Jessica A., GG8.5
Slater, Ben, GG9.33
Sleight, A.W., GG8.7
Smith, M.D., GG1.4
Spinu, Leonard, GG8.5
Srdanov, Vojislav, GG11.8
Starrost, F., GG4.9
Steunou, Nathalie, GG8.3
Stitzer, K.E., GG1.4
Stone, William E.E., GG3.36

Stoneham, A. Marshall, GG9.33
Suárez, Donají Y., GG11.9
Subramanian, M.A., GG3.4
Sugahara, Yoshiyuki, GG6.24
Sugi, Yoshihiro, GG6.28
Sugimoto, Wataru, GG6.24
Sugiyama, Katsumasa, GG1.10

Takei, T., GG8.7
Talham, Daniel R., GG5.2
Tarascon, J.-M., GG10.1
Thompson, J.R., GG4.4
Tiedje, O., GG4.9
Tirado, José L., GG9.7
Tomar, M.S., GG6.19
Tsuji, Masamichi, GG6.25
Tsunoda, Yu, GG6.24

Ukyo, Yoshio, GG9.6

v. Toussaint, U., O5.31
Valladares, Ariel A., GG6.15
Valladares, R.M., GG6.15
Viciu, Marilena L., GG8.5
Vogt, Frederick G., GG3.22

Waldron, J.E.L., GG4.5
Wang, Francis, GG9.11, GG10.9
Wang, Shuqiang, GG6.1
Wang, Xiqu, GG8.1
Watanabe, Yasuhiro, GG5.10
Wedel, Boris, GG1.10
Wei, B., GG6.33
Weinberg, Henry, GG11.8
Weller, Mark T., GG3.31, GG7.2,
    GG9.5
Whangbo, M.-H., GG5.3
White, T.J., GG6.33
Whittingham, M. Stanley, GG6.16,
    GG9.16, GG10.7
Wiley, John B., GG8.5
Williams, David E., GG9.33
Wilson, C.C., GG3.31
Wilson, Mark, GG10.9
Woods, L.M., GG4.4
Wright, Jon P., GG2.6
Wu, Jennifer L., GG11.8

Yokogawa, Yoshiyuki, GG6.14

Yoo, Han-Ill, GG3.25
Yuen, Tan, GG6.12

Zavalij, Peter Y., GG6.16, GG9.16,
 GG10.7
Zhang, Xiao, GG8.5

Zheng, J.P., GG9.17
Zhou, Weilie L., GG8.5
Zhou, Yuanzhong, G9.19
Zuhr, R.A., O5.31
zur Loye, H.-C., GG1.4

SUBJECT INDEX

ab initio
    calculations, GG4.9
    simulations, GG6.15
acid treatment, GG6.24
air-water interface, GG5.2
Al-doped antimonic acids, GG9.1
alkaline batteries, GG10.1
aluminum titanate, GG3.17
antimony chalcogenides, GG4.9
apatite, GG6.33
Aurivillius phase(s), GG6.24, GG11.9

Ba-Sr-Bi oxides, GG4.7
battery, GG9.6
    material(s), GG6.19, GG9.5
Bi, GG9.1
bimetallic coordination polymer,
    GG6.17
bismuth oxide(s), GG8.7, GG10.1
boron, GG3.22
BSTO, GG3.4

calcium strontium barium bismuth
    oxide, GG9.14
capacitors, GG9.17
carbon clusters, GG6.15
$Cd_2Os_2O_7$, GG4.4
CDW/SDW, GG4.1
ceria, G9.19
cerium(IV) oxides, GG6.14
charge ordering, GG1.7, GG4.5
chemical deintercalation, GG9.7
clathrates, GG11.1
cobalt nickel oxide, GG8.12
combined QM-Pot calculation, GG7.4
conductivity, GG4.9
contact shift, GG9.9
copper oxides, GG2.2
critical phenomena, GG4.3
cross effect, GG3.25
crystal chemistry, GG1.10, GG8.12

d^1, GG4.1
dielectric ceramics, GG3.14
$DyCo_5$, GG5.10

electrical
    measurements, GG4.3

    properties, GG9.1
electrolyte, GG9.19
electrotransport, GG3.25
ESCA, GG6.25
EXAFS, GG3.36

Fe_3O_4, GG2.6, GG3.25
ferroelectric, GG3.4, GG3.5
ferroelectricity, GG11.9
Fe-Si colloids, GG3.36
film structure, GG3.27
first principles, GG9.33
fission plasma-wall interactions, O5.31
fluoride ion conductor, GG10.9
formation reaction, GG3.17
framework, GG7.2
frustration, GG2.4
FTIR, GG3.36
    spectroscopy, GG6.25
fused salt electrolysis, GG5.6

gallium disphosphonate, GG6.31
Gd-doped CeO_2, GG6.1
glasses, GG3.20
graphite intercalation compounds,
    GG6.18

heavy metals, GG6.33
high pressure synthesis, GG8.1
homogeneous precipitation method,
    GG6.14
hybrid material, GG6.17
hydrothermal
    reaction, GG8.7
    synthesis, GG8.3
    technique, GG9.16
H-ZSM-5, GG7.4

image analysis, GG6.18
intercalation, GG8.5
ion
    exchange, GG8.5
    implantation and sputtering, O5.31
ionic electronic conductor, GG9.14
IOS-1, GG9.29
isotopic substitution, GG3.31

$LaM_xFe_{1-x}O_3$ (M= Ni, Co), GG6.22

Langmuir-Blodgett film, GG5.2
lanthanide coordination polymers,
    GG6.12
lanthanum
    manganite, GG9.36
    strontium manganate crystals,
        GG5.6
large crystal, GG8.1
layered
    compounds, GG1.7
    perovskite, GG6.24
lead, GG3.22
    zirconate, GG3.21
        titanate, GG3.21
Li-intercalation, GG6.19
liquid crystal, GG7.5
lithium, GG9.11
    batteries, GG9.7, GG10.7
    conductivity, GG2.3
    ions battery, GG9.9
    nickel iron oxide, GG9.7
    sodium cuprate, GG9.5
luminescence, GG11.8

magnetic, GG2.4
    Compton
        profile, GG5.10
        scattering, GG5.10
    force microscopy, GG5.6
    solids, GG5.3
magnetism, GG4.5, GG6.4, GG6.22,
    GG6.32
magnetite, GG2.6
manganese
    dioxide, GG6.25
    oxides, GG5.9
Mars van Krevelen, GG9.33
mesoporous silicate, GG7.5
mesostructure, GG6.14
metal-insulator transition, GG4.4
microporous, GG9.17
microstructure, GG2.3
mixed
    alkali effect, GG3.20
    ionic-electronic conductor, GG4.7
model, GG9.19
molecular
    dynamics simulation, GG10.9

sieve, GG6.28

nanocomposite, GG6.16
nanoparticle, GG6.32
neodymium nickel oxide, GG3.27
neutron diffraction, GG9.5
nickel, GG9.6
    hydroxide, GG10.1
    $NiS_{2-x}Se_x$ alloys, GG4.3
NMR, GG9.11
    spectroscopy, GG3.21
NO decomposition, GG9.36

open
    framework, GG10.7
    structured compounds, GG11.1
optical
    properties and electronic structure,
        GG6.15
    waveguide sensor, GG9.29
organic linkers, GG7.8
oxalate, GG6.12
oxide, GG2.4

perovskite(s)(-), GG8.5
    oxides, GG1.4
    type phases, GG4.2
phase
    equilibria, GG9.14
    transition(s), GG2.2, GG5.9
pillared structure, GG6.17
pnictide-oxide, GG4.1
polarization microscope, GG3.17
polymeric precursor solution method,
    GG6.1
polyoxovanadate, GG8.3
powder
    diffraction, GG6.28
        neutron diffraction, GG3.31
precursor plasma spray, GG6.29
proton mobility, GG7.4
pulsed laser deposition, GG3.27
pyridinedicarboxylate, GG6.12
pyrochlore, GG4.4
pyrolusite, GG9.11

Raman scattering, GG6.19
rare earth, GG5.9

redox processes, GG4.2
(RE/Li)titanates and niobates, GG2.3
reversible intercalation, GG9.16
rf induction plasma technique, GG6.29
Rietveld refinement, GG6.31
ruthenium oxide, GG9.17

samarium nickelate, GG3.31
SAXS, GG3.36
SiC nanotubes, GG6.4
single crystal structure, GG1.4
soft chemistry synthesis, GG8.12
sol-gel, GG9.29
solid state
    NMR, GG9.9, GG10.9
    nuclear magnetic resonance,
        GG3.22
spin
    dimers, GG5.3
    exchange interactions, GG5.3
stabilization, GG6.33
structure(s)(-), GG4.5, GG7.2, GG7.8
    determination, GG6.28
    property relation, GG4.2
superconductivity, GG2.2
superspace group formalism, GG1.4
supramolecular chemistry, GG5.2
surfactant, GG6.16
    templation, GG7.5
synthesis and characterization, GG6.1

tellurium oxide compounds, GG1.10
TEM, GG3.14, GG11.9
temperature programmed reduction
    (TPR), GG6.22
thermoelectrics, GG11.1
tin dioxide, GG9.33

TiO_2, GG10.4
transition
    metal, GG7.2
    frameworks, GG7.8
transmission electron microscopy,
    GG6.18
TRAPDOR NMR, GG3.20
triglycine sulphate, GG3.5
tunability, GG3.4
tungsten ions in carbon, O5.31
2-probe AC and 4-probe DC
    conductivity measurement, GG4.7

unprecedented coercive field, GG6.4

^{51}V NMR spectroscopy, GG8.3
vanadium
    oxide, GG6.16 GG9.16, GG10.7
    phosphate, GG1.7
Verwey transition, GG2.6

water adsorption behavior, GG9.1
white LED, GG11.8

XAFS, GG9.6
x-ray
    diffraction (XRD), GG3.5,
        GG3.14, GG6.31,
        GG6.32
    structure investigation, GG1.10

Y, GG9.1
YAG:Ce, GG11.8
yttria stabilized zirconia, GG9.36
yttrium aluminum garnet films, GG6.29

zeolite ZSM-5, GG8.1